SOLUTIONS MANUAL ACCOMPANYING

Elements of Electromagnetics

Third Edition

Matthew N.O. Sadiku
Jerry Sagliocca
Oladega Soriyan

New York Oxford
OXFORD UNIVERSITY PRESS
2001

Oxford University Press

Oxford New York
Athens Auckland Bangkok Bogotá Buenos Aires Calcutta
Cape Town Chennai Dar es Salaam Delhi Florence Hong Kong Istanbul
Karachi Kuala Lumpur Madrid Melbourne Mexico City Mumbai
Nairobi Paris São Paulo Shanghai Singapore Taipei Tokyo Toronto Warsaw

and associated companies in
Berlin Ibadan

Published by Oxford University Press, Inc.
198 Madison Avenue, New York, New York 10016
http://www.oup-usa.org

ISBN 0-19-514497-X (paper)

Printing (last digit): 9 8 7 6 5 4 3 2 1

Printed in the United States of America
on acid-free paper.

Table of Contents

CHAPTER 1

P. E. 1.1

(a) $\boldsymbol{A}+\boldsymbol{B}=(1,0,3)+(5,2,-6)=(6,2,-3)$

$$|\boldsymbol{A}+\boldsymbol{B}|=\sqrt{36+4+9}=\underline{\underline{7}}$$

(b) $5\boldsymbol{A}-\boldsymbol{B}=(5,0,15)-(5,2,-6)=\underline{\underline{(0,-2,21)}}$

(c) The component of $\boldsymbol{A}$ along $\mathbf{a}_y$ is $\mathbf{A}_y=\underline{\underline{0}}$

(d) $3\boldsymbol{A}+\boldsymbol{B}=(3,0,9)+(5,2,-6)=(8,2,3)$

A unit vector parallel to this vector is

$$\boldsymbol{a}_{11}=\frac{(8,2,3)}{\sqrt{64+4+9}}$$
$$=\underline{\underline{\pm(0.9117\boldsymbol{a}_x+0.2279\boldsymbol{a}_y+0.3419\boldsymbol{a}_z)}}$$

P. E. 1.2 (a) The distance vector

$$\boldsymbol{r}_{QR}=\boldsymbol{r}_R-\boldsymbol{r}_Q=(0,3,8)-(2,4,6)$$
$$=\underline{\underline{-2\boldsymbol{a}_x-\boldsymbol{a}_y+2\boldsymbol{a}_z}}$$

(b) The distance between Q and R is

$$|\boldsymbol{r}_{QR}|=\sqrt{4+1+4}=\underline{\underline{3}}$$

(c) Vector $\boldsymbol{r}_{QP}=\boldsymbol{r}_P-\boldsymbol{r}_Q=(1,-3,5)-(2,4,6)=(-1,-7,-1)$

$$\cos\theta_{PQR}=\frac{\boldsymbol{r}_{QR}\cdot\boldsymbol{r}_{QP}}{|\boldsymbol{r}_{QR}||\boldsymbol{r}_{QP}|}=\frac{7}{3\sqrt{51}}$$

$$\theta_{PQR}=\underline{\underline{70.93^\circ}}$$

(d) Area $=\frac{1}{2}|\boldsymbol{r}_{QR}\times\boldsymbol{r}_{QP}|=\frac{1}{2}|(15,-4,13)|$

$$=\underline{\underline{10.12}}$$

P. E. 1.3 Consider the figure shown below:

$$U_Z = U_P + U_W = -350a_X + \frac{40}{\sqrt{2}}(-a_X + a_Y)$$

$$= -378a_X + 28.28a_Y$$

or

$$\mathbf{u} = 379.3\angle 175.72^\circ$$

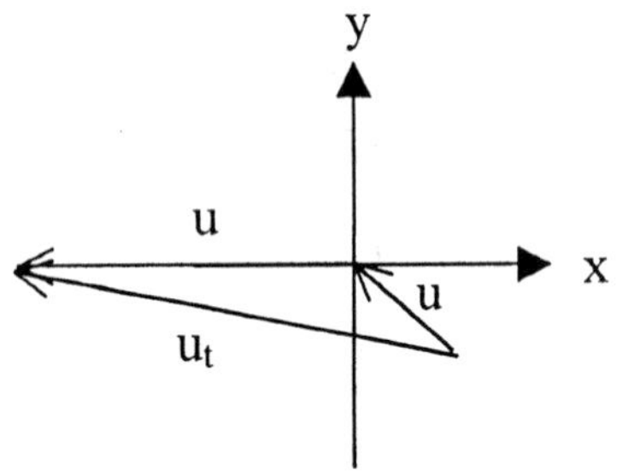

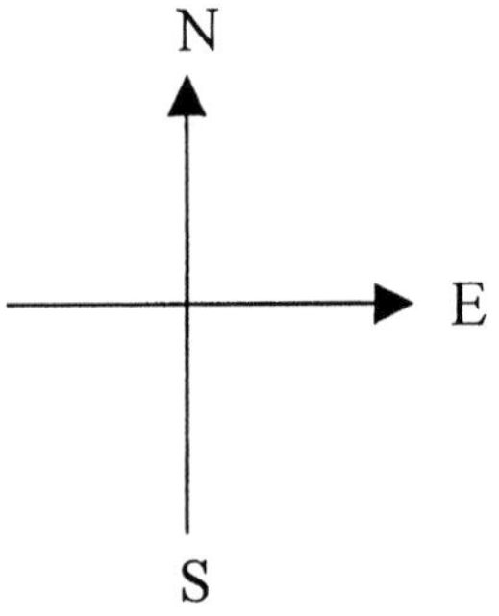

P. E. 1.4

At point (1,0), $\mathbf{G} = \mathbf{a}_y$;
at point (0,1), $\mathbf{G} = -\mathbf{a}_x$;
at point (2,0), $\mathbf{G} = \mathbf{a}_y$;
at point (1,1), $\mathbf{G} = \frac{-\mathbf{a}_x + \mathbf{a}_y}{\sqrt{2}}$; and so on.

It is evident that **G** is a unit vector at each point. Thus the vector field **G** is as sketched in Fig.1.8.

P. E. 1.5

Using the dot product,

$$\cos\theta_{AB} = \frac{\mathbf{A}\bullet\mathrm{B}}{AB} = \frac{-13}{\sqrt{10}\sqrt{65}} = -\sqrt{\frac{13}{50}}$$

or using the cross product,

$$\sin\theta_{AB} = \frac{|\mathrm{A}\,x\,\mathbf{B}|}{AB} = \sqrt{\frac{481}{650}}$$

Either way,

$$\underline{\underline{\theta_{AB} = 120.66^\circ}}$$

P. E. 1.6

(a) $$E_F = (E \cdot a_F)a_F = \frac{(E \cdot F)F}{|F|^2} = \frac{-10(4,-10,5)}{141}$$

$$= -0.2837a_x + 0.7092a_y - 0.3546a_z$$

(b) $$E \times F = \begin{vmatrix} a_x & a_y & a_z \\ 0 & 3 & 4 \\ 4 & -10 & 5 \end{vmatrix} = (55,16,-12)$$

$$a_{E \times F} = \pm(0.9398,0.2734,-0.205)$$

P. E. 1.7 $a + b + c = 0$ showing that a, b, and c form the sides of a triangle.

$a \cdot b = 0$,
hence it is a right angle triangle.

$$\text{Area} = \frac{1}{2}|a \times b| = \frac{1}{2}|b \times c| = \frac{1}{2}|c \times a|$$

$$\frac{1}{2}|a \times b| = \frac{1}{2}\begin{vmatrix} 4 & 0 & -1 \\ 1 & 3 & 4 \end{vmatrix} = \frac{1}{2}|(3,-17,12)|$$

$$\text{Area} = \frac{1}{2}\sqrt{9 + 289 + 144} = 10.51$$

P. E. 1.8

(a) $$P_1P_2 = \sqrt{(x_2 - x_1)^2 + (y_2 - y_1)^2 + (z_2 - z_1)^2}$$

$$= \sqrt{25 + 4 + 64} = 9.644$$

(b) $$r_P = r_{P_1} + \lambda(r_{P_2} - r_{P_1})$$

$$= (1,2,-3) + \lambda(-5,-2,8)$$

$$= (1 - 5\lambda, 2 - 2\lambda, -3 + 8\lambda).$$

(c) The shortest distance is

$$d = P_1P_3 \sin\theta = |P_1P_3 \times a_{P_1P_2}|$$

$$= \frac{1}{\sqrt{93}}\begin{vmatrix} 6 & -3 & 5 \\ -5 & -2 & 8 \end{vmatrix}$$

$$= \frac{1}{\sqrt{93}}|(-14,-73,-27)| = 8.2$$

Prob.1.1

$$\boldsymbol{r} = (-3,2,2) - (2,4,4) = (-5,-2,-2)$$

$$\boldsymbol{a}_r = \frac{\boldsymbol{r}}{|\boldsymbol{r}|} = \frac{(-5,-2,-2)}{\sqrt{25+4+4}} = -0.8703\boldsymbol{a}_x - 0.3482\boldsymbol{a}_y - 0.3482\boldsymbol{a}_z$$

Prob. 1.2

(a) $\boldsymbol{A} + 2\boldsymbol{B} = (2,5,-3) + (6,-8,0) = \underline{\underline{8\boldsymbol{a}_x - 3\boldsymbol{a}_y - 3\boldsymbol{a}_z}}$

(b) $\boldsymbol{A} - 5\boldsymbol{C} = (2,5,-3) - (5,5,5) = (-3,0,-8)$

$|\boldsymbol{A} - 5\boldsymbol{C}| = \sqrt{9+0+64} = \underline{\underline{8.544}}$

(c) $k\boldsymbol{B} = 3k\boldsymbol{a}_x - 4k\boldsymbol{a}_y$

$|k\boldsymbol{B}| = \sqrt{9k^2 + 16k^2} = \pm 5k = 2$

$\Rightarrow \underline{\underline{k = \pm 0.4}}$

(d) $\boldsymbol{A} \cdot \boldsymbol{B} = (2,5,-3) \cdot (3,-4,0) = 6 - 20 + 0 = 14$

$$\boldsymbol{A} \times \boldsymbol{B} = \begin{vmatrix} 2 & 5 & -3 \\ 3 & -4 & 0 \end{vmatrix} = (-12,-9,-23)$$

$$\frac{\boldsymbol{A} \times \boldsymbol{B}}{\boldsymbol{A} \cdot \boldsymbol{B}} = \left(\frac{12}{14}, \frac{9}{14}, \frac{23}{14}\right) = \underline{\underline{0.8571\boldsymbol{a}_x + 0.6428\boldsymbol{a}_y + 1.642\boldsymbol{a}_z}}$$

Prob. 1.3

(a) $\boldsymbol{A} - 2\boldsymbol{B} = (2,1,-3) - (0,2,-2) = (2,-1,-1)$

$\boldsymbol{A} - 2\boldsymbol{B} + \boldsymbol{C} = \underline{\underline{5\boldsymbol{a}_x + 4\boldsymbol{a}_y + 6\boldsymbol{a}_z}}$

(b) $\boldsymbol{A} + \boldsymbol{B} = (2,2,-4)$

$\boldsymbol{C} - 4(\boldsymbol{A} + \boldsymbol{B}) = (3,5,7) - (8,8,-16) = \underline{\underline{-5\boldsymbol{a}_x - 3\boldsymbol{a}_y + 23\boldsymbol{a}_z}}$

(c) $2\boldsymbol{A} - 3\boldsymbol{B} = (4,2,-6) - (0,3,-3) = (4,-1,-3)$

$|\boldsymbol{C}| = \sqrt{9+25+49} = 9.11$

$$\frac{2\boldsymbol{A} - 3\boldsymbol{B}}{|\boldsymbol{C}|} = \underline{\underline{0.439\boldsymbol{a}_x - 0.11\boldsymbol{a}_y - 0.3293\boldsymbol{a}_z}}$$

(d) $\boldsymbol{A}\cdot\boldsymbol{C} = 6+5-21 = -10,$

$$|\boldsymbol{B}| = \sqrt{2}$$

$$\boldsymbol{A}\cdot\boldsymbol{C} - |\boldsymbol{B}|^2 = -10+2 = \underline{\underline{-8}}$$

(e) $\frac{1}{3}\boldsymbol{A} + \frac{1}{4}\boldsymbol{C} = \left(\frac{2}{3},\frac{1}{3},-1\right) + \left(\frac{3}{4},\frac{5}{4},\frac{7}{4}\right) = (1.4167, 1.5833, 0.75)$

$$\frac{1}{2}\boldsymbol{B}\times\left(\frac{1}{3}\boldsymbol{A}+\frac{1}{4}\boldsymbol{C}\right) = \frac{1}{2}\begin{vmatrix} 0 & 1 & -1 \\ 1.4167 & 1.5833 & 0.75 \end{vmatrix} = \underline{\underline{1.1667\boldsymbol{a}_x - 0.7084\boldsymbol{a}_y - 0.7084\boldsymbol{a}_z}}$$

Prob.1.4

(a) T = $\underline{(3, -2, 1)}$ and S = $\underline{(4, 6, 2)}$

(b) $r_{TS} = r_s - r_t = (4, 6, 2) - (3, -2, 1) = \underline{a_x + 8a_y + a_z}$

(c) distance = $|r_{TS}| = \sqrt{1+64+1} = \underline{\underline{8.124}}$ m

Prob. 1.5

Let $\boldsymbol{D} = \alpha\boldsymbol{A} + \beta\boldsymbol{B} + \boldsymbol{C}$

$$= (5\alpha - \beta + 8)\boldsymbol{a}_x + (3\alpha + 4\beta + 2)\boldsymbol{a}_y + (-2\alpha + 6\beta)\boldsymbol{a}_z$$

$$\boldsymbol{D}_x = 0 \rightarrow 5\alpha - \beta + 8 = 0 \qquad (1)$$

$$\boldsymbol{D}_z = 0 \rightarrow -2\alpha + 6\beta = 0 \rightarrow \alpha = 3\beta \qquad (2)$$

Substituting (2) into (1),

$$15\beta - \beta + 8 = 0 \rightarrow \beta = -\frac{8}{14} = -\frac{4}{7}$$

Thus

$$\underline{\underline{\alpha = -\frac{12}{7}, \beta = -\frac{4}{7}}}$$

Prob. 1.6

$$\boldsymbol{A}\cdot\boldsymbol{B}=0\rightarrow 0=3\alpha+\beta-24 \qquad (1)$$
$$\boldsymbol{A}\cdot\boldsymbol{C}=0\rightarrow 0=5\alpha-2+4\gamma \qquad (2)$$
$$\boldsymbol{B}\cdot\boldsymbol{C}=0\rightarrow 0=15-2\beta-6\gamma \qquad (3)$$

In matrix form,

$$\begin{bmatrix}24\\2\\15\end{bmatrix}=\begin{bmatrix}3&1&0\\5&0&4\\0&2&6\end{bmatrix}\begin{bmatrix}\alpha\\\beta\\\gamma\end{bmatrix}$$

$$\Delta=\begin{vmatrix}3&1&0\\5&0&4\\0&2&6\end{vmatrix}=3(0-8)-1(30-0)+0(10-0)=-24-30=-54$$

$$\Delta_1=\begin{vmatrix}24&1&0\\2&0&4\\15&2&6\end{vmatrix}=-24\times 8-(12-60)=-144$$

$$\Delta_2=\begin{vmatrix}3&24&0\\5&2&4\\0&15&6\end{vmatrix}=3(12-60)-24\times 30=-864$$

$$\Delta_3=\begin{vmatrix}3&1&24\\5&0&2\\0&2&15\end{vmatrix}=-12-75+240=153$$

$$\alpha=\frac{\Delta_1}{\Delta}=\frac{-144}{-54}=\underline{\underline{2.667}}$$

$$\beta=\frac{\Delta_2}{\Delta}=\frac{-864}{-54}=\underline{\underline{16}}$$

$$\gamma=\frac{\Delta_3}{\Delta}=\frac{153}{-54}=\underline{\underline{-2.833}}$$

Prob. 1.7

(a) $\boldsymbol{A}\cdot\boldsymbol{B}=\boldsymbol{AB}\cos\theta_{AB}$

$\boldsymbol{A}\times\boldsymbol{B}=\boldsymbol{AB}\sin\theta_{AB}\boldsymbol{a}_n$

$(\boldsymbol{A}\cdot\boldsymbol{B})^2+|\boldsymbol{A}\times\boldsymbol{B}|^2=(\boldsymbol{AB})^2(\cos^2\theta_{AB}+\sin^2\theta_{AB})=(\boldsymbol{AB})^2$

(b) $\boldsymbol{a}_x \cdot (\boldsymbol{a}_y \times \boldsymbol{a}_z) = \boldsymbol{a}_x \cdot \boldsymbol{a}_x = 1$. Hence,

$$\frac{\boldsymbol{a}_y \times \boldsymbol{a}_z}{\boldsymbol{a}_x \cdot \boldsymbol{a}_y \times \boldsymbol{a}_z} = \frac{\boldsymbol{a}_x}{1} = \boldsymbol{a}_x$$

$$\frac{\boldsymbol{a}_z \times \boldsymbol{a}_x}{\boldsymbol{a}_x \cdot \boldsymbol{a}_y \times \boldsymbol{a}_z} = \frac{\boldsymbol{a}_y}{1} = \boldsymbol{a}_y$$

$$\frac{\boldsymbol{a}_x \times \boldsymbol{a}_y}{\boldsymbol{a}_x \cdot \boldsymbol{a}_y \times \boldsymbol{a}_z} = \frac{\boldsymbol{a}_z}{1} = \boldsymbol{a}_z$$

Prob. 1.8

(a) $\boldsymbol{P} + \boldsymbol{Q} = (2,2,0), \boldsymbol{P} + \boldsymbol{Q} - \boldsymbol{R} = (3,1,-2)$

$$|\boldsymbol{P} + \boldsymbol{Q} - \boldsymbol{R}| = \sqrt{9+1+4} = \sqrt{14} = \underline{\underline{3.742}}$$

(b) $$\boldsymbol{P} \cdot \boldsymbol{Q} \times \boldsymbol{R} = \begin{vmatrix} -2 & -1 & -2 \\ 4 & 3 & 2 \\ -1 & 1 & 2 \end{vmatrix} = -2(6-2) + (8+2) - 2(4+3) = -8 + 10 - 14 = \underline{\underline{-12}}$$

$$\boldsymbol{Q} \times \boldsymbol{R} = \begin{vmatrix} 4 & 3 & 2 \\ -1 & 1 & 2 \end{vmatrix} = (4,-10,7)$$

$$\boldsymbol{P} \cdot \boldsymbol{Q} \times \boldsymbol{R} = (-2,-1,-2) \cdot (4,-10,7) = -8 + 10 - 14 = \underline{\underline{-12}}$$

(c) $$\boldsymbol{Q} \times \boldsymbol{P} = \begin{vmatrix} 4 & 3 & 2 \\ -2 & -1 & -2 \end{vmatrix} = (-4,4,2)$$

$$\boldsymbol{Q} \times \boldsymbol{P} \cdot \boldsymbol{R} = (-4,4,2) \cdot (-1,1,2) = 4 + 4 + 4 = \underline{\underline{12}}$$

or $$\boldsymbol{Q} \times \boldsymbol{P} \cdot \boldsymbol{R} = \boldsymbol{R} \cdot \boldsymbol{Q} \times \boldsymbol{P} = \begin{vmatrix} -1 & 1 & 2 \\ 4 & 3 & 2 \\ -2 & -1 & -2 \end{vmatrix} = -(-6+2) - (-8+4) + 2(-4+6) = \underline{\underline{12}}$$

(d) $$(\boldsymbol{P} \times \boldsymbol{Q}) \cdot (\boldsymbol{Q} \times \boldsymbol{R}) = (4,-4,2) \cdot (4,-10,7) = 16 + 40 - 14 = \underline{\underline{42}}$$

(e) $$(\boldsymbol{P} \times \boldsymbol{Q}) \times (\boldsymbol{Q} \times \boldsymbol{R}) = \begin{vmatrix} 4 & -4 & 2 \\ 4 & -10 & 7 \end{vmatrix} = \underline{\underline{-48\boldsymbol{a}_x - 36\boldsymbol{a}_y - 24\boldsymbol{a}_z}}$$

(f) $$\cos\theta_{PR} = \frac{\boldsymbol{P} \cdot \boldsymbol{R}}{|\boldsymbol{P}||\boldsymbol{R}|} = \frac{(2-1-4)}{\sqrt{4+1+4}\sqrt{1+1+4}} = \frac{-3}{3\sqrt{6}} = \frac{-1}{\sqrt{6}}$$

$$\underline{\underline{\theta_{PR} = 114.1^\circ}}$$

$$(g)\ \sin\theta_{PQ} = \frac{|\boldsymbol{P}\times\boldsymbol{Q}|}{|\boldsymbol{P}||\boldsymbol{Q}|} = \frac{\sqrt{16+16+4}}{3\sqrt{16+9+4}} = \frac{6}{3\sqrt{29}}$$

$$\theta_{PQ} = \underline{\underline{21.8^\circ}}$$

Prob. 1.9

$$(a)\ \boldsymbol{T}_s = \boldsymbol{T}\cdot\boldsymbol{a}_s = \frac{\boldsymbol{T}\cdot\boldsymbol{S}}{|\boldsymbol{S}|} = \frac{(2,-6,-3)\cdot(1,2,1)}{\sqrt{6}} = \frac{-7}{\sqrt{6}} = \underline{\underline{-2.8577}}$$

$$(b)\ \boldsymbol{S}_T = (\boldsymbol{S}\cdot\boldsymbol{a}_T)\boldsymbol{a}_T = \frac{(\boldsymbol{S}\cdot\boldsymbol{T})\boldsymbol{T}}{\boldsymbol{T}^2} = \frac{-7(2,-6,3)}{7^2}$$

$$= \underline{\underline{-0.2857\boldsymbol{a}_x + 0.8571\boldsymbol{a}_y - 0.4286\boldsymbol{a}_z}}$$

$$(c)\ \sin\theta_{TS} = \frac{|\boldsymbol{T}\times\boldsymbol{S}|}{|\boldsymbol{T}||\boldsymbol{S}|} = \begin{vmatrix} 2 & -6 & 3 \\ 1 & 2 & 1 \end{vmatrix} = \frac{|(-12,1,10)|}{7\sqrt{6}} = \frac{\sqrt{245}}{7\sqrt{6}} = 0.9129$$

$$\Rightarrow \theta_{TS} = \underline{\underline{65.91^\circ}}$$

Prob. 1.10

$$(a)\ \boldsymbol{A}_B = \boldsymbol{A}\cdot\boldsymbol{a}_B = \frac{\boldsymbol{AB}}{|\boldsymbol{B}|} = \frac{-1+12+15}{\sqrt{1+4+9}} = \frac{26}{\sqrt{14}} = \underline{\underline{6.95}}$$

$$(b)\ \boldsymbol{B}_A = (\boldsymbol{B}\cdot\boldsymbol{a}_A)\boldsymbol{a}_A = \frac{(\boldsymbol{B}\cdot\boldsymbol{A})\boldsymbol{A}}{|\boldsymbol{A}|^2} = \frac{26(-1,6,5)}{(1+36+25)}$$

$$= \underline{\underline{-0.4193\boldsymbol{a}_x + 2.516\boldsymbol{a}_y + 2.097\boldsymbol{a}_z}}$$

$$(c)\ \cos\theta_{AB} = \frac{\boldsymbol{A}\cdot\boldsymbol{B}}{|\boldsymbol{A}||\boldsymbol{B}|} = \frac{26}{\sqrt{62}\sqrt{1+4+9}} = \frac{26}{\sqrt{62}\sqrt{14}}$$

$$\theta_{AB} = \underline{\underline{28.05^\circ}}$$

$$(d)\ \boldsymbol{A}\times\boldsymbol{B} = \begin{vmatrix} -1 & 6 & 5 \\ 1 & 2 & 3 \end{vmatrix} = 8\boldsymbol{a}_x + 8\boldsymbol{a}_y - 8\boldsymbol{a}_z$$

A unit vector perpendicular to both $\boldsymbol{A}$ and $\boldsymbol{B}$ is

$$\boldsymbol{a}_{A\times B} = \frac{8\boldsymbol{a}_x + 8\boldsymbol{a}_y - 8\boldsymbol{a}_z}{8\sqrt{1+1+1}} = \frac{\boldsymbol{a}_x + \boldsymbol{a}_y - \boldsymbol{a}_z}{\sqrt{3}} = \underline{\underline{0.577\boldsymbol{a}_x + 0.577\boldsymbol{a}_y - 0.577\boldsymbol{a}_z}}$$

Prob. 1.11

$$\cos\theta = \frac{\boldsymbol{H}\cdot\boldsymbol{a}_x}{|\boldsymbol{H}|} = \frac{3}{\sqrt{9+25+64}} = \frac{3}{98}$$

$$\theta_x = \underline{\underline{72.36^\circ}}$$

$$\cos\theta = \frac{\boldsymbol{H}\cdot\boldsymbol{a}_y}{|\boldsymbol{H}|} = \frac{5}{\sqrt{9+25+64}} = \frac{5}{98}$$

$$\theta_y = \underline{\underline{59.66^\circ}}$$

$$\cos\theta = \frac{\boldsymbol{H}\cdot\boldsymbol{a}_z}{|\boldsymbol{H}|} = \frac{-8}{\sqrt{9+25+64}} = \frac{-8}{98}$$

$$\theta_z = \underline{\underline{143.91^\circ}}$$

Prob. 1.12

$$\boldsymbol{Q}\times\boldsymbol{R} = \begin{vmatrix} 1 & 1 & 1 \\ 2 & 0 & 3 \end{vmatrix} = (3,-1,-2)$$

$$\boldsymbol{P}\cdot(\boldsymbol{Q}\times\boldsymbol{R}) = (2,-1,1)\cdot(3,-1,2) = 6+1-2 = \underline{\underline{5}}$$

Prob. 1.13

(a) Using the fact that

$$(\boldsymbol{A}\times\boldsymbol{B})\times\boldsymbol{C} = (\boldsymbol{A}\cdot\boldsymbol{C})\boldsymbol{B} - (\boldsymbol{B}\cdot\boldsymbol{C})\boldsymbol{A},$$

we get

$$\boldsymbol{A}\times(\boldsymbol{A}\times\boldsymbol{B}) = -(\boldsymbol{A}\times\boldsymbol{B})\times\boldsymbol{A} = \underline{\underline{(\boldsymbol{B}\cdot\boldsymbol{A})\boldsymbol{A} - (\boldsymbol{A}\cdot\boldsymbol{A})\boldsymbol{B}}}$$

(b) $\boldsymbol{A}\times(\boldsymbol{A}\times(\boldsymbol{A}\times\boldsymbol{B})) = \boldsymbol{A}\times[(\boldsymbol{A}\cdot\boldsymbol{B})\boldsymbol{A} - (\boldsymbol{A}\cdot\boldsymbol{A})\boldsymbol{B}]$

$$= \underline{\underline{(\boldsymbol{A}\cdot\boldsymbol{B})(\boldsymbol{A}\times\boldsymbol{A}) - (\boldsymbol{A}\cdot\boldsymbol{A})(\boldsymbol{A}\times\boldsymbol{B})}}$$

Prob. 1.14

$$\mathrm{A}\cdot(\mathrm{B}\times\mathrm{C}) = \begin{vmatrix} A_x & A_y & A_z \\ B_x & B_y & B_z \\ C_x & C_y & C_z \end{vmatrix}, \qquad (\mathrm{A}\times\mathrm{B})\cdot\mathrm{C} = \begin{vmatrix} A_x & A_y & A_z \\ B_x & B_y & B_z \\ C_x & C_y & C_z \end{vmatrix}$$

Hence, $\mathrm{A}\cdot(\mathrm{B}\times\mathrm{C}) = (\mathrm{A}\times\mathrm{B})\cdot\mathrm{C}$

Prob. 1.15

$$P_1P_2 = r_{P_2} - r_{P_1} = (-6,0,-3)$$
$$P_1P_3 = r_{P_3} - r_{P_1} = (1,5,-6)$$
$$P_1P_2 \times P_1P_3 = \begin{vmatrix} -6 & 0 & -3 \\ 1 & 5 & -6 \end{vmatrix} = (15,39,-30)$$

$$\text{Area of the triangle} = \frac{1}{2}\left|P_1P_2 \times P_1P_3\right| = \frac{1}{2}\sqrt{15^2 + 39^2 + 30^2} = \underline{\underline{25.72}}$$

Prob. 1.16

Let $P_1 = (4,1,-3)$, $P_2 = (-2, 5, 4)$, and $P_3 = (0, 1, 6)$

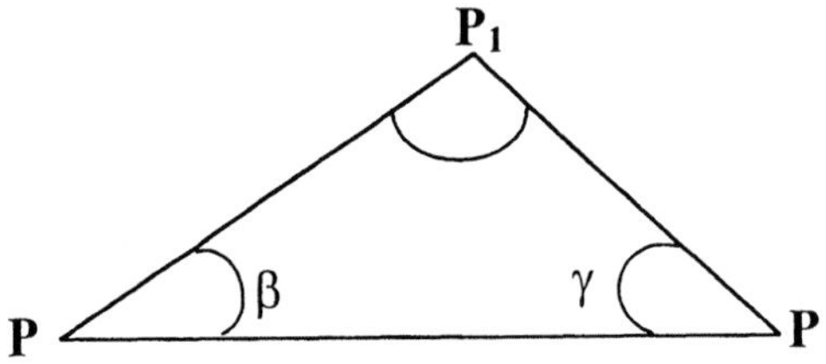

$$a = r_{P_2} - r_{P_1} = (-2,5,4) - (4,1,-3) = (-6,4,7)$$
$$b = r_{P_3} - r_{P_2} = (0,1,6) - (-2,5,4) = (2,-4,2)$$
$$c = r_{P_1} - r_{P_3} = (4,1,-3) - (0,1,6) = (4,0,-9)$$

Note that $a + b + c = 0$

$$a \cdot b = ab\cos(180 - \gamma) \rightarrow -\cos\gamma = \frac{a \cdot b}{|a||b|} = \frac{-12 - 16 + 14}{\sqrt{101}\sqrt{24}}$$

$$\gamma = \cos^{-1}\frac{14}{\sqrt{101}\sqrt{24}} = \underline{\underline{73.47^\circ}}$$

$$b \cdot c = bc\cos(180 - \beta) \rightarrow -\cos\beta = \frac{b \cdot c}{|b||c|} = \frac{8 + 0 - 18}{\sqrt{24}\sqrt{97}}$$

$$\beta = \cos^{-1}\frac{10}{\sqrt{24}\sqrt{97}} = \underline{\underline{78.04^\circ}}$$

$$a \cdot c = ac\cos(180 - \alpha) \rightarrow -\cos\alpha = \frac{a \cdot c}{|a||c|} = \frac{-24 + 0 - 63}{\sqrt{101}\sqrt{97}}$$

$$\alpha = \cos^{-1}\frac{87}{\sqrt{101}\sqrt{97}} = \underline{\underline{28.48^\circ}}$$

Prob. 1.17

(a) $\boldsymbol{r}_{PQ} = \boldsymbol{r}_Q - \boldsymbol{r}_P = (2,-1,3)-(-1,4,8) = (3,-5,-5)$

$r_{PQ} = |\boldsymbol{r}_{PQ}| = \sqrt{9+25+25} = \underline{\underline{7.681}}$

(b) $\boldsymbol{r}_{PR} = \boldsymbol{r}_R - \boldsymbol{r}_P = (-1,2,3)-(-1,4,8) = (0,-2,-5) = \underline{\underline{-2\boldsymbol{a}_y - 5\boldsymbol{a}_z}}$

(c) $\boldsymbol{r}_{QP} = -\boldsymbol{r}_{PQ} = -3\boldsymbol{a}_x + 5\boldsymbol{a}_y + 5\boldsymbol{a}_z$

$\boldsymbol{r}_{QR} = \boldsymbol{r}_Q - \boldsymbol{r}_R = (2,-1,3)-(-1,2,3) = 3\boldsymbol{a}_x - 3\boldsymbol{a}_y$

$$\cos\theta = \frac{\boldsymbol{r}_{QP}\cdot\boldsymbol{r}_{QR}}{|\boldsymbol{r}_{QP}||\boldsymbol{r}_{QR}|} = \frac{-9-15}{\sqrt{9+25+25}\sqrt{9+9}} = \frac{-24}{\sqrt{18}\sqrt{59}}$$

$\underline{\underline{\theta = 137.43^\circ}}$

(d) Area $= \frac{1}{2}|\boldsymbol{r}_{QP}\times\boldsymbol{r}_{QR}|$

$$\boldsymbol{r}_{QP}\times\boldsymbol{r}_{QR} = \begin{vmatrix} -3 & 5 & 5 \\ 3 & -3 & 0 \end{vmatrix} = 15\boldsymbol{a}_x + 15\boldsymbol{a}_y - 6\boldsymbol{a}_z$$

Area $= \frac{1}{2}\sqrt{15^2+15^2+6^2} = \underline{\underline{11.02}}$

(e) Perimeter $= QP + PR + RQ = \boldsymbol{r}_{QP} + \boldsymbol{r}_{PR} + \boldsymbol{r}_{QR}$

$= \sqrt{59} + \sqrt{4+25} + \sqrt{18}$

$= 7.681 + 5.385 + 4.243$

$= \underline{\underline{17.31}}$

Prob. 1.18

(a) Let $\boldsymbol{A} = (A,B,C)$ and $\boldsymbol{r} = (x,y,z)$

$(\boldsymbol{r}-\boldsymbol{A})\cdot\boldsymbol{A} = (x-A)A + (y-B)B + (z-C)C$

$= Ax + By + Cz + D$

where $D = -A^2 - B^2 - C^2$. Hence,

$(\boldsymbol{r}-\boldsymbol{A})\cdot\boldsymbol{A} = 0 \rightarrow Ax + By + Cz + D = 0$

which is the equation of a plane.

(b) $(\boldsymbol{r}-\boldsymbol{A})\cdot\boldsymbol{r} = (x-A)x + (y-B)y + (z-C)z$

If $(\boldsymbol{r}-\boldsymbol{A})\cdot\boldsymbol{r} = 0$, then

$$x^2 + y^2 + z^2 - Ax - By - Cz = 0$$

which is the equation of a sphere whose surface touches the origin.

(c) See parts (a) and (b).

Prob. 1.19

(a) Let P and Q be as shown below:

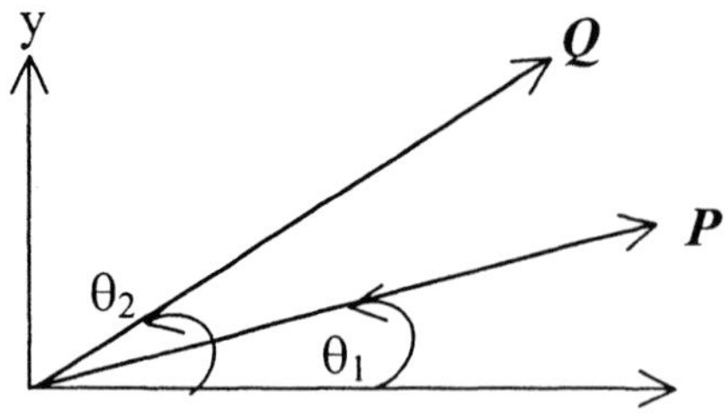

$|\boldsymbol{P}| = \cos^2\theta_1 + \sin^2\theta_1 = 1, |\boldsymbol{Q}| = \cos^2\theta_2 + \sin^2\theta_2 = 1,$

Hence **P** and **Q** are unit vectors.

(b) $\boldsymbol{P}\cdot\boldsymbol{Q} = (1)(1)\cos(\theta_2 - \theta_1)$

But $\boldsymbol{P}\cdot\boldsymbol{Q} = \cos\theta_1\cos\theta_2 + \sin\theta_1\sin\theta_2$. Thus,

$\underline{\underline{\cos(\theta_2 - \theta_1) = \cos\theta_1\cos\theta_2 + \sin\theta_1\sin\theta_2}}$

Let $\boldsymbol{P}_1 = \boldsymbol{P} = \cos\theta_1\boldsymbol{a}_x + \sin\theta_1\boldsymbol{a}_y$ and

$\boldsymbol{Q}_1 = \cos\theta_2\boldsymbol{a}_x - \sin\theta_2\boldsymbol{a}_y$.

$\boldsymbol{P_1}$ and $\boldsymbol{Q_1}$ are unit vectors as shown below:

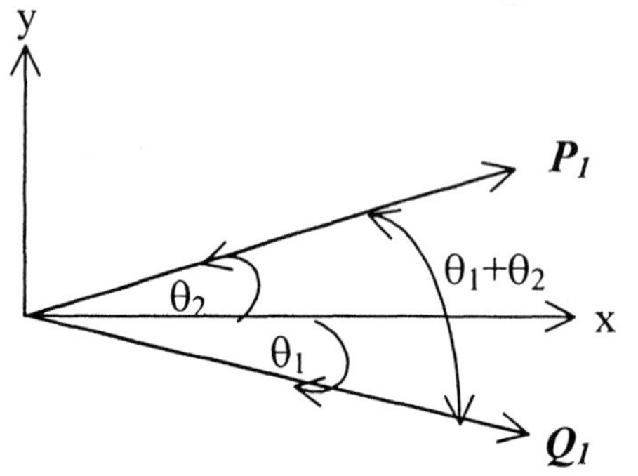

$\boldsymbol{P}_1\cdot\boldsymbol{Q}_1 = (1)(1)\cos(\theta_1 + \theta_2)$

But $\boldsymbol{P}_1\cdot\boldsymbol{Q}_1 = \cos\theta_1\cos\theta_2 - \sin\theta_1\sin\theta_2$,

$\underline{\underline{\cos(\theta_2 + \theta_1) = \cos\theta_1\cos\theta_2 - \sin\theta_1\sin\theta_2}}$

Alternatively, we can obtain this formula from the previous one by replacing θ_2 by $-\theta_2$ in $\boldsymbol{Q}$.

(c)

$$\frac{1}{2}|P-Q| = \frac{1}{2}|(\cos\theta_1 - \cos\theta_2)a_x + (\sin\theta_1 - \sin\theta_2)a_y$$

$$= \frac{1}{2}\sqrt{\cos^2\theta_1 + \sin^2\theta_1 + \cos^2\theta_2 + \sin^2\theta_2 - 2\cos\theta_1\cos\theta_2 - 2\sin\theta_1\sin\theta_2}$$

$$= \frac{1}{2}\sqrt{2 - 2(\cos\theta_1\cos\theta_2 + \sin\theta_1\sin\theta_2)} = \frac{1}{2}\sqrt{2 - 2\cos(\theta_2 - \theta_1)}$$

Let $\theta_2 - \theta_1 = \theta$, the angle between **P** and **Q.**

$$\frac{1}{2}|P-Q| = \frac{1}{2}\sqrt{2 - 2\cos\theta}$$

But $\cos 2A = 1 - 2\sin^2 A$.

$$\frac{1}{2}|P-Q| = \frac{1}{2}\sqrt{2 - 2 + 4\sin^2\theta/2} = \sin\theta/2$$

Thus,

$$\underline{\frac{1}{2}|P-Q| = |\sin\frac{\theta_2 - \theta_1}{2}|}$$

Prob. 1.20

$$w = \frac{w(1,-2,2)}{3} = (1,-2,2), \quad \boldsymbol{r} = \boldsymbol{r}_p - \boldsymbol{r}_o = (1,3,4) - (2,-3,1) = (-1,6,3)$$

$$\boldsymbol{u} = \boldsymbol{w} \times \boldsymbol{r} = \begin{vmatrix} 1 & -2 & 2 \\ -1 & 6 & 3 \end{vmatrix} = (-18,-5,4)$$

$$\underline{\underline{\boldsymbol{u} = -18\boldsymbol{a}_x - 5\boldsymbol{a}_y + 4\boldsymbol{a}_z}}$$

Prob. 1.21

(a) At T, $\boldsymbol{A} = (-4,3,-9)$

$$|A| = \sqrt{16+9+81} = \sqrt{106} = \underline{\underline{10.3}}$$

(b) Let $\boldsymbol{r}_{TS} = \boldsymbol{B} = B\boldsymbol{a}_B$

$$B = 5.6, \boldsymbol{a}_B = \boldsymbol{a}_A = \frac{(-4,3,-9)}{10.3}$$

$$\boldsymbol{r}_{TS} = \boldsymbol{B} = \frac{5.6(-4,3,9)}{10.3}$$

$$= -2.175\boldsymbol{a}_x + 1.631\boldsymbol{a}_y - 4.893\boldsymbol{a}_z$$

(c) $\boldsymbol{r}_{TS} = \boldsymbol{r}_S - \boldsymbol{r}_T \rightarrow \boldsymbol{r}_S = \boldsymbol{r}_T + \boldsymbol{r}_{TS}$

$$\therefore \boldsymbol{r}_S = -0.175\boldsymbol{a}_x + 0.631\boldsymbol{a}_y - 1.893\boldsymbol{a}_Z$$

Prob. 1.22

(a) At (1,2,3), $\boldsymbol{E}$ = (2,1,6)

$$|\boldsymbol{E}| = \sqrt{4+1+36} = \sqrt{41} = 6.403$$

(b) At (1,2,3), $\boldsymbol{F}$ = (2,-4,6)

$$\boldsymbol{E}_F = (\boldsymbol{E}\cdot\boldsymbol{a}_F)\boldsymbol{a}_F = \frac{(\boldsymbol{E}\cdot\boldsymbol{F})\boldsymbol{F}}{|\boldsymbol{F}|^2} = \frac{36}{56}(2,-4,6)$$

$$= 1.286\boldsymbol{a}_x - 2.571\boldsymbol{a}_y + 3.857\boldsymbol{a}_z$$

(c) At (0,1,-3), $\boldsymbol{E}$ = (0,1,-3), $\boldsymbol{F}$ = (0,-1,0)

$$\boldsymbol{E}\times\boldsymbol{F} = \begin{vmatrix} 0 & 1 & -3 \\ 0 & -1 & 0 \end{vmatrix} = (-3,0,0)$$

$$\boldsymbol{a}_{E\times F} = \pm\frac{\boldsymbol{E}\times\boldsymbol{F}}{|\boldsymbol{E}\times\boldsymbol{F}|} = \pm\boldsymbol{a}_x$$

CHAPTER 2

P. E. 2.1

(a) At P(1,3,5), x = 1, y = 3, z =5,

$\rho = \sqrt{x^2 + y^2} = \sqrt{10}$, z = 5, $\phi = \tan^{-1} y/x = 3$

$P(\rho,\phi,z) = P(\sqrt{10}, \tan^{-1} 3,5) = \underline{\underline{P(3.162, 71.6^o, 5)}}$

Spherical system:

$r = \sqrt{x^2 + y^2 + z^2} = \sqrt{35} = 5.916$

$r = \sqrt{x^2 + y^2 + z^2} = \sqrt{35} = 5.916$

$\theta = \tan^{-1} \sqrt{x^2 + y^2}/z = \tan^{-1} \sqrt{10}/5 = \tan^{-1} 0.6325 = 32.31°$

$P(r,\theta,\varphi) = \underline{\underline{P(5.916, 32.31°, 71.56°)}}$

At T(0,-4,3), x = 0 y =-4, z =3;

$\rho = \sqrt{x^2 + y^2} = 4, z = 3, \varphi = \tan^{-1} y/x = \tan^{-1} -4/0 = 270°$

$\underline{\underline{T(\rho,\varphi,z) = T(4, 270°, 3).}}$

Spherical system:

$r = \sqrt{x^2 + y^2 + z^2} = 5, \theta = \tan^{-1} \rho/z = \tan^{-1} 4/3 = 53.13°.$

$\underline{\underline{T(r,\theta,\varphi) = T(5, 53.13°, 270°).}}$

At S(-3-4-10), x =-3, y =-4, z =-10;

$\rho = \sqrt{x^2 + y^2} = 5,\ \phi = \tan^{-1} -4/-3 = 233.1°$

$\underline{\underline{S(\rho,\phi,z) = S(5, 233.1, -10).}}$

Spherical system:

$r = \sqrt{x^2 + y^2 + z^2} = 5\sqrt{5} = 11.18.$

$\theta = \tan^{-1} \rho/z = \tan^{-1} 5/-10 = 153.43°;$

$\underline{\underline{S(r,\theta,\phi) = S(11.18, 153.43°, 233.1°).}}$

(b) In Cylindrical system, $\rho = \sqrt{x^2 + y^2}$; $yz = z\rho\sin\theta$,

$Q_x = \dfrac{\rho}{\sqrt{\rho^2 + z^2}}$; $Q_y = 0$; $Q_z = \dfrac{z\rho\sin\phi}{\sqrt{\rho^2 + z^2}}$;

$$\begin{bmatrix} Q_\rho \\ Q_\phi \\ Q_z \end{bmatrix} = \begin{bmatrix} \cos\phi & \sin\phi & 0 \\ -\sin\phi & \cos\phi & 0 \\ 0 & 0 & 1 \end{bmatrix} \begin{bmatrix} Q_x \\ 0 \\ Q_z \end{bmatrix};$$

$$Q_\rho = Q_x \cos\phi = \frac{\rho\cos\phi}{\sqrt{\rho^2 + z^2}}, \qquad Q_\phi = -Q_x \sin\phi = \frac{-\rho\sin\phi}{\sqrt{\rho^2 + z^2}}$$

Hence,

$$\bar{Q} = \frac{\rho}{\sqrt{x^2 + z^2}}(\cos\phi\, \bar{a}_\rho, -\sin\phi\, a_\phi, -z\sin\phi\, \bar{a}_z).$$

In Spherical coordinates:

$$Q_x = \frac{r\sin\phi}{r} = \sin\phi;$$

$$Q_z = -r\sin\phi\sin\theta r\cos\theta\frac{1}{r} = -r\sin\theta\cos\theta\sin\phi.$$

$$\begin{bmatrix} Q_r \\ Q_\theta \\ Q_\phi \end{bmatrix} = \begin{bmatrix} \sin\theta\cos\phi & \sin\theta\sin\phi & \cos\theta \\ \cos\theta\cos\phi & \cos\theta\sin\phi & -\sin\phi \\ -\sin\phi & \cos\phi & 0 \end{bmatrix} \begin{bmatrix} Q_x \\ 0 \\ Q_z \end{bmatrix};$$

$$Q_r = Q_x \sin\theta\cos\phi + Q_z\cos\theta = \sin^2\theta\cos\phi - r\sin\theta\cos^2\theta\sin\phi.$$

$$Q_\theta = Q_x\cos\theta\cos\phi - Q_z\sin\theta = \sin\theta\cos\theta\cos\phi + r\sin^2\theta\cos\theta\sin\phi.$$

$$Q_\phi = -Q_x\sin\phi = -\sin\theta\sin\phi.$$

$$\therefore\ \bar{Q} = \sin\theta(\sin\theta\cos\phi - r\cos^2\theta\sin\phi)\bar{a}_r + \sin\theta\cos\phi(\cos\phi + r\sin\theta\sin\phi)\bar{a}_\theta - \sin\theta\sin\phi\bar{a}_\phi\ .$$

At T :

$$\bar{Q}(x,y,z) = \frac{4}{5}\bar{a}_x + \frac{12}{5}\bar{a}_z = 0.8\bar{a}_x + 2.4\bar{a}_z\ ;$$

$$\bar{Q}(\rho,\phi,z) = \frac{4}{5}(\cos 270^\circ\, \bar{a}_\rho - \sin 270^\circ\, \bar{a}_\phi - 3\sin 270^\circ\, \bar{a}_z$$

$$= 0.8\bar{a}_\phi + 2.4\bar{a}_z\ ;$$

$$\bar{Q}(r,\theta,\phi) = \frac{4}{5}(0 - \frac{45}{25}(-1))\bar{a}_r + \frac{4}{5}(\frac{3}{5})(0 - \frac{20}{5}(-1))\bar{a}_\theta - \frac{4}{5}(-1)\bar{a}_\phi$$

$$= \frac{36}{25}\bar{a}_r + \frac{48}{25}\bar{a}_\theta + \frac{4}{5}\bar{a}_\phi = 1.44\bar{a}_r + 1.92\bar{a}_\theta + 0.8\bar{a}_\phi\ ;$$

Note, that the magnitude of vector Q = 2.53 in all 3 cases above.

P.E. 2.2 (a)

$$\begin{bmatrix} A_x \\ A_y \\ A_z \end{bmatrix} = \begin{bmatrix} \cos\phi & -\sin\phi & 0 \\ \sin\phi & \cos\phi & 0 \\ 0 & 0 & 1 \end{bmatrix} \begin{bmatrix} \rho z \sin\phi \\ 3\rho\cos\phi \\ \rho\cos\phi\sin\phi \end{bmatrix}$$

$$\bar{A} = (\rho z\cos\phi\sin\phi - 3\rho\cos\phi\sin\phi)\bar{a}_x + (\rho z\sin^2\phi + 3\rho\cos^2\phi)\bar{a}_y + \rho\cos\phi\sin\phi\,\bar{a}_z.$$

But $\rho = \sqrt{x^2+y^2}$, $\tan\phi = \dfrac{y}{x}$, $\cos\phi = \dfrac{x}{\sqrt{x^2+y^2}}$, $\sin\phi = \dfrac{y}{\sqrt{x^2+y^2}}$;

Substituting all this yields:

$$\bar{A} = \frac{1}{\sqrt{x^2+y^2}}[(xyz - 3xy)\bar{a}_x + (zy^2 + 3x^2)\bar{a}_y + xy\bar{a}_z].$$

$$\begin{bmatrix} B_x \\ B_y \\ B_z \end{bmatrix} = \begin{bmatrix} \sin\theta\cos\phi & \cos\theta\cos\phi & -\sin\phi \\ \sin\theta\sin\phi & \cos\theta\sin\phi & \cos\phi \\ \cos\theta & -\sin\theta & 0 \end{bmatrix} \begin{bmatrix} r^2 \\ 0 \\ \sin\theta \end{bmatrix}$$

Since $r = \sqrt{x^2+y^2+z^2}$, $\tan\theta = \dfrac{\sqrt{x^2+y^2}}{z}$, $\tan\phi = \dfrac{y}{z}$;

and $\sin\theta = \dfrac{\sqrt{x^2+y^2}}{\sqrt{x^2+y^2+z^2}}$, $\cos\theta = \dfrac{z}{\sqrt{x^2+y^2+z^2}}$;

and $\sin\phi = \dfrac{y}{\sqrt{x^2+y^2}}$, $\cos\phi = \dfrac{x}{\sqrt{x^2+y^2}}$;

$$B_x = r^2\sin\theta\cos\phi - \sin\theta\sin\phi = rx - \frac{y}{r} = \frac{1}{r}(r^2x - y).$$

$$B_y = r^2\sin\theta\sin\phi + \sin\theta\cos\phi = ry + \frac{y}{x} = \frac{1}{r}(r^2y + x).$$

$$B_z = r^2\cos\theta = rz = \frac{1}{r}(r^2z).$$

Hence,

$$B = \frac{1}{\sqrt{x^2+y^2+z^2}}[\{x(x^2+y^2+z^2) - y\}a_x + \{y(x^2+y^2+z^2) + x\}\bar{a}_y + z(x^2+y^2+z^2)\bar{a}_z]$$

P.E.2.3 (a) At:

$(1, \pi/3, 0), \quad H = (0, 0.5, 1)$

$$\bar{a}_x = \cos\phi\, \bar{a}_\rho - \sin\phi\, \bar{a}_\phi = \frac{1}{2}(\bar{a}_\rho - \sqrt{3}\,\bar{a}_\phi)$$

$$\bar{H} \bullet \bar{a}_x = -\frac{\sqrt{3}}{4} = \underline{\underline{-0.433.}}$$

(b) At:

$$(1, \pi/3, 0), \quad \bar{a}_\theta = \cos\theta\, \bar{a}_\rho - \sin\theta\, \bar{a}_z = -\bar{a}_z.$$

$$\bar{H} \times \bar{a}_\theta = \begin{vmatrix} \bar{a}_\rho & \bar{a}_\phi & \bar{a}_z \\ 0 & \frac{1}{2} & 1 \\ 0 & 0 & -1 \end{vmatrix} = \underline{\underline{-0.5\, \bar{a}_\rho.}}$$

(c) $$(H \bullet \bar{a}_\rho)\bar{a}_\rho = \underline{\underline{0\, \bar{a}_\rho.}}$$

(d)

$$\bar{H} \times \bar{a}_z = \begin{vmatrix} \bar{a}_\rho & \bar{a}_\phi & \bar{a}_z \\ 0 & 1/2 & 1 \\ 0 & 0 & 1 \end{vmatrix} = 0.5\, \bar{a}_\rho.$$

$$\left| H \times \bar{a}_z \right| = \underline{\underline{0.5}}$$

P.E. 2.4

(a)

$$\bar{A} \bullet \bar{B} = (3, 2, -6) \bullet (4, 0, 3) = \underline{\underline{-6.}}$$

(b)

$$\left| \bar{A} \times \bar{B} \right| = \begin{vmatrix} 3 & 2 & -6 \\ 4 & 0 & 3 \end{vmatrix} = \left| 6\, \bar{a}_r - 33\, \bar{a}_\theta - 8\, \bar{a}_\phi \right|.$$

Thus the magnitude of $A \times B = \underline{\underline{34.48.}}$

(c)

At $(1, \pi/3, 5\pi/4), \quad \theta = \pi/3,$

$$\bar{a}_z = \cos\theta\, \bar{a}_r - \sin\theta\, \bar{a}_\theta = \frac{1}{2}\bar{a}_r - \frac{\sqrt{3}}{2}\bar{a}_\theta.$$

$$(\bar{A} \bullet \bar{a}_z)\bar{a}_z = (\frac{3}{2} - \sqrt{3})(\frac{1}{2}\bar{a}_r - \frac{\sqrt{3}}{2}\bar{a}_\theta)$$

$$= \underline{\underline{-0.116\,\bar{a}_r + 0.201\,\bar{a}_\theta}}.$$

Prob. 2.1

(a)

$x = \rho\cos\phi = 1\cos 60^\circ = 0.5;$

$y = \rho\sin\phi = 1\sin 120^\circ = 0.866;$

$z = 2;$

$P(x, y, z) = \underline{\underline{P(0.5, 0.866, 2)}}.$

(b)

$x = 2\cos 90^\circ = 0; \quad y = 2\sin 90^\circ = 1; \quad z = -10.$

$Q = \underline{\underline{Q(0, 1, -4)}}.$

(c)

$x = r\sin\theta\cos\phi = 3\sin 45^\circ\cos 210^\circ = -1.837;$

$y = r\sin\theta\sin\phi = 10\sin 135^\circ\sin 90^\circ = -1.061;$

$z = r\cos\theta = 10\cos 135^\circ = 2.121.$

$R(x, y, z) = \underline{\underline{R(-1.837, -1.061, 2.121)}}.$

(d)

$x = 4\sin 90^\circ\cos 30^\circ = 3.464.$

$y = 3\sin 30^\circ\sin 240^\circ = 2.$

$z = r\cos\theta = 4\cos 90^\circ = 0.$

$T(x, y, z) = \underline{\underline{T(3.464, 2, 0)}}.$

Prob.2.2

(a) Given P(1,-4,-3), convert to cylindrical and spherical values;

$$\rho = \sqrt{x^{2=} + y^2} = \sqrt{1^2 + (-4)^2} = \sqrt{17} = 4.123.$$

$$\phi = \tan^{-1}\frac{y}{x} = \tan^{-1}\frac{-4}{1} = 284.04^\circ.$$

$$\therefore\ P(\rho,\phi,z) = \underline{\underline{(4.123,\ 284.04^\circ,\ -3)}}.$$

Spherical:

$$r = \sqrt{x^2 + y^2 + z^2} = \sqrt{1 + 16 + 9} = 5.099.$$

$$\theta = \tan^{-1}\frac{\rho}{z} = \tan^{-1}\frac{4.123}{-3} = 126.04^\circ.$$

$$P(r,\theta,\phi) = \underline{\underline{P(5.099,\ 126.04^\circ,\ 284.04^\circ)}}.$$

Prob . 2.3

(a)

$$x = \rho\cos\phi, \qquad y = \rho\sin\phi,$$

$$V = \underline{\underline{\rho z\cos\phi - \rho^2\sin\phi\cos\phi + \rho z\sin\phi}}$$

(b)

$$U = x^2 + y^2 + z^2 + y^2 + 2z^2$$

$$= r^2 + r^2\sin^2\theta\sin^2\phi + 2r^2\cos^2\theta$$

$$= \underline{\underline{r^2[1 + \sin^2\theta\sin^2\phi + 2\cos^2\theta}}$$

Prob. 2.4

(a)

$$\begin{bmatrix} D_\rho \\ D_\phi \\ D_z \end{bmatrix} = \begin{bmatrix} \cos\phi & \sin\phi & 0 \\ -\sin\phi & \cos\phi & 0 \\ 0 & 0 & 1 \end{bmatrix} \begin{bmatrix} 0 \\ x+z \\ 0 \end{bmatrix}$$

$$D_\rho = (x+z)\sin\phi = (\rho\cos\phi + z)\sin\phi$$

$$D_\phi = (x+z)\cos\phi = (\rho\cos\phi + z)\cos\phi$$

$$\bar{D} = \underline{\underline{(\rho\cos\phi + z)[\sin\phi\ \bar{a}_\rho + \cos\phi\ \bar{a}_\phi]}}$$

Spherical:

$$\begin{bmatrix} D_r \\ D_\theta \\ D_\phi \end{bmatrix} = \begin{bmatrix} \ldots & \sin\theta\sin\phi & \ldots \\ \ldots & \cos\theta\sin\phi & \ldots \\ \ldots & \cos\phi & \ldots \end{bmatrix} \begin{bmatrix} 0 \\ x+z \\ 0 \end{bmatrix}$$

$D_r = (x+z)\sin\theta\cos\phi = r(\sin\theta\cos\phi + \cos\theta)\sin\theta\sin\phi.$

$D_\theta = (x+z)\cos\theta\sin\phi = r(\sin\theta\sin\phi + \cos\theta)\cos\theta\sin\phi.$

$D_\phi = (x+z)\cos\phi \quad = r(\sin\theta\cos\phi + \cos\theta)\cos\phi.$

$$\underline{\bar{D} = r(\sin\theta\cos\phi + \cos\theta)[\sin\theta\sin\phi\,\bar{a}_r + \cos\theta\sin\phi\,\bar{a}_\theta + \cos\phi\,\bar{a}_\phi].}$$

(b) Cylindrical:

$$\begin{bmatrix} E_\rho \\ E_\phi \\ E_z \end{bmatrix} = \begin{bmatrix} \cos\phi & \sin\phi & 0 \\ -\sin\phi & \cos\phi & 0 \\ 0 & 0 & 1 \end{bmatrix} \begin{bmatrix} y^2 - x^2 \\ xyz \\ x^2 - z^2 \end{bmatrix}$$

$$E_\rho = (y^2 - x^2)\cos\phi + xyz\sin\phi$$

$$= \rho^2(\sin^2\phi - \cos^2\phi)\cos\phi + \rho^2 z\cos\phi\sin^2\phi$$

$$= -\rho^2\cos 2\phi\cos\phi + \rho^2 z\sin^2\phi\cos\phi.$$

$$E_\phi = -(y^2 - x^2)\sin\phi + xyz\cos\phi$$

$$= \rho^2\cos 2\phi\sin\phi + \rho^2\cos 2\phi\sin\phi + \rho^2 z\sin\phi\cos^2\phi.$$

$$E_z = x^2 - z^2 = \rho^2\cos^2\phi - z^2.$$

$$\underline{\bar{E} = \rho^2\cos\phi(z\sin^2\phi - \cos 2\phi)\bar{a}_\rho + \rho^2\sin\phi(2\cos^2\phi + \cos 2\phi)\bar{a}_\phi + (\rho^2\cos\phi - z^2)\bar{a}_z.}$$

In spherical:

$$\begin{bmatrix} E_r \\ E_\theta \\ E_\phi \end{bmatrix} = \begin{bmatrix} \sin\theta\cos\phi & \sin\theta\sin\phi & \cos\theta \\ \cos\theta\cos\phi & \cos\theta\sin\phi & -\sin\theta \\ -\sin\phi & \cos\phi & 0 \end{bmatrix} \begin{bmatrix} y^2 - x^2 \\ xyz \\ x^2 - z^2 \end{bmatrix}$$

$$E_r = (y^2 - x^2)\sin\theta\cos\phi + xyz\sin\theta\sin\phi + (x^2 - z^2)\cos\theta;$$

but $x = r\sin\theta\cos\phi$, $\quad y = r\sin\theta\sin\phi$, $\quad z = r\cos\theta$;

$$E_r = r^2\sin^2\theta(\sin^2\phi - \cos^2\phi)\cos\phi + r^3\sin^3\theta\cos\theta\sin^2\phi\cos\phi + r^2(\sin^2\theta\cos^2\phi)\cos\theta;$$

$$E_\theta = (y^2 - x^2)\cos\theta\cos\phi + xyz\cos\theta\sin\phi - (x^2 - z^2)\sin\theta;$$

$$= -r^2\sin^2\theta\cos 2\phi\cos\theta\cos\phi + r^3\sin^2\theta\cos^2\theta\sin^2\phi\cos\phi - r^2(\sin^2\theta\cos^2\phi - \cos^2\theta)\sin\theta;$$

$$E_\phi = (x^2 - y^2)\sin\phi + xyz\cos\phi$$

$$= r^2\sin^2\theta\cos 2\phi\sin\phi + r^3\sin^2\theta\cos^2\phi\sin\phi\cos\theta;$$

$$\bar{E} = [-r^2 \sin^3\theta \cos 2\phi + r^3 \sin^3\theta \cos\theta \sin^2\phi \cos\phi + r^2(\sin^2\theta \cos^2\phi - \cos^2\theta)\cos\theta]\bar{a}_r +$$

$$[-r^2 \sin^2\theta \cos 2\phi \cos\theta \cos\phi + r^3 \sin^2\theta \cos^2\theta \sin^2\phi \cos\phi - r^2 \sin\theta(\sin^2\theta \cos^2\phi - \cos^2\theta)]\bar{a}_\theta +$$

$$+ \quad [r^2 \sin^2\theta \cos 2\phi \sin\phi + r^3 \sin^2\theta \cos^2\phi \sin\phi \cos\theta]\bar{a}_\phi$$

Prob. 2.5 (a)

$$\begin{bmatrix} F_\rho \\ F_\phi \\ F_z \end{bmatrix} = \begin{bmatrix} \cos\phi & \sin\phi & 0 \\ -\sin\phi & \sin\phi & 0 \\ 0 & 0 & 1 \end{bmatrix} \begin{bmatrix} \dfrac{x}{\sqrt{\rho^2 + z^2}} \\ \dfrac{y}{\sqrt{\rho^2 + z^2}} \\ \dfrac{4}{\sqrt{\rho^2 + z^2}} \end{bmatrix}$$

$$F_\rho = \frac{1}{\sqrt{\rho^2 + z^2}}[\rho\cos^2\phi + \rho\sin^2\phi] = \frac{\rho}{\sqrt{\rho^2 + z^2}};$$

$$F_\phi = \frac{1}{\sqrt{\rho^2 + z^2}}[-\rho\cos\phi\sin\phi + \rho\cos\phi\sin\phi] = 0;$$

$$F_z = \frac{4}{\sqrt{\rho^2 + z^2}};$$

$$\bar{F} = \frac{1}{\sqrt{\rho^2 + z^2}}(\rho\bar{a}_\rho + 4\bar{a}_z).$$

In Spherical:

$$\begin{bmatrix} F_r \\ F_\theta \\ F_\phi \end{bmatrix} = \begin{bmatrix} \sin\theta\cos\phi & \sin\theta\sin\theta & \cos\theta \\ \cos\theta\cos\phi & \cos\theta\sin\phi & -\sin\theta \\ -\sin\theta & \cos\phi & 0 \end{bmatrix} \begin{bmatrix} \dfrac{x}{r} \\ \dfrac{y}{r} \\ \dfrac{4}{r} \end{bmatrix}$$

$$F_r = \frac{r}{r}\sin^2\theta\cos^2\theta + \frac{r}{r}\sin^2\theta\sin^2\theta + \frac{4}{r}\cos\theta \quad = \sin^2\theta + \frac{4}{r}\cos\theta;$$

$$F_\theta = \sin\theta\cos\theta\cos^2\phi + \sin\theta\cos\theta\sin^2\phi - \frac{4}{r}\sin\theta = \sin\theta\cos\theta - \frac{4}{r}\sin\theta;$$

$$F_\phi = -\sin\theta\cos\phi\sin\phi + \sin\theta\sin\phi\cos\phi = 0;$$

$$\therefore\ \bar{F} = (\sin^2\theta + \frac{4}{r}\sin\theta)\bar{a}_r + \sin\theta(\cos\theta - \frac{4}{r})\bar{a}_\theta.$$

(b)

$$\begin{bmatrix} G_\rho \\ G_\phi \\ G_z \end{bmatrix} = \begin{bmatrix} \cos\phi & \sin\phi & 0 \\ -\sin\phi & \sin\phi & 0 \\ 0 & 0 & 1 \end{bmatrix} \begin{bmatrix} \frac{x\rho^2}{\sqrt{\rho^2+z^2}} \\ \frac{y\rho^2}{\sqrt{\rho^2+z^2}} \\ \frac{z\rho^2}{\sqrt{\rho^2+z^2}} \end{bmatrix}$$

$$G_\rho = \frac{\rho^2}{\sqrt{\rho^2+z^2}}[\rho\cos^2\phi + \rho\sin^2\phi] = \frac{\rho^3}{\sqrt{\rho^2+z^2}};$$

$$G_\phi = 0;$$

$$G_z = \frac{z\rho^2}{\sqrt{\rho^2+z^2}};$$

$$\bar{G} = \frac{\rho^2}{\sqrt{\rho^2+z^2}}(\rho\bar{a}_\rho + z\bar{a}_z).$$

Spherical :

$$\begin{bmatrix} G_r \\ G_\theta \\ G_\phi \end{bmatrix} = \begin{bmatrix} \sin\theta\cos\phi & \sin\theta\sin\phi & \cos\theta \\ \cos\theta\cos\phi & \cos\theta\sin\phi & -\sin\theta \\ -\sin\phi & \cos\phi & 0 \end{bmatrix} \begin{bmatrix} \frac{xr\sin\theta}{r} \\ y\sin\theta \\ z\sin\theta \end{bmatrix}$$

$$G_r = r\sin^2\theta\cos^2\phi + r\sin^2\theta\sin^2\phi + r\cos^2\theta\sin\theta$$

$$= r\sin^3\theta + r\cos^2\sin\theta = r\sin\theta.$$

$$G_\theta = r\sin^2\theta\cos\theta\cos^2\phi + r\sin^2\theta\cos\sin^3\phi - r\sin^3\theta\cos\theta$$

$$= r\sin^2\theta\cos\theta - r\sin^2\theta\cos\theta = 0.$$

$$G_\phi = -r\sin^2\theta\sin\phi\cos\phi + r\sin^2\theta\cos\phi\sin\phi = 0.$$

$$\therefore \bar{G} = \underline{\underline{r\sin\theta\,\bar{a}_r.}}$$

Prob. 2.6 (a)

$$\begin{bmatrix} A_x \\ A_y \\ A_z \end{bmatrix} = \begin{bmatrix} \cos\phi & -\sin\phi & 0 \\ \sin\phi & \cos\phi & 0 \\ 0 & 0 & 1 \end{bmatrix} \begin{bmatrix} \rho(z^2+1) \\ -\rho z\cos\phi \\ 0 \end{bmatrix}$$

$$A_x = \rho(z^2+1)\cos\phi + \rho z\sin\phi\cos\phi$$

$$= \sqrt{x^2+y^2}\,(z^2+1)\frac{x}{\sqrt{x^2+y^2}} + \sqrt{x^2+y^2}\left(\frac{zxy}{x^2+y^2}\right)$$

$$= x(z^2+1) + \frac{xyz}{\sqrt{x^2+y^2}}.$$

$$A_y = \rho(z^2+1)\sin\phi - \rho z\cos^2\phi$$

$$= \sqrt{x^2+y^2}\,(z^2+1)\frac{y}{\sqrt{x^2+y^2}} - \frac{x^2z}{\sqrt{x^2+y^2}}$$

$$= y(z^2+1) - \frac{x^2z}{\sqrt{x^2+y^2}};$$

$$A_z = 0;$$

$$\therefore \bar{A} = \underline{\underline{\left[x(z^2+1) + \frac{xyz}{\sqrt{x^2+y^2}}\right]\bar{a}_x + \left[y(z^2+1) - \frac{x^2z}{\sqrt{x^2+y^2}}\right]\bar{a}_y.}}$$

$$\begin{bmatrix} B_x \\ B_y \\ B_z \end{bmatrix} = \begin{bmatrix} \sin\theta\cos\phi & \cos\theta\cos\phi & -\sin\phi \\ \sin\theta\sin\phi & \cos\theta\sin\phi & \cos\phi \\ \cos\theta & -\sin\theta & 0 \end{bmatrix} \begin{bmatrix} 2x \\ r\cos\theta\cos\theta \\ -r\sin\phi \end{bmatrix}$$

$$B_x = 2x\sin\theta\cos\phi + r\cos^2\theta\cos^2\phi + r\sin^2\phi$$

$$= \frac{2x^2\sqrt{x^2+y^2}}{\sqrt{x^2+y^2}\sqrt{x^2+y^2+z^2}} + \frac{\sqrt{x^2+y^2+z^2}}{x^2+y^2+z^2}\left(\frac{xz}{x^2+y^2}\right) + \sqrt{x^2+y^2+z^2}\left(\frac{y^2}{x^2+y^2}\right)$$

$$= \frac{2x^2}{\sqrt{x^2+y^2+z^2}} + \frac{xz}{(x^2+y^2)\sqrt{x^2+y^2+z^2}} + \frac{y^2\sqrt{x^2+y^2+z^2}}{x^2+y^2};$$

$$B_y = 2x\sin\theta\sin\phi + r\cos^2\theta\sin\phi\cos\phi - r\sin\phi\cos\phi$$

$$= \frac{2xy\sqrt{x^2+y^2}}{\sqrt{x^2+y^2}\sqrt{x^2+y^2+z^2}} + \frac{\sqrt{x^2+y^2+z^2}\,(xyz^2)}{x^2+y^2+z^2} - \sqrt{x^2+y^2+z^2}\left(\frac{xy}{x^2+y^2}\right)$$

$$= \frac{2xy}{\sqrt{x^2+y^2+z^2}} + \frac{xyz^2}{x^2+y^2\sqrt{x^2+y^2+z^2}} - \frac{xy\sqrt{x^2+y^2+z^2}}{\sqrt{x^2+y^2}};$$

$$B_z = 2x\cos - r\sin\theta\cos\theta\cos\phi$$

$$= \frac{2xz}{\sqrt{x^2+y^2+z^2}} - \frac{\sqrt{x^2+y^2+z^2}\quad (xy)\sqrt{x^2+y^2}}{(x^2+y^2+z^2)\quad \sqrt{x^2+y^2}}$$

$$= \frac{2xz}{\sqrt{x^2+y^2+z^2}} - \frac{xz}{\sqrt{x^2+y^2+z^2}} = \frac{xz}{\sqrt{x^2+y^2+z^2}};$$

$$\therefore \bar{B} = \left[\frac{2x^2}{\sqrt{x^2+y^2+z^2}} + \frac{xz}{(x^2+y^2)\sqrt{x^2+y^2+z^2}} + \frac{y^2\sqrt{x^2+y^2+z^2}}{x^2+y^2}\right]\bar{a}_x +$$

$$\left[\frac{2xy}{\sqrt{x^2+y^2+z^2}} + \frac{xyz^2}{(x^2+y^2)\sqrt{x^2+y^2+z^2}} - \frac{xy\sqrt{x^2+y^2+z^2}}{x^2+y^2}\right]\bar{a}_y +$$

$$\left[\frac{xz}{\sqrt{x^2+y^2+z^2}}\right]\bar{a}_z.$$

Prob 2.7 (a)

$$\begin{bmatrix} C_x \\ C_y \\ C_z \end{bmatrix} = \begin{bmatrix} \cos\phi & -\sin\phi & 0 \\ \sin\phi & \cos\phi & 0 \\ 0 & 0 & 1 \end{bmatrix} \begin{bmatrix} z\sin\phi \\ -\rho\cos\phi \\ 2\rho z \end{bmatrix}$$

$$C_x = z\sin\phi\cos\phi + \rho\sin\phi\cos\phi = \frac{xyz}{x^2+y^2} + \frac{xy\sqrt{x^2+y^2}}{x^2+y^2};$$

$$C_y = z\sin^2\phi - \rho\cos^2\phi = \frac{y^2 z}{x^2+y^2} - \frac{x^2\sqrt{x^2+y^2}}{x^2+y^2};$$

$$C_z = 2\rho z = 2z\sqrt{x^2+y^2};$$

$$\therefore\ \bar{C} = \left(\frac{xyz}{x^2+y^2} + \frac{xy}{\sqrt{x^2+y^2}}\right)\bar{a}_x + \left(\frac{y^2 z}{x^2+y^2} - \frac{x^2}{\sqrt{x^2+y^2}}\right)\bar{a}_y + 2z\sqrt{x^2+y^2}\,\bar{a}_z.$$

(b)

$$\begin{bmatrix} D_x \\ D_y \\ D_z \end{bmatrix} = \begin{bmatrix} \sin\theta\cos\phi & \cos\theta\cos\phi & -\sin\phi \\ \sin\theta\sin\phi & \cos\theta\sin\phi & \cos\phi \\ \cos\theta & -\sin\phi & 0 \end{bmatrix} \begin{bmatrix} \frac{\sin\theta}{r^2} \\ \frac{\cos\theta}{r^2} \\ 0 \end{bmatrix}$$

$$D_x = \frac{\sin^2\theta\cos\phi}{r^2} + \frac{\cos^2\theta\cos\phi}{r^2} = \frac{\cos\phi}{r^2} = \frac{x}{\sqrt{x^2+y^2}\,(x^2+y^2+z^2)};$$

$$D_y = \frac{\sin^2\theta\sin\phi}{r^2} + \frac{\cos^2\theta\sin\phi}{r^2} = \frac{\sin\phi}{r^2} = \frac{y}{\sqrt{x^2+y^2}\,(x^2+y^2+z^2)};$$

$$D_z = \frac{\sin\theta\cos\theta}{r^2} - \frac{\sin\theta\cos\theta}{r^2} = 0;$$

$$\therefore\ \bar{D} = \frac{1}{\sqrt{x^2+y^2}\,(x^2+y^2+z^2)}(x\bar{a}_x + y\bar{a}_y)$$

Prob. 2.8 (a)

$$\bar{a}_x \bullet \bar{a}_\rho = (\cos\phi \bar{a}_\rho - \sin\phi \bar{a}_\phi) \bullet \bar{a}_\rho = \cos\phi$$

$$\bar{a}_x \bullet \bar{a}_\phi = (\cos\phi \bar{a}_\rho - \sin\phi \bar{a}_\phi) \bullet \bar{a}_\phi = -\sin\phi$$

$$\bar{a}_y \bullet \bar{a}_\rho = (\sin\phi \bar{a}_\rho + \cos\phi \bar{a}_\phi) \bullet \bar{a}_\rho = \sin\phi$$

$$\bar{a}_y \bullet \bar{a}_\phi = (\sin\phi \bar{a}_\rho + \sin\phi \bar{a}_\phi) \bullet \bar{a}_\phi = \cos\phi$$

(b)

Since $\bar{a}_\rho$, $\bar{a}_\phi$, and $\bar{a}_z$ are mutually orthogonal

$$\bar{a}_z \bullet \bar{a}_z = 1; \qquad \bar{a}_z \bullet \bar{a}_\rho = 0; \qquad \bar{a}_z \bullet \bar{a}_\phi = 0.$$

Also, $\bar{a}_x \bullet \bar{a}_z = 0; \qquad \bar{a}_y \bullet \bar{a}_z = 0.$

$$\begin{bmatrix} \cos\phi & -\sin\phi & 0 \\ \sin\phi & \cos\phi & 0 \\ 0 & 0 & 1 \end{bmatrix} = \begin{bmatrix} \bar{a}_x \bullet \bar{a}_\rho & \bar{a}_x \bullet \bar{a}_\phi & \bar{a}_z \bullet \bar{a}_z \\ \bar{a}_y \bullet \bar{a}_\rho & \bar{a}_y \bullet \bar{a}_\phi & \bar{a}_y \bullet \bar{a}_z \\ \bar{a}_z \bullet \bar{a}_\rho & \bar{a}_z \bullet \bar{a}_\phi & \bar{a}_z \bullet \bar{a}_z \end{bmatrix}$$

(c)

In spherical system:

$$\bar{a}_x = \sin\theta\cos\phi\,\bar{a}_r + \cos\theta\cos\phi\,\bar{a}_\theta - \sin\phi\,\bar{a}_\phi.$$

$$\bar{a}_y = \sin\theta\sin\phi\,\bar{a}_r + \cos\theta\sin\phi\,\bar{a}_\theta - \cos\phi\,\bar{a}_\phi.$$

$$\bar{a}_z = \cos\theta\,\bar{a}_r - \sin\theta\,\bar{a}_\theta.$$

Hence,

$$\bar{a}_x \bullet \bar{a}_r = \sin\theta\cos\phi;$$

$$\bar{a}_x \bullet \bar{a}_\theta = \cos\theta\cos\phi;$$

$$\bar{a}_y \bullet \bar{a}_r = \sin\theta\sin\phi;$$

$$\bar{a}_y \bullet \bar{a}_\theta = \cos\theta\sin\phi;$$

$$\bar{a}_z \bullet \bar{a}_r = \cos\theta;$$

$$\bar{a}_z \bullet \bar{a}_\theta = -\sin\theta;$$

Prob 2.9 (a)

$$r = \sqrt{x^2 + y^2 + z^2} = \sqrt{\rho^2 + z^2}.$$

$$\theta = \tan^{-1}\frac{\rho}{z}; \qquad \phi = \phi.$$

or

$$\rho = \sqrt{x^2 + y^2} = \sqrt{r^2 \sin^2\theta\cos^2\phi + r^2\sin^2\theta\sin^2\phi}.$$

$$= r\sin\theta;$$

$$z = r\cos\theta; \qquad \phi = \phi.$$

(a) From the figures below,

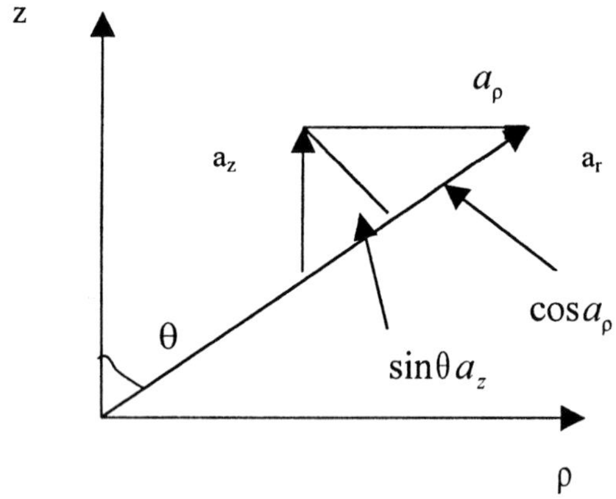

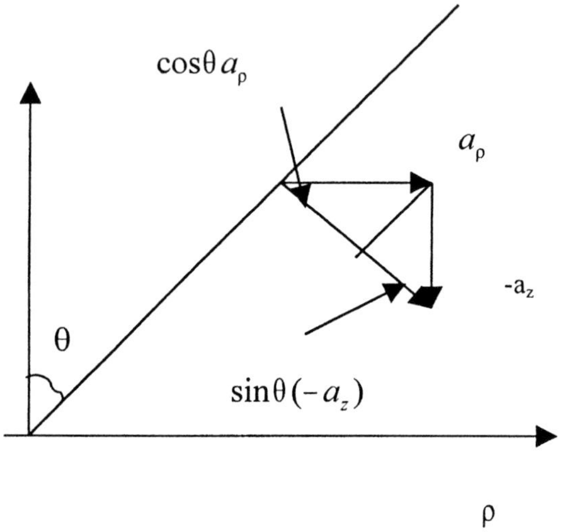

$\bar{a}_r = \sin\theta\, \bar{a}_z + \cos\theta\, \bar{a}_\rho; \quad \bar{a}_\theta = \cos\theta\, \bar{a}_\rho - \sin\theta\, \bar{a}_z; \quad \bar{a}_\phi = \bar{a}_\phi.$

Hence,

$$\begin{bmatrix} \bar{a}_r \\ \bar{a}_\theta \\ \bar{a}_\phi \end{bmatrix} = \begin{bmatrix} \sin\theta & 0 & \cos\theta \\ \cos\theta & 0 & -\sin\theta \\ 0 & 1 & 0 \end{bmatrix} \begin{bmatrix} \bar{a}_\rho \\ \bar{a}_\phi \\ \bar{a}_z \end{bmatrix}$$

From the figures below,

$\bar{a}_\rho = \cos\theta\, \bar{a}_\theta + \sin\theta\, \bar{a}_x; \quad \bar{a}_z = \cos\theta\, \bar{a}_x - \sin\theta\, \bar{a}_\theta; \quad \bar{a}_\phi = \bar{a}_\phi.$

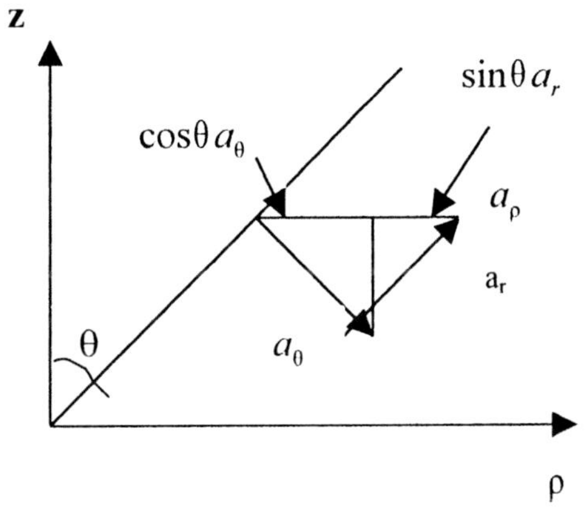

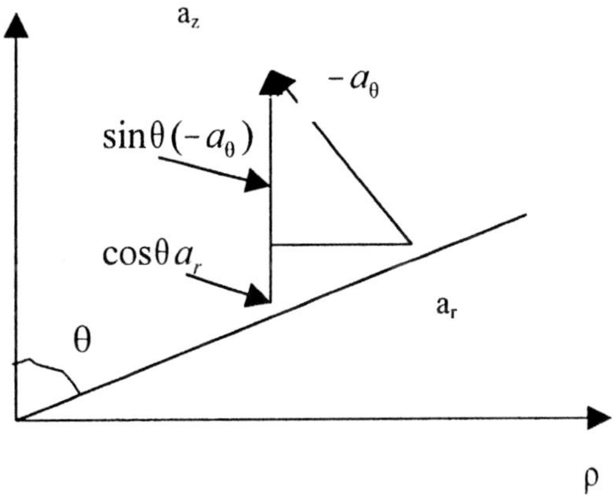

$$\begin{bmatrix} \bar{a}_\rho \\ \bar{a}_\phi \\ \bar{a}_z \end{bmatrix} = \begin{bmatrix} \sin\theta & \cos\theta & 0 \\ 0 & 0 & 1 \\ \cos\theta & -\sin\theta & 0 \end{bmatrix} \begin{bmatrix} \bar{a}_r \\ \bar{a}_\theta \\ \bar{a}_z \end{bmatrix}$$

Prob. 2.10 (a)

$$\begin{bmatrix} H_\rho \\ H_\phi \\ H_z \end{bmatrix} = \begin{bmatrix} \cos\phi & \sin\phi & 0 \\ -\sin\phi & \cos\phi & 0 \\ 0 & 0 & 1 \end{bmatrix} \begin{bmatrix} xy^2z \\ x^2yz \\ xyz^2 \end{bmatrix}$$

$$H_\rho = xy^2z\cos\phi + x^2yz\sin\phi = \rho^3 z\cos^2\phi\sin^2\phi + \rho^3 z\cos^2\phi\sin^2\phi.$$

$$= \frac{1}{2}\rho^3 z\sin^2 2\phi$$

$$H_\phi = -xy^2z\sin\phi + x^2yz\cos\phi = -\rho^3 z\cos\phi\sin^3\phi + \rho^3 z\cos\phi\sin\phi$$

$$= \rho^3 z\cos\phi\sin\phi\cos 2\phi.$$

$$H_z = xyz^2 = \rho^2 z^2 \sin\phi\cos\phi.$$

$$\bar{H} = \frac{1}{2}\rho^3 z\sin^2 2\phi\, \bar{a}_\rho + \frac{1}{2}\rho^3 z\sin 2\phi\cos 2\phi\, \bar{a}_\phi + \frac{1}{2}\rho^3 z\sin 2\phi\, \bar{a}_z.$$

$$\begin{bmatrix} H_r \\ H_\theta \\ H_\phi \end{bmatrix} = \begin{bmatrix} \sin\theta\cos\phi & \sin\theta\sin\phi & \cos\theta \\ \cos\theta\cos\phi & \cos\theta\sin\phi & -\sin\theta \\ -\sin\phi & \cos\phi & 0 \end{bmatrix} \begin{bmatrix} xy^2z \\ x^2yz \\ xyz^2 \end{bmatrix}$$

$x = r\sin\theta\cos\phi, \qquad y = r\sin\theta\sin\phi, \qquad z = r\cos\theta.$

$H_r = xyz[y\sin\theta\cos\phi + x\sin\theta\sin\phi + z\cos\theta$

$= r^3\sin^2\theta\cos\theta\sin\phi[r\sin^2\theta\sin\phi\cos\phi + r\sin^2\theta\sin\phi\cos\phi + r\cos^2\theta]$

$H_\theta = xyz[y\cos\theta\cos\phi + x\cos\theta\sin\phi - z\sin\theta]$

$= r^3\sin^2\theta\cos\theta\sin\phi\cos\phi\,[r\sin\theta\cos\theta\sin\phi\cos\phi + r\sin\theta\cos\theta\sin\phi\cos\phi - r\cos\theta\sin\theta]$

$H_\phi = xyz[-y\sin\phi + x\cos\phi]$

$= r^3\sin^2\theta\cos\theta\sin\phi\cos\phi[-r\sin\theta\sin^2\phi + r\sin\theta\cos^2\phi]$

$= r^4\sin^3\theta\cos\theta\sin\phi\cos 2\phi.$

$$\bar{H} = r^4\sin^2\theta\cos\theta\sin\phi\cos\phi[(\sin^2\theta\sin 2\phi + \cos^2\theta)\bar{a}_r + (\sin\theta\cos\theta\sin 2\phi - \cos\theta\sin\theta)\bar{a}_\theta + \sin\theta\cos 2\phi\,\bar{a}_\phi].$$

(b)

$At\,(3 - 45),\ \bar{H}(x,y,z) = -60(-4,3,5)$

$\left|\bar{H}(x,y,z)\right| = 424.3$

This will help check $H(\rho,\phi,z)$ and $H(r,\theta,\phi)$

$\rho = 5,\ z = 5,\ \phi = 360° - \tan^{-1}\frac{4}{3} = 306.87°$

$$\bar{H} = \frac{1}{2}(125)(5)(-0.96)\bar{a}_\rho + \frac{1}{2}(125)(5)(-0.90)\big(-0.277\big)\bar{a}_\phi + \frac{1}{2}(25)(5)(-0.96)\bar{a}_z$$

$$= 288\,\bar{a}_\rho + 84\,\bar{a}_\phi - 300\,\bar{a}_z$$

Spherical,

$r = \sqrt{50} = 5\sqrt{2}; \quad \sin\theta = \frac{5}{5\sqrt{2}} = \frac{1}{\sqrt{2}}; \quad \cos\theta = \frac{5}{5\sqrt{2}} = \frac{1}{\sqrt{2}}.$

$\& \quad \sin\phi = -\frac{4}{5}, \quad \cos\phi = \frac{3}{5}.$

$$\therefore\ \bar{H} = 2500(\frac{1}{2})(\frac{1}{\sqrt{2}})(-\frac{12}{25})[\{\frac{1}{2}*2(-\frac{12}{28}) + \frac{1}{2}\}\bar{a}_r + \{\frac{1}{2}*2(-\frac{12}{25}) - \frac{1}{2}\}\bar{a}_\theta + \frac{1}{\sqrt{2}}\{\frac{9}{12} - \frac{16}{25}\}\bar{a}_\phi]$$

$$= -8.485\bar{a}_r + 415.8\bar{a}_\theta + 84\bar{a}_\phi.$$

Prob 2.11 (a)

$$\begin{bmatrix} A_x \\ A_y \\ A_y \end{bmatrix} = \begin{bmatrix} \cos\phi & -\sin\phi & 0 \\ \sin\phi & \cos\phi & 0 \\ 0 & 0 & 1 \end{bmatrix} \begin{bmatrix} \rho\cos\phi \\ 0 \\ \rho z^2\sin\phi \end{bmatrix}$$

$$A_x = \rho\cos^2\phi = \sqrt{x^2+y^2}\,\frac{x^2}{x^2+y^2} = \frac{x^2}{\sqrt{x^2+y^2}}$$

$$A_y = \rho\sin\phi\cos\phi = \sqrt{x^2+y^2}\,\frac{xy}{x^2+y^2} = \frac{xy}{\sqrt{x^2+y^2}}$$

$$A_z = \frac{1}{\sqrt{x^2+y^2}}[x^2\bar{a}_x + xy\bar{a}_y + yz\bar{a}_z].$$

At (3,-4,0) x=3, y=-4, z=0;

$$\bar{A} = \frac{1}{5}[9\bar{a}_x - 12\bar{a}_y].$$

$$|\bar{A}| = 3$$

(b)

$$\begin{bmatrix} A_r \\ A_\theta \\ A_\phi \end{bmatrix} = \begin{bmatrix} \sin\theta\cos\phi & \sin\theta\sin\phi & \cos\theta \\ \cos\phi\cos\phi & \cos\theta\sin\phi & -\sin\theta \\ -\sin\phi & \cos\phi & 0 \end{bmatrix} \begin{bmatrix} \frac{x^2}{\rho} \\ \frac{xy}{\rho} \\ \frac{yz^2}{\rho} \end{bmatrix}$$

$$x = r\sin\theta\cos\phi,\quad y = r\sin\theta\sin\phi,\quad z = r\cos\theta,\quad \rho = r\sin\theta.$$

$$A_r = \frac{r^2\sin^2\theta\cos^2\phi}{r\sin\theta}\sin\theta\cos\phi + \frac{r^2\sin^2\theta\cos\phi\sin\phi}{r\sin\theta}\sin\theta\sin\phi + \frac{r^3\sin\theta\cos^2\phi}{r\sin\theta}\sin\phi\cos\theta$$

$$= r\sin^2\theta\cos\phi + r^2\cos^3\theta\sin\theta$$

$$A_\theta = r\sin\theta\cos^2\phi\cos\theta\cos\phi + r\sin\theta\cos\phi\sin\phi\cos\theta\sin\phi - r^2\cos^2\theta\sin\phi\sin\theta$$

$$= r\sin\theta\cos\theta\cos\phi - r^2\sin\theta\cos^2\sin\phi$$

$$= r\sin\theta\cos\theta[\cos\phi - r\cos\theta\sin\phi]$$

$$A_\phi = -r\sin\theta\cos^2\phi\sin\phi + r\sin\theta\cos\phi\sin\phi\cos\phi = 0.$$

$\therefore$

$$\bar{A} = r[\sin^2\theta\cos\phi + r\cos^3\theta\sin\phi]\bar{a}_r + r\sin\theta\cos\theta[\cos\phi - r\cos\theta\sin\phi]\bar{a}_\theta.$$

$$At\ (3-4,0),\ r=5,\ \theta=\pi/2,\quad \phi=306.83$$

$$\cos\phi = 3/5, \qquad \sin\phi = -4/5.$$

$$\bar{A} = 5[1^2 * \frac{3}{5} + 5(0)(-4/5)]\bar{a}_r + 5(1)(0)a_\theta$$

$$= 3\bar{a}_r.$$

$$|\bar{A}| = 3.$$

Prob 2.12

$$\begin{bmatrix} A_x \\ A_y \\ A_z \end{bmatrix} = \begin{bmatrix} \cos\phi & -\sin\phi & 0 \\ \sin\phi & \cos\phi & 0 \\ 0 & 0 & 1 \end{bmatrix} \begin{bmatrix} A_\rho \\ A_\phi \\ A_z \end{bmatrix}$$

$$= \begin{bmatrix} \frac{x}{\sqrt{x^2+y^2}} & -\frac{y}{\sqrt{x^2+y^2}} & 0 \\ \frac{y}{\sqrt{x^2+y^2}} & \frac{x}{\sqrt{x^2+y^2}} & 0 \\ 0 & 0 & 1 \end{bmatrix} \begin{bmatrix} A_\rho \\ A_\phi \\ A_z \end{bmatrix}$$

$$\begin{bmatrix} A_x \\ A_y \\ A_z \end{bmatrix} = \begin{bmatrix} \sin\theta\cos\phi & \cos\theta\cos\phi - & \sin\phi \\ \sin\theta\sin\phi & \cos\theta\sin\phi & \cos\phi \\ \cos\theta & -\sin\theta & 0 \end{bmatrix} \begin{bmatrix} A_r \\ A_\theta \\ A_\phi \end{bmatrix}$$

$$= \begin{bmatrix} \frac{x}{\sqrt{x^2+y^2+z^2}} & = \frac{xz}{\sqrt{x^2+y^2}\sqrt{x^2+y^2+z^2}} & \frac{-y}{\sqrt{x^2+y^2}} \\ \frac{y}{\sqrt{x^2+y^2+z^2}} & \frac{yz}{\sqrt{x^2+y^2}\sqrt{x^2+y^2+z^2}} & \frac{x}{\sqrt{x^2+y^2}} \\ \frac{z}{\sqrt{x^2+y^2+z^2}} & -\frac{\sqrt{x^2+y^2}}{\sqrt{x^2+y^2+z^2}} & 0 \end{bmatrix} \begin{bmatrix} A_r \\ A_\phi \\ A_\phi \end{bmatrix}$$

Prob 2.13 (a) Using the results in Prob.2.9,

$$A_\rho = \rho z \sin\phi = r^2 \sin\theta\cos\theta\sin\phi$$
$$A_\phi = 3\rho\cos\phi = 3r\sin\theta\cos\phi$$
$$A_z = \rho\cos\phi\sin\phi = r\sin\theta\cos\phi\sin\phi$$

Hence,

$$\begin{bmatrix} A_r \\ A_\theta \\ A_\phi \end{bmatrix} = \begin{bmatrix} \sin\theta & 0 & \cos\theta \\ \cos\theta & 0 & -\sin\theta \\ 0 & 1 & 0 \end{bmatrix} \begin{bmatrix} r^2 \sin\theta\cos\theta\sin\phi \\ 3r\sin\theta\cos\phi \\ r\sin\theta\cos\phi\sin\phi \end{bmatrix}$$

$$A(r,\theta,\phi) = \underline{\underline{r\sin\theta\left[\sin\phi\cos\theta(r\sin\theta + \cos\phi)a_r + \sin\phi(r\cos^2\theta - \sin\theta\cos\phi)a_\theta + 3\cos\phi\, a_\phi\right]}}$$

At $(10, \pi/2, 3\pi/4)$, $\quad r = 10, \theta = \pi/2, \phi = 3\pi/4$

$$\bar{A} = 10(0a_r + 0.5a_\theta - \frac{3}{\sqrt{2}}a_\phi) = \underline{\underline{5a_\theta - 21.21a_\phi}}$$

(b) $\quad B_r = r^2 = (\rho^2 + z^2), \quad B_\theta = 0, \quad B_\phi = \sin\theta = \dfrac{\rho}{\sqrt{\rho^2 + z^2}}$

$$\begin{bmatrix} B_\rho \\ B_\phi \\ B_z \end{bmatrix} = \begin{bmatrix} \sin\theta & \cos\theta & 0 \\ 0 & 0 & 1 \\ \cos\theta & -\sin\theta & 0 \end{bmatrix} \begin{bmatrix} B_r \\ B_\theta \\ B_\phi \end{bmatrix}$$

$$B(\rho,\phi,z) = \sqrt{\rho^2 + z^2}\left(\rho a_\rho + \frac{\rho}{\rho^2 + z^2}a_\phi + z a_z\right)$$

At $(2, \pi/6, 1)$, $\quad \rho = 2, \phi = \pi/6, z = 1$

$$B = \sqrt{5}(2a_\rho + 0.4a_\phi + a_z) = \underline{\underline{4.472a_\rho + 0.8944a_\phi + 2.236a_z}}$$

Prob 2.14

(a) $d = \sqrt{(6-2)^2 + (-1-1)^2 + (2-5)^2} = \sqrt{29} = \underline{\underline{5.385}}$

(b)
$$d^2 = 3^2 + 5^2 - 2(3)(5)\cos\pi + (-1-5)^2 = 100$$
$$d = \sqrt{100} = \underline{\underline{10}}$$

(c)

$$d^2 = 10^2 + 5^2 - 2(10)(5)\cos\frac{\pi}{4}\cos\frac{\pi}{6}$$

$$d^2 = (10)(5)\sin\frac{\pi}{4}\sin\frac{\pi}{6}\cos 7\frac{\pi}{4} - \frac{3\pi}{4}$$

$$d = \sqrt{99.12} = \underline{\underline{9.956.}}$$

Prob 2.15

(a) An infinite line parallel to the z-axis.

(b) Point (2,-1,10).

(c) A circle of radius $r\sin\theta = 5$, i.e. the intersection of a cone and a sphere.

(d) An infinite line parallel to the z-axis.

(e) A semi-infinite line parallel to the x-y plane.

(f) A semi-circle of radius 5 in the x-y plane.

Prob.2.16

At $T(2,3,-4)$

$$\theta = \tan^{-1}\sqrt{x^2+y^2} = \tan^{-1}\frac{\sqrt{13}}{-4} = 137.97$$

$$\cos\theta = \frac{-4}{\sqrt{29}} = -0.7428, \sin\theta = \frac{\sqrt{13}}{\sqrt{29}} = 0.6695$$

$$\phi = \tan^{-1}\frac{y}{x} = \tan^{-1}\frac{3}{2} = 56.31$$

$$\cos\phi\ \frac{2}{\sqrt{13},}\quad \sin\phi = \frac{3}{\sqrt{13}}$$

$$\bar{a}_z = \cos\theta\, \bar{a}_r - \sin\theta\, \bar{a}_\theta = \underline{\underline{-0.7428\,\bar{a}_r - 0.6695\,\bar{a}_\theta.}}$$

$$\bar{a}_r = \sin\theta\cos\phi\ \bar{a}_x + \sin\theta\sin\phi\ \bar{a}_y + \cos\theta\ \bar{a}_z.$$

$$= \underline{\underline{0.3714\bar{a}_x + 0.5571\bar{a}_y - 0.7428\bar{a}_z.}}$$

Prob.2.17

At $P(0,2,-5)$, $\phi = 90°$;

$$\begin{bmatrix} B_x \\ B_y \\ B_z \end{bmatrix} = \begin{bmatrix} \cos\phi & -\sin\phi & 0 \\ \sin\phi & \cos\phi & 0 \\ 0 & 0 & 1 \end{bmatrix} \begin{bmatrix} B_\rho \\ B_\phi \\ B_z \end{bmatrix}$$

$$= \begin{bmatrix} 0 & -1 & 0 \\ 1 & 0 & 0 \\ 0 & 0 & 1 \end{bmatrix} \begin{bmatrix} -5 \\ 1 \\ -3 \end{bmatrix}$$

$$\bar{B} = -\bar{a}_x - 5\bar{a}_y - 3\bar{a}_z$$

(a) $\bar{A} + \bar{B} = (2,4,10) + (-1,-5,-3)$

$$= \underline{\underline{\bar{a}_x - \bar{a}_y + 7\bar{a}_z.}}$$

(b) $\cos\theta_{AB} = \dfrac{\bar{A} \bullet \bar{B}}{\|A\|\,\|B\|} = \dfrac{-52}{\sqrt{4200}}$

$$\theta_{AB} = \cos^{-1}\left(\frac{-52}{\sqrt{4200}}\right) = \underline{\underline{143.26°}}.$$

(c) $A_B = \bar{A} \bullet \bar{a}_B = \dfrac{\bar{A} \bullet \bar{B}}{B} = -\dfrac{52}{\sqrt{35}} = \underline{\underline{-8.789.}}$

Prob. 2.18

At $P(8, 30°, 60°) = P(r,\theta,\phi)$,

$$x = r\sin\theta\cos\phi = 8\sin 30°\cos 60° = 2.$$

$$y = r\sin\theta\sin\phi = 8\sin 30°\sin 60° = 2\sqrt{3}$$

$$z = r\cos\theta = 8\left(\frac{1}{2}\sqrt{3}\right) = 4\sqrt{3}.$$

$$\bar{G} = 14\bar{a}_x + 8\sqrt{3}\,\bar{a}_y + (48+24)\bar{a}_z = (14, 13.86, 72);$$

$$\bar{a}_\phi = -\sin\phi\,\bar{a}_x + \cos\phi\,\bar{a}_y = -\frac{\sqrt{3}}{2}\bar{a}_x + \frac{1}{2}\bar{a}_y;$$

$$\bar{G}_\phi = (\bar{G} \bullet \bar{a}_\phi)\bar{a}_\phi = (-7\sqrt{3} + 4\sqrt{3})\frac{1}{2}(-\sqrt{3}\,\bar{a}_x + \bar{a}_y)$$

$$= \underline{\underline{4.5\bar{a}_x - 2.598\bar{a}_y.}}$$

Prob. 2.19

(*a*) $\bar{J}_z = (\bar{J} \bullet \bar{a}_z)\bar{a}_z.$

$At(2, \ \pi/2, \ 3\pi/2), \ \bar{a}_z = \cos\theta \bar{a}_r - \sin\theta \bar{a}_\theta = -\bar{a}_\theta.$

$\bar{J}_z = -\cos 2\theta \sin\phi \bar{a}_\theta = -\cos\pi \sin(3\pi/2)\bar{a}_\theta = -\bar{a}_\theta.$

(*b*) $\bar{J}_\theta = \tan\frac{\theta}{2}\ln r\,\bar{a}_\phi = \tan\frac{\pi}{4}\ln 2\,\bar{a}_\phi = \ln 2\,\bar{a}_\phi = 0.6931\bar{a}_\phi.$

(*c*) $\bar{J}_t = \bar{J} - \bar{J}_n = \bar{J} - \bar{J}_r = -\bar{a}_\theta + \ln 2\,\bar{a}_\phi = \underline{\underline{-\bar{a}_\theta + 0.6931\bar{a}_\phi}}\ .$

(*d*) $\bar{J}_P = (\bar{J} \bullet \bar{a}_x)\bar{a}_x$

$\bar{a}_x = \sin\theta\cos\phi\,\bar{a}_r + \cos\theta\cos\phi\,\bar{a}_\theta - \sin\phi\,\bar{a}_\phi = \bar{a}_\phi.$

At $(2, \ \pi/2, \ 3\pi/2)$,

$\bar{J}_P = \underline{\underline{\ln 2\,\bar{a}_\phi}}.$

Prob 2.20

$At\ P, \quad \rho = 2, \ \phi = 30°, \quad z = -1$

$\bar{H} = 10\sin 30\bar{a}_\rho + 2\cos 30°\,\bar{a}_\phi - 4\bar{a}_z.$

$= 5\bar{a}_\rho + 1.732\bar{a}_\phi - 4\bar{a}_z.$

$$\bar{a}_{//} = \frac{(5,\ 1.732,\ -4)}{\sqrt{5^2 + 1.732^2 + 4^2}} = \underline{\underline{0.7538\bar{a}_\rho + 0.2611\bar{a}_\phi - 0.603\bar{a}_z}}.$$

(*b*) $H_x = H_\rho\cos\phi - H_\phi\sin\phi = 5\rho\sin\phi\cos\phi - \rho z\cos\phi\sin\phi$

or P at $\rho = 5, \ \phi = 30, \ z = 1;$

$\bar{H}_x = H_x\bar{a}_x = (25\sin 30°\cos 30° + 5\sin 30°\cos 30°)\bar{a}_x.$

$= \underline{\underline{13\bar{a}_x}}$

(*c*) Normal to $\rho = 2$ *is* $\bar{H}_n = \bar{H}_\rho\bar{a}_\rho;$

i.e. $\bar{H}_n = \underline{\underline{0.7538\ \bar{a}_\rho}}.$

(*d*) Tangential to $\phi = 30°$.

$\bar{H}_t = H_\rho\bar{a}_\rho + H_z\bar{a}_z = \underline{\underline{0.7538\bar{a}_\rho - 0.603\bar{a}_z}}$

Prob.2.21

(a) At $T, x = 3, y = -4, z = 1, \rho = 5, \cos\phi = -\frac{3}{5}$

$$\bar{A} = 0\bar{a}_\rho - 5(1)(-\frac{3}{5})\bar{a}_\rho + 25(1)\bar{a}_z$$

$$= 3\bar{a}_\phi + 25\bar{a}_z$$

$$r = \sqrt{26}, \qquad \sin\theta = \frac{5}{\sqrt{26},} \qquad \cos\theta = \frac{1}{\sqrt{26}}$$

$$\bar{B} = 26(\frac{-3}{5})\bar{a}_r + 2(\sqrt{26})\frac{5}{\sqrt{26}}\bar{a}_\phi$$

$$= -15.6\bar{a}_r + 10\bar{a}_\phi$$

(b) In cylindrical coordinates,

$$\begin{bmatrix} B_\rho \\ B_\phi \\ B_z \end{bmatrix} = \begin{bmatrix} \sin\theta & \cos\theta & 0 \\ 0 & 0 & 1 \\ \cos\theta & -\sin\theta & 0 \end{bmatrix} \begin{bmatrix} -15.6 \\ 0 \\ 10 \end{bmatrix}$$

$$B_\rho = 15.6\sin\theta = 26(-\frac{3}{5})(\frac{5}{\sqrt{26}}) = 15.3$$

$$B_\phi = 10, \quad B_z = 15.6\cos\theta = -3.059$$

$$\bar{B}(\rho,\phi,z) = (-15.3, 10, -3.059)$$

$$\bar{A}_B = (\bar{A}\bullet\bar{a}_B)\bar{a}_B = (\bar{A}\bullet\bar{B})\bar{B}\frac{1}{|\bar{B}|^2} = \frac{(30-76.485)(-15.3,10,-3.059)}{343.36}$$

$$= 2.071\bar{a}_\rho - 1.354\bar{a}_\phi + 0.4141\bar{a}_z.$$

(c) In spherical coordinates,

$$\begin{bmatrix} A_r \\ A_0 \\ A_\phi \end{bmatrix} = \begin{bmatrix} \sin\theta & 0 & \cos\theta \\ \cos\theta & 0 & -\sin\theta \\ 0 & 1 & 0 \end{bmatrix} \begin{bmatrix} 0 \\ 3 \\ 25 \end{bmatrix}$$

$$A_r = 25\cos\theta = \frac{25}{\sqrt{26}} = 4.903$$

$$A_\theta = 25\sin\theta = -25\left(\frac{5}{\sqrt{26}}\right) = -24.51$$

$$A_\phi = 0.$$

$$\bar{A}\times\bar{B} = \begin{vmatrix} \bar{a}_r & \bar{a}_\theta & \bar{a}_\phi \\ 4.903 & -24.51 & 0 \\ -15.6 & 0 & 10 \end{vmatrix} = -245.1a_r + 49.03a_\theta - 382.43a_\phi$$

$$\bar{a}_{AxB} = \frac{\pm\bar{A}\times\bar{B}}{456.87} = \underline{\underline{\pm(0.5365\bar{a}_r - 0.1073\bar{a}_\theta + 0.8371\bar{a}_\phi.}}$$

Prob 2.22

(*a*) For $(x,y,z) = (2,3,6)$,

$$r = \sqrt{x^2+y^2+z^2} = 7$$

$$x = r\cos\alpha \quad \cos\alpha = \frac{x}{r} = \frac{-2}{7}, \alpha = 106.6°$$

$$y = r\cos\beta \quad \cos\beta = \frac{y}{r} = \frac{3}{7}, \beta = 64.6°$$

$$z = r\cos\gamma \quad \cos\gamma = \frac{z}{r} = \frac{6}{7}, \gamma = 31°$$

Hence,

$$(r,\alpha,\beta,\gamma) = \underline{\underline{(7, 106.6°, 64.6°, 31°)}}$$

(*b*) For $(\rho,\phi,z) = (4, 30°, -3)$,

$$r = \sqrt{\rho^2 + z^2 = 5},$$

$$\cos y = \frac{z}{r} = \frac{-3}{5} \; y = 126.9°$$

$$\cos\alpha = \frac{x}{r} = \rho\frac{\cos\phi}{r} = \frac{4\cos 30°}{5} \; \alpha = 46.15°$$

$$\cos B = \frac{y}{r} = \frac{\rho\sin\phi}{r} = \frac{4}{5}\sin 30° \quad B = 66.42°$$

$$(r,\alpha,B,y) = \underline{\underline{(5, 46.15°, 66.42°, 126.9°)}}$$

(*c*) For $(r,\theta,\phi) = (3, 30°, 60°)$,

$r = 3,\ y = \theta = 30°,$

$$\cos\alpha = \frac{x}{r} = \frac{r\sin\theta\cos\phi}{r} = \frac{1}{4} \quad \alpha = 75.52°,$$

$$\cos B = \frac{y}{r} = \sin\theta\sin\phi = 0.433 \quad B = 64.34°,$$

$$(r,\alpha,B,y) = \underline{\underline{(3, 75.52°, 64.34°, 30°)}}.$$

Prob 2.23

$$\bar{G} = \cos y\,\bar{a}_y + \frac{2r\cos\theta\sin\phi}{r\sin\theta}\bar{a}_y + (1-\cos^2\phi)\bar{a}_z$$

$$= \cos\phi\,\bar{a}_x + 2\tan\theta\sin\phi\,\bar{a}_y + \sin\phi\,\bar{a}_z$$

$$\begin{pmatrix} Gr \\ G_\theta \\ G_\phi \end{pmatrix} = \begin{bmatrix} \sin\theta\cos\phi & \sin\theta\cos\phi & \cos\theta \\ \sin\theta\cos\phi & \cos\theta\sin\phi & -\sin\theta \\ -\sin\phi & \cos\phi & 0 \end{bmatrix} \begin{bmatrix} \cos^2\phi \\ 2\tan\theta\sin\phi \\ \sin^2\phi \end{bmatrix}$$

$$Gr = \sin\theta\cos\phi + 2\cos\theta\sin^2\phi + \cos\theta\sin^2\phi$$

$$= \sin\theta\cos^2\phi + 3\cos\theta\sin^2\phi$$

$$G_\theta = \cos\theta\cos^2\phi + 2\tan\theta\cos\theta\sin^2\phi - \sin\theta\sin^2\phi$$

$$G_\phi = -\sin\phi\cos^2\phi + \sin^2\phi\cos\phi = \sin\phi\cos\phi(\sin\phi - \cos\phi)$$

$$\bar{G} = [\sin\theta\cos^2\phi + 3\cos\theta\sin^2\phi]\bar{a}_x$$

$$+ [\cos\theta\cos^2\phi + 2\tan\theta\cos\theta\sin^2\phi - \sin\theta\sin^2\phi]\bar{a}_\theta$$

$$\underline{\underline{+ \sin\phi\cos\phi(\cos\phi - \cos\phi)\,\bar{a}_\phi}}$$

CHAPTER 3

P. E. 3.1

(a) $DH = \int_{\phi=45^\circ}^{\phi=60^\circ} r\sin\phi \, d\phi \Big|_{r=3,\theta=90^\circ} = 3(1)[\frac{\pi}{3} - \frac{\pi}{4}] = \frac{\pi}{4} = \underline{\underline{0.7854.}}$

(b) $FG = \int_{\theta=60^\circ}^{\theta=90^\circ} r d\theta \Big|_{r=5} = 5(\frac{\pi}{2} - \frac{\pi}{3}) = \frac{5\pi}{6} = \underline{\underline{2.618}}.$

(c)

$$AEHD = \int_{\theta=60^\circ}^{\theta=90^\circ} \int_{\phi=45^\circ}^{\phi=60^\circ} r^2 \sin\theta \, d\theta \, d\phi \Big|_{r=3} = 9(-\cos\theta)\Big|_{\theta=60^\circ}^{\theta=90^\circ} \phi \Big|_{\phi=45^\circ}^{\phi=60^\circ}$$

$$= 9(\frac{1}{2})(\frac{\pi}{12}) = \frac{3\pi}{8} = \underline{\underline{1.178.}}$$

(d)

$$ABCD = \int_{r=3}^{r=5} \int_{\theta=60}^{\theta=90} r d\theta \, dr = \frac{r^2}{2}\Big|_{r=3}^{r=5} (\frac{\pi}{2} - \frac{\pi}{3}) = \frac{4\pi}{3} = \underline{\underline{4.189}}.$$

(e)

$$\text{Volume} = \int_{r=3}^{r=5} \int_{\phi=45^\circ}^{\phi=60^\circ} \int_{\theta=60}^{\theta=90} r^2 \sin\theta \, d\theta \, d\phi = \frac{r^3}{3}\Big|_{r=3}^{r=5} (-\cos\theta)\Big|_{\theta=60^\circ}^{\theta=90^\circ} \phi \Big|_{\phi=45^\circ}^{\phi=60^\circ} = \frac{1}{3}(98)(\frac{1}{2})\frac{\pi}{12}$$

$$= \frac{49\pi}{36} = \underline{\underline{4.276}}.$$

P.E. 3.2

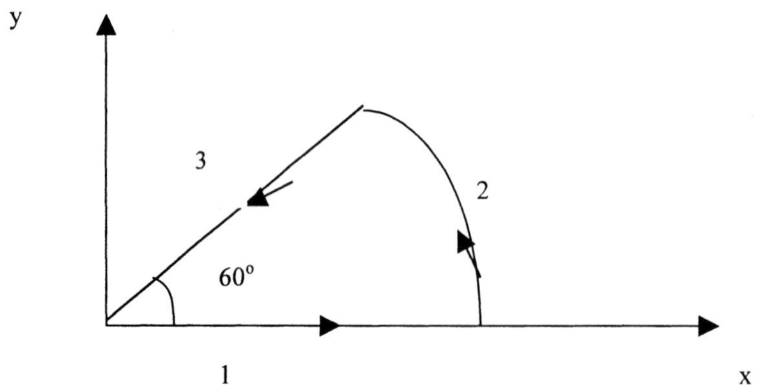

$$\oint_L \bar{A}\bullet d\bar{l} = \left(\int_1 + \int_2 + \int_3\right)\bar{A}\bullet d\bar{l} = C_1 + C_2 + C_3$$

Along (*1*), $C_1 = \int \bar{A}\bullet d\bar{l} = \int_0^2 \rho\cos\phi\, d\rho|_{\phi=0} = \frac{\rho^2}{2}\Big|_0^2 = 2.$

Along (*2*), $d\bar{l} = \rho\, d\phi\, \bar{a}_\phi,\ \bar{A}\bullet d\bar{l} = 0,\ \ C_2 = 0$

Along (*3*), $C_3 = \int_2^0 \rho\cos\phi\, d\rho_{\ \phi=60^\circ} = \frac{\rho^2}{2}\Big|_2^0 \left(\frac{1}{2}\right) = -1$

$$\oint_l \bar{A}\bullet d\bar{l} = C_1 + C_2 + C_3 = 2 + 0 - 1 = \underline{\underline{1}}$$

P.E. 3.3

(*a*) $\nabla U = \frac{\partial U}{\partial x}\bar{a}_x + \frac{\partial U}{\partial y}\bar{a}_y + \frac{\partial U}{\partial z}\bar{a}_z$

$= \underline{\underline{y(2x+z)\bar{a}_x + x(x+z)\bar{a}_y + xy\,\bar{a}_z}}$

(b) $\nabla V = \frac{\partial V}{\partial \rho}\bar{a}_\rho + \frac{1}{\rho}\frac{\partial V}{\partial \phi}\bar{a}_\phi + \frac{\partial V}{\partial z}\bar{a}_z$

$= \underline{\underline{(z\sin\phi + 2\rho)\bar{a}_\rho + (z\cos\phi - \frac{z}{\rho}\sin 2\phi)\bar{a}_\phi + (\rho\cos\phi + 2z\cos^2\phi)\bar{a}_z}}$

(c)

$$\nabla f = \frac{\partial f}{\partial r}\bar{a}_r + \frac{1}{r}\frac{\partial f}{\partial \theta}\bar{a}_\theta + \frac{1}{r\sin\theta}\frac{\partial f}{\partial \phi}\bar{a}_\phi$$

$= (\cos\theta\sin\phi + 2r\phi)\bar{a}_r - \sin\theta\sin\phi\ln r\,\bar{a}_\theta$

$\underline{\underline{+ \left(\cos\theta\cos\phi\ln r + r\,\mathrm{cosec}\,\theta\right)\bar{a}_\phi}}$

P.E. 3.4

$\nabla\Phi = (x+y)\bar{a}_x + (x+z)\bar{a}_y + (y+z)\bar{a}_z$

At $(1,2,3)$ $\nabla\Phi = \underline{\underline{(5,4,3)}}$

$\nabla\Phi\bullet\bar{a}_l = (5,4,3)\bullet\frac{(2,2,1)}{3} = \frac{21}{3} = \underline{\underline{7}},$

where $(2,2,1) = (3,4,4) - (1,2,3)$

P.E. 3.5

Let $f = x^2y + z - 3, \quad g = x\log z - y^2 + 4,$

$$\nabla f = 2xy\bar{a}_x + x^2\bar{a}_y + \bar{a}_z,$$

$$\nabla g = \log z\,\bar{a}_x - 2y\bar{a}_y + \frac{x}{z}\bar{a}_z$$

At P(-1,2,1),

$$\bar{n}_f = \pm\frac{\nabla f}{\nabla f} = -\frac{(-4\bar{a}_x + \bar{a}_y + \bar{a}_z)}{\sqrt{18}}$$

$$\cos\theta = \bar{n}_f.\bar{n}_g = \pm\frac{(-5)}{\sqrt{18x17}}$$

$$\theta = \cos^{-1}\frac{5}{17.493} \quad = 73.39°$$

P.E. 3.6

(a) $\nabla\bullet\bar{A} = \frac{\partial A_x}{\partial x} + \frac{\partial A_y}{\partial y} + \frac{\partial A_z}{\partial z} = 0 + 4 + 0 = \underline{\underline{4x}}.$

At (1,-2,3), $\nabla\bullet\bar{A} = \underline{\underline{4}}.$

(b)

$$\nabla\bullet\bar{B} = \frac{1}{\rho}\frac{\partial}{\partial\rho}(\rho B_\rho) + \frac{1}{\rho}\frac{\partial B_\phi}{\partial\phi} + \frac{\partial B_z}{\partial\rho}$$

$$= \frac{1}{\rho}2\rho z\sin\phi - \frac{1}{\rho}3\rho z^2\sin\phi + 2z\sin\phi - 3z^2\sin\phi$$

$$= \underline{\underline{(2-3z)z\sin\phi}}.$$

$At\,(5,\frac{\pi}{2},1), \quad \nabla\bullet\bar{B} = (2-3)(1) = \underline{\underline{-1}}.$

(c)

$$\nabla\bullet\bar{C} = \frac{1}{r^2}\frac{\partial}{\partial r}(r^2 A_r) + \frac{1}{r\sin\theta}\frac{\partial}{\partial\theta}(A_\theta\sin\theta) + \frac{1}{r\sin\theta}\frac{\partial A_\phi}{\partial\phi}$$

$$= \frac{1}{r^2}6r^2\cos\theta\cos\phi$$

$$= \underline{\underline{6\cos\theta\cos\phi}}$$

$At\,(1,\frac{\pi}{6},\frac{\pi}{3}), \quad \nabla\bullet\bar{C} = 6\cos\frac{\pi}{6}\cos\frac{\pi}{3} = \underline{\underline{2.598}}.$

P.E. 3.7 This is similar to Example 3.7.

$$\Psi = \oint_S \bar{A} \bullet d\bar{s} = \Psi_t + \Psi_b + \Psi_c$$

$$\Psi_t = 0 = \Psi_b \text{ since } \bar{A} \text{ has no z-component}$$

$$\Psi_c = \iint \rho^2 \cos^2\phi \, \rho \, d\phi \, dz = \rho^3 \int_{\phi=0}^{\phi=2\pi} \cos^2\phi \, d\phi \int_{z=0}^{z=1} dz \Bigg|_{\rho=4}$$

$$= (4)^3 \pi (1) = 64\pi$$

$$\Psi = 0 + 0 + 64\pi = \underline{\underline{64\pi}}$$

By the divergence theorem,

$$\oint_S \bar{A} \bullet d\bar{s} = \oint_V \nabla \bullet \bar{A} \, dv$$

$$\nabla \bullet \bar{A} = \frac{1}{\rho}\frac{\partial}{\partial \rho}(\rho^3 \cos^2\phi) + \frac{1}{\rho}\frac{\partial}{\partial \phi} z \sin\phi + \frac{\partial A_z}{dz}$$

$$= 3\rho \cos^2\phi + \frac{1}{\rho}\cos\phi .$$

$$\Psi = \int_V \nabla \bullet \bar{A} \, dv = \int_V (3\rho \cos^2\phi + \frac{1}{\rho}\cos\phi)\rho \, d\phi \, dz \, d\rho$$

$$= 3\int_0^4 \rho^2 d\rho \int_0^{2\pi} \cos^2\phi \, d\phi \int_0^1 dz + \int_0^4 d\rho \int_0^{2\pi} \cos\phi \, d\phi \int_0^1 dz$$

$$= 3(\frac{4^3}{3})\pi(1) = \underline{\underline{64\pi}} .$$

P.E. 3.8

(a)

$$\nabla \times \bar{A} = \bar{a}_x(1-0) + \bar{a}_y(y-0) + \bar{a}_z(4y-z)$$

$$= \underline{\underline{\bar{a}_x + y\bar{a}_y + (4y-z)\bar{a}_z}}$$

At $(1,-2,3)$, $\nabla \times \bar{A} = \underline{\underline{\bar{a}_x - 2\bar{a}_y - 11\bar{a}_z}}$

(b)

$$\nabla \times \bar{B} = \bar{a}_\rho(0 - 6\rho z \cos\phi) + \bar{a}_\phi(\rho \sin\phi - 0) + \bar{a}_z \frac{1}{\rho}(6\rho z^2 \cos\phi - \rho z \cos\phi)$$

$$= \underline{\underline{-6\rho z \cos\phi \, \bar{a}_\rho + \rho \sin\phi \, a_\phi + (6z-1)z\cos\phi \, \bar{a}_z}}$$

At $(5, \frac{\pi}{2}, -1)$, $\nabla \times \bar{B} = \underline{\underline{5\bar{a}_\phi}}$.

(c)

$$\nabla \times \bar{C} = \bar{a}_r \frac{1}{r\sin\theta}(r^{-1/2}\cos\theta - 0) + \frac{\bar{a}_\theta}{r}(-\frac{2r\cos\theta\sin\phi}{\sin\theta} - \frac{3}{2}r^{1/2}) + \frac{\bar{a}_\phi}{r}(0 - 2r\sin\theta\cos\phi)$$

$$= r^{-1/2}\cot\theta\,\bar{a}_r - (2\cot\theta\sin\phi + \frac{3}{2}r^{-1/2})\bar{a}_\theta - 2\sin\theta\cot\phi\,\bar{a}_\phi$$

At $(1, \frac{\pi}{6}, \frac{\pi}{3})$, $\nabla \times C = \underline{1.732\bar{a}_r - 4.5\bar{a}_\theta - 0.5\bar{a}_\phi}$

P.E. 3.9

$$\oint_L \bar{A}\bullet d\bar{l} = \int_S (\nabla \times \bar{A})\bullet d\bar{S}$$

But $(\nabla \times \bar{A}) = \sin\phi\,\bar{a}_z + \frac{z\cos\phi}{\rho}\bar{a}_\rho$ and $d\bar{S} = \rho\,d\phi\,d\rho\,\bar{a}_z$

$$\int_S (\nabla \times \bar{A})\bullet d\bar{S} = \iint \rho\sin\phi\,d\phi\,d\rho$$

$$= \frac{\rho}{2}\Big|_0^2 (-\cos\phi)\Big|_0^{60°}$$

$$= 2(-\frac{1}{2} + 1) = \underline{1.}$$

P.E. 3.10

$$\nabla \times \nabla V = \begin{vmatrix} \frac{\partial}{\partial x} & \frac{\partial}{\partial y} & \frac{\partial}{\partial z} \\ \frac{\partial V}{\partial x} & \frac{\partial V}{\partial y} & \frac{\partial V}{\partial z} \end{vmatrix} =$$

$$= (\frac{\partial^2 V}{\partial y\partial z} - \frac{\partial^2 V}{\partial y\partial z})\bar{a}_x + (\frac{\partial^2 V}{\partial x\partial z} - \frac{\partial^2 V}{\partial z\partial x})\bar{a}_y + (\frac{\partial^2 V}{\partial x\partial y} - \frac{\partial^2 V}{\partial y\partial x})\bar{a}_z = 0$$

P.E. 3.11

(a)

$$\nabla^2 U = \frac{\partial}{\partial x}(2xy + yz) + \frac{\partial}{\partial x}(x^2 + xz) + \frac{\partial}{\partial x}(xy)$$

$$= \underline{2y.}$$

(b)

$$\nabla^2 V = \frac{1}{\rho}\frac{\partial}{\partial \rho}\rho(z\sin\phi + 2\rho) + \frac{1}{\rho^2}(-\rho z\sin\phi - 2z^2\frac{\partial}{\partial \rho}\sin\phi\cos\phi) + \frac{\partial}{\partial z}(\rho\sin\phi + 2z\cos^2\phi)$$

$$= \frac{1}{\rho}(z\sin\phi + 4\rho) - \frac{1}{\rho^2}(z\rho\sin\phi + 2z^2\cos 2\phi) + 2\cos^2\phi.$$

$$= 4 + 2\cos^2\phi - \frac{2z^2}{\rho^2}\cos 2\phi.$$

(c)

$$\nabla^2 f = \frac{1}{r^2}\frac{\partial}{\partial r}[\frac{1}{r^2}\frac{1}{r}\cos\theta\sin\phi + 2r^2\phi] + \frac{1}{r^2\sin\theta}\frac{\partial}{\partial\theta}[-\sin^2\theta\sin\phi\ln r]$$

$$+ \frac{1}{r^2\sin^2\theta}[-\cos\theta\sin\theta\ln r]$$

$$= \frac{1}{r^2}\cos\theta\sin\phi(1 - 2\ln r - \csc^2\theta\ln r) + 6\theta$$

P.E. 3.12

If $\bar{B}$ is conservative, $\nabla\times\bar{B} = 0$ must be satisfied.

$$\nabla\times\bar{B} = \begin{vmatrix} \frac{\partial}{\partial x} & \frac{\partial}{\partial y} & \frac{\partial}{\partial z} \\ y + z\cos xz & x & x\cos xz \end{vmatrix}$$

$$= 0\bar{a}_x + (\cos xz - xz\sin xz - \cos xz + xz\sin xz)\bar{a}_y + (1-1)\bar{a}_z \qquad = 0$$

Hence $\bar{B}$ is a conservative field.

Prob. 3.1

(a)

$$dl = \rho d\phi; \qquad \rho = 3$$

$$L = \int dl = 3\int_{\pi/4}^{\pi/2} d\phi = 3(\frac{\pi}{2} - \frac{\pi}{4}) = \frac{3\pi}{4} = \underline{\underline{2.356}}$$

(b)

$$dl = r\sin\theta d\phi; \qquad r = 1, \quad \theta = 30°;$$

$$L = \int dl = r\sin\theta \int_0^{\pi/3} d\phi = (1)\sin 30° [(\frac{\pi}{3}) - 0] = \underline{\underline{0.5236.}}$$

(c)

$$dl = r d\phi$$

$$L = \int dl = r\int_{\pi/6}^{\pi/2} d\theta = 4(\frac{\pi}{2} - \frac{\pi}{6}) = \frac{4\pi}{3} = \underline{\underline{4.189}}$$

Prob. 3.2

(a)

$$dS = \rho d\phi dz$$

$$S = \int dS = \rho\iint d\phi dz = 2\int_0^5 dz \int_{\pi/3}^{\pi/2} d\phi = 2(5)[\frac{\pi}{2} - \frac{\pi}{3}] = \frac{10\pi}{6} = \underline{\underline{5.236}}$$

(b)

In cylindrical, $dS = \rho\, d\rho\, d\phi$

$$S = \int dS = \int_1^3 \rho\, d\rho \int_0^{\pi/4} d\phi = \frac{\pi}{4}(\frac{\rho}{2})_1^3 = \underline{\underline{3.142}}$$

(c) In spherical, $dS = r^2 \sin\theta\, d\phi\, d\theta$

$$S = \int dS = 100\int_{\pi/4}^{2\pi/3} \sin\theta\, d\theta \int_0^{2\pi} d\phi = 100(2\pi)(-\cos\theta)\Big|_{\pi/4}^{2\pi/3} = 200\pi(0.5 - 0.7071) = \underline{\underline{7.58.4}}$$

(d)

$$dS = r\, dr\, d\theta$$

$$S = \int dS = \int_0^4 r dr \int_{\pi/3}^{\pi/2} d\theta = \frac{r^2}{2}\Big|_0^4 (\frac{\pi}{2} - \frac{\pi}{3}) = \frac{8\pi}{6} = \underline{\underline{4.189}}$$

Prob.3.3

(a) $dV = dxdydz$

$$V = \int dxdydz = \int_0^1 dx \int_1^2 dy \int_{-3}^3 dz = (1)\,(2-1)(3--3) = \underline{\underline{6}}$$

(b) $dV = \rho\, d\phi\, d\rho\, dz$

$$V = \int_2^5 \rho\, d\rho \int_1^4 dz \int_{\pi/3}^{\pi} d\phi = \frac{\rho^2}{2}\Big|_2^5 (4--1)(\pi - \frac{\pi}{3}) = \frac{1}{2}(25-4)(5)(\frac{2\pi}{3}) = 35\pi = \underline{\underline{110}}$$

(c) $dV = r^2 \sin\theta\, dr d\theta\, d\phi$

$$V = \int_1^3 r^2 dr \int_{\pi/2}^{2\pi/3} \sin\theta \int_{\pi/6}^{\pi/2} d\phi = \frac{r^3}{3}\Big|_1^3 (-\cos\theta)\Big|_{\pi/2}^{\pi/3} (\frac{\pi}{2} - \frac{\pi}{6})$$

$$= \frac{1}{3}(27-1)(\frac{1}{2})(\frac{\pi}{3}) = \frac{26\pi}{18} = \underline{\underline{4.538}}$$

Prob 3.4

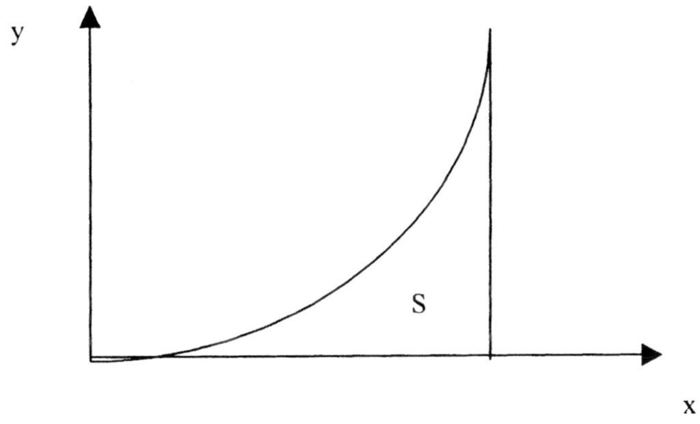

$$\int \rho_s dS = \int_{x=0}^{x=1} \int_{y=0}^{y=x^2} (x^2 + xy)dy\,dx$$

$$= \int_0^1 (x^2 y + \frac{xy^2}{2}\Big|_0^{x^2})dx = \int_0^1 (x^4 + \frac{x^5}{2})dx$$

$$= \frac{1}{5} + \frac{1}{12} = \frac{17}{60} = \underline{\underline{0.2833}}$$

Prob. 3.5

$$\int_L \bar{H}\bullet \bar{dl} = \int (x^2 dx + y^2\, dy)$$

But on L, $y = x^2 \quad dy = 2xdx$

$$\int_L \bar{H}\bullet \bar{dl} = \int_0^1 (x^2 + x^4 .2x)dx = \frac{x^3}{3} + 2\frac{x^6}{6}\Big|_0^1 = \frac{1}{3} + \frac{1}{3} = \underline{\underline{0.6667}}$$

Prob. 3.6

$$V = \int_{\phi=0}^{2\pi} \int_{\theta=0}^{\alpha} \int_{r=o}^{a} r^2 \sin\theta\, d\theta\, dr\, d\phi = \frac{2\pi\, a^3}{3}(1 - \cos\alpha)$$

$$V(\alpha = \frac{\pi}{3}) = \frac{2\pi\, a^3}{3}(1 - \frac{1}{2}) = \underline{\underline{\frac{\pi\, a^3}{3}}}$$

$$V(\alpha = \frac{\pi}{2}) = \frac{2\pi\, a^3}{3}(1 - 0) = \underline{\underline{\frac{2\pi\, a^3}{3}}}$$

Prob.3.7

(a)

$$\int \bar{F}\bullet d\bar{l} = \int_{y=0}^{1} (x^2 - z^2)dy\Big|_{z=0}^{x=0} + \int_{x=0}^{x=2} 2xydx\Big|_{z=0}^{y=1} + \int_{z=0}^{z=3} (-3xz^2)dz\Big|_{y=1}^{x=0}$$

$$= 0 + 2(1)\frac{x^2}{2}\Big|_0^2 - 3(2)\frac{z^3}{3}\Big|_0^3$$

$$= 0 + 4 - 54 = \underline{\underline{-50}}$$

(b)

Let $x = 2t. \quad y = t, \quad z = 3t$

$dx = 2dt, \quad dy = dt, \quad dz = 3dt;$

$$\int \bar{F}\bullet d\bar{l} = \int_0^1 (8t^2 - 5t^2 - 162t^3)\, dt = -\frac{79}{2} = \underline{\underline{-39.5}}$$

Prob.3.8

$$\int \bar{H}\bullet d\bar{l} = \int_{x=0}^{1}(x-y)dx\Big|_{y=0} + \int_{z=0}^{1}(x^2+zy)dz\Big|_{z=0}^{y=0}$$

$$+ \int (x^2+zy)dy + 5yzdz\Big|_{z=1-\frac{y}{2}}^{x=0}$$

$$= \frac{x^2}{2}\Big|_0^1 + 0 + \int_{y=0}^{2} y(1-\frac{y}{2})dy + 5y(1-\frac{y}{2})(-\frac{dy}{2})$$

$$= \frac{1}{2} + \int_0^2 (-\frac{3}{2}y + \frac{3y^2}{4})dy$$

$$= \frac{1}{2} + (\frac{-3}{4}y^2 + \frac{y^4}{4})\Big|_0^2 = \frac{1}{2} - 3 + 4$$

$$= \underline{\underline{1.5}}$$

Prob. 3.9

The surface S can be divided into 5 parts as shown below:

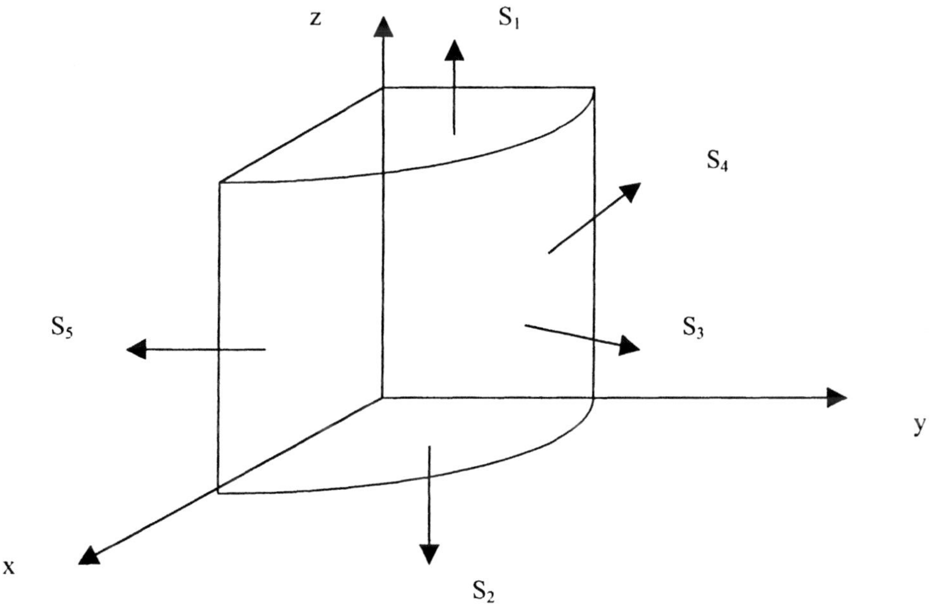

$$V = (x+y)z = \rho z(\cos\phi + \sin\phi)$$

Let

$$\bar{A} = \int V d\bar{S} = \left(\int_{S_1} + \int_{S_2} + \int_{S3} + \int_{S4} + \int_{S5}\right) V d\bar{S} = \overline{A_1} + \overline{A_2} + \overline{A_3} + \overline{A_4} + \overline{A_5}$$

For $\overline{A_1}$, $z = 2$, $d\bar{S} = \rho d\phi d\rho \bar{a}_z$,

$$\bar{A}_1 = \int_{\rho=0}^{1}\int_{\phi=0}^{\pi/2} \rho^2 z(\cos\phi + \sin\phi)d\phi\, d\rho\, \bar{a}_z = (2)\frac{\rho^3}{3}\Big|_0^1 \left(\sin\phi - \cos\phi\right)\Big|_0^{\pi/2}\bar{a}_z$$

$$= \frac{2}{3}(1 - 0 - 0 + 1)\bar{a}_z = \frac{4}{3}\bar{a}_z$$

For $\bar{A}_2$, $z = 0$, $d\bar{S} = \rho\, d\phi(-\bar{a}_z)$,

$$\bar{A}_2 = -\int_{\rho=0}^{1}\int_{\phi=0}^{\pi/2}\rho^2 z(\cos\phi + \sin\phi)d\phi\, d\rho\, \bar{a}_z = 0$$

For A_3 $\rho = 1$, $d\bar{S} = \rho\, d\phi\, dz\, \bar{a}_\rho$

$$\bar{A}_3 = \int_{z=0}^{z=2}\int_{\phi=0}^{\frac{\pi}{2}}\rho^2 z(\cos\phi + \sin\phi)\, d\phi\, dz\, \bar{a}_\rho$$

$$= (1^2)\, \frac{z^2}{2}\Big|_0^2 (1 + 1)\, \bar{a}_\rho = 4\bar{a}_\rho$$

For A_4, $\phi = \frac{\pi}{2}$, $d\bar{S} = d\rho\, dz\, \bar{a}_\phi$

$$A_4 = \int_{\rho=0}^{1}\int_{z=0}^{2}\rho\, z\, (\cos\phi + \sin\phi)\, d\rho\, dz\, \bar{a}_\phi$$

$$= \frac{\rho^2}{2}\Big|_0^1 \frac{z}{2}\Big|_0^2 (0 + 1)\bar{a}_\phi = 1\bar{a}_\phi$$

For A_5, $\phi = 0$, $d\bar{S} = d\rho\, dz)(-\bar{a}_\phi)$

$$A_5 = -\bar{a}_\phi$$

Thus, $\bar{A} = \int V d\bar{S} = \frac{4}{3}\bar{a}_z + 0 + 4\bar{a}_\phi - \bar{a}_\phi = \underline{\underline{4\bar{a}_\rho + 1.333\bar{a}_z}}$

Prob 3.10

(a) $\int \bar{A}\, dv = \int 2xy\, dxdydz\, \bar{a}_x + \int x\, z\, dxdydz\, \bar{a}_y - \int y\, dxdydz\, \bar{a}_z$

$$= 2\int_0^2 xdx\int_0^2 ydy\int_0^2 dz\,\bar{a}_x + \int_0^2 xdx\int_0^2 dy\int_0^2 zdz\,\bar{a}_y + \int_0^2 dx\int_0^2 ydy\int_0^2 dz\,\bar{a}_z$$

Since $\int_0^2 x\,dx = \frac{x^2}{2}\Big|_0^2 = 2$ and $\int_0^2 dx = 2$, we get

$$\int \bar{A}\,dv = 2(2)(2)(2)\bar{a}_x + (2)(2)(2)\bar{a}_y - (2)(2)(2)\bar{a}_z$$

$$= \underline{\underline{16\bar{a}_x + 8\bar{a}_y - 8\bar{a}_z}}$$

$$\begin{bmatrix} A_\rho \\ A_\phi \\ A_z \end{bmatrix} = \begin{bmatrix} \cos\phi & \sin\phi & 0 \\ -\sin\phi & \cos\phi & 0 \\ 0 & 0 & 1 \end{bmatrix} \begin{bmatrix} 2xy \\ xz \\ -y \end{bmatrix}$$

$$A_\rho = 2xy\cos\phi + xz\sin\phi = 2\rho^2\cos^3\phi\sin\phi + \rho z\cos\phi\sin\phi$$

$$A_\phi = -2xy\sin\phi + xz\cos\phi = -2\rho^2\cos\phi\sin^2\phi + \rho z\cos^2\phi$$

$$A_z = -y = -\rho\cos\phi$$

$$dv = \rho\,d\phi\,d\rho\,dz$$

$$\int \bar{A}\,dv = \iiint 2\rho^3\cos^3\phi\,d(-\cos\phi)\,d\rho\,dz\,\bar{a}_\rho + \iiint \rho^2 z\cos\phi\,d(-\cos\phi)\,d\rho\,dz\,\bar{a}_\rho$$

$$-2\iiint \rho^3\sin^2\phi\,d(\sin\phi)\,d\rho\,dz\,\bar{a}_\phi + \iiint \rho^2 z\cos^2\phi\,d\phi\,d\rho\,dz\,\bar{a}_\phi$$

$$-\iiint \rho^2\cos\phi\,d\phi\,d\rho\,dz\,\bar{a}_z$$

Since $\int_0^{2\pi}\cos\phi\,d\phi = 0$,

$$\int \bar{A}\,dv = -2\frac{\rho^4}{4}\Big|_0^3 \frac{4}{4}\cos\phi\Big|_0^{2\pi} z\Big|_0^5\bar{a}_\rho - \frac{\rho^3}{3}\Big|_0^3\frac{z^2}{2}\Big|_0^5\frac{\cos^2}{2}\phi\Big|_0^{2\pi}\bar{a}_\rho$$

$$-\frac{2\rho^4}{4}\Big|_0^3 z\Big|_0^3\frac{\sin^3\phi}{3}\Big|_0^{2\pi}\bar{a}_\rho + \frac{\rho^3}{3}\Big|_0^3\frac{z^2}{2}\Big|_0^5\left(\frac{1}{2}+\frac{1}{4}\sin 2\phi\right)\Big|_0^{2\pi}\bar{a}_\phi$$

$$= 0 + 0 + 0 + (9)\left(\frac{25}{2}\right)\left(\frac{1}{2}\right)\bar{a}_\phi = \underline{\underline{56.25\,\bar{a}_\phi}}$$

(c)

$$\begin{bmatrix} A_r \\ A_\theta \\ A_\phi \end{bmatrix} = \begin{bmatrix} \sin\theta\cos\phi & \cos\theta\cos\phi & -\sin\theta \\ \sin\theta\sin\phi & \cos\theta\sin\phi & \cos\phi \\ \cos\theta & -\sin\theta & 0 \end{bmatrix} \begin{bmatrix} 2xy \\ xz \\ -y \end{bmatrix}$$

$$\int \bar{A}\,dv = \iiint 2r^4 \sin^4\theta \cos^2\phi \; d(\cos\phi)\,d\theta\, d\phi\; dr\, \bar{a}_r$$

$$+ \iiint r^4 \sin^2\theta \cos^2\theta \cos^2\phi \; d\theta\, d\phi\, dr\, \bar{a}_r$$

$$+ \iiint r^4 \sin^2\theta \cos\phi \sin\phi \;\; d\theta\, d\phi\, dr\, \bar{a}_r$$

$$+ \iiint 2\, r^4 \sin^4\theta \sin^2\phi \;\; d(\sin\phi)\,d\theta\, d\phi\, dr\, \bar{a}_\theta$$

$$+ \iiint r^4 \sin^2\theta \cos^2\theta \cos\phi \;\; \sin\phi \;\; d\theta\, d\phi\, dr\, \bar{a}_\theta$$

$$- \iiint r^3 \sin^2\theta \cos\phi \sin\phi \; d\theta\, d\phi\; dr\, \bar{a}_\theta$$

$$+ \iiint 2\, r^4 \sin^2\theta \cos\theta \sin\phi \cos\phi \; d\theta\, d\phi\; dr\, \bar{a}_\phi$$

$$- \iiint r^4 \sin^3\theta \cos^2\theta \cos\phi \; d\theta\, d\phi\; dr\, \bar{a}_\phi$$

$$= \frac{r^5}{5}\Big|_0^4 \left(\frac{1}{2} + \frac{1}{4}\cos 2\phi\right)\Big|_0^{2\pi} \int_0^{\pi} \;\; \cos\theta\,(1 - \cos^2\theta)\,d\theta\, \bar{a}_r$$

$$= 204.8\,\left(\frac{1}{2}\right) \left[\int_0^{\pi} \cos^2\theta\, d\theta - \int_0^{\pi} \cos^4\theta\, d\theta\right] \bar{a}_r$$

But $\displaystyle\int_0^{\pi} \cos^2\theta\, d\theta = \left(\frac{\theta}{2} + \frac{\sin 2\theta}{4}\right)\Big|_0^{\pi} = \frac{\pi}{2}$

Since $\cos 4\theta = 8\cos^4\theta - 8\cos^2\theta - 1$

$$\int_0^{\pi} \cos^4\theta\, d\theta = \frac{\pi}{2} + \frac{1}{8}\int_0^{\pi} \;\; \cos 4\theta\, d\theta - \frac{1}{8}\int_0^{\pi} d\theta$$

$$= \frac{\pi}{2} + \frac{1}{8}\frac{1}{4}\sin 4\theta\,\Big|_0^{\pi} - \frac{\pi}{8} = \frac{\pi}{2} - \frac{\pi}{8}$$

$$\int \bar{A}\, dv = 102.4\left(\frac{\pi}{2} - \frac{\pi}{2} + \frac{\pi}{8}\right)\bar{a}_r = \underline{\underline{40.21\,\bar{a}_r}}$$

Prob 3.11

$$\bar{a} = \left(\frac{dV_x}{dt}, \frac{dV_y}{dt}, \frac{dV_z}{dt}\right) = 2.4\bar{a}_z$$

$$\frac{dV_x}{dt} = 0 \quad \longrightarrow \quad V_x = A$$

$$\frac{dV_y}{dt} = 0 \quad \longrightarrow \quad V_y = B$$

$$\frac{dV_z}{dt} = 2.4 \quad \longrightarrow \quad V_z = 2.4t + C$$

$At\ t = 0,\ (V_x, V_y, V_z) = (-2, 0, 5)$. Hence,

$$A = -2, \quad B = 0, \quad C = 5$$

$$V_x = \frac{dx}{dt} = -2 \quad \longrightarrow \quad x = -2t + D$$

$$V_y = \frac{dy}{dt} = 0 \quad \longrightarrow \quad y = E$$

$$V_z = \frac{dt}{dt} = 2.4t + 5 \quad \longrightarrow \quad z = 1.2t^2 + 5t + F$$

$At\ t = 0 x = 0, y = 0, 2 = 0$. Hence, $D = 0 = E = F$

$x = -2t, y = 0, z = 1.2t^2 + 5t$

(a) At $t = 1,\ x = -2,\ y = 0,\ z = 6.2.$ Thus the particle is at $\underline{\underline{(-2, 0, 6.2)}}$

(b) $\bar{V} = (V_x, V_y, V_z) = \underline{\underline{-2\bar{a}_x + (2.4t + 5)\bar{a}_z \ \text{m/s}}}$

Prob 3.12

(a)

$$\nabla U = \frac{\partial U}{\partial x}\bar{a}_x + \frac{\partial U}{\partial y}\bar{a}_y + \frac{\partial U}{\partial z}\bar{a}_z$$

$$= \underline{\underline{4z^2\bar{a}_x + 3z\bar{a}_y + (8xz + 3y)\bar{a}_z}}$$

(b)

$$\nabla T = \frac{\partial T}{\partial \rho}\bar{a}_\rho + \frac{1}{\rho}\frac{\partial T}{\partial \phi}\bar{a}_\phi + \frac{\partial T}{\partial z}\bar{a}_z$$

$$= \underline{\underline{5e^{-2z}\sin\phi\,\bar{a}_\rho + 5e^{-2z}\cos\phi\,\bar{a}_\phi - 10\rho e^{-2z}\sin\phi\,\bar{a}_z}}$$

(c)

$$\nabla H = \frac{\partial H}{\partial r}\bar{a}_r + \frac{1}{r}\frac{\partial H}{\partial \theta}\bar{a}_\theta + \frac{1}{r\sin\theta}\frac{\partial H}{\partial \phi}\bar{a}_\phi$$

$$= \underline{\underline{2r\cos\theta\cos\phi\,\bar{a}_r - r\sin\theta\cos\phi\,\bar{a}_\theta - r\cos\theta\sin\phi\,\bar{a}_\phi}}$$

Prob 3.13

$$(a)\ \nabla V = \frac{\partial V}{\partial x}\bar{a}_x + \frac{\partial V}{\partial y}\bar{a}_y + \frac{\partial V}{\partial z}\bar{a}_z$$

$$= \underline{\underline{2e^{(2x+3y)}\cos 5z\,\bar{a}_z + 3e^{(2x+3y)}\cos 5z\,\bar{a}_y - 5e^{(2x+3y)}\sin 5z\,\bar{a}_z.}}$$

At $(0.1,-0.2,0.4)$

$$e^{(2x+3y)} = e^{0.2-0.6} = 0.6703,\quad \cos 5z = \cos 2 = -0.4161,\quad \sin 5z = 0.9092$$

$$\nabla V = 2(0.6073)(-0.4161)\bar{a}_x + 3(0.6703)(-0.4161)\bar{a}_y - 5(0.6203)(0.9092)$$

$$= \underline{\underline{-0.5578\,\bar{a}_x - 0.8367\,\bar{a}_y - 3.047\,\bar{a}_z}}$$

(b)

$$\nabla T = \underline{\underline{5e^{-2z}\sin\phi\,\bar{a}_\rho + 5e^{-z}\cos\phi\,\bar{a}_\phi - 10\rho e^{-2z}\sin\phi\,\bar{a}_z}}$$

At $(2, \frac{\pi}{3}, 0)$,

$$\nabla T = (5)(1)(0.5)\bar{a}_\rho + 5(1)(0.5)\bar{a}_\phi - 10(2)(1)(0.866)\bar{a}_z$$

$$= \underline{\underline{2.5\,\bar{a}_\rho + 2.5\,\bar{a}_\phi - 17.32\,\bar{a}_z}}$$

(c)

$$\nabla Q = \underline{\underline{\frac{-2\sin\theta\sin\phi}{r^3}\bar{a}_r + \frac{\cos\theta\sin\phi}{r^3}\bar{a}_\theta + \frac{\cos\phi}{r^3}\bar{a}_\phi}}$$

At $(1, 30°, 90°)$,

$$\nabla Q = \frac{-2(0.5)(1)}{1}\bar{a}_r + \frac{(0.86)(1)}{1}\bar{a}_\theta + 0 = \underline{\underline{-\bar{a}_r + 0.866\,\bar{a}_\theta}}$$

Prob 3.14

$$\nabla S = 2x\,\bar{a}_x + 2y\,\bar{a}_y - \bar{a}_z$$

At $(1,3,0)$,

$$\nabla S = 2\bar{a}_x + 6\bar{a}_y - \bar{a}_z \text{ and } \quad \bar{a}_n = \frac{\nabla S}{|\nabla S|} = \frac{(2,6,-1)}{\sqrt{4+36+1}}$$

$$\bar{a}_n = \underline{\underline{0.3123\bar{a}_x + 0.937\bar{a}_y - 0.1562\bar{a}_z}}$$

Prob 3.15

$$\nabla T = 2x\bar{a}_x + 2y\bar{a}_y - \bar{a}_z$$

At $(1,1,2)$, $\nabla T = (2,2,-1)$. The mosquito should move in the direction of

$$\underline{\underline{2\bar{a}_x + 2\bar{a}_y - \bar{a}_z}}$$

Prob 3.16 (a)

$$\nabla \bullet \bar{A} = \underline{\underline{ye^{xy} + x\cos xy - 2x\cos zx \sin zx}}$$

$$\nabla \times \bar{A} = \begin{vmatrix} \dfrac{\partial}{\partial x} & \dfrac{\partial}{\partial y} & \dfrac{\partial}{\partial z} \\ e^{xy} & \sin xy & \cos^2 xz \end{vmatrix}$$

$$= (0-0)\bar{a}_x + (0 + 2z\cos xz \sin xz)\bar{a}_y + (y\cos xy - xe^{xy})\bar{a}_z$$

$$= \underline{\underline{z\sin 2xz\,\bar{a}_y + (y\cos xy - xe^{xy})\bar{a}_z.}}$$

(b)

$$\nabla \bullet \bar{B} = \frac{1}{\rho}\frac{\partial}{\partial \rho}(\rho^2 z^2 \cos\phi) + 0 + \sin^2\phi$$

$$= \underline{\underline{2z^2\cos\phi + \sin^2\phi}}$$

$$\nabla \times \bar{B} = \left(\frac{1}{\rho}\frac{\partial \bar{B}_z}{\partial \phi} - 0\right)\bar{a}_\rho + \left(\frac{\partial B_\rho}{\partial z} - \frac{\partial B_z}{\partial \rho}\right)\bar{a}_\phi + \frac{1}{\rho}\left(0 - \frac{\partial B_e}{\partial \phi}\right)\bar{a}_z$$

$$= \underline{\underline{\frac{z\sin 2\phi}{\rho}\bar{a}_\rho + 2\rho z\cos\phi\,\bar{a}_\phi + z^2\sin\phi\,\bar{a}_z}}$$

(c)

$$\nabla \bullet \bar{C} = \frac{1}{r^2}\frac{\partial}{\partial r}(r^3\cos\theta) + \frac{1}{r\sin\theta}\frac{\partial}{\partial \theta}\left(-\frac{1}{r}\sin^2\theta\right) + 0$$

$$= \underline{\underline{3\cos\theta - \frac{2\cos\theta}{r^2}}}$$

$$\nabla \times \bar{C} = \frac{1}{r\sin\theta}[\frac{\partial}{\partial\theta}(2r^2\sin^2\theta) - 0]\bar{a}_r + \frac{1}{r}[0 - \frac{\partial}{\partial r}(2r^3\sin\theta)]\bar{a}_\theta$$

$$+ \frac{1}{r}[\frac{\partial}{\partial r}(-\sin\theta) + r\sin\theta]\bar{a}_\phi$$

$$= 4r\cos\theta\,\bar{a}_r - 6r\sin\theta\,\bar{a}_\theta + \sin\theta\,\bar{a}_\phi$$

Prob 3.17 (a)

$$\nabla \times \bar{A} = \begin{vmatrix} \frac{\partial}{\partial x} & \frac{\partial}{\partial y} & \frac{\partial}{\partial z} \\ x^2y & y^2z & -2xz \end{vmatrix} = \underline{\underline{-y^2\bar{a}_x + 2z\bar{a}_y - y^2\bar{a}_z.}}$$

$$\nabla \bullet \nabla \times \bar{A} = \underline{\underline{0}}$$

(b)

$$\nabla \times \bar{A} = (\frac{1}{\rho}\frac{\partial A_z}{\partial\phi} - \frac{\partial A_\phi}{\partial z})\bar{a}_\rho + (\frac{\partial A_\rho}{\partial z} - \frac{\partial A_z}{\partial\rho})\bar{a}_\phi + \frac{1}{\rho}(\frac{\partial(\rho A_\rho)}{\partial\rho} - \frac{\partial A_\rho}{\partial\phi})\bar{a}_z$$

$$= (0 - 0)\bar{a}_\rho + (\rho^2 - 3z^2)\bar{a}_\phi + \frac{1}{\rho}(4\rho^3 - 0)\bar{a}_z$$

$$= \underline{\underline{(\rho^2 - 3z^2)\bar{a}_\phi + 4\rho^2\bar{a}_z}}$$

$$\nabla \bullet \nabla \times \bar{A} = \underline{\underline{0}}$$

(c)

$$\nabla \times \bar{A} = \frac{1}{r\sin\theta}\frac{\partial}{\partial\theta}(-\sin\theta\cos\phi)\bar{a}_r + [\frac{1}{\sin\theta}\frac{\cos\phi}{r^2} - \frac{\partial}{\partial r}(r^{-1}\cos\theta)]\bar{a}_\theta + \frac{1}{r}(0 - 0)\bar{a}_\phi$$

$$= \frac{-\cos\theta\cos\phi}{r\sin\theta}\bar{a}_r + \frac{1}{r}[\frac{\cos\phi}{r^2\sin\theta} + r^{-2}\cos\theta]\bar{a}_\theta$$

$$= \frac{-1}{r}\cot\theta\cos\phi\,\bar{a}_r + \frac{1}{r^3}(\frac{\cos\phi}{\sin\theta} + \cos\theta)\,\bar{a}_\theta$$

$$\nabla \bullet \nabla \times \bar{A} = \underline{\underline{0}}$$

Prob 3.18

$$\bar{\nabla} \bullet \bar{H} = k\bar{\nabla} \bullet \bar{\nabla} T = k\bar{\nabla}^2 T$$

$$\bar{\nabla}^2 T = \frac{\partial^2 T}{\partial x^2} + \frac{\partial^2 T}{\partial y^2} = 50\sin\frac{\pi x}{2}\cosh\frac{\pi y}{2}(-\frac{\pi^2}{4} + \frac{\pi^2}{4}) = 0$$

Hence, $\bar{\nabla} \bullet \bar{H} = 0$

Prob 3.19

(a)

$$\nabla \bullet (V\bar{A}) = \frac{\partial}{\partial x}(V A_x) + \frac{\partial}{\partial y}(V A_y) + \frac{\partial}{\partial z}(V A_z)$$

$$= (A_x \frac{\partial V}{\partial x} + V\frac{\partial A_x}{\partial x}) + (A_y \frac{\partial V}{\partial y} + V\frac{\partial A_y}{\partial y}) + (A_z \frac{\partial V}{\partial z} + V\frac{\partial A_z}{\partial z})$$

$$= V(\frac{\partial A_x}{\partial x} + \frac{\partial A_y}{\partial y} + \frac{\partial A_z}{\partial z}) + A_x\frac{\partial V}{\partial x} + A_y\frac{\partial V}{\partial y} + A_z\frac{\partial V}{\partial z}$$

$$= \underline{\underline{V\nabla \bullet \bar{A} + \bar{A} \bullet \nabla V}}$$

(b)

$$\nabla \bullet A = 2 + 3 - 4 = 1; \quad \nabla V = yz\bar{a}_x + xz\bar{a}_y + xy\bar{a}_z$$

$$\nabla \bullet (V\bar{A}) = V\nabla \bullet \bar{A} + \bar{A} \bullet \nabla V$$

$$= xyz + 2xyz + 3xyz - 4xyz = \underline{\underline{2xyz}}$$

Prob 3.20 (a)

$$\nabla \times V\bar{A} = \begin{vmatrix} \frac{\partial}{\partial x} & \frac{\partial}{\partial y} & \frac{\partial}{\partial z} \\ VA_x & VA_y & VA_z \end{vmatrix}$$

$$= [\frac{\partial}{\partial y}(VA_z) - \frac{\partial}{\partial z}(VA_y)]\bar{a}_x + [\frac{\partial}{\partial z}(VA_x) - \frac{\partial}{\partial x}(VA_z)]\bar{a}_y$$

$$+ [\frac{\partial}{\partial x}(VA_y) - \frac{\partial}{\partial y}(VA_x)]\bar{a}_z$$

$$= [A_z\frac{\partial V}{\partial x} + V\frac{\partial A_z}{\partial y} - A_y\frac{\partial V}{\partial z} + V\frac{\partial A_y}{\partial z}]\bar{a}_x$$

$$+ [A_x\frac{\partial V}{\partial z} + V\frac{\partial A_x}{\partial z} - A_z\frac{\partial V}{\partial x} + V\frac{\partial A_z}{\partial x}]\bar{a}_y$$

$$+ [A_y\frac{\partial V}{\partial x} + V\frac{\partial A_y}{\partial x} - A_x\frac{\partial V}{\partial y} + V\frac{\partial A_x}{\partial y}]\bar{a}_z$$

$$= V[(\frac{\partial A_z}{\partial y} - \frac{\partial A_y}{\partial z})\bar{a}_x + (\frac{\partial A_x}{\partial z} - \frac{\partial A_z}{\partial x})\bar{a}_y$$

$$+ (\frac{\partial A_y}{\partial x} + \frac{\partial A_x}{\partial y})\bar{a}_z]$$

$$+ (\frac{\partial V}{\partial x}\bar{a}_x + \frac{\partial V}{\partial y}\bar{a}_y + \frac{\partial V}{\partial z}\bar{a}_z) \times (A_x\bar{a}_y + A_y\bar{a}_y + A_z\bar{a}_z)$$

$$= \underline{\underline{V(\nabla \times \bar{A}) + \nabla V \times \bar{A}}}$$

(b)

$$V\bar{A} = \frac{1}{r}\cos\theta\,\bar{a}_r + \frac{1}{r}\sin\theta\,\bar{a}_\theta + \frac{1}{r^2}\sin\theta\cos\phi\,\bar{a}_\phi$$

$$\nabla \times (V\bar{A}) = \frac{1}{r\sin\theta}[\frac{2}{r^2}\sin\theta\cos\theta\cos\phi - 0)\bar{a}_r + \frac{1}{r}(0 + \frac{1}{r^2}\sin\theta\cos\phi)\bar{a}_\theta +$$

$$\frac{1}{r}(0 + \frac{1}{r}\sin\theta)\bar{a}_\phi$$

$$= \underline{\underline{\frac{2\cos\theta\cos\phi}{r^3}\bar{a}_r + \frac{\sin\theta\cos\phi}{r^3}\bar{a}_\theta + \frac{\sin\theta}{r^2}\bar{a}_\phi}}$$

Prob 3.21

$$grad\ U = \frac{\partial U}{\partial x}\bar{a}_x + \frac{\partial U}{\partial y}\bar{a}_y + \frac{\partial U}{\partial z}\bar{a}_z$$

$$= (z - 2xy)\bar{a}_x + (2yz^2 - x^2)\bar{a}_y + (x - 2y^2z)\bar{a}_z$$

$$Div\ grad U = \nabla \bullet \nabla U = \frac{\partial}{\partial x}(z - 2xy) + \frac{\partial}{\partial y}(2yz^2 - x^2) + \frac{\partial}{\partial z}(x - 2y^2z)$$

$$= -2y + 2z^2 - 2y^2$$

$$= \underline{\underline{2(z^2 - y^2 - y)}}$$

Prob 3.22

$$\nabla \ln\rho = (\frac{\partial}{\partial x}\ln\rho)\bar{a}_x + (\frac{\partial}{\partial y}\ln\rho)\bar{a}_y + (\frac{\partial}{\partial z}\ln\rho)\bar{a}_z$$

$$= \frac{x}{\rho^2}\bar{a}_x + \frac{y}{\rho^2}\bar{a}_y$$

$$\nabla \times \phi \bar{a}_z = \nabla \times \tan^{-1}\frac{y}{x}\bar{a}_z$$

$$= \begin{vmatrix} \frac{\partial}{\partial x} & \frac{\partial}{\partial y} & \frac{\partial}{\partial z} \\ 0 & 0 & \tan^{-1}\frac{y}{x} \end{vmatrix}$$

$$= \frac{x}{x^2+y^2}\bar{a}_x + \frac{y}{x^2+y^2}\bar{a}_y$$

$$= \frac{x}{\rho^2}\bar{a}_x + \frac{y}{\rho^2}\bar{a}_y$$

$$= \underline{\underline{\nabla \ln\rho, \quad \text{as expected !}}}$$

Prob 3.23

$$\nabla\phi = \frac{1}{r\sin\phi}\bar{a}_\phi, \qquad \nabla\theta = \frac{1}{r}\bar{a}_\theta$$

$$\frac{r\nabla\theta}{\sin\theta} = \frac{\bar{a}_\theta}{\sin\theta}$$

$$\nabla \times \left(\frac{r\nabla\theta}{\sin\theta}\right) = \frac{1}{r}\sin\theta\,\bar{a}_\theta$$

Thus, $\nabla\phi = \underline{\underline{\nabla \times \left(\frac{r\nabla\theta}{\sin\theta}\right)}}$

Prob 3.24

(a) $\nabla V = \underline{\underline{(6xy+z)\bar{a}_x + 3x^2\bar{a}_y + x\bar{a}_z}}$

$\nabla \bullet \nabla V = \underline{\underline{6y}}$

$$\nabla \times \nabla V = \begin{vmatrix} \frac{\partial}{\partial x} & \frac{\partial}{\partial y} & \frac{\partial}{\partial z} \\ 6xy+z & 3x^2 & x \end{vmatrix} = \underline{\underline{0}}$$

(b) $\nabla V = \underline{\underline{z\cos\phi\,\bar{a}_\rho - z\sin\phi\,\bar{a}_\phi + \rho\cos\phi\,\bar{a}_z}}$

$$\nabla \bullet \nabla V = \frac{1}{\rho}\frac{\partial}{\partial\rho}(\rho z\cos\phi) + \frac{z}{\rho}\cos\theta + 0 = \frac{z}{\rho}\cos\phi - \frac{z}{\rho}\cos\phi = \underline{\underline{0}}$$

$\nabla \times \nabla V = \underline{\underline{0}}$

$$(c)\ \bar{\nabla} V = \frac{1}{r^2}(24r^2)\cos\theta\sin\phi + \frac{4r\cos\phi}{r\sin\theta}(\cos^2\theta\sin^2\theta)$$

$$- \frac{4}{r^2\sin^2\theta}\cos\theta\sin\phi$$

$$= \underline{\underline{24r\cos\theta\sin\phi + \frac{4\cos\phi}{\sin\theta} - 8\cos\phi\sin\theta - \frac{4\cos\theta\sin\phi}{\sin^2\theta}}}$$

$$\nabla \times \nabla V = \underline{\underline{0}}$$

Prob. 3.25

(a)

$$(\bar{\nabla} \bullet \bar{r})\bar{T} = 3\bar{T} = 6yz\bar{a}_y + 3xy^2\bar{a}_y + 3x^2yz\bar{a}_z$$

(b)

$$x\frac{\partial \bar{T}}{\partial x} + y\frac{\partial \bar{T}}{\partial y} + z\frac{\partial \bar{T}}{\partial z} = x\ (y^2\bar{a}_y + 2xyz\bar{a}_z) + y(2z\bar{a}_x + 2xy\bar{a}_y + x^2z\bar{a}_x)$$

$$+ z(2y\bar{a}_x + x^2y\bar{a}_z)$$

$$= \underline{\underline{4yz\bar{a}_x + 3xy^2\bar{a}_y + 4x^2yz\bar{a}_z}}$$

(c)

$$\nabla \bullet \bar{r}(\bar{r} \bullet \bar{T}) = 3\,(2xyz + xy^3 + x^2yz^2)$$

$$= \underline{\underline{6xyz + 3xy^3 + 3x^2yz^2}}$$

(d)

$$(\bar{r} \bullet \nabla)\bar{r} = (x\frac{\partial}{\partial x} + y\frac{\partial}{\partial y} + z\frac{\partial}{\partial z})(x^2 + y^2 + z^2)$$

$$= x(2x) + y(2y) + z(2z)$$

$$= \underline{\underline{2(x^2 + y^2 + z^2) = 2r^2}}$$

Prob. 3.26

$$(a)\nabla r^n\bar{r} = \frac{\partial}{ox}(xr^n) + \frac{\partial}{oy}(yr^n) + \frac{\partial}{oz}(zr^n)$$

where $r^n = (x^2 + y^2 + z^2)^{n/2}$

$$\nabla \cdot r^n \bar{r} = 2x^2(\frac{n}{2})\ (x^2 + y^2 + z^2)^{\frac{n}{2}-1} + 2y^2(\frac{n}{2})\ (x^2 + y^2 + z^2)^{\frac{n}{2}-1}$$

$$+ 2z^2(\frac{n}{2})\ (x^2 + y^2 + z^2)^{\frac{n}{2}-1} + r^n + r^n + r^n$$

$$= n(x^2 + y^2 + z^2)\ (x^2 + y^2 + z^2)^{\frac{n}{2}-1} + 3r^n$$

$$= nr^n + 3r^n \quad = \underline{\underline{(n+3)r^n}}$$

$$(b) \nabla \times r^n \bar{r} = \begin{vmatrix} \frac{\partial}{\partial x} & \frac{\partial}{\partial y} & \frac{\partial}{\partial z} \\ r^n x & r^n y & r^n z \end{vmatrix}$$

$$= [2y(\frac{n}{2})z(x^2 + y^2 + z^2)^{\frac{n}{2}-1} - 2z(\frac{n}{2})y(x^2 + y^2 + z^2)^{\frac{n}{2}-1}]a_x + \ldots$$

$$= 0$$

Prob. 3.27

(*a*) Let $V = \ln r = \ln\sqrt{x^2 + y^2 + z^2}$

$$\frac{\partial V}{\partial x} = \frac{1}{r}\ \frac{1}{2}(2x)\ (x^2 + y^2 + z^2)^{-\frac{1}{2}} = \frac{x}{r^2}$$

$$\nabla V = \frac{\partial V}{\partial x}\bar{a}_x + \frac{\partial V}{\partial y}\bar{a}_y + \frac{\partial V}{\partial z}\bar{a}_z = \frac{x\bar{a}_x + y\bar{a}_y + z\bar{a}_z}{r^2} = \underline{\underline{\frac{\bar{r}}{r^2}}}$$

(*b*) Let $\nabla V = \bar{A} = \frac{\bar{r}}{r^2} = \frac{1}{r}\bar{a}_x$ in spherical coordinates.

$$\nabla^2\ (\ln r) = \nabla \cdot \nabla\ (\ln r) = \nabla \cdot \bar{A} = \frac{1}{r^2}\ \frac{\partial}{\partial r}(r^2 A r) = \frac{1}{r^2}\frac{d}{dr}(r)$$

$$= \underline{\underline{\frac{1}{r^2}}}$$

Prob 3.28

(a)

$$V_1 = x^3 + y^3 + z^3$$

$$\nabla^2 V_1 = \frac{\partial^2 V_1}{\partial x^2} + \frac{\partial^2 V_1}{\partial y^2} + \frac{\partial^2 V_1}{\partial z^2}$$

$$= \frac{\partial}{\partial x}(3x^2) + \frac{\partial}{\partial y}\left(3y^2\right) + \frac{\partial}{\partial x}(3z^2)$$

$$= 6x + 6y + 6z = \underline{\underline{6(x + y + z)}}$$

(b)

$$V_2 = \rho z^2 \sin 2\phi$$

$$\nabla^2 V_2 = \frac{1}{\rho}\frac{\partial}{\partial \rho}(\rho z^2 \sin 2\phi) - \frac{4z^2}{\rho}\sin 2\phi + \frac{\partial}{\partial z}(2\rho z \sin 2\phi)$$

$$= \frac{z}{\rho}\sin 2\phi - \frac{4z^2}{\rho}\sin 2\phi + 2\rho \sin 2\phi$$

$$= (\frac{-3z^2}{\rho} + 2\rho)\sin 2\phi$$

(c)

$$V_3 = r^2(1 + \cos\theta \sin\phi)$$

$$\nabla^2 V_3 = \frac{1}{r^2}\frac{\partial}{\partial r}[2r^3(1 + \cos\theta \sin\phi)] + \frac{1}{r^2 \sin\theta}\frac{\partial}{\partial \theta}\left(-\sin\theta \sin\phi\right)r^2$$

$$+ \frac{1}{r^2 \sin\theta}\frac{\partial}{\partial \theta}(-\sin^2\theta \sin\phi)r^2 + \frac{1}{r^2 \sin^2\theta}r^2(-\cos\theta \sin\phi)$$

$$= 6(1 + \cos\theta \sin\phi) - \frac{2\sin\theta}{\sin\theta}\cos\theta \sin\phi - \frac{\cos\theta \sin\phi}{\sin^2\theta}$$

$$= 6 + 4\cos\theta \sin\phi - \frac{\cos\theta \sin\phi}{\sin^2\theta}$$

Prob 3.29

(a)

$$U = x^3 y^2 e^{xz}$$

$$\nabla^2 U = \frac{\partial}{\partial x}(3x^2 y^2 e^{xz}) + \frac{\partial}{\partial y}(2x^2 y e^{xz}) + \frac{\partial}{\partial z}(x^4 y^2 e^{xz})$$

$$= 6xy^2 e^{xz} + 2x^2 e^{xz} + x^5 y^2 e^{xz} \qquad = (6xy^2 + 2x^2 + x^5 y^2)e^{xz}$$

At $(1,-1,1)$,

$$\nabla^2 U = e^1\ (6 + 2 + 1) = 9e = 24.46$$

(b)

$$V = \rho^2 z(\cos\phi + \sin\phi)$$

$$\nabla^2 V = \frac{1}{\rho}\frac{\partial}{\partial \rho}[2\rho^2 z(\cos\phi + \sin\phi)] - z(\cos\phi + \sin\phi) + 0$$

$$= 4z(\cos\phi + \sin\phi) - z(\cos\phi + \sin\phi)$$

$$= 3z(\cos\phi + \sin\phi)$$

At $(5, \frac{\pi}{6}, -2)$, $\nabla^2 V = -6(0.866 + 0.5) = -8.196$

(c)

$$W = e^{-r}\sin\theta\cos\phi$$

$$\nabla^2 W = \frac{1}{r^2}\frac{\partial}{\partial r}\ (-r^2 e^{-r}\sin\theta\cos\phi) + \frac{e^{-r}}{r^2\sin\theta}\cos\phi\frac{\partial}{\partial\theta}(\sin\theta\cos\theta)$$

$$-\frac{e^{-r}\sin\theta\cos\phi}{r^2\sin^2\theta}$$

$$= \frac{1}{r^2}(-2re^{-r}\sin\theta\cos\phi) + e^{-r}\sin\theta\cos\phi$$

$$+\frac{e^{-r}\cos\phi}{r^2\sin\theta}(\cos^2\theta - \sin^2\theta) - \frac{-e^{-r}\cos\theta}{r^2\sin\theta}$$

$$\underline{\underline{\nabla^2 W = e^{-r}\sin\theta\cos\phi(1-\frac{4}{r})}}$$

At $(1, 60^\circ, 30^\circ)$,

$$\nabla^2 W = e^{-1}\sin 60\cos 30(1-4) = -2.25e^{-1} = \underline{\underline{-0.8277}}$$

Prob 3.30

(a)

$$\nabla^2 V = \frac{\partial^2 V}{\partial x^2} + \frac{\partial^2 V}{\partial y^2} + \frac{\partial^2 V}{\partial y^2}$$

$$= \underline{\underline{2(y^2z^2 + x^2z^2 + x^2y^2)}}$$

(b)

$$\nabla^2 \bar{A} = \nabla^2 A_x\ \bar{a}_x + \nabla^2 A_y \bar{a}_y + \nabla^2 A_z \bar{a}_z$$

$$= (2y + 0 + 0)\bar{a}_x + (0 + 0 + 6xz)\bar{a}_y + (0 - 2z^2 - 2y^2)\bar{a}_z$$

$$= \underline{\underline{2y\bar{a}_x + 6xz\bar{a}_y - 2(y^2 + z^2)\bar{a}_z}}$$

(c)

$$grad\ div\ A = \nabla(\nabla \bullet \bar{A}) = \nabla(2xy + 0 - 2y^2z)$$

$$= \underline{\underline{2y\bar{a}_x + 2(x - 2yz)\bar{a}_y - 2y^2\bar{a}_z}}$$

(d)

$$curl\ curl\ \bar{A} = \nabla x \nabla x \bar{A} = \nabla(\nabla \bullet \bar{A}) - \nabla^2 \bar{A}$$

From parts (b) and (c),

$$\nabla x \nabla x \bar{A} = \underline{\underline{2(x - 2yz - 3xz)\bar{a}_y + 2z^2\bar{a}_z}}$$

Prob. 3.31

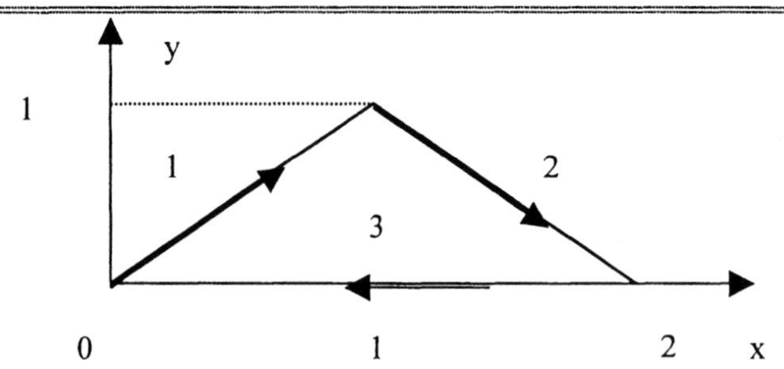

(a)

$$\oint_L \bar{F} \bullet d\bar{l} = \left(\int_1 + \int_2 + \int_3\right) \bar{F} \bullet d\bar{l}$$

For $1, y = x \quad dy = dx, d\bar{l} = dx\,\bar{a}_x + dy\,\bar{a}_y,$

$$\int_1 \bar{F} \bullet d\bar{l} = \int_0^1 x^3 dx - x dx = -\frac{1}{4}$$

For $2,\ y = -x + 2, dy = -dx, d\bar{l} = dx\,\bar{a}_x + dy\,\bar{a}_y,$

$$\int_2 \bar{F}\, d\bar{l} = \int_1^2 (-x^3 + 2x^2 - x + 2) dx = \frac{17}{12}$$

For 3,

$$\int_3 \bar{F} \bullet d\bar{l} = \int_2^0 x^2 y dx \Big|_{y=0} = 0$$

$$\oint_L \bar{F} \bullet d\bar{l} = -\frac{1}{4} + \frac{17}{12} + 0 = \frac{7}{6}$$

(b)

$\nabla \times \bar{F} = -x^2\,\bar{a}_z\ ; \qquad d\bar{S} = dxdy(-\bar{a}_z)$

$$\int (\nabla \times \bar{F}) \bullet d\bar{S} = -\iint (-x^2) dxdy = \int_0^1 \int_0^x x^2 dydx + \int_1^2 \int x^2 dydx$$

$$= \int_0^1 x^2 y \Big|_0^x dx + \int_1^2 x^2 y \Big|_0^{-x+2} dx = \frac{x}{4}\Big|_0^1 + \int_1^2 x^2(-x+2) dx = \frac{7}{6}$$

(c) Yes

Prob 3.32

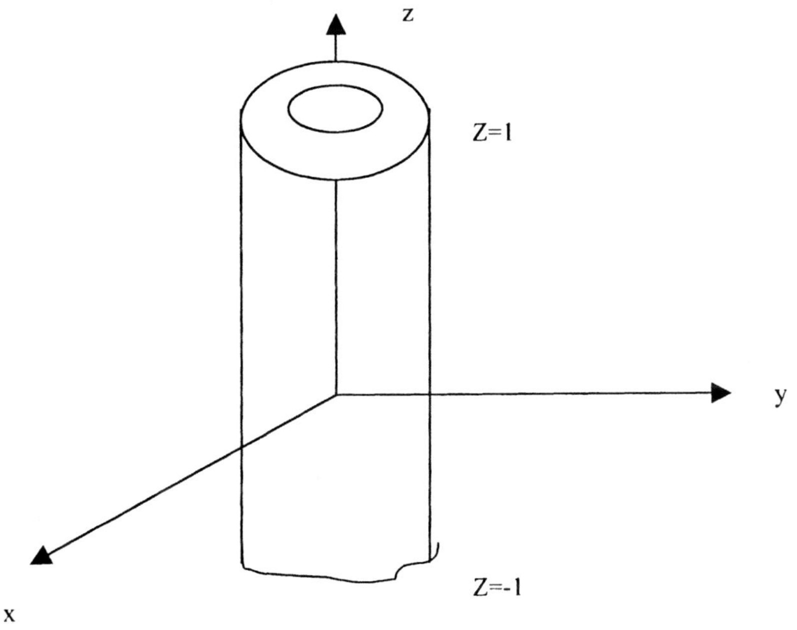

(a)

$$\oint \bar{D} \bullet d\bar{s} = \left[\iint_{z=-1} + \iint_{z=1} + \iint_{\rho=2} + \iint_{\rho=5} \right] \bar{D} \bullet d\bar{s}$$

$$= -\iint \rho^2 \cos^2\phi \, d\phi \, d\rho + \iint \rho^2 \cos^2\phi \, d\phi \, d\rho - \iint 2\rho^2 z^2 d\phi \, dz \Big|_{\rho=2} + \iint 2\rho^2 z^2 d\phi \, dz \Big|_{\rho=5}$$

$$= -2(2)^2 \int_0^{2\pi} d\phi \int_{-1}^{1} z^2 dz + 2(5)^2 \int_0^{2\pi} d\phi \int_{-1}^{1} z^2 dz$$

$$= -8(2\pi)\left(\frac{z^3}{3}\Big|_{-1}^{1}\right) + 50(2\pi)\left(\frac{z^3}{3}\Big|_{-1}^{1}\right)$$

$$= \frac{-32\pi}{3} + \frac{200\pi}{3} = \underline{\underline{176}}$$

(b) $$\nabla \bullet \bar{D} = \frac{1}{\rho}\frac{\partial}{\partial \rho}(2\rho^2 z^2) = 4z^2$$

$$\int \nabla \bullet D dv = \iiint 4z^2 \rho \, d\rho \, d\phi \, dz = 4\int_{-1}^{1} z^2 dz \int_2^5 \rho \, d\rho \int_0^{2\pi} d\phi$$

$$= 4x \frac{z^3}{3}\Bigg|_{-1}^{1} \frac{\rho^2}{2}\Bigg|_2^5 (2\pi) = 56\pi = \underline{\underline{176}}$$

Prob 3.33

Transform $\bar{F}$ into cylindrical system.

$$\begin{bmatrix} F_\rho \\ F_\phi \\ F_z \end{bmatrix} = \begin{bmatrix} \cos\phi & \sin\phi & 0 \\ -\sin\phi & \cos\phi & 0 \\ 0 & 0 & 1 \end{bmatrix} \begin{bmatrix} x^2 \\ y^2 \\ z^2 - 1 \end{bmatrix}$$

$$F_\rho = x^2 \cos\phi + y^2 \sin\phi = \rho^2 \cos^3\phi + \rho^2 \sin^3\phi, F_z = z^2 - 1$$

$$F_\phi = -x^2 \sin\phi + y^2 \cos\phi = -\rho^2 \cos^2\phi \sin\phi + \rho^2 \sin^2\phi \cos\phi$$

$$\nabla \bullet \bar{F} = \frac{1}{\rho}\frac{\partial}{\partial\rho}(\rho^3 \cos^3\phi + \rho^3 \sin^3\phi) + 2z - \rho\cos^3\phi - 2\rho\cos\phi\sin^2\phi$$

$$+ 2\rho\sin\phi\cos^2\phi + \rho\sin^3\phi$$

$$= 2\rho\cos^3\phi + 4\rho\sin^3\phi - 2\rho\cos\phi\sin^2\phi + 2\rho\cos^2\phi\sin\phi + 2z$$

$$\int \bar{F} \bullet d\bar{S} = \int \nabla \bullet \bar{F}\, dv$$

Due to the fact that we are integrating $\sin\phi$ and $\cos\phi$ over $0 < \phi < 2\pi$, all terms involving $\cos\phi$ *and* $\sin\phi$ will vanish. Hence,

$$\int \bar{F} d\bar{S} = \iiint 2z\, \rho\, d\rho\, d\phi\, dz = 2\int_0^{2/\pi} d\phi \int_0^2 z dz \int_0^2 \rho\, d\rho$$

$$= 2(2\pi)(\frac{2^2}{2}\Big|_0^2) = 16\pi$$

$$= \underline{\underline{50.26}}$$

Prob 3.34

(a)

$$\oint \bar{A} \bullet d\bar{S} = \int_V \nabla \bullet \bar{A} dv, \quad \nabla \bullet \bar{A} = y + z + x$$

$$\oint \bar{A} \bullet d\bar{S} = \int_0^1\int_0^1\int_0^1 (x + y + z) dxdydz$$

$$= 3\int_0^1 xdy \int_0^1 dy \int_0^1 dz = 3(\frac{x^2}{2}\Big|_0^1)(1)(1)$$

$$= \underline{\underline{1.5}}$$

(b)

$\nabla \bullet \bar{A} = 0.$ Hence, $\oint \bar{A} \bullet d\bar{S} = \underline{\underline{0}}$

Prob 3.35

(a)

$$\nabla \bullet \bar{A} = y^2 + 3y^2 + y^2 = 5y^2$$

$$\int \nabla \bullet \bar{A}\, dv = \int\int\int 5y^2 dxdydz$$

$$= 5\int_0^1 dx \int_0^1 y^2 dy \int_0^1 dz = 5(1)(1)(\frac{y^3}{3}\Big|_0^1) = \underline{\underline{1.667}}$$

$$\oint \bar{A} \bullet d\bar{S} = [\iint_{x=0} + \iint_{x=1} + \iint_{y=0} + \iint_{y=1} + \iint_{z=0} + \iint_{z=1}]\bar{A} \bullet d\bar{S}$$

$$= -\int\int xy^2 dydz\Big|_{x=0} + \int\int xy^2 dydz\Big|_{x=1} - \int\int y^3 dxdz\Big|_{y=0}$$

$$+ \int\int y^3 dxdz\Big|_{y=1} - \iint y^2 zdxdy\Big|_{z=0} + \iint y^2 zdxdy\Big|_{z=1}$$

$$= (1)(1)(\frac{y^3}{3}\Big|_0^1) + (1)(1)(1) + (1)(1)(\frac{y^3}{3}\Big|_0^1) = \underline{\underline{1.667}}$$

(b)

$$\nabla \bullet \bar{A} = \frac{1}{\rho}\frac{\partial}{\partial \rho}(2\rho^2 z) + \frac{3z\cos\phi}{\rho} - 0 = 4z + \frac{3z\cos\phi}{\rho}$$

$$\int_V \nabla \bullet \bar{A} = \iiint (4z + \frac{3z}{\rho}\cos\phi)\rho\, d\rho\, d\phi\, dz$$

$$= 4\int_0^2 \rho\, d\rho \int_0^5 zdz \int_0 d\phi + 3\int_0^2 d\rho \int_0^5 zdz \int^{45^\circ} \cos\phi\, d\phi$$

$$= 4(\frac{4}{2})(\frac{25}{2})(\frac{11}{4}) + 3(2)(\frac{25}{2})\sin 45^\circ$$

$$= 25\pi + 75\sin 45^\circ = \underline{\underline{131.57}}$$

$$\oint_s \bar{A} \bullet d\bar{S} = [\iint_{\rho=2} + \iint_{z=0} + \iint_{z=5} + \iint_{\phi=0} + \iint_{\phi=45^\circ}]\bar{A} \bullet d\bar{S}$$

$$= J_1 + J_2 + J_3 + J_4 + J_5$$

where $J_1 = \iint 2\rho z \rho \, d\phi \, dz \Big|_{\rho=2} = (2)(2)^2 \int_0^5 z dz \int_0^{\frac{\pi}{4}} d\phi = 25\pi$

$$J_2 = \iint 4\rho \cos\phi \, d\rho \, \rho \, d\phi \Big|_{z=0} = -\frac{32}{3}\sin\frac{\pi}{4}$$

$$J_3 = -\iint 4\rho \cos\phi \, d\rho \, dd \Big|_{z=5} = \frac{32}{3}\sin\frac{\pi}{4}$$

$$J_4 = \iint 3z \sin\phi \, d\rho \, dz \Big|_{\phi=0} = 0$$

$$J_5 = \iint 3z \sin\phi \, d\rho \, d\phi \Big|_{\phi=\frac{\pi}{4}} = 75\sin\frac{\pi}{4}$$

$$\oint \bar{A} \bullet d\bar{S} = 25\pi + 75\sin\frac{\pi}{4} = \underline{\underline{131.57}}$$

(c)

$$\nabla \bullet \bar{A} = \frac{1}{r^2}\frac{\partial}{\partial r}(r^4) + \frac{1}{\sin\theta}\frac{\partial}{\partial\theta}(r\sin^2\theta\cos\phi)$$
$$= 4r + 2\cos\theta\cos\phi$$

$$\int \nabla \bullet \bar{A} \, dv = \iiint 4r^3 \sin\theta \, d\theta \, d\phi \, dr + \iiint 2r^2 \sin\theta \cos\theta \cos\phi \, d\theta \, d\phi \, dr$$
$$= 4\frac{r^4}{4}\Big|_0^3 (-\cos\theta)\Big|_0^{\pi/2} (\frac{\pi}{2}) + \frac{2r^3}{3}\Big|_0^3 (-\frac{\cos\theta}{2})\Big|_0^{\pi/2} \sin\phi \Big|_0^{\frac{\pi}{2}}$$
$$= 81(1)(\frac{\pi}{2}) + 18(0 + \frac{1}{2})(1 - 0)$$
$$= \frac{81\pi}{2} + 9 = \underline{\underline{136.23}}$$

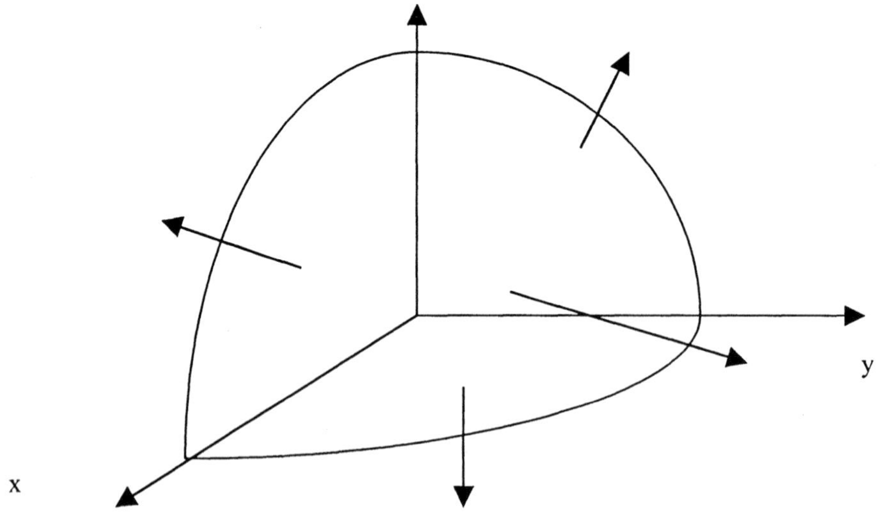

$$\int \bar{A} \bullet d\bar{S} = [\iint_{\phi=0} + \iint_{\phi=\pi/2} + \iint_{r=3} + \iint_{\theta=\pi/2}] \bar{A} \bullet d\bar{S}$$

Since $\bar{A}$ has no ϕ – component, the first two integrals vanish.

$$\int \bar{A} \bullet d\bar{S} = \int_{\phi=0}^{\pi/2} \int_{\phi=0}^{\pi/2} r^4 \sin\theta \, d\theta \, d\phi \Big|_{r=3} + \int_{r=0}^{3} \int_{\phi=0}^{\pi/2} r^2 \sin^2\theta \cos\phi \, dr d\phi \Big|_{\theta=\pi/2}$$

$$= 81 \left(\frac{\pi}{2}\right) (-\cos\theta) \Big|_0^{\pi/2} + 9(1) \sin\phi \Big|_0^{\pi/2}$$

$$= \frac{81\pi}{2} + 9 \quad = \underline{\underline{136.23}}$$

Prob. 3.36

$$\int \rho_v \, dv = \oint_S \bar{A} \bullet d\bar{S} \qquad \text{(divergence theorem)}$$

where $\rho_v = \nabla \bullet \bar{A} = x^2 + y^2$

$$\frac{\partial A_x}{\partial x} = x^2 \longrightarrow A_x = \frac{x^3}{3} + C_1$$

$$\frac{\partial A_y}{\partial y} = y^2 \longrightarrow A_y = \frac{y^3}{3} + C_2$$

Hence,

$$\bar{A} = \underline{\underline{\left(\frac{x^3}{3} + C_1\right)\bar{a}_x + \left(\frac{y^3}{3} + C_2\right)\bar{a}_y}}$$

Prob. 3.37

(a)

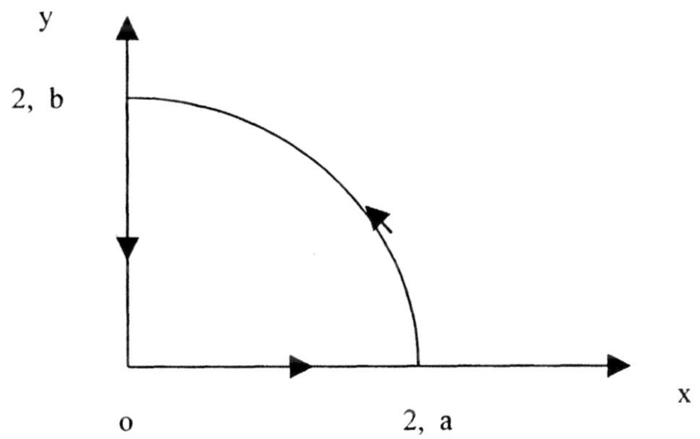

$d\bar{l} = d\rho\, \bar{a}_\rho + \rho d\phi\, \bar{a}_\phi$

$\bar{A} \bullet d\bar{l} = \rho \sin\phi\, d\rho + \rho^3 d\phi$

Along *oa*: $d\phi = 0, \quad \phi = 0, \quad \bar{A} \bullet d\bar{l} = 0, \quad \int_0^a \bar{A} \bullet d\bar{l} = 0$

Along *ob*: $d\rho = 0, \quad \rho = 2, \quad \int_a^b \bar{A} \bullet d\bar{l} = 0, \quad \int_0^{\frac{\pi}{2}} 8 d\phi = 4\pi$

Along *bo*: $d\phi = 0, \quad \phi = \frac{\pi}{2}, \quad \int_b^0 \bar{A} \bullet d\bar{l} = \int_2^0 \rho\, d\rho = -2$

Hence, $\oint \bar{A} \bullet d\bar{l} = \underline{\underline{4\pi - 2}}$

(b)

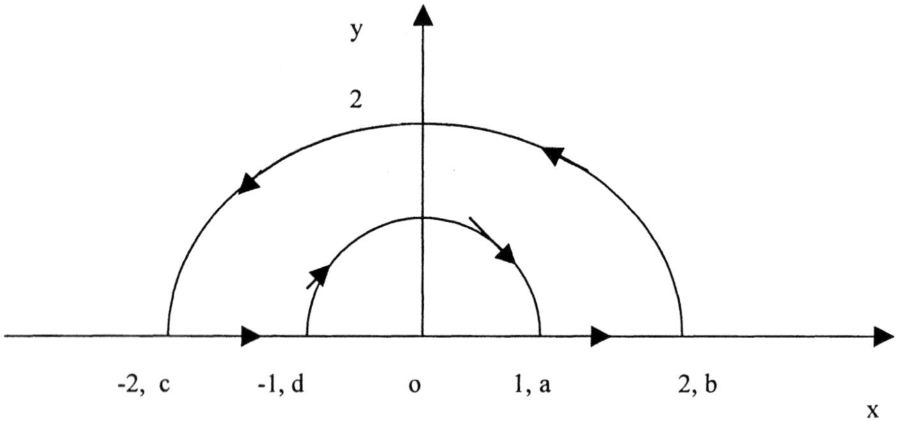

Along *ab*, $d\phi = 0, \ \phi = 0, \ \bar{A} \bullet d\bar{l} = 0, \ \int_a^b \bar{A} \bullet d\bar{l} = 0.$

Along *bc*, $d\rho = 0, \ \bar{A} \bullet d\bar{l} = \rho^3 d\phi,$

$$\int_b^c \bar{A} \bullet d\bar{l} = \int \rho^3 d\phi = (2)^3(\pi - 0) = 8\pi$$

Along *cd*, $d\phi = 0, \ \phi = \pi, \ \bar{A} \bullet d\bar{l} = 0, \ \int_c^d \bar{A} \bullet d\bar{l} = 0$

Along *da*, $d\rho = 0, \ \bar{A} \bullet d\bar{l} = \rho^3 d\phi,$

$$\int_d^a \bar{A} \bullet d\bar{l} = \rho^3 \int_\pi^0 d\phi = (1)^3(0 - \pi) = -\pi.$$

Hence, $\oint \bar{A} \bullet d\bar{l} = 0 + 8\pi + 0 - \pi = \underline{\underline{7\pi}}$.

This may be checked by using Stokes' theorem.

Prob. 3.38

Let $\psi = \oint \bar{F} \bullet d\bar{S} = \psi_t + \psi_b + \psi_o + \psi_i$

where ψ_t, ψ_b, ψ_o, ψ_i are the fluxes through the top surface, bottom surface, outer surface ($\rho = 3$), and inner surface respectively.

For the top surface, $d\bar{S} = \rho \, d\phi \, d\rho \, \bar{a}_z$, $\quad z = 5$;

$$\bar{F} \bullet d\bar{S} = \rho^2 \, z \, d\phi \, dz. \text{ Hence:}$$

$$\psi_t = \int_{\rho=2}^{3} \int_{\phi=0}^{2\pi} \rho^2 \, z \, d\phi \, dz \Big|_{z=5} = \frac{190 \, \pi}{3}$$

For the bottom surface, $\quad z = 0, \; d\bar{S} = \rho \, d\phi \, d\rho(-\bar{a}_z)$

$$\bar{F} \bullet d\bar{S} = \rho z \, d\phi \, d\rho = 0. \quad \text{Hence, } \psi_b = 0.$$

For the outer curved surface, $\rho = 3, \quad d\bar{S} = \rho \, d\phi \, dz \, \bar{a}_\rho$

$$\bar{F} \bullet d\bar{S} = \rho^3 \sin\phi \, d\phi \, dz. \text{ Hence,}$$

$$\psi_a = \int_{z=0}^{5} dz \, \rho^3 \int_{\phi=0}^{2\pi} \sin\phi \, d\phi \Big|_{\rho=3} = 0$$

For the inner curved surface, $\rho = 2, \quad d\bar{S} = \rho \, d\phi \, dz(-\bar{a}_z)$

$$\bar{F} \bullet d\bar{S} = \rho^3 \sin\phi \, d\phi \, dz. \text{ Hence,}$$

$$\psi_a = \int_{z=0}^{5} dz \, \rho^3 \int_{\phi=0}^{2\pi} \sin\phi \, d\phi \Big|_{\rho=2} = 0$$

$$\psi = \frac{190\pi}{3} + 0 + 0 + 0 = \underline{\underline{\frac{190 \, \pi}{3}}}$$

$$\psi = \oint \bar{F} \bullet d\bar{S} = \int \nabla \bullet \bar{F} \, dV$$

$$\nabla \bullet \bar{F} = \frac{1}{\rho}\frac{\partial}{\partial \rho}(\rho^3 \sin\phi) + \frac{1}{\rho}\frac{\partial}{\partial \phi}(z\cos\phi) + \rho$$

$$= 3\rho \sin\phi - \frac{z}{\rho}\sin\phi + \rho$$

$$\int_V \nabla \bullet \bar{F}\, dv = \iiint (3\rho \sin\phi - \frac{z}{\rho}\sin\phi + \rho)\, \rho\, d\phi\, d\rho\, dz$$

$$= 0 + 0 + \int_0^5 dz \int_0^{2\pi} d\phi \int_2^3 \rho^2 d\rho$$

$$= \underline{\underline{\frac{190\,\pi}{3}}}$$

Prob. 3.39

Let $\bar{B} = \nabla \times \bar{T}$

$$\psi = \oint_S \bar{B} \bullet d\bar{S} = \int \nabla \bullet \bar{B}\, dv = \int \nabla \bullet \nabla \times \bar{T}\, dv = 0$$

Prob 3.40

$$\bar{Q} = \frac{r}{r\sin\theta} r\sin\theta[(\cos\phi - \sin\phi)\bar{a}_x + (\cos\phi + \sin\phi)\bar{a}_y$$

$$= r(\cos\phi - \sin\phi)\bar{a}_x + r(\cos\phi + \sin\phi)\bar{a}_y$$

$$\begin{bmatrix} Q_r \\ Q_\theta \\ Q_\phi \end{bmatrix} = \begin{bmatrix} \sin\theta\cos\phi & \sin\theta\sin\phi & \cos\theta \\ \cos\theta\cos\phi & \cos\theta\sin\phi & -\sin\theta \\ -\sin\phi & \cos\phi & 0 \end{bmatrix} \begin{bmatrix} Q_x \\ Q_y \\ Q_z \end{bmatrix}$$

$$\bar{Q} = r\sin\theta\bar{a}_r + r\cos\theta\bar{a}_\theta + r\bar{a}_\phi$$

(a)

$$d\bar{l} = \rho d\phi \bar{a}_\phi, \quad \rho = r\sin 30° = 2(\frac{1}{2}) = 1$$

$$z = r\cos 30° = \sqrt{3}$$

$$Q_\phi = r = \sqrt{\rho^2 + z^2}$$

$$\oint \bar{Q} \bullet d\bar{l} = \int_0^{2\pi} \sqrt{\rho^2 + z^2}\, \rho d\phi = 2(1)(2\pi) = \underline{\underline{4\pi}}$$

(b)

$$\nabla \times \bar{Q} = \cot\theta\, \bar{a}_r - 2\bar{a}_\theta + \cos\theta\, \bar{a}_\phi$$

For S_1, $d\bar{S} = r^2 \sin\theta\, d\theta\, d\phi\, \bar{a}_r$

$$\int_{S_1} (\nabla \times \bar{Q}) \bullet d\bar{S} = \int r^2 \sin\theta \cot\theta\, d\theta\, d\phi \Big|_{r=2}$$

$$= 4\int_0^{2\pi} d\phi \int_0^{30°} \cos\theta\, d\theta = \underline{\underline{4\pi}}$$

(c)

For S_2, $d\bar{S} = r\sin\theta\, d\theta\, dr\, \bar{a}_\theta$

$$\int_{S_2} (\nabla \times \bar{Q}) \bullet d\bar{S} = -2\int r\sin\theta\, d\phi\, dr \Big|_{\theta=30°}$$

$$= -2\sin 30 \int_0^2 r dr \int_0^{2\pi} d\phi$$

$$= -4\pi$$

(d)

For S_1, $d\bar{S} = r^2 \sin\theta\, d\phi\, d\theta\, \bar{a}_r$

$$\int_{S_1} \bar{Q} \bullet d\bar{S} = r^3 \int \sin^2\theta\, d\theta\, d\phi \Big|_{r=2}$$

$$= 8\int_0^{2\pi} d\phi \int_0^{30°} \sin^2\theta\, d\theta$$

$$= 4\pi\left[\frac{\pi}{3} - \frac{\sqrt{3}}{2}\right]$$

(e)

For S_2, $d\bar{S} = r\sin\theta\, d\phi\, dr\, \bar{a}_\theta$

$$\int_{S_2} \bar{Q} \bullet d\bar{S} = \int r^2 \sin\theta\cos\theta\, d\phi\, dr \Big|_{\theta=30°}$$

$$= \frac{4\pi\sqrt{3}}{3}$$

(f)

$$\nabla \bullet \bar{Q} = \frac{1}{r^2}\frac{\partial}{\partial r}(r^3 \sin\theta) + \frac{r}{r\sin\theta}\frac{\partial}{\partial\theta}(\sin\theta\cos\theta) + 0$$

$$= 2\sin\theta + \cos\theta\cot\theta$$

$$\int \nabla \bullet \bar{Q} dV = \int (2\sin\theta + \cos\theta\cot\theta) r^2 \sin\theta\, d\theta\, d\phi\, dr$$

$$= \frac{r^3}{3}\Big|_0^2 (2\pi) \int_0^{30} (1 + \sin^2\theta) d\theta$$

$$= \frac{4\pi}{3}\left(\pi - \frac{\sqrt{3}}{2}\right)$$

Check: $\int \nabla \bullet \bar{Q} dV = (\int_{S_1} + \int_{S_2})(\nabla \times \bar{Q}) \bullet d\bar{S}$

$$= 4\pi[\frac{\pi}{3} - \frac{\sqrt{3}}{2} + \frac{\sqrt{3}}{3}]$$

$$= \frac{4\pi}{3}[\pi - \frac{\sqrt{3}}{2}] \quad \text{(It checks.)}$$

Prob. 3.41

Since $\bar{u} = \bar{\omega} \times \bar{r}$, $\nabla \times \bar{u} = \nabla \times (\bar{\omega} \times \bar{r})$. From Appendix A.10,

$$\nabla \times (\bar{A} \times \bar{B}) = \bar{A}(\nabla \bullet \bar{B}) - \bar{B}(\nabla \bullet \bar{A}) + (\bar{B} \bullet \nabla)\bar{A} - (\bar{A} \bullet \nabla)\bar{B}$$

$$\nabla \times \bar{u} = \nabla \times (\bar{\omega} \times \bar{r})$$

$$\nabla \times (\bar{\omega} \times \bar{r}) = \bar{\omega}(\nabla \bullet \bar{r}) - \bar{r}(\nabla \bullet \bar{\omega}) + (\bar{r} \bullet \nabla)\bar{\omega} - (\bar{\omega} \bullet \nabla)\bar{r}$$

$$= \bar{\omega}(3) - \bar{\omega} = 2\bar{\omega}$$

or $\bar{\omega} = \frac{1}{2}\nabla \times \bar{u}$.

Alternatively, let $x = r\cos\omega t$, $y = r\sin\omega t$

$$\bar{u} = \frac{\partial x}{\partial t}\bar{a}_x + \frac{\partial y}{\partial t}\bar{a}_y$$

$$= -\omega r \sin\omega t\, \bar{a}_x + \omega r \cos\omega t\, \bar{a}_y$$

$$= -\omega y \bar{a}_x + \omega x \bar{a}_y$$

$$\nabla \times \bar{u} = \begin{vmatrix} \frac{\partial}{\partial x} & \frac{\partial}{\partial y} & \frac{\partial}{\partial z} \\ -\omega y & \omega x & 0 \end{vmatrix} = 2\omega \bar{a}_z = 2\bar{\omega}$$

i.e., $\bar{\omega} = \frac{1}{2}\nabla \times \bar{u}$

Prob 3.42

Let $\bar{A} = U\nabla V$ and apply Stokes' theorem.

$$\int_L U\nabla V \bullet d\bar{l} = \int \nabla X(U\nabla V) \bullet d\bar{S}$$

$$= \int_s (\nabla U X \nabla V) d\bar{S} + \int_s U\, (\nabla X \nabla V) \bullet d\bar{S}$$

But $\nabla X \nabla V = 0$. Hence,

$$\int_L U\nabla V \bullet d\bar{l} = \int_s (\nabla U X \nabla V) \bullet d\bar{S}$$

Similarly, we can show that

$$\int_L V\nabla U \bullet d\bar{l} = \int (\nabla V X \nabla U) \bullet d\bar{S} - \int (\nabla U X \nabla V) \bullet d\bar{S}$$

Thus, $\int_L U\nabla V \bullet d\bar{l} = -\int_L V\nabla U \bullet d\bar{l}$

Prob. 3.43

Let $\bar{A} = r^n \bar{r} = (x^2 + y^2 + z^2)^{n/2}\ (x\bar{a}_x + y\bar{a}_y + z\bar{a}_z)$

By divergence theorem,

$$\int \bar{A} \bullet d\bar{S} = \int \nabla \bullet \bar{A}\, dv$$

$$\nabla \bullet \bar{A} = \frac{\partial Ax}{\partial x} + \frac{\partial Ay}{\partial y} + \frac{\partial Az}{\partial z}$$

$$= \frac{\partial}{\partial x}(xr^n) + \frac{\partial}{\partial y}(yr^n) + \frac{\partial}{\partial z}(zr^n)$$

$$= r^n + 2x^2(\frac{n}{2})(x^2 + y^2 + z^2)^{n/2-1}$$

$$+ r^n + 2y^2(\frac{n}{2})(x^2 + y^2 + z^2)^{n/2-1}$$

$$+ r^n + 2z^2(\frac{n}{2})(x^2 + y^2 + z^2)^{n/2-1}$$

$$= 3r^n + n(x^2 + y^2 + z^2)r^{n-1}$$

$$= (3 + n)r^n$$

Thus, $\oint r^n \bar{r}\, d\bar{s} = \int (3 + n) r^n dV$

or $\int r^n dv = \frac{1}{n+3} \oint r^n \bar{r}\, d\bar{s}$

Prob 3.44

(a)

$$\nabla \times \bar{G} = \begin{vmatrix} \frac{\partial}{\partial x} & \frac{\partial}{\partial y} & \frac{\partial}{\partial z} \\ 16xy - z & 8x^2 & -x \end{vmatrix}$$

$$= 0\bar{a}_x + (-1 + 1)\bar{a}_y + (16x - 16x)\bar{a}_z = 0$$

Thus, G is irrotational.

(b)

$$4 = \oint \bar{G} \bullet d\bar{s} = \int \nabla \bullet \bar{G} dv$$

$$\nabla \bullet \bar{G} = 16y + 0 + 0 = 16y$$

$$4 = \iiint 16y dx dy dz = 16 \int_0^1 dx \int_0^1 dz \int_0^1 y dy = 16(1)(1)(\frac{y^2}{2} \Big|_0^1) = \underline{\underline{8}}$$

(c)

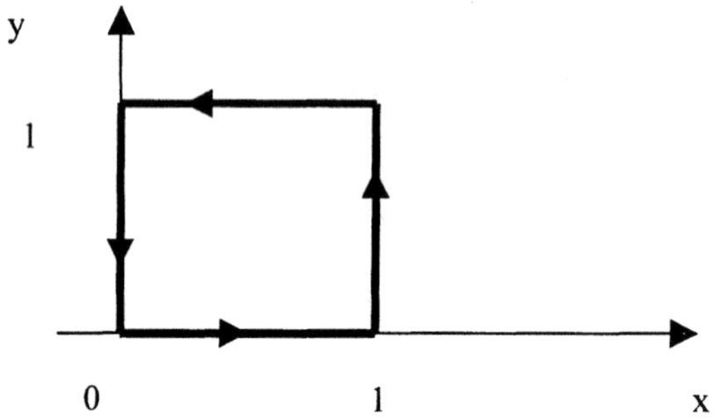

$$\oint_L \bar{G} \bullet d\bar{l} = \int_{x=0}^{x=1} (16xy - z)dx \Big|_{\substack{y=0 \\ z=0}} + \int_{y=0}^{y=1} 8x^2 dy \Big|_{\substack{x=1 \\ z=0}} + \int_{x=1}^{x=0} (16xy - z)dx \Big|_{\substack{y=1 \\ z=0}} + \int_{y=1}^{y=0} 8x^2 dy \Big|_{\substack{x=0 \\ z=0}}$$

$$= 0 + 8(1)y \Big|_0^1 + 16(1)\frac{x^2}{2} \Big|_1^0 + 0$$

$$= 8 - 8 = \underline{\underline{0}}$$

This is expected since G is irrotational, i.e.

$$\oint \bar{G} \bullet d\bar{l} = \int (\nabla \times \bar{G}) \bullet d\bar{S} = 0$$

Prob 3.45

$$\nabla \times \bar{T} = \begin{vmatrix} \frac{\partial}{\partial x} & \frac{\partial}{\partial y} & \frac{\partial}{\partial z} \\ \alpha x + \beta z^2 & 3x^2 - \gamma z & 3xz^2 - y \end{vmatrix}$$

$$= (-1 + \gamma)\bar{a}_x + (3\beta z^2 - 3z^2)\bar{a}_y + (6x - \alpha x)\bar{a}_z$$

If $\bar{T}$ is irrotational, $\nabla \times \bar{T} = 0$, *i.e.*

$$\underline{\underline{\alpha = 1 = \beta = \gamma}}$$

$$\nabla \bullet \bar{T} = \frac{\partial \bar{T}_x}{\partial x} + \frac{\partial \bar{T}_y}{\partial y} + \frac{\partial \bar{T}_z}{\partial z} = \alpha\, y + 0 + 6xz$$

At $(2, -1, 0)$,

$$\nabla \bullet \bar{T} = -1 + 0 = \underline{\underline{-1}}$$

CHAPTER 4

P. E. 4.1

$$(a)\ \bar{F} = \frac{1x10^{-9}}{4\pi\left(\frac{10^{-9}}{36\pi}\right)}\left[\frac{5x10^{-9}[(1,-3,7)-(2,0,4)]}{[(1,-3,7)-2,0,4)]^3}\right.$$

$$\left.\frac{-2x10^{-9}[(1,-3,7)-(-3,0,5)]}{[(1,-3,7)-(-3,0,5)]^3}\right.$$

$$= \left[\frac{45(-1,-3,3)}{19^{3/2}} - \frac{18(4,-3,2)}{29^{3/2}}\right] \text{ nN}$$

(a) $$= -1.004\bar{a}_x - 1.284\bar{a}_y + 1.4\bar{a}_z \text{ nN}$$

$$(b)\quad \bar{E} = \frac{\bar{F}}{Q} = -1.004\bar{a}_x - 1.284\bar{a}_y + 1.4\bar{a}_z \text{ V/m}$$

P. E. 4.2

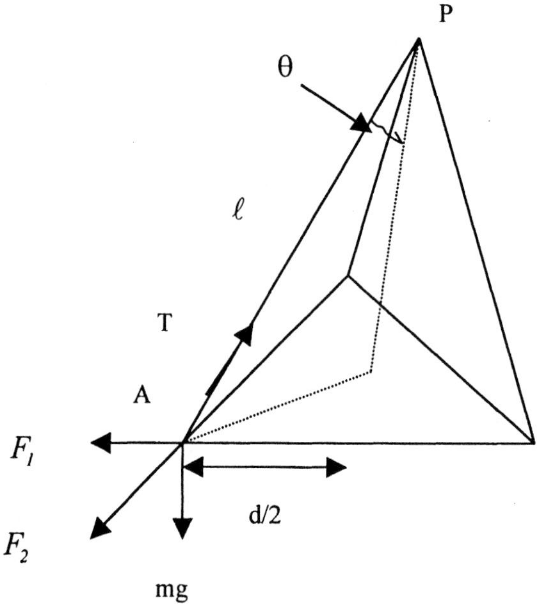

At point A,

$$T\sin\theta\cos 30° = F_1 + F_2\cos 60°$$

$$= \frac{q^2}{4\pi\varepsilon_0 d^2} + \frac{q^2}{4\pi\varepsilon_0 d^2}\left(\frac{1}{2}\right)$$

$$= \frac{3q^2}{8\pi\varepsilon_0 d^2}$$

$$T\cos\theta = mg$$

Hence, $\tan\theta \cos 30° = \dfrac{3q^2}{8\pi\varepsilon_0 d^2} mg$

But $\sin\theta = \dfrac{h}{l} = \dfrac{d}{\sqrt{3}\, l}$ $\tan\theta = \dfrac{\frac{d}{\sqrt{3}}}{\sqrt{L^2 - \frac{d^2}{3}}}$

Thus, $\dfrac{\frac{d}{\sqrt{3}}(\frac{\sqrt{3}}{2})}{\sqrt{L^2 - \frac{d^2}{3}}} = \dfrac{3q^2}{8\pi\varepsilon_0 d^2 mg}$

or $q^2 = \dfrac{4\pi\varepsilon_0 d^3 mg}{3\sqrt{l^2 - \frac{d^2}{3}}}$

but $q = \dfrac{Q}{3} \longrightarrow q^2 = \dfrac{Q}{9}$. Hence,

$$Q^2 = \frac{12\pi\varepsilon_0 d^3 mg}{\sqrt{l^2 - \frac{d^2}{3}}}$$

P.E. 4.3

$$e\bar{E} = m\frac{d^2\bar{l}}{dt^2}$$

$$eE_0(-2\bar{a}_x + \bar{a}_y) = m(\frac{d^2x}{dt^2}\bar{a}_x + \frac{d^2y}{dt^2}\bar{a}_y + \frac{d^2z}{dt^2}\bar{a}_z)$$

where $E_0 = 200$ kV / m

$$\frac{d^2z}{dt^2} = 0 \longrightarrow z = ct + c_2$$

$$m\frac{d^2x}{dt^2} = -2eE_0 \longrightarrow x = \frac{-2eE_0t^2}{2m} + c_3t + c_4$$

$$m\frac{d^2y}{dt^2} = eE_0 \longrightarrow y = \frac{eE_0t^2}{2m} + c_5t + c_6$$

At $t = 0$, $(x,y,z) = (0,0,0)$ $\quad c_1 = 0 = c_4 = c_6$

Also, $(\frac{dx}{dt}, \frac{dy}{dt}, \frac{dz}{dt}) = (0,0,0)$

At $t = 0 \longrightarrow \quad c_1 = 0 = c_3 = c_5$

Hence, $(x,y) = \frac{eE_0 t^2}{2m} (-2,1)$

i.e. $2\,|y| = |x|$

Thus the largest horizontal distance is

80 cm = $\underline{\underline{0.8 \text{ m}}}$

P.E. 4.4

(a)

Consider an element of area ds of the disk.

The contribution due to $ds = \rho\, d\phi\, d\rho$ is

$$dE = \frac{\rho_s ds}{4\pi\varepsilon_0 r^2} = \frac{\rho_s ds}{4\pi\varepsilon_0(\rho^2 + h^2)}$$

The sum of the contribution along ρ gives zero.

$$E_z = \frac{\rho_s}{4\pi\,\varepsilon_0} \int_{\rho=0}^{a} \int_{\phi=0}^{2\pi} \frac{z\rho\, d\rho\, d\phi}{(\rho^2 + h^2)^{3/2}} = \frac{h\rho_s}{2\varepsilon_0} \int_{\rho-0}^{a} \frac{\rho\, d\rho}{(\rho^2 + h^2)^{3/2}}$$

$$= \frac{h\rho_s}{4\varepsilon_0} \int_0^a (\rho^2 + h^2)^{3/2} d(\rho^2) = \frac{h\rho_s}{2\varepsilon_0} (-2(\rho^2 + h^2)^{-1/2} \Big|_0^a$$

$$= \frac{\rho_s}{2\varepsilon_0} [1 - \frac{h}{(h^2 + a^2)^{1/2}}]$$

(b)

As $a \longrightarrow \infty$,

$$\underline{\underline{\bar{E} = \frac{\rho_s}{2\varepsilon_0} \bar{a}_z}}$$

P.E. 4.5

$$Q_S = \int \rho_S dS = \int_{-2}^{2} \int_{-2}^{2} 12|y|\, dx\, dy$$

$$= 12(4) \int_0^2 2y\, dy = \underline{\underline{192 \text{ mC}}}$$

$$\bar{E} = \int \frac{\rho_s dS}{4\pi\varepsilon r^2} \bar{a}_r = \int \frac{\rho_s dS|\bar{r} - \bar{r}'|}{4\pi\varepsilon_o |\bar{r} - \bar{r}'|^3}$$

where $\bar{r} - \bar{r}' = (0,0,10) - (x,y,z) = (-x,-y,10)$.

$$\bar{E} = \int_{x=2}^{2}\int_{y=2}^{2} \frac{12\,|y|10^{-3}(-x,-y,10)}{4\pi(\frac{10^{-9}}{36\pi})(x^2+y^2+100)^{3/2}}$$

$$= 108(10^{-6})[\int_{-2}^{2}|y|\int_{-2}^{2}\frac{-x dx\, dy\, \bar{a}_x}{(x^2+y^2+100)^{3/2}} + \int_{-2}^{2}|y|\int_{-2}^{2}\frac{-y|y|\; dy\, dx\, \bar{a}_y}{(x^2+y^2+100)^{3/2}}$$

$$+ 10\bar{a}_z \int_{-2}^{2}\int_{-2}^{2}\frac{-|y| dx\, dy}{(x^2+y^2+100)^{3/2}}]$$

$$\bar{E} = 108(10^7)\bar{a}_z \int_{-2}^{2}[2\int_{0}^{2}\frac{\frac{1}{2}d(y^2)}{(x^2+y^2+100)^{3/2}}]dx$$

$$= -216(10^7)\bar{a}_z \int_{-2}^{2}[\frac{1}{(x^2+104)^{1/2}} - \frac{1}{(x^2+100)^{1/2}}]dx$$

$$= -216(10^7)\bar{a}_z\ \ln\left|\frac{x+\sqrt{x^2+104}}{x+\sqrt{x^2+100}}\right|_{-2}^{2}$$

$$= -216(10^7)\bar{a}_z(\ln(\frac{2+\sqrt{108}}{2+\sqrt{104}}) - \ln(\frac{-2+\sqrt{108}}{-2+\sqrt{104}}))$$

$$= -216(10^7)\bar{a}_z(-7.6202\,(10^{-3}))$$

$$\bar{E} = \underline{\underline{16.46\,\bar{a}_z \quad \text{mV/m}}}$$

P.E. 4.6

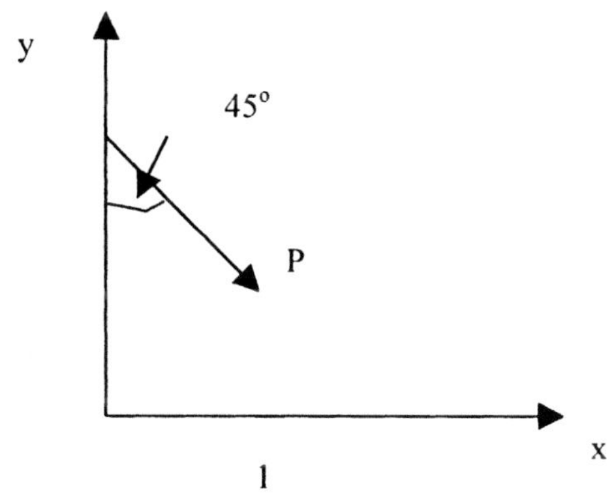

$$\bar{E} = \frac{\rho_L}{2\pi\varepsilon_0\rho}\bar{a}_\rho$$

To get $\bar{a}_\rho$, consider the $z = -1$ plane. $\rho = \sqrt{2}$

$$\bar{a}_\rho = \bar{a}_x \cos 45° - \bar{a}_y \sin 45°$$

$$= \frac{1}{\sqrt{2}}(\bar{a}_x - \bar{a}_y)$$

$$\bar{E}_3 = \frac{10(10^{-9})}{2\pi(\frac{10^{-9}}{36\pi})}\frac{1}{2}(\bar{a}_x - \bar{a}_y)$$

$= 90\pi\,(\bar{a}_x - \bar{a}_y)$. Hence,

$$\bar{E} = \bar{E}_1 + \bar{E}_2 + \bar{E}_3$$

$$= -180\pi\,\bar{a}_x + 270\pi\,\bar{a}_y + 90\pi\,\bar{a}_x - 90\pi\,\bar{a}_y.$$

$$= \underline{\underline{-282.7\,\bar{a}_x + 565.5\bar{a}_y \text{ V/m}}}$$

P.E. 4.7

$$\bar{D} = \bar{D}_Q + \bar{D}_\rho = \frac{Q}{4\pi r^2}\bar{a}_r + \frac{\rho_s}{2}\bar{a}_n$$

$$= \frac{30x10^{-9}}{4\pi(5)^2}\frac{[(0,4,3) - (0,0,0)]}{5} + \frac{10x10^{-9}}{2}\bar{a}_y$$

$$= \frac{30}{500\pi}(0,4,3) + 5\bar{a}_y \text{ nC/m}^2$$

$$= \underline{\underline{5.076\bar{a}_y + 0.0573\bar{a}_z \text{ nC/m}^2}}$$

P.E. 4.8

(a) $\rho v = \nabla \bullet \bar{D} = 4x$

$\rho v(-1,0,3) = \underline{\underline{-4 \text{ C/m}^3}}$

(b) $4 = Q = \int \rho v dv = \int_0^1\int_0^1\int_0^1 4x dx dy dz$

$= 4(1)(1)(1/2) = \underline{\underline{2 \text{ C}}}$

(c) $Q = 4 = \underline{\underline{2 \text{ C}}}$

P.E. 4.9

$$Q = \int \rho_v dv = \psi = \oint \bar{D} \bullet d\bar{s}$$

For $0 \le r \le 10$,

$$D_r (4\pi r^2) = \iiint 2r \ (r^2) \sin\theta \, d\theta \, dr \, d\phi$$

$$D_r (4\pi r^2) = 4\pi (\frac{2r^4}{4} \Big|_0^r) = 2\pi r^4$$

$$Dr = \frac{r^2}{2} \qquad \bar{E} = \frac{r^2}{2\varepsilon_0} \bar{a}_r \ \text{nV/m}$$

$$\bar{E}(r = 2) = \frac{4(10^{-9})}{2(\frac{10^{-9}}{36\pi})} \bar{a}_r = 72\pi \, \bar{a}_r = \underline{\underline{226 \ \bar{a}_r \ \text{V/m}}}$$

For $r \le 10$,

$$D_r (4\pi r^2) = 2\pi r_0^4, \qquad r_0 = 10\text{m}$$

$$D_r = \frac{r_0^4}{2r^2} \longrightarrow \bar{E} = \frac{r_0^4}{2\varepsilon_0 r^2} \bar{a}_r \ \text{nV/m}$$

$$\bar{E}(r = 12) = \frac{10^4 (10^{-9})}{2(\frac{10^{-9}}{36\pi})(144)} \bar{a}_r = 1250\pi \, \bar{a}_r$$

$$= \underline{\underline{3.927 \, \bar{a}_r \ \text{kV/m}}}$$

P. E. 4.10

$$V(\bar{r}) = \sum_{k=1}^{3} \frac{Q_k}{4\pi\varepsilon_0 |\bar{r} - \bar{r}_k|} + C$$

At $V(\infty) = 0, \quad C = 0$

$$|\bar{r} - \bar{r}_1| = |(-1,5,2) - (2,-1,3)| = \sqrt{46}$$

$$|\bar{r} - \bar{r}_2| = |(-1,5,2) - (0,4,-2)| = \sqrt{18}$$

$$|\bar{r} - \bar{r}_3| = |(-1,5,2) - (0,0,0)| = \sqrt{30}$$

$$V(-1,5,2) = \frac{10^{-6}}{4\pi(\frac{10^{-9}}{36\pi})} [\frac{-4}{\sqrt{46}} + \frac{5}{\sqrt{18}} + \frac{3}{\sqrt{30}}]$$

$$= \underline{\underline{10.3 \ k\text{V}}}$$

P.E. 4.11

$$V = \frac{Q}{4\pi\varepsilon_0 r} + C$$

If $V(0,6,-8) = V(r = 10) = 2;$

$$2 = \frac{5(10^{-9})}{4\pi(\frac{10^{-9}}{36\pi})} + C \quad \longrightarrow \quad C = -2.5$$

(a)

$$V_A = \frac{5(10^{-9})}{4\pi(\frac{10^{-9}}{36\pi})|(-3,2,6)-(0,0,0)|} - 2.5$$

$$= \underline{\underline{3.929\ V}}$$

(b)

$$V_B = \frac{45}{\sqrt{7^2+1^2+5^2}} - 2.5 = \underline{\underline{2.696\ V}}$$

(b) $V_{AB} = V_B - V_A = 2.696 - 3.929 = \underline{\underline{-1.233\ V}}$

P.E. 4.12

(a)

$$\frac{-W}{Q} = \int \bar{E} \bullet d\bar{l} = \int (3x^2 + y)dx + xdy$$

$$= \int_0^2 (3x^2 + y)dx \Big|_{y=5} + \int_5^{-1} x\,dy \Big|_{x=2}$$

$$= 18 - 12 = 6$$

$$W = -6Q = \underline{\underline{12}}\ \text{mJ}$$

(b)

$$dy = -3\,dx$$

$$\frac{-W}{Q} = \int \bar{E} \bullet d\bar{l} = \int_0^2 (3x^2 + 5 - 3x)dx + x(-3)dx$$

$$= \int_0^2 (3x^2 - 6x + 5)dx = 8 - 12 + 10 = 6$$

$$W = \underline{\underline{12}}\ \text{nJ}$$

P.E. 4.13

(a)

$$(0,0,10) \longrightarrow (r = 10, \theta = 0, \phi = 0)$$

$$V = \frac{100\cos 0}{4\pi\varepsilon_0(10)}(10^{-12}) = \frac{10^{-12}}{4\pi(\frac{10^{-9}}{36\pi})} = \underline{\underline{9 \text{ mV}}}$$

$$\bar{E} = \frac{100(10^{-12})}{4\pi(\frac{10^{-9}}{36\pi})10^3}[2\cos 0\,\bar{a}_r = \sin 0\,\bar{a}_\theta]$$

$$= \underline{\underline{1.8\,\bar{a}_r \text{ mV/m}}}$$

(b)

$$At\ (1, \frac{\pi}{3}, \frac{\pi}{2}),$$

$$V = \frac{100\cos\frac{\pi}{3}(10^{-12})}{4\pi(\frac{10^{-9}}{36\pi})(1)^2} = \underline{\underline{0.45 V}}$$

$$\bar{E} = \frac{100(10^{-12})}{4\pi(\frac{10^{-9}}{36\pi})(1)^2}(2\cos\frac{\pi}{3}\bar{a}_r + \sin\frac{\pi}{3}\bar{a}_\theta)$$

$$= \underline{\underline{0.9\bar{a}_r + 0.7794\,\bar{a}_\theta \text{ V/m}}}$$

P.E. 4.14

After Q_1, $W_1 = 0$

After Q_2, $W_2 = Q_2 V_{21} = \dfrac{Q_2\, Q_1}{4\pi\varepsilon_0|(1,0,0) - (0,0,0)|}$

$$= \frac{1(-2)(10^{-18})}{4\pi(10^{-9})\frac{1}{36\pi}} = \underline{\underline{-18 \text{ nJ}}}$$

After Q_3,

$$W_3 = Q_3(V_{31} + V_{32}) + Q_2 V_{21}$$

$$= 3(9)(10^{-9})\left\{\frac{1}{|(0,0,-1) - (0,0,0)|} + \frac{-2}{|(0,0,-1) - (1,0,0)|}\right\} - 18 \text{nJ}$$

$$= 27(1 - \frac{2}{\sqrt{2}}) - 18$$

$$= \underline{\underline{-29.18 \text{ nJ}}}$$

P.E. 4.15

After Q_4,

$$W_4 = Q_4(V_{41} + V_{42} + V_{43}) + Q_3(V_{31} + V_{32}) + Q_2V_{21}$$

$$= -4(9)(10^{-9})\left\{\frac{1}{|(0,0,1)-(0,0,0)|} + \frac{-2}{|(0,0,1)-(1,0,0)} + \frac{3}{|(0,0,1)-(0,0,-1)}\right\} + W_3$$

$$= -36\left(1 - \frac{2}{\sqrt{2}} + \frac{3}{2}\right) + W_3$$

$$= -39.09 - 29.18 \text{ nJ} = \underline{\underline{-68.27 \text{ nJ}}}$$

Prob. 4.1

(a)

$$\bar{F}_{Q_1} = \frac{Q_1Q_2(\bar{r}_{Q_1} - \bar{r}_{Q_1})}{4\pi\varepsilon\left|\bar{r}_{Q_1} - \bar{r}_{Q_1}\right|^3} = \frac{-20(10^{-12})[(3,2,1)-(-4,0,0)]}{4\pi\dfrac{10^{-9}}{36\pi}|(3,2,1)-(-4,0,0)|^3} = -0.5655\frac{(7,2,5)}{688.88}$$

$$= \underline{\underline{-5.746\bar{a}_x - 1.642\bar{a}_y + 4.104\bar{a}_z \text{ mN}}}$$

Prob 4.2

(a)

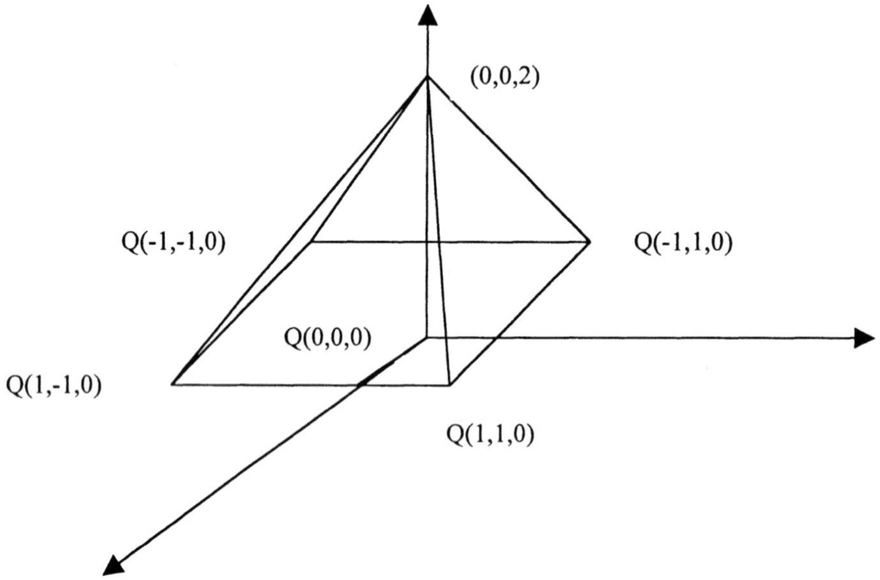

$$\bar{F} = \frac{qQ}{4\pi\varepsilon_0}\frac{[(0,0,2)-(0,0,0)]}{|(0,0,2)-(0,0,0)|^3} + \frac{qQ}{4\pi\varepsilon_0}\frac{[(0,0,2)-(1,1,0)]}{|(0,0,2)-(1,1,0)|^3} + \frac{qQ}{4\pi\varepsilon_0}\frac{[(0,0,2)-(-1,1,0)]}{|(0,0,2)-(-1,1,0)|^3}$$

$$+ \frac{qQ}{4\pi\varepsilon_0}\frac{[(0,0,2)-(1,-1,0)]}{|(0,0,2)-(1,-1,0)|^3} + \frac{qQ}{4\pi\varepsilon_0}\frac{[(0,0,2)-(-1,-1,0)]}{|(0,0,2)-(-1,-1,0)|^3}$$

But $\frac{qQ}{4\pi\varepsilon_0} = \frac{15(10)(10^{-12})}{4\pi(10^{-9}/36\pi)} = 1.35$

Factoring and dividing by 1.35 yields

$$\frac{\bar{F}}{1.35} = \frac{(0,0,2)}{8} + \frac{(-1,-1,2)}{6^{3/2}} + \frac{(1,-1,2)}{6^{3/2}} + \frac{(-1,1,2)}{6^{3/2}} + \frac{(1,1,2)}{6^{3/2}}$$

$$\bar{F} = 1.35(0.25 + \frac{8}{6^{3/2}})\bar{a}_z = \underline{\underline{1.072\,\bar{a}_z\,N}}$$

(b)

$$\bar{E} = \frac{\bar{F}}{q} = \frac{1.072\bar{a}_z}{10(10^{-6})} = 107.2\,\bar{a}_z \quad k\text{V/m}$$

Prob 4.3 (a)

$$\bar{E}(5,0,6) = \frac{qQ}{4\pi\varepsilon_0}\frac{[(5,4,6)-(4,0,-3)]}{|(5,4,6)-(4,0,-3)|^3} + \frac{qQ}{4\pi\varepsilon_0}\frac{[(5,0,6)-(2,0,1)]}{|(5,0,6)-(2,0,1)|^3}$$

$$= \frac{qQ}{4\pi\varepsilon_0}\frac{(1,0,9)}{(\sqrt{82})^3} + \frac{qQ}{4\pi\varepsilon_0}\frac{(3,0,5)}{(61)^{3/2}}$$

If $\bar{E}_z = 0$, then

$$\frac{9\,qQ}{4\pi\varepsilon_0}\frac{1}{(82)^{3/2}} + \frac{5\,qQ}{4\pi\varepsilon_0}\frac{1}{(61)^{3/2}} = 0$$

$$\bar{Q}_1 = -\frac{5}{9}Q_2(\frac{82}{61})^{3/2} = -\frac{5}{9}4(\frac{82}{61})^{3/2}\ \text{nC}$$

$$= \underline{\underline{-3.463\ \text{nC}}}$$

(b)

$$\bar{F}(5,0,6) = q\,\bar{E}(5,0,6)$$

If $E_x = 0$, then

$$\frac{qQ_1}{4\pi\varepsilon_0(82)^{3/2}} + \frac{3qQ_2}{4\pi\varepsilon_0(61)^{3/2}} = 0$$

$$Q_1 = -3Q_2(\frac{82}{61})^{3/2} = -12(\frac{82}{61})^{3/2}\ nC$$

$$Q_1 = \underline{\underline{-18.7\ nC}}$$

Prob 4.4

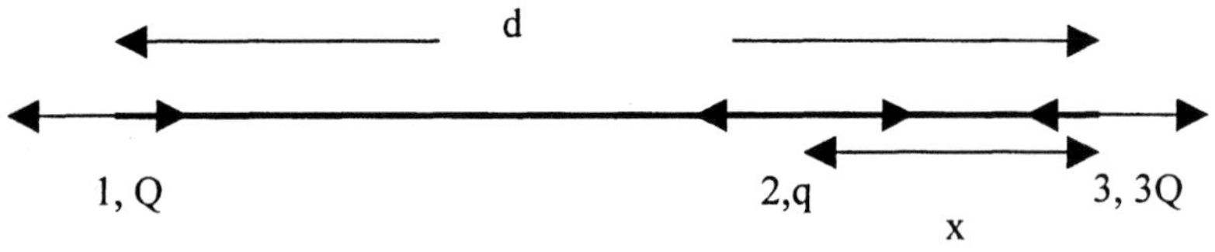

For the system to be in equilibrium, q must be negative and

$$\bar{F}_{12} = \bar{F}_{23} = \bar{F}_{13}$$

or
$$\frac{-1\,Qq}{4\pi\,(d-x)^2} = \frac{-3\,Qq}{4\pi\,x^2} = \frac{4Q^2}{4\pi\,d^2}$$

that is, $3(d-x)^2 = x^2 \quad \longrightarrow \quad 3d^2 - 6dx + 3x^2 = x^2$

$$2x^2 - 6dx + 3d^2 = 0$$

$$x = \frac{6d \pm \sqrt{36d^2 - 24d^2}}{4} = \frac{6d \pm d\sqrt{12}}{4}$$

$$x = 3 \pm \sqrt{3} = \underline{\underline{4.732\text{m},\;\; 1.268 \text{ m}}}$$

Prob 4.5

(a) $Q = \int \rho_L dl = \int_0^5 2x^2 dx = 4x^3 \Big|_0^5 mC = \underline{\underline{0.5\,C}}$

(b)
$$Q = \int \rho_S dS = \int_{z=0}^{4} \int_{\phi=0}^{2\pi} \rho z^2 \rho\, d\phi\, dz \Big|_{\rho=3} = 9(2\pi)\frac{z^3}{3}\Big|_0^4 nC$$
$$= \underline{\underline{1.206\ \mu C}}$$

(c)
$$Q = \int \rho_V dV = \iiint \frac{10}{r\sin\theta}\, r^2 \sin\theta\, d\theta\, d\phi\, dr$$
$$= 10\int_0^{2\pi} d\phi \int_0^{\pi} d\theta \int_0^{4} r dr = 10(2\pi)(\pi)\frac{4^2}{2}$$
$$= \underline{\underline{157.91\ C}}$$

Prob. 4.6

$$Q_A = \int \rho_L dl = \rho_L \int_{-5}^{0} dl = 5\rho_L = 10\ mC$$

$$\bar{Q}_B = \int \rho_S dS = \rho_S \int dS = \rho_S \iint \rho\, d\phi\, d\rho$$
$$= \rho_S \int_0^4 \rho\, d\rho \int_{\phi=0}^{\pi/2} d\phi = \rho_S \frac{\rho^2}{2}\Big|_0^4 \left(\frac{\pi}{2}\right)$$
$$= \rho_S (8)\left(\frac{\pi}{2}\right) = 20\pi\ mC = \underline{\underline{62.83\ mC}}$$

Prob 4.7

$$\bar{E} = \int \frac{\rho dl\, \bar{R}}{4\pi\varepsilon_o R^3} \; ; \quad dl = dy; \quad \bar{R} = (5,0,0) - (0,y,0) = 5\bar{a}_x - y\bar{a}_y$$

$$\bar{E} = \rho_L \int \frac{5\bar{a}_x - y\bar{a}_y}{4\pi\varepsilon_0 (y^2 + 25)^{3/2}}$$

$$= \frac{2(10^{-3})}{4\pi(10^{-9}/36\pi)} \int_0^{-5} (5\bar{a}_x + y\bar{a}_y) \frac{1}{(y^2+25)^{3/2}} dy$$

$$= 18(10^6)[k_x \bar{a}_x + k_y \bar{a}_y]$$

where $k_x = \int_0^{-5} \frac{dy}{(y^2+25)^{3/2}} = \frac{5(y/25)}{\sqrt{y^2+25}}\Big|_0^{-5} = -\frac{1}{\sqrt{50}} = -0.1414$

where $k_y = \int_0^{-5} \frac{y}{(y^2+25)^{3/2}} dy = \frac{1}{\sqrt{y^2+25}}\Big|_0^{-5} = - -\frac{1}{\sqrt{50}} + \frac{1}{5} = 0.05858$

$$\bar{E} = \underline{\underline{-2.545\bar{a}_x + 1.054\bar{a}_y \text{ mV/m}}}$$

Prob. 4.8

$$\bar{E} = \int \frac{\rho\, dS\, \bar{R}}{4\pi\varepsilon_0 R^3} \; ; \quad dS = \rho\, d\phi\, d\rho; \quad R = \sqrt{\rho^2 + h^2}$$

$$\bar{R} = -\rho\,\bar{a}_\rho + h\bar{a}_z$$

$$\bar{E} = \frac{\rho_S}{4\pi\varepsilon_0} \int \frac{(-\rho\,\bar{a}_\rho + h\bar{a}_z)\rho\, d\phi\, d\rho}{(\rho^2 + h^2)^{3/2}}$$

$$= \frac{5(10^{-3})}{4\pi(10^{-9}/36\pi)} [-\int_{\phi=0}^{\pi/2} \int_{\rho=0}^{4} \frac{\rho^2 d\phi\, d\rho}{(\rho^2+h^2)^{3/2}} \bar{a}_\rho + h \int_{\phi=0}^{\pi/4} \int_{\rho=0}^{4} \frac{\rho\, d\phi\, d\rho}{(\rho^2+h^2)^{3/2}} \bar{a}_z]$$

$$= 45(10^6 [-\frac{\pi}{2} \int \frac{\rho^2 d\rho}{(\rho^2+h^2)^{3/2}} \bar{a}_\rho + \frac{\pi h}{2} \int \frac{\rho\, d\rho}{(\rho^2+h^2)^{3/2}} \bar{a}_z]$$

But $\int \frac{x^2\,dx}{(x^2+a^2)^{3/2}} = \ln(\frac{\sqrt{x^2+a^2}}{a} + \frac{x}{a}) - \frac{x}{\sqrt{x^2+a^2}} + C$

and $\int \frac{x\,dx}{(x^2+a^2)} = -\frac{1}{\sqrt{x^2+a^2}} + C$

Let $\bar{E} = 45\ (10^2)\ [\frac{-\pi}{2} k_\rho \bar{a}_\rho + \frac{\pi}{2} h k_z \bar{a}_z]$

$$k_\rho = [\ln(\frac{\sqrt{\rho^2+h^2}}{h} + \frac{\rho}{h}) - \frac{\rho}{\sqrt{\rho^2+h^2}}]\Big|_{\rho=0}^{4} = \ln 2 - \frac{4}{5} = -\ 0.1068$$

$$k_z = \frac{-1}{\sqrt{\rho^2+h^2}}\Big|_0^4 = -\frac{1}{5} + \frac{1}{3} = 0.1338$$

$$\bar{E} = \frac{45}{4}(10^6)[0.671\bar{a}_\rho + 2.5126\bar{a}_z]$$

$$= \underline{\underline{7.549\bar{a}_\rho + 28.27\bar{a}_z\ \ \text{mV/m}}}$$

(b)

The result is the same as that in (a) except that instead of

$\int_{\phi=0}^{\pi/2} d\phi = \frac{\pi}{2}$, we now have $\int_{\phi=0}^{\pi/2} \sin\phi\ d\phi = -\cos\phi\Big|_0^{\pi/2} = 1$

That is, we replace $\pi/2$ by 1

$$\bar{E} = 45(10^6)\ [-k_\rho \bar{a}_\rho + h k_z \bar{a}_z]$$

$$= \underline{\underline{4.806\bar{a}_\rho + 18\ \bar{a}_z\ \text{mV/m}}}$$

Prob 4.9

$V = \int_S \frac{\rho_S dS}{4\pi\varepsilon_0 r}; \qquad \rho_S = \frac{1}{\rho}; \quad dS = \rho\, d\phi\, d\rho; \quad r = \sqrt{\rho^2+h^2}$

$$V = \frac{1}{4\pi\varepsilon_0}\iint \frac{\frac{1}{\rho}(\rho\, d\phi\, d\rho)}{(\rho^2+h^2)^{1/2}} = \frac{1}{4\pi\varepsilon_0}\int_0^{2\pi} d\phi \int_{\rho=0}^{a} \frac{d\rho}{(\rho^2+h^2)}$$

$$= \frac{2\pi}{4\pi\varepsilon_0}\ln(\rho + \sqrt{\rho^2+h^2})\Big|_{\rho=0}^{a} = \frac{1}{2\varepsilon_0}[\ln(a + \sqrt{\rho^2+h^2}) - \ln h]$$

$$= \underline{\underline{\frac{1}{2\varepsilon_0}\ln\frac{a+\sqrt{\rho^2+h^2}}{h}}}$$

Prob. 4.10 (a)

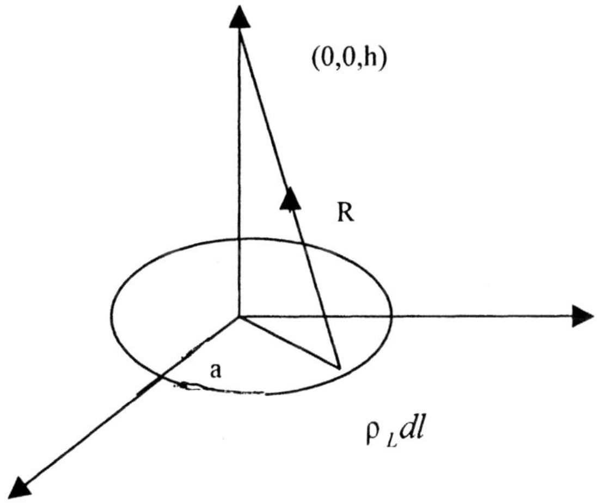

$$\bar{D} = \int \frac{\rho_L dl\, \bar{R}}{4\pi R^3}, \qquad \bar{R} = -a\bar{a}_\rho + h\bar{a}_z$$

$$\bar{D} = \frac{\rho_L}{4\pi} \int_{\phi=0}^{\phi=2\pi} \frac{a d\phi(-a\bar{a}_\rho + h\bar{a}_z}{(a^2+h^2)^{3/2}}$$

Due to symmetry, the ρ component varies.

$$\bar{D} = \frac{\rho_L a(2\pi h)\bar{a}_z}{4\pi(a^2+h^2)^{3/2}} = \frac{\rho_L a h \bar{a}_z}{2(a^2+h^2)^{3/2}}$$

$a = 2,\ h = 3,\ \rho_L = 5\mu\text{C/m}$

Since the ring is placed in $x = 0$, $\bar{a}_z$ becomes $\bar{a}_x$.

$$\bar{D} = \frac{2(6)(5)\bar{a}_x}{2(4+9)^{3/2}} = \underline{\underline{0.64\,\bar{a}_x\ \mu\text{C/m}^2}}$$

(b)

$$\bar{D}_Q = \frac{Q}{4\pi}\frac{[(3,0,0)-(0,-3,0)]}{|(3,0,0)-(0,-3,0)|^3} + \frac{Q}{4\pi}\frac{[(3,0,0)-(0,3,0)]}{|(3,0,0)-(0,3,0)|^3}$$

$$= \frac{Q(3,3,0)}{4\pi(18)^{3/2}} + \frac{Q(3,-3,0)}{4\pi(18)^{3/2}} = \frac{6Q}{4\pi(18)^{3/2}}$$

$$\bar{D} = \bar{D}_R + \bar{D}_Q = 0$$

$$0.64(10^{-6}) + \frac{6Q}{4\pi(18)^{3/2}} = 0$$

$$\therefore\ Q = -0.64(4\pi)(18^{3/2})10^{-6}\frac{1}{6} = \underline{\underline{-102.4\mu C}}$$

Prob. 4.11

Due to symmetry, $\bar{E}$ has only z – component given by

$$dE_z = dE\cos\alpha$$

$$= \frac{\rho_S dx\,dy}{4\pi\varepsilon_0(x^2+y^2+h^2)}\frac{h}{(x^2+y^2+h^2)^{1/2}}$$

$$E_z = \frac{\rho_S h}{4\pi\varepsilon_0}\int_{-a}^{a}\int_{-b}^{b}\frac{dxdy}{(x^2+y^2+h^2)^{3/2}}$$

$$= \frac{\rho_S h}{\pi\varepsilon_0}\int_{-a}^{a}\int_{-b}^{b}\frac{dxdy}{(x^2+y^2+h^2)^{3/2}}$$

$$= \frac{\rho_S h}{\pi\varepsilon_0}\int_{0}^{a}\frac{y\,dx}{(x^2+h^2)(x^2+y^2+h^2)^{1/2}}\Bigg|_0^b$$

$$= \frac{\rho_S h}{\pi\varepsilon_0}\int_{0}^{a}\frac{b\,dx}{(x^2+h^2)(x^2+b^2+h^2)^{1/2}}$$

By changing variables, we finally obtain

$$E_z = \frac{\rho_s}{\pi\ \varepsilon_0}\tan^{-1}\left\{\frac{ab}{h(a^2+b^2+h^2)^{1/2}}\right\}\bar{a}_z$$

$$= 36(10^{-3})(0.0878 \text{ radians})\ \bar{a}_z = \underline{\underline{31.61\ \mu\text{V/m}}}$$

Prob 4.12

$$\bar{E} = \bar{E}_1 + \bar{E}_2 + \bar{E}_3$$

$$= \frac{Q}{4\pi\varepsilon_0 r^2}\bar{a}_r + \frac{\rho_L}{2\pi\varepsilon_0\rho}\bar{a}_\rho + \frac{\rho_S}{2\varepsilon_0}\bar{a}_n$$

$$= \frac{100(10^{-12})}{4\pi(\frac{10^{-9}}{36\pi})}\left\{\frac{(1,1,1)-(4,1,-3)}{|(1,1,1)-(4,1,-3)|^3}\right\} + \frac{2(10^{-9})}{2\pi(\frac{10^{-9}}{36\pi})}\left\{\frac{(1,1,1)-(1,0,0)}{|(1,1,1)-(1,0,0)|^2}\right\} + \frac{5(10^{-9})}{2\pi(\frac{10^{-9}}{36\pi})}\bar{a}_z$$

$$= (-0.0216,0,0.0288) + (0,18,18) - 90\pi(0,0,1)$$

$$= \underline{\underline{-0.0216\,\bar{a}_x + 18\,\bar{a}_y - 264.7\,\bar{a}_z\ \ \text{V/m}}}$$

Prob 4.13

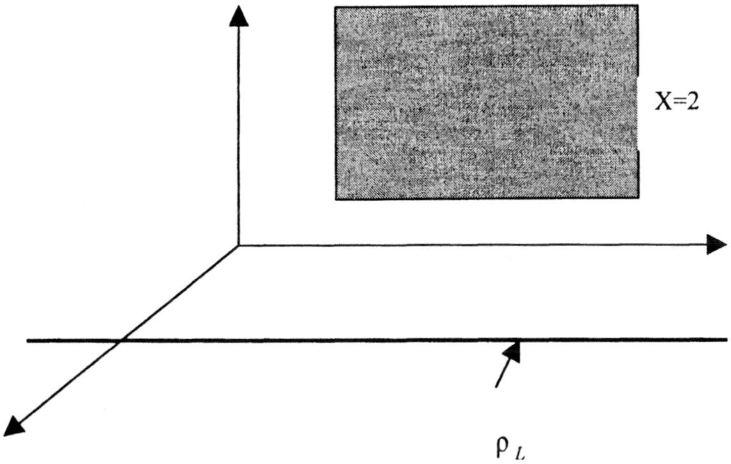

$$\bar{E} = \frac{\rho_S}{2\varepsilon_0}\bar{a}_n + \frac{\rho_L}{2\pi\varepsilon_0\rho}\bar{a}_\rho$$

$$\bar{\rho} = (0,0,0) - (3,0,-1) = -3\bar{a}_x + \bar{a}_z$$

$$\bar{E} = \frac{4(10^{-9})}{2(10^{-9}/36\pi)}(\bar{a}_x) + \frac{20(10^{-9})}{2\pi(10^{-9}/36\pi)}\frac{(-3\bar{a}_x + \bar{a}_z)}{(9+1)}$$

$$= 72\pi\,\bar{a}_x + 36(-3\bar{a}_x + \bar{a}_z)$$

$$\bar{F} = q\,\bar{E} = -5(36)\ [(2\pi - 3)\bar{a}_x + \bar{a}_z]\ mN$$

$$= \underline{\underline{-0.591\bar{a}_x - 0.18\,\bar{a}_z\ \text{N}}}$$

Prob 4.14

$$\bar{D} = \sum_{k=1}^{4}\frac{\bar{Q}_k(\bar{r} - \bar{r}_k)}{4\pi|(\bar{r} - \bar{r}_k)|^3}$$

$$\bar{D} = \frac{Q}{4\pi}\left\{\frac{2[(0,0,0)-(2,2,0)]}{|(0,0,0)-(2,2,0)|^3} - \frac{2[(0,0,0)-(-2,-2,0)]}{|(0,0,0)-(-2,-2,0)|^3} + \frac{[(0,0,6)-(-2,2,0)]}{|(0,0,6)-(-2,2,0)|^3}\right.$$

$$\left. - \frac{[(0,0,6)-(2,-2,0)]}{|(0,0,6)-(2,-2,0)|^3} - \right\}$$

$$= \frac{15}{4\pi}\left\{\frac{2(-2,-2,6)}{44^{3/2}} - \frac{2(2,2,6)}{44^{3/2}} + \frac{2(2,-2,6)}{44^{3/2}} - \frac{2(-2,2,6)}{44^{3/2}} - \right\}$$

$$= \frac{15}{4\pi(44)^{3/2}}(-4,-12,0)\ \mu\text{C/m}^2$$

$$= \underline{\underline{-16.36\,\bar{a}_x - 49.08\bar{a}_y\ \ \text{nC/m}^2}}$$

Prob 4.15

Let Q_l be located at the origin. At the spherical surface of radius r,

$$Q_l = \oint \bar{D} \bullet d\bar{S} = \varepsilon E_r (4\pi r^2)$$

or

$$\bar{E} = \frac{Q_l}{4\pi\varepsilon r^2}\bar{a}_r \quad \text{by Gauss's law.}$$

If a second charge Q_2 is placed on the spherical surface, Q_2 experiences a force:

$$\bar{F} = Q_2\bar{E} = \frac{Q_l Q_2}{4\pi\varepsilon r^2}\bar{a}_r$$

which is Columb's law.

Prob. 4.16
(a)

$$\rho_V = \nabla \bullet \bar{D} = \frac{\partial D_x}{\partial x} + \frac{\partial D_y}{\partial y} + \frac{\partial D_z}{\partial z} = 8y + 0 = \underline{\underline{8y \ \ \text{C/m}^3}}$$

(b)

$$\rho_V = \nabla \bullet \bar{D} = \frac{1}{\rho}\frac{\partial}{\partial\rho}(\rho^2 \sin\phi) + \frac{1}{\rho}\frac{\partial}{\partial\phi}(2\rho\cos\phi) + \frac{\partial}{\partial z}(2z^2)$$

$$= 2\sin\phi - 2\sin\phi + 4z = 4z \ \ \text{C/m}^3$$

(c)

$$\rho_V = \nabla \bullet \bar{D} = \frac{1}{r^2}\frac{\partial}{\partial r}\left(\frac{2}{r}\cos\theta\right) + \frac{1}{r^4\sin\theta}\frac{\partial}{\partial\theta}(\sin^2\theta)$$

$$= \frac{-2}{r^3}\cos\theta + \frac{1}{r^4\sin\theta}(2\sin\theta\cos\theta) = \underline{\underline{0}}$$

Prob 4.17
(a)

$$\bar{D} = \varepsilon_0(\bar{E}) = 10^{-9}\frac{1}{36\pi}(xy\bar{a}_x + x^2\bar{a}_y)$$

$$\bar{D} = \underline{\underline{8.84\,xy\bar{a}_x + 8.84x^2\bar{a}_y \ \text{pC/m}^2}}$$

(b)

$$\rho_V = \nabla \bullet \bar{D} = \frac{\partial D_x}{\partial x} + \frac{\partial D_y}{\partial y} + \frac{\partial D_z}{\partial z}$$

$$= \underline{\underline{8.84y\,\text{pC/m}^3}}$$

Prob 4.18

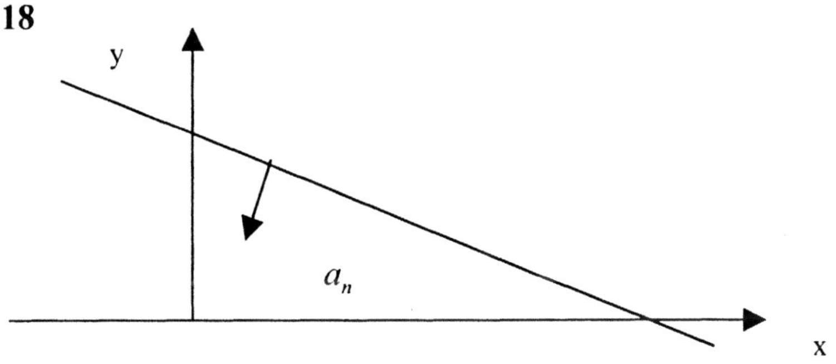

Let $f(x,y) = x + 2y - 5; \quad \nabla f = \bar{a}_x + 2\bar{a}_y$

$$\bar{a}_n = \pm \frac{\nabla f}{|\nabla f|} = \pm \frac{(\bar{a}_x + 2\bar{a}_y)}{\sqrt{5}}$$

Since point $(-1,0,1)$ is below the plane,

$$\bar{a}_n = -\frac{(\bar{a}_x + 2\bar{a}_y)}{\sqrt{5}} .$$

$$\bar{E} = \frac{\rho_s}{2\varepsilon_0}\bar{a}_n = \frac{6(10^{-9})}{2(10^{-9}/36\pi)}\left(-\frac{(\bar{a}_x + 2\bar{a}_y)}{\sqrt{5}}\right)$$

$$= \underline{\underline{-151.7\,\bar{a}_x - 303.5\bar{a}_y \ \text{V/m}}}$$

Prob 4.19

$$W = \frac{1}{2}\int \bar{D}\bullet \bar{E}\,dV = \frac{1}{2\varepsilon_0}\int |\bar{D}|^2\,dV \ \text{nJ}$$

$$2\varepsilon_0 W = \iiint (4y^4 + 16x^2y^2 + 1)\,dx\,dy\,dz$$

$$= 4\int_{x=0}^{2} dx \int_{y=1}^{2} y^4\,dy \int_{z=-1}^{4} dz + 16\int_{x=1}^{2} x^2 dx \int_{y=1}^{2} y^2 dy \int_{z=1}^{4} dz + \int_{x=1}^{2} dx \int_{x=-1}^{2} dy \int_{x=-1}^{4} dz$$

$$= 4(3)\frac{y^5}{5}\Big|_1^4 (5) + 16\left(\frac{x^3}{3}\Big|_1^2\right)^2 (5) + (3)(3)(5)$$

$$= 372 + 435.56 + 45 = 852.56$$

Thus,

$$W = \frac{10^{-9}}{2(10^{-9}/36\pi)}(852.56) = 853.56 = \underline{\underline{5.357 \ \text{kJ}}}$$

Prob 4.20

(a)

$$\rho_V = \nabla \bullet \bar{D} = \frac{1}{\rho}\frac{\partial}{\partial \rho}(\rho D_\rho) + \frac{1}{\rho}\frac{\partial D_\phi}{\partial \phi} + \frac{\partial D_z}{\partial z}$$

$$\rho_V = 4(z+1)\cos\phi - (z+1)\cos\phi + 0$$

$$\rho_V = \underline{\underline{3(z+1)\cos\phi \quad \mu C/m^2}}$$

$$Q_{enc} = \int \rho_V dv = \iiint 3(z+1)\cos\phi \; \rho d\phi \; d\rho dz$$

(b)

$$= 3\int_0^2 \rho d\rho \int_0^4 (z+1) \int_0^{\pi/2} \cos\phi d\phi = 3(2)(\frac{z^2}{2}+z)\Big|_0^4 (\sin\phi \Big|_0^{\pi/2})$$

$$= 6(8+4)(1-0) = \underline{\underline{72\mu C}}$$

(c)

Let $\psi = \psi_1 + \psi_2 + \psi_3 + \psi_4 + \psi_5 = \oint D \bullet d\bar{S}$

where ψ_1, ψ_2, ψ_3, ψ_4, ψ_5 respectively correspond witn surfaces S_1, S_2, S_3, S_4, S_4 (in the figure below) respectively.

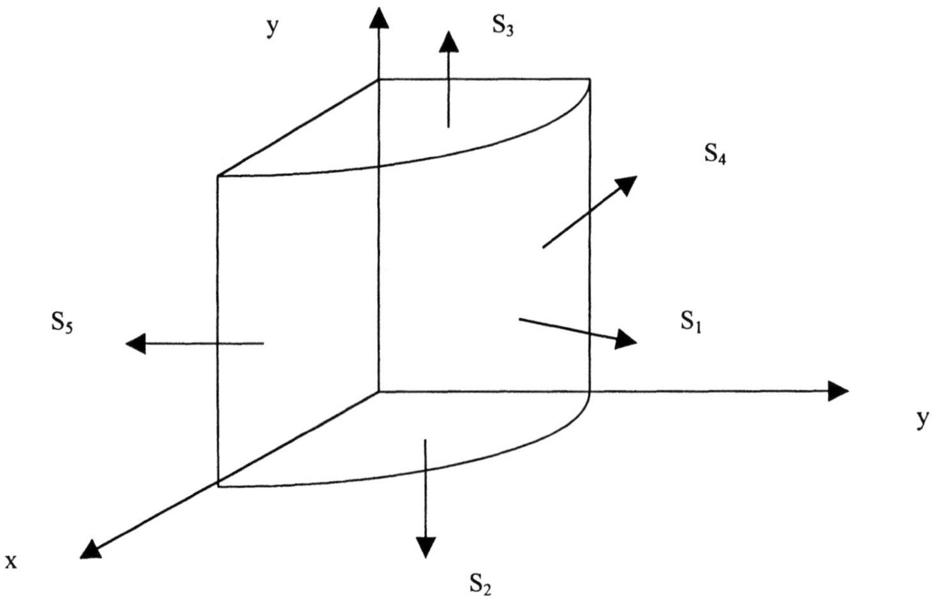

For S_1 $\rho = 2, dS = \rho d\phi dz \bar{a}\rho$

$$\psi_1 = \iint 2\rho(z+1)\cos\phi \Big|_{\rho=2} = 2(2)\int_0^4 (z+1)dz \int_0^{\pi/2} \cos\phi d\phi$$

$$= 4(12)(1) = 48$$

For S_2, $z = 0, dS = \rho d\phi d\rho(-\bar{a}_z)$

$$\psi_2 = -\iint \rho^2 \cos\phi \;\; \rho d\phi \;\; d\rho = -\int_0^2 \rho^3 d\rho \int_0^{\pi/2} \cos\phi d\phi$$

$$= -\frac{\rho^4}{4}\Big|_0^2 (1) = -4$$

For $S_{3,}$ $z = 1, d\bar{S} = \rho d\phi d\rho \bar{a} z, \;\; \psi_3 = +4$

For $S_4, d = \pi/2, \; d\bar{S} = d\rho dz a\phi$

$$\psi_4 = -\iint \rho(z = 1)\sin\phi \;\; d\rho dz\Big|_{d=\pi/2} = (11\int_0^2 \rho d\rho \int_0^4 (z+1)dz$$

$$= -\frac{\rho^2}{2}\Big|_0^2 (12) = -(2)(12) = -24$$

For S_5, $d = 0, d\bar{S} = d\rho dz(-\bar{a}_\phi), \psi_5 = \iint \rho(z+1)\sin\phi \;\; d\rho dz\Big|_{\phi=0} = 0$

$$\psi = 48 - 4 + 4 - 24 + 0 = \underline{\underline{24\mu\text{C}}}$$

Prob. 4.21

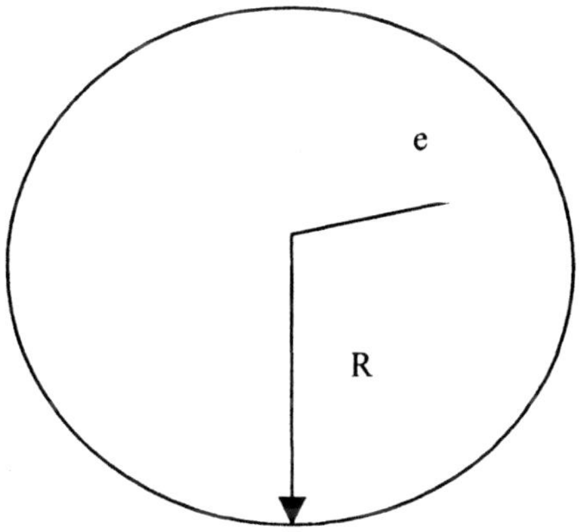

$$F = eE$$

$$\rho_0 = \frac{e}{4\pi \frac{R^3}{3}} = \frac{3e}{4\pi R^3}$$

$$\rho_V = \begin{cases} \rho_0, & 0 < r < R \\ 0, & \text{elsewhere} \end{cases}$$

$$\oint \bar{D} \bullet d\bar{S} = Q_{enc} = \int \rho_V dV = \frac{3e}{4\pi R^3} \frac{4\pi r^3}{3} = D_r(4\pi r^2)$$

$$E_r = \frac{3e\,r}{12\pi\varepsilon_0 R^3}$$

$$F = eE = \frac{e^2 r}{4\pi\varepsilon_0 R^3}$$

Prob 4.22

(a)

$$\psi = Q_{enc}$$

For $r = 1.5$m,

$$Q_{enc} = \int \rho_{S1} ds = \rho_{s1} \int ds = \rho_{s1}(4\pi R^2\ )$$
$$= 2(10^{-6})4\pi(1^2) = 8\pi(10^{-6})$$
$$\psi = Q_{enc} = \underline{\underline{25.13\quad \mu\text{C}}}$$

For $r = 2.5$m,

$$Q_{enc} = \rho_{s1}(4\pi R_1^{\,2}) + \rho_{s2}(\,4\pi R_2^{\,2})$$
$$= 8\pi(10^{-6}) + (-4)10^{-6}(4\pi 2^2)$$
$$= (8\pi - 64\pi)10^{-6}$$
$$\psi = Q_{enc} = \underline{\underline{-175.93\quad \mu\text{C}}}$$

(b)

$$\psi = Q_{enc}, \qquad \int \bar{D} \bullet d\bar{S} = Q_{enc}$$

$$D_r(\,4\pi r^2) = Q_{enc}$$

$$D_r = \frac{Q_{enc}}{4\pi r^2}$$

For $r = 0.5$, $\quad Q_{enc} = 0 \quad \longrightarrow \quad \bar{D} = 0$

For $r = 2.5$, $\quad Q_{enc} = -175.93\mu C = -56\pi(10^{-6})$

$$D_r = -\frac{56\pi(10^{-6})}{4\pi(25)} = -2.24\bar{a}_r \ \mu\text{C}/\text{m}^2$$

For $r = 3.5$, $\quad Q_{enc} = \rho_{s1} 4\pi R_1^2 + \rho_{s2} 4\pi R_2^2 + \rho_{s3} 4\pi R_3^2$

$$= -56\pi + 5(4\pi(3^3)) \ \ \mu\text{C}$$

$$= 124\pi \ \ \mu\text{C}$$

$$D_r = \frac{124\pi}{4\pi(3-5)^2} \mu\text{C}/\text{m}^2 = 2.531\bar{a}_r \ \ \mu\text{C}/\text{m}^2$$

Prob 4.23

For $\rho < 1$, $\ Q_{enc.} = 0 \quad \longrightarrow \quad \bar{D} = 0$

For $1 < \rho < 2$,

$$Q_{enc.} = \int_{\phi=0}^{2\pi} \int_{\rho=1} \int_{z=0}^{L} 12\rho \, d\phi \, d\rho \, dz$$

$$= 12(2\pi) L \frac{\rho^3}{3}\Big|_1^{\rho} = 8\pi L(\rho^3 - 1)$$

$$\psi = \int \bar{D} \bullet d\bar{S} = D_\rho \int_{\rho=0}^{2} \int_{\phi=0}^{2\pi} \rho \, d\phi \, dz = D_\rho(2\pi\rho L)$$

Hence,

$$8\pi L(\rho^3 - 1) = D_\rho(2\pi\rho L)$$

$$D_\rho = \frac{8(\rho^3 - 1)}{2\rho}$$

For $\rho > 2$, $\quad \psi = D_\rho(2\pi\rho L)$

$$Q_{enc} = 8\pi L\rho^3\Big|_1^2 = 56\pi L$$

$$56\pi L = D_\rho(2\pi\rho L)$$

$$D_\rho = \frac{28}{\rho}$$

$$\text{Thus,}\quad D_\rho = \begin{cases} 0, & \rho < 1, \quad 1 < \rho < 2 \\ \dfrac{8(\rho^3 - 1)}{2\rho}, & \rho > 2 \\ \dfrac{28}{\rho} \end{cases}$$

Prob 4.24

(a)

$\psi = Q_{enc}$ at $r = 2$

$$Q_{enc} = \int \rho_V \, dV = \iiint \frac{10}{r^2} r^2 \sin\theta \, d\theta \, dr \, d\phi$$

$$= 10 \int_{r=0}^{2} \int_{\phi=0}^{2\pi} \int_{\theta=0}^{\pi} \sin\theta \, d\theta$$

$$= 10(2)(2\pi)(2) = (80\pi) \text{ mC}$$

Thus, $\psi = \underline{\underline{251.3 \text{ mC}}}$

At $r = 6$;

$$Q_{enc.} = 10 \int_{r=0}^{4} dr \int_{\phi=0}^{2\pi} d\phi \int_{\theta=0}^{\pi} \sin\theta \, d\theta$$

$$= 10(4)(2\pi)(2) = 160\pi$$

$\psi = \underline{\underline{502.6 \text{ mC}}}$

(b)

$$\psi = Q_{enc}$$

But $\psi = \oint \bar{D} \bullet d\bar{S} = D_r \oint dS = D_r (4\pi r^2)$

At $r = 1$,

$$Q_{enc} = 10 \int_{r=0}^{1} dr \int_{\phi=0}^{2\pi} d\phi \int_{\theta=0}^{\pi} \sin\theta \, d\theta$$

$$Q_{enc} = 10(1)(2\pi)(2) = 40\pi$$

Thus,

$$D_r = \frac{Q_{enc}}{4\pi r^2} = \frac{40\pi}{4\pi(1)} = 10$$

$\bar{D} = \underline{\underline{10 \, \bar{a}_r \quad nC/m^2}}$

At $r = 5$, $Q_{enc} = 160\pi$

$$D_r = \frac{Q_{enc}}{4\pi r^2} = \frac{160\pi}{4\pi (5)^2} = 1.6$$

$$= \underline{\underline{1.6\,\bar{a}_r \ \text{nC/m}^3}}$$

Prob. 4.25

Break up path $P(1,2,-4) \longrightarrow R(3,-5,6)$

$P(1,2,-4)$ (↓ to P') $\qquad$ $R(3,-5,6)$ (↑ from R')

$P'(3,2,-4) \longrightarrow R'(3,-5,-4)$

$$\frac{-W}{Q} = \int \bar{E}\bullet d\bar{l} = \left\{ \int_P^{P'} + \int_{P'}^{R'} + \int_{R'}^{R} \right\} \bar{E}\bullet d\bar{l}$$

$$= \int_{x=1}^{3} dx + \int_{y=2}^{-5} z^2 dy \,\Big|_{z=-4} + \int_{z=-4}^{6} 2yzdz \,\Big|_{y=-5}$$

$$= 2 + 16(-7) + 2(-5)\frac{z^2}{2}\Big|_{-4}^{6} = 2 - 112 - 100 = -210$$

$$W = 210\,Q = 210(5) = \underline{\underline{1050 \text{ J}}}$$

Prob 4.26

(a)

$$W_{AB} = q\int \bar{E}\bullet d\bar{l}, \qquad d\bar{l} = d\rho\, \bar{a}_\rho$$

$$\frac{-W_{AB}}{q} = \int (z+1)\sin\phi \, d\rho \,\Big|_{\phi=0, z=0} = 0$$

$$W_{AB} = 0$$

(b)

$$\frac{-W_{BC}}{q} = \int_{\phi=0}^{30} (z+1)\cos\phi \,\rho\, d\phi \,\Big|_{\rho-4, z=0} = 4\sin\phi \Big|_0^{30^\circ} = 2$$

$$W_{BC} = -2q = \underline{\underline{-8 \text{ nJ}}}$$

(c)

$$\frac{-W_{CD}}{q} = \int_{z=0}^{-2} \rho \sin\phi \, dz \Big|_{\substack{\phi=30^\circ \\ \rho=4}} = 4\sin 30^\circ (z \Big|_0^{-2}) = -4$$

$$W_{CD} = 4q = \underline{\underline{16 \text{ nJ}}}$$

(d)

$$W_{AD} = W_{AB} + W_{BC} + W_{CD} = 0 - 8 + 16 = \underline{\underline{8 \text{ nJ}}}$$

Prob. 4.27

(a)

From A to B, $d\bar{l} = rd\theta \, \bar{a}_\theta$,

$$W_{AB} = -Q \int_{\theta=30^\circ}^{90^\circ} 10r\cos\theta \, r\,d\theta \Big|_{r=5} = \underline{\underline{-1250 \text{ nJ}}}$$

(b)

From A to C, $d\bar{l} = dr\,\bar{a}_r$,

$$W_{AC} = -Q \int_{r=5}^{10} 20r\sin\theta \, dr \Big|_{\theta=30^\circ} = \underline{\underline{-3750 \text{ nJ}}}$$

(c)

From A to D, $d\bar{l} = r\sin\theta \, d\phi \, \bar{a}_\phi$,

$$W_{AD} = -Q\int 0(r\sin\theta)\,d\phi = \underline{\underline{0 \text{ J}}}$$

(d)

$$W_{AE} = W_{AD} + W_{DF} + W_{FE}$$

where F is $(10,30,60)$. Hence,

$$W_{AE} = -Q\left\{ \int_{r=5}^{10} 20r\sin\theta \, dr \Big|_{\theta=30^\circ} + 10 \int_{\theta=30}^{90} 10r\cos\theta \, r\,d\theta \Big|_{r=10} \right\}$$

$$= -100\left[\frac{75}{2} + \frac{100}{2}\right] \text{ nJ} = \underline{\underline{-8750 \text{ nJ}}}$$

Prob 4.28

$$W = qV_{AB} = q(V_B - V_A)$$

$$= 2(10^{-6})[2(1)(-3) - 1(1)(2)] = \underline{\underline{-16 \ \mu\text{J}}}$$

Prob 4.29
(a)

$$\bar{E} = -\nabla V = -(2x\bar{a}_x + 4y\bar{a}_y + 8z\bar{a}_z)$$

$$= \underline{\underline{-2x\bar{a}_x + 4y\bar{a}_y + 8z\bar{a}_z \text{ V/m}}}$$

(b)

$$-\bar{E} = \frac{\partial V}{dx}\bar{a}_x + \frac{\partial V}{dy}\bar{a}_y + \frac{\partial V}{dz}\bar{a}_z$$

$$= \cos(x^2+y^2+z^2)^{1/2}[2x\bar{a}_x + 2y\bar{a}_y + 2z\bar{a}_z](\frac{1}{2})$$

$$= \underline{\underline{-(x\bar{a}_x + y\bar{a}_y + z\bar{a}_z)\cos(x^2+y^2+z^2)^{1/2} \quad \text{V/m}}}$$

(c)

$$-\bar{E} = \frac{\partial V}{\partial \rho}\bar{a}_\rho + \frac{1}{\rho}\frac{\partial V}{\partial \phi}\bar{a}_\phi + \frac{\partial V}{\partial z}\bar{a}_z$$

$$= 2\rho(z+1)\sin\phi\bar{a}_\rho + \rho(z+1)\cos\phi\bar{a}_\rho + \rho^2\sin\phi\bar{a}_z$$

$$= \underline{\underline{-2\rho(z+1)\sin\phi\bar{a}_\rho - \rho(z+1)\cos\phi\bar{a}_\rho - \rho^2\sin\phi\bar{a}_z}}$$

(d)

$$\bar{E} = \frac{\partial V}{\partial}\bar{a}_z + \frac{1}{r}\frac{\partial V}{\partial\theta}\bar{a}_\theta + \frac{1}{r\sin\theta}\frac{\partial V}{\partial\phi}\bar{a}_\phi$$

$$-\bar{E} = -e^x\sin\theta\cos 2\phi\,\bar{a}_r + \frac{1}{r}e^{-r}\cos\theta\cos 2\phi\,\bar{a}_\theta + \frac{e^{-r}}{r}(-2\sin 2\phi)\bar{a}_\phi$$

$$\bar{E} = \underline{\underline{e^x\sin\theta\cos 2\phi\,\bar{a}_r - \frac{1}{r}e^{-r}\cos\theta\cos 2\phi\,\bar{a}_\theta + \frac{2e^{-r}}{r}(\sin 2\phi)\bar{a}_\phi}} \quad \text{V/m}$$

Prob 4.30 (a)

$$V_P = \sum \frac{Q_k}{4\pi|\bar{r}_P - \bar{r}_k|}$$

$$4\pi\varepsilon_o V_p = \frac{10^{-3}}{|(-1,1,2)-(0,0,4)|} + \frac{-2(10^{-3})}{|(-1,1,2)-(-2,5,1)|} + \frac{3(10^{-3})}{|(-1,1,2)-(3,-4,6)|}$$

$$4\pi\varepsilon_0(10^3)V_p = \frac{1}{|(-1,1,-2)|} - \frac{2}{|(1,-4,1)} + \frac{3}{|(-4,5,-4)|} = \frac{1}{\sqrt{6}} - \frac{2}{\sqrt{18}} + \frac{3}{\sqrt{5}}$$

$$4\pi\frac{10^{-9}}{36\pi}(10^3)V_p = 0.3542$$

$$\therefore \quad V_p = \underline{\underline{3\,(10^6)\ \text{V}}}$$

(b)

$$V_Q = \sum \frac{Q_k}{4\pi\varepsilon_o|\bar{r}_p - \bar{r}_k|}$$

$$4\pi\varepsilon_o V_Q = \frac{10^{-3}}{|(1,2,3)-(0,0,4)|} + \frac{-2(10^{-3})}{|(1,2,3)-(-2,5,1)|} + \frac{3(10^{-3})}{|(1,2,3)-(3,-4,6)|}$$

$$4\pi\varepsilon_0(10^3)V_p = \frac{1}{|(1,2,-1)|} - \frac{2}{|(3,-3,2)} + \frac{3}{|(-2,6,-3)|} = \frac{1}{\sqrt{6}} - \frac{2}{\sqrt{22}} + \frac{3}{\sqrt{7}}$$

$$4\pi\frac{10^{-9}}{36\pi}(10^3)V_p = 0.410$$

$$V_Q = \underline{\underline{3.694\,(10^6)\text{ V}}}$$

$$\therefore V_{PQ} = V_Q - V_P = 0.69(10^6) = \underline{\underline{694\text{ kV}}}$$

Prob 4.31

(a)

$$\bar{E} = -(\frac{\partial V}{\partial x}\bar{a}_x + \frac{\partial V}{\partial y}\bar{a}_y + \frac{\partial V}{\partial z}\bar{a}_z)$$

$$= -2xy(z+3)\bar{a}_x - x^2(z+3)\bar{a}_y - x^2y\bar{a}_z$$

At $(3,4,-6)$, $x = 2,\ y = 4,\ z = -6,$

$$\bar{E} = -2(3)(4)(-3)\bar{a}_x - 9(-3)\bar{a}_y - 9(4)\bar{a}_z$$

$$= \underline{\underline{72\bar{a}_x + 27\bar{a}_y - 36\bar{a}_z\text{ V/m}}}$$

(b)

$$\rho_V = \nabla\bullet\bar{D} = \varepsilon_0\nabla\bullet\bar{E} = -\varepsilon_0(2y)(z+3)$$

$$\psi = Q_{enc} = \int\rho_V dV = -2\varepsilon_0\iiint y(z+3)dx\,dy\,dz$$

$$= -2\varepsilon_0\int_0^1 dx\int_0^1 y\,dy\int_0^1(z+3)dz = -2\varepsilon_0(1)(1/2)(\frac{z^2}{2} + 3z)\Big|_0^1$$

$$= -\varepsilon_0(\frac{1}{2} + 3) = \frac{-7}{2}(\frac{10^{-9}}{36\pi})$$

$$Q_{enc} = \underline{\underline{-30.95\ pC}}$$

Prob 4.32

$$\bar{E} = \begin{cases} \dfrac{\rho_0 a^3}{4\varepsilon_0 r^2}\bar{a}_r, & r > a \\ \dfrac{\rho_0 r^2}{4\varepsilon_0 a}\bar{a}_r, & r < a \end{cases}$$

Since $V = -\int \bar{E} \bullet d\bar{l} = -\int E\,dr$,

$$V = \begin{cases} \dfrac{-\rho r^3}{12\varepsilon_0 a} + C_1, & r < a \\ \dfrac{\rho a^3}{4\varepsilon_0 r} + C_2, & r > a \end{cases}$$

But $V(\infty) = 0 \quad \longrightarrow \quad C_2 = 0;$

$$V(r = a) = \frac{\rho_0 a^2}{4\varepsilon_0} = \frac{-\rho_0 a^2}{12\varepsilon_0} + C_1 \quad \longrightarrow \quad C_1 = \frac{\rho_0}{3C_0}$$

Thus, $$V = \begin{cases} \dfrac{-\rho_0 r^3}{12\varepsilon_0 a} + \dfrac{\rho_0 a^2}{3\varepsilon_0}, & r < a \\ \dfrac{\rho_0}{4\varepsilon_0 r}, & r > a \end{cases}$$

Prob 4.33
(a)

$$\nabla \times \bar{E} = \begin{vmatrix} \dfrac{\partial}{\partial x} & \dfrac{\partial}{\partial y} & \dfrac{\partial}{\partial z} \\ yz & xz & xy \end{vmatrix}$$

$$= (x - x)\bar{a}_x + (y - y)\bar{a}_y + (z - z)\bar{a}_z = 0$$

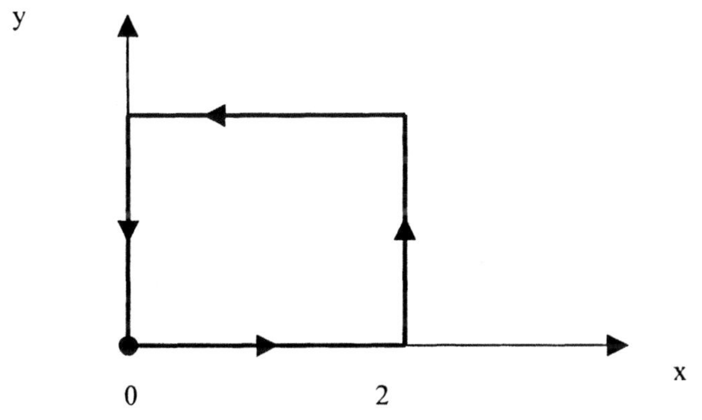

(b)

$$\oint \bar{E}\bullet d\bar{l} = \int_{x=0}^{2} yz\,dx \Big|_{\substack{y=0\\ z=1}} + \int_{y=0}^{2} xz\,dy \Big|_{\substack{z=1\\ x=2}} + \int_{x=2}^{0} yz\,dx \Big|_{\substack{y=2\\ z=1}} + \int_{y=2}^{0} xz\,dy \Big|_{\substack{x=0\\ z=1}}$$

$$= 2y\Big|_0^2 + 2x\Big|_0^2 = 4-4 = \underline{\underline{0}}$$

Prob. 4.34 (a)

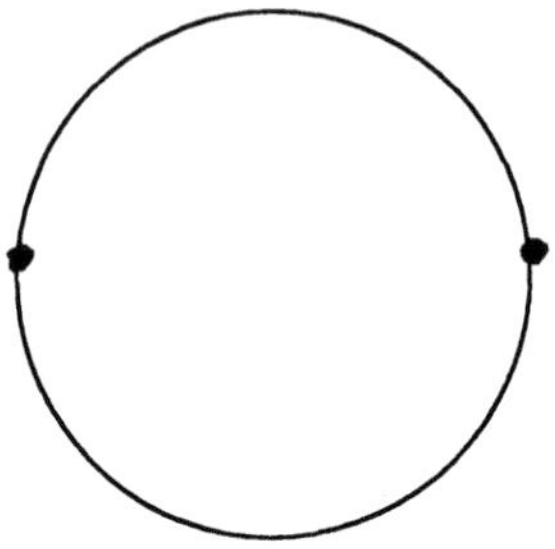

$$V = \frac{2\frac{Q}{2}}{4\pi\varepsilon_0 r} = \frac{Q}{4\pi\varepsilon_0 r}$$

$$= \frac{60(10^{-6})}{4\pi(10^{-9})\frac{1}{36\pi}} = \underline{\underline{15\ k\text{V}}}$$

(b)

$$V = \frac{3\,(\frac{Q}{3})}{4\pi\varepsilon_0 r} = \underline{\underline{15\ k\text{V}}}$$

(c)

$$V = \int \frac{\rho_L}{4\pi\varepsilon_0 r} = \frac{Q}{4\pi\varepsilon_0 r} = \underline{\underline{15\ k\text{V}}}$$

Prob 4.35 (a)

For $r \geq a$,

$$Q_{enc} = \int \rho_V dV = \iiint \rho_0 (a^2 - r^2) r^2 \sin\theta\, d\theta\, d\phi\, dr$$

$$Q_{enc} = \rho_0 \int_0^{\pi} \sin\theta\, d\theta \int_0^{2\pi} d\phi \int_0^{a} (a^2 r^2 - r^4)\, dr$$

$$Q_{enc} = 4\pi\rho_0(a^2\frac{r^3}{3} - \frac{r^5}{5})\Big|_0^2$$

$$Q_{enc} = \frac{8\pi}{15}\rho_0$$

$$\psi = \int \bar{D}\bullet d\bar{S} = \varepsilon_0 E_r(4\pi r^2)$$

$$\psi = Q_{enc} :$$

$$\varepsilon_0 E_r(4\pi r^2) = \frac{8\pi}{15}\rho_0$$

$$E_r = \frac{2\rho_0}{15\varepsilon_0 r^2} \quad \text{or}$$

$$\bar{E} = \frac{2\rho_0}{15\varepsilon_0 r^2}\bar{a}_r$$

$$V = \int \bar{E}\bullet d\bar{l} = -\frac{2\rho_0}{15\varepsilon_0}\int r^{-2}dr = \frac{2\rho_0}{15\varepsilon_0 r} + C_1$$

Since $V(r->0) = 0, \quad C_1 = 0;$

$$V = \frac{2\rho_0}{15\varepsilon_0 r}$$

(b)

For $r \le a$,

$$Q_{enc} = \rho_0(4\pi)(\frac{a^2r^3}{3} - \frac{r^5}{5})\Big|_0^r = 4\pi\rho_0(\frac{a^2r^3}{3} - \frac{r^5}{5})$$

$$E_r = \frac{Q_{enc}}{4\pi\varepsilon_0 r^2} = \frac{\rho_0}{\varepsilon_0}(\frac{a^2r}{3} - \frac{r^3}{5})$$

$$\bar{E} = \frac{\rho_0}{\varepsilon_0}(\frac{a^2r^3}{3} - \frac{r^5}{5})$$

$$V = -\int \bar{E}\bullet d\bar{l} = -\frac{\rho_0}{\varepsilon_0}(\frac{a^2r^2}{6} - \frac{r^4}{20}) + C_2$$

$$= \frac{\rho_0}{\varepsilon_0}(\frac{r^4}{20} - \frac{a^2r^2}{6}) + C_2$$

Since $V(r = a^+) = V(r = a^-)$,

$$\frac{2\rho_0}{15\varepsilon_0 a} = \frac{\rho_0}{\varepsilon_0}(\frac{a^4}{20} - \frac{a^4}{6}) + C_2 \quad \longrightarrow \quad C_2 = \frac{2\rho_0}{15\varepsilon_0 a} + \frac{7\rho_0 a^4}{60\varepsilon_0}$$

$$V = \frac{\rho_0}{\varepsilon_0}(\frac{r^4}{20} - \frac{a^2 r^2}{6}) + \frac{2\rho_0}{15\varepsilon_0} + \frac{7\rho_0 a^4}{60\varepsilon_0}$$

(c)

The total charge is found in part (a) as

$$Q = \frac{8\pi\rho_0}{15}$$

(d)

For $r \geq a, \bar{E}$ decays to zero with no maxima.

For $r \leq a$,

$$E_r = \frac{\rho_0}{\varepsilon_0}(\frac{a^2 r}{3} - \frac{r^3}{5})$$

$$\frac{\partial E_r}{dr} = \frac{\rho_0}{\varepsilon_0}(\frac{a^2}{3} - \frac{3r^2}{5}) = 0 \quad \longrightarrow \quad r = \frac{a\sqrt{5}}{3}$$

$$r = 0.7453\,a$$

Prob 4.36

$m\dfrac{d^2y}{dt^2} = eE$; divide by m, and integrate once, one obtains:

$$u\,\frac{dy}{dt} = \frac{eEt}{m} + c_0$$

$$y = \frac{eEt^2}{2\,m} + c_0 t + c_1 \qquad (1)$$

"From rest" implies $c_1 = 0 = c_0$

At $t = t_0$, $y = d$, $E = \dfrac{V}{d}$ or $V = Ed$.

Substituting this in (1) yields:

$$t^2 = \frac{2\,m\,d}{e\,E}$$

Hence:

$$u = \frac{eE}{m}\sqrt{\frac{2md}{eE}} = \sqrt{\frac{2eE}{m}}$$

that is, $u \,\alpha\, \sqrt{V}$

or $\quad u = k\sqrt{V}$

(b)

$$k = \sqrt{\frac{2e}{m}} = \sqrt{\frac{2\,(1.603)\,10^{-19}}{9.1066\;\;(10^{-31})}}$$

$$= \underline{\underline{5.933\,(10^5)}}$$

(c)

$$V = \frac{u^2\,m}{2e} = \frac{9(10^{16})\,\dfrac{1}{100}}{2(176)\,(10^{11})} = \underline{\underline{2557\;\,k\,\mathrm{V}}}$$

Prob 4.37

(a)

This is similar to Example 4.3.

$$u_y = \frac{eEt}{m}, \quad u_x = u_0$$

$$y = \frac{eEt^2}{2m}, \quad x = u_0 t$$

$$t = \frac{x}{u_0} = \frac{10(10^{-2})}{10^7} = 10 \text{ ns}$$

Since $x = 10$ cm when $y = 1$cm,

$$E = \frac{2my}{et^2} = \frac{2(10^{-2})}{1.76(10^{11})(10^{-16})} = 1.136\ \mathrm{kV/m}$$

$$E = \underline{\underline{-\,1.136\,\bar{a}_y\ \mathrm{kV/m}}}$$

(b)

$$u_x = u_0 = 10^7,$$

$$u_y = \frac{2000}{1.76}(1.76)10^{11}(10^{-8}) = 2(10^6)$$

$$\bar{u} = \underline{\underline{(\bar{a}_x + 0.2\bar{a}_y)(10^7)\ \ \mathrm{m/s}}}$$

Prob 4.38

$$V = \frac{p\cos\theta}{4\pi\varepsilon_0 r} = \frac{k\cos\theta}{r}$$

$At\ (0, 1\text{nm}),\quad \theta = 0,\quad r = 1\text{ nm},\quad V = 9;$

$$\text{that is,}\quad 9 = \frac{k(1)}{1(10^{-18})},\quad \therefore\ k = 9(10^{-18})$$

$$V = 9(10^{-18})\frac{\cos\theta}{r}$$

At $(1,1)$ nm, $\quad r = \sqrt{2}$ nm, $\quad \theta = 45°$,

$$V = \frac{9(10^{-18})\cos 45°}{10^{-18}\sqrt{2}} = \frac{9}{2\sqrt{2}} = \underline{\underline{3.182\ \text{V}}}$$

Prob 4.39

The dipole is oriented along y- axis.

$$V = \frac{\bar{p}\bullet\bar{r}}{4\pi\varepsilon_0 r^2};\qquad \bar{p}\bullet\bar{r} = Qd\,\bar{a}_y\bullet\bar{a}_r = Qd\sin\theta\sin\phi$$

$$V = \frac{Qd\sin\theta\sin\phi}{4\pi\varepsilon_0 r^2}$$

$$\bar{E} = -\nabla V = -\frac{\partial V}{\partial r}\bar{a}_r - \frac{1}{r}\frac{\partial V}{\partial\theta}\bar{a}_\theta - \frac{1}{r\sin\theta}\frac{\partial V}{\partial\phi}\bar{a}_\phi$$

$$= \frac{Qd}{4\pi\varepsilon_0}\left\{\frac{2\sin\theta\sin\phi}{r^3}\bar{a}_r - \frac{\cos\theta\sin\phi}{r^3}\bar{a}_\theta - \frac{\cos\theta}{r^3}\bar{a}_\phi\right\}$$

$$\bar{E} = \underline{\underline{\frac{Qd}{4\pi\varepsilon_0}(2\sin\theta\sin\phi\,\bar{a}_r - \cos\theta\sin\phi\,\bar{a}_\theta - \cos\phi\,\bar{a}_\phi)}}$$

Prob 4.40

$$W = Q_2 V_{21} = Q_2\frac{Q_1}{4\pi\varepsilon_0|\bar{r}_2 - \bar{r}_1|}$$

$$= \frac{-2(1)(10^{-6})}{4\pi(\frac{10^{-9}}{36\pi})\ |(5,-10,-1)|} = \frac{-18(10^{-3})}{\sqrt{126}}$$

$$W = \underline{\underline{-1.604}}$$

Prob 4.41

$$W = \frac{1}{2}\int \bar{D} \bullet \bar{E}\, dv = \frac{\varepsilon_0}{2}\int |\bar{E}| dv,$$

$$\bar{E} = \frac{Q}{4\varepsilon_0 r^2}\bar{a}_r,$$

$$W = \frac{\varepsilon_0}{2}\iiint \frac{Q^2}{16\pi^2\varepsilon_0 r^4}(r^2 \sin\theta\, dr\, d\theta\, d\phi)$$

$$W = \frac{Q^2}{32\pi^2\varepsilon_0} 4\pi \int_a^\infty \frac{1}{r^2} dr = \underline{\underline{\frac{Q^2}{8\pi\varepsilon_0 a}}}$$

Prob 4.42

$$W = \frac{1}{2}\varepsilon_0 \int |\bar{E}|^2 dv = \frac{1}{2}\varepsilon_0 \iiint (4r^2 \sin^2\theta \cos^2\phi + r^2 \cos^2\theta \cos^2\phi + r^2 \sin^2\phi) r^2 \sin\theta\, d\theta\, d\phi$$

$$= \frac{1}{2}\varepsilon_0 \int r^4\, dr \left\{ 4\int_0^{2\pi} \cos^2\phi\, d\phi \int_0^{\pi} \sin^3\theta\, d\theta + \int_0^{2\pi} \cos^2\phi\, d\phi \int_0^{\pi} \cos^2\theta \sin\theta\, d\theta \right.$$

$$\left. + \int_0^{2\pi} \sin^2\phi\, d\phi \int_0^{2\pi} \sin\theta\, d\theta \right\}$$

$$= \frac{1}{2}\varepsilon_0 \left.\frac{r^5}{5}\right|_0^2 \left\{ 4\left(\frac{1}{2}\right)(2\pi)\int_0^{\pi}(1 - \cos^2\theta)\, d(-\cos\theta) \right.$$

$$\left. + \frac{1}{2}(2\pi)\int_0^{\pi} \cos^2\theta\, d(-\cos\theta) + \frac{1}{2}(2\pi)(-\cos\theta)\Big|_0^{\pi} \right\}$$

$$= 3.2\varepsilon_0 [4\pi(\frac{\cos^3\theta}{3} - \cos\theta)\Big|_0^{\pi} + \pi(-\frac{\cos^3\theta}{3})\Big|_0^{\pi} + 2\pi]$$

$$= 3.2\,\varepsilon_0 (8\pi) = 25.6\pi\, \frac{10^{-9}}{36\pi}$$

$$W = \underline{\underline{0.7111\ nJ}}$$

Prob 4.43

$$\bar{E} = -\nabla V = -(\frac{\partial V}{\partial \rho}\bar{a}_\rho + \frac{1}{\rho}\frac{\partial V}{\partial \rho}\bar{a}_\phi + \frac{\partial V}{\partial z}\bar{a}_z)$$

$$\bar{E} = -(2\rho z \sin\phi\, \bar{a}_\rho + \rho z \cos\phi\, \bar{a}_\phi + \rho^2 \sin\phi\, \bar{a}_z)$$

$$W = \frac{1}{2}\varepsilon_0 \int |\bar{E}|^2 dv = \iiint (4\rho^2 z^2 \sin^2\phi + \rho^2 z^2 \cos^2\phi + \rho^4 \sin^2\phi)\rho\, d\phi\, dz d\rho$$

$$\frac{2W}{\varepsilon_0} = 4\int_1^4 \rho^3 dz \int_{-2}^{2} z^2 dz \int_0^{\pi/3} \sin^2\phi\, d\phi + \int_1^4 \rho^3 d\rho \int_{-2}^{2} z^2 dz \int_0^{\pi/3} \cos^2\phi\, d\phi$$

$$+ \int_1^4 \rho^5 d\rho \int_{-2}^{2} dz \int_0^{\pi/3} \sin^2\phi\, d\phi$$

But $\int_0^{\pi/3} \cos^2\phi\, d\phi = \frac{1}{2}\int_0^{\pi/3} [1 + \cos 2\phi]\, d\phi = \frac{\pi}{6} + \frac{1}{4}\sin\frac{2\pi}{3} = 0.7401$

$$\int_0^{\pi/2} \sin^2\phi\, d\phi = \frac{1}{2}\int_0^{2} (1 - \cos 2\phi) d\phi = \frac{\pi}{6} - \frac{1}{4}\sin^2\frac{\pi}{3} = 0.3071$$

$$\frac{2W}{\varepsilon_0} = \frac{4}{4}\rho^4 \Big|_0^1 \frac{2z^2}{3}\Big|_0^2 (0.3071) + \frac{\rho^4}{4}\Big|_0^1 \frac{2z^3}{3}\Big|_0^2 (0.7401) + \frac{\rho^6}{6}\Big|_0^4 (4)(0.3071)$$

$$= 255(\frac{16}{3})(0.3071) + \frac{255}{4}(\frac{16}{3})(0.7041) + \frac{4096}{6}(0.3071)$$

$$= 417.67 + 239.394 + 838.59 = 1495.6$$

$$W = \frac{1495.6}{2}(\frac{10^{-9}}{36\pi})$$

$$= \underline{\underline{6.612 \text{ nJ}}}$$

CHAPTER 5

P. E. 5.1 $\quad d\mathbf{S} = \rho d\phi dz a_\rho$

$$I = \int J \bullet dS = \int_{\phi=0}^{2\pi} \int_{z=1}^{5} 10z \sin^2\phi \rho dz d\phi|_{\rho=2} = 10(2)\frac{z^2}{2}\Big|_1^5 \int_0^{2\pi} \frac{1}{2}(1-\cos 2\phi) d\phi = 240\pi$$

I = 754 A

P. E. 5.2

$$I = \rho_s wu = 0.5x10^{-6} x0.1x10 = 0.5\mu \text{ A}$$

$$V = IR = 10^{14} x0.5x10^{-6} = \underline{50 \text{ MV}}$$

P. E. 5.3 $\sigma = 5.8x10^7$ S/m

$$J = \sigma E \quad \longrightarrow \quad E = \frac{J}{\sigma} = \frac{8x10^6}{5.8x10^7} = \underline{0.138 \text{ V/m}}$$

$$J = \rho_v u \quad \longrightarrow \quad u = \frac{J}{\rho_v} = \frac{8x10^6}{1.81x10^{10}} = \underline{4.42x10^{-4} \text{ m/s}}$$

P. E. 5.4 The composite bar can be modeled as a parallel combination of resistors as shown below.

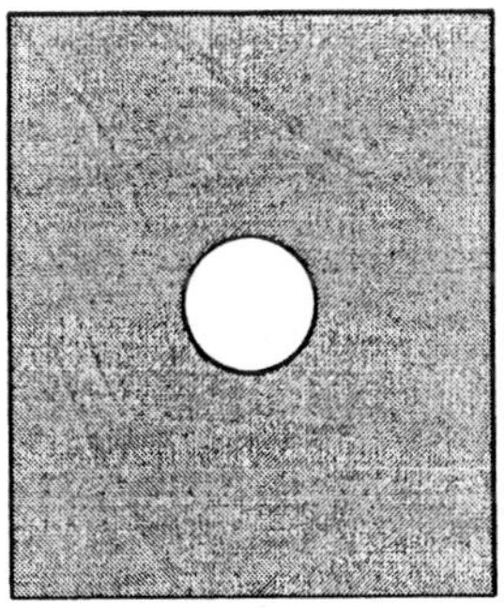

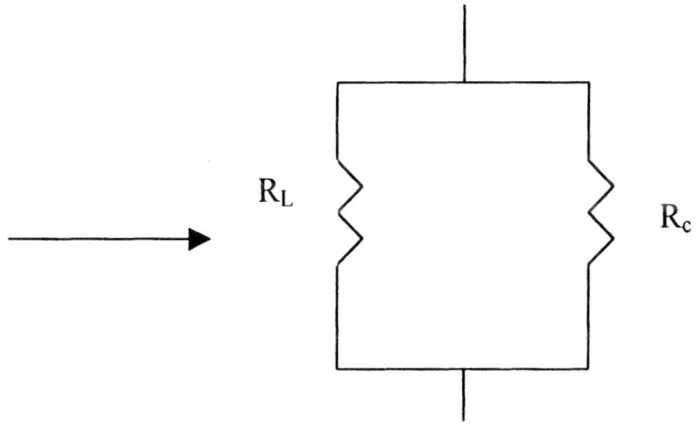

For the lead, $\quad R_L = \dfrac{l}{\sigma_L S_L}, \quad S_L = d^2 - \pi r^2 = 9 - \dfrac{\pi}{4}$ cm^2

$R_L = 0.974$ mΩ

For copper, $\quad R_c = \dfrac{l}{\sigma_c S_c}, \quad S_c = \pi r^2 = \dfrac{\pi}{4}$ cm^2

$$R_c = \frac{4}{5.8x10^7 x \frac{\pi}{4} x10^{-4}} = 0.8781 \ \text{m}\Omega$$

$$R_c = \frac{R_L R_c}{R_L + R_c} = \frac{0.974x0.8781}{0.974 + 0.8781} = \underline{\underline{461.7\mu\Omega}}$$

P. E. 5.5 $\rho_{Ps} = P \bullet a_x = ax^2 + b$

$$\rho_{ps}\Big|_{x=0} = P \bullet (-a_x)\Big|_{x=0} = \underline{\underline{-b}}$$

$$\rho_{ps}\Big|_{x=L} = P \bullet a_x \Big|_{x=L} = \underline{\underline{aL^2 + b}}$$

$$Q_s = \int \rho_{ps} dS = -bA + (aL^2 + b)A = AaL^2$$

$$\rho_{pv} = -\nabla \cdot P = -\frac{d}{dx}(ax^2 + b) = -2ax$$

$$\rho_{pv}\Big|_{x=0} = \underline{\underline{0}}, \qquad \rho_{pv}\Big|_{x=L} = \underline{\underline{-2aL}}$$

$$Q_v = \int \rho_{pv} dv = \int_0^L (-2ax) A dx = -AaL^2$$

Hence,

$$Q_T = Q_v + Q_s = -AaL^2 + AaL^2 = 0$$

P. E. 5.6

$$E = \frac{V}{d} a_x = \frac{10^3}{2x10^{-3}} a_x = 500 a_x \ \text{kV/m}$$

$$P = \chi_e \varepsilon_o E = (2.25 - 1)x \frac{10^{-9}}{36\pi} x0.5x10^6 a_x = \underline{\underline{6.853 a_x \mu C / \text{m}^2}}$$

$$\rho_{ps} = P \bullet a_x = \underline{\underline{6.853 \mu \text{C} / \text{m}^2}}$$

P. E. 5.7 (a) Since $P = \varepsilon_o \chi_e E, \quad P_x = \varepsilon_o \chi_e E_x$

$$\chi_e = \frac{P_x}{\varepsilon_o E_x} = \frac{3x10^9}{10\pi} \frac{1}{5} x36\pi x10^9 = \underline{\underline{2.16}}$$

(b) $E = \dfrac{P}{\chi_e \varepsilon_o} = \dfrac{36\pi x 10^9}{2.16} \dfrac{1}{10\pi}(3,-1,4)10^{-9} = \underline{\underline{5a_x - 1.67a_y + 6.67a_z}}$ V/m

(c)

$$D = \varepsilon_o \varepsilon_r E = \frac{\varepsilon_r P}{\chi_e} = \frac{3.16}{2.16}\left(\frac{1}{10\pi}\right)(3,-1,4) \quad \text{nC/m}^2 = \underline{\underline{139.7a_x - 46.6a_y + 186.3a_z}} \quad \text{pC/m}^2$$

P. E. 5.8 From Example 5.8,

$$F = \frac{\rho_s^2 S}{2\varepsilon_o} \longrightarrow \rho_s^2 = \frac{2\varepsilon_o F}{S}$$

But $\rho_s = \varepsilon_o E = \varepsilon_o \dfrac{V_d}{d}$. Hence

$$\rho_s^2 = \frac{2\varepsilon_o F}{S} = \frac{\varepsilon_o^2 V_d^2}{d^2} \longrightarrow V_d^2 = \frac{2Fd^2}{\varepsilon_o S}$$

i.e.

$$V_d = V_1 - V_2 = \sqrt{\frac{2Fd^2}{\varepsilon_o S}}$$

as required.

P. E. 5.9 (a) Since $a_n = a_x$,

$$D_{1n} = 12a_x, \qquad D_{1t} = -10a_x + 4a_z, \qquad D_{2n} = D_{1n} = 12a_x$$

$$E_{2t} = E_{1t} \longrightarrow D_{2t} = \frac{\varepsilon_2 D_{1t}}{\varepsilon_1} = \frac{1}{2.5}(-10a_y + 4a_z) = -4a_y + 1.6a_z$$

$D_2 = D_{2n} + D_{2t} = \underline{\underline{-12a_x - 4a_y + 1.6a_z}}$ nC/m^2.

(b) $\tan\theta_2 = \dfrac{D_{2t}}{D_{2n}} = \dfrac{\sqrt{(-4)^2 + (1.6)^2}}{12} = 0.359 \longrightarrow \underline{\underline{\theta_2 = 19.75^o}}$

(c) $E_{1t} = E_{2t} = E_2 \sin\theta_2 = 12\sin 60^o = 10.392$

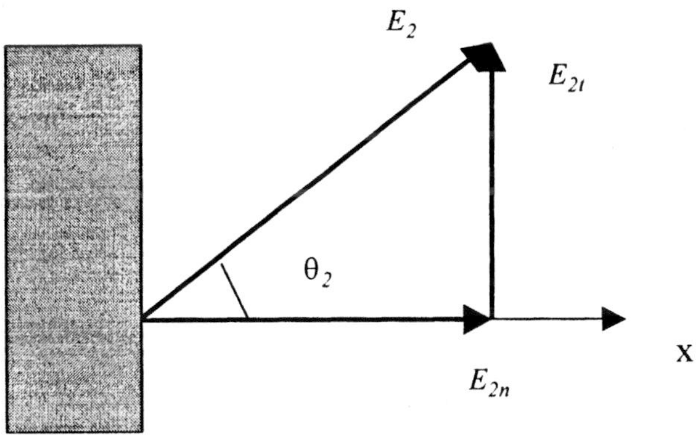

$$E_{1n} = \frac{\varepsilon_{r2}}{\varepsilon_{r1}} E_{2n} = \frac{1}{2.5} 12\cos 60^o = 2.4$$

$$E_1 = \sqrt{E_{1t}^2 + E_{1n}^2} = \underline{\underline{10.67}}$$

$$\tan\theta_1 = \frac{\varepsilon_{r1}}{\varepsilon_{r2}} \tan\theta_2 = \frac{2.5}{1} \tan 60^o = 4.33 \quad \longrightarrow \quad \underline{\underline{\theta_1 = 77^o}}$$

Note that $\theta_1 > \theta_2$.

P. E. 5.10

$$D = \varepsilon_o E = \frac{10^{-9}}{36\pi}(60,20,-30)x10^{-3} = \underline{\underline{0.531a_x + 0.177a_y - 0.265a_z}} \text{ pC/m}^2$$

$$\rho_s = D_n = |D| = \frac{10^{-9}}{36\pi}(10)\sqrt{36+4+9}(10^{-3}) = \underline{\underline{0.619}} \text{ pC/m}^2$$

Prob. 5.1

$$I = \int J \bullet dS, \qquad dS = r\sin\theta d\phi dr a_\theta$$

$$I = -\int_{r=0}^{2}\int_{\phi=0}^{2\pi} r^3 \sin^2\theta d\phi dr \Big|_{\theta=30^o} = -(\sin 30^o)^2 \left.\frac{r^4}{4}\right|_0^2 (2\pi) = -2\pi = \underline{\underline{-6.283}} \text{ A}$$

Prob. 5.2

$$I = \int J \bullet dS = \int_{\rho=0}^{a}\int_{\phi=0}^{2\pi} \frac{500}{\rho}\rho d\phi d\rho = 500(2\pi a) = 1000\pi x1.6x10^{-3} = 1.6\pi = \underline{\underline{5.026}} \text{ A}$$

Prob. 5.3

$$I = \int J \bullet dS = 10\int_{\rho=0}^{a}\int_{\phi=0}^{2\pi} e^{-(1-\rho/a)}\rho d\phi d\rho = 20\pi \int_{\rho=0}^{a} \rho e^{-(1-\rho/a)} d\rho$$

But $\int xe^{ax}dx = \frac{e^{ax}}{a^2}(ax-1),$

$$I = 20\pi e^{-1}a^2(\frac{\rho}{a} - 1)e^{\rho/a}\Big|_0^a = \frac{20\pi a^2}{e}(1+0) = \underline{\underline{23.11a^2}} \text{ A}$$

Prob. 5.4

$$I = \frac{dQ}{dt} = -3x10^{-4}e^{-3t}$$

I(t=0) = - 0.3 mA. I(t=2.5) = -0.3 $e^{-7.5}$ = -166 nA

Prob. 5.5 (a) $\nabla^2 V = -\rho_v / \varepsilon$

$$\nabla^2 V = \frac{\partial}{\partial x}(2xy^2z) + \frac{\partial}{\partial y}(2x^2yz) = 6xyz$$

$$\rho_v = -8xyz(2\varepsilon_o) = \underline{\underline{-16xyz\varepsilon_o}}$$

(b) $J = \rho_v u = -16xy^2z\varepsilon_o(10^4)a_y$

$$I = \int J \bullet dS = -16(10^4)\frac{10^{-9}}{36\pi}\int_0^{0.5} xdx \int_0^{0.5} zdz = -16(36\pi)(10^{-5})\left(\frac{x^2}{2}\Big|_0^{0.5}\right)^2$$

$$I = -4(36\pi)(10^{-5})(0.5)^2 = \underline{\underline{-1.131}} \text{ mA}$$

Prob. 5.6 (a) $R = \frac{l}{\sigma S} = \frac{8x10^{-2}}{3x10^4 x\pi x25x10^{-6}} = \frac{8}{75\pi} = \underline{\underline{33.95\text{m}\Omega}}$

(b) $I = V/R = 9x\frac{75\pi}{8} = \underline{\underline{265.1}}$ A

(c) P = IV = 2.386 kW

Prob. 5.7 (a) $R = \frac{\rho l}{S} \longrightarrow \rho = RS/l = \frac{4.04}{10^3}\frac{\pi d^2}{4} = 2.855x10^{-8}$

$\sigma = 1/\rho = \underline{\underline{3.5x10^7}}$ S/m (Aluminum)

(b) $J = I/S = \frac{40}{\frac{\pi}{4}x90x10^{-6}} = \underline{\underline{5.66x10^6 \text{A/m}^2}}$

or

$$J = \sigma E = 3.5x0.1616x10^7 = 5.66x10^6 \text{A/m}^2$$

Prob. 5.8

$$R = \frac{l}{\sigma S}, \qquad S = \pi r^2 = \pi d^2 / \pi, d = 0.4\text{mm}, \qquad l = N 2\pi R = N\pi D, \quad D = 6.5 \text{ mm}$$

$$R = \frac{150x\pi(6.5)x10^{-3}}{5.8x10^{7} x\pi \frac{(0.4)^2}{4} x10^{-6}} = \underline{\underline{0.42\Omega}}$$

Prob. 5.9 (a) $R = \frac{\rho_c l}{S}, \qquad S_i = \pi r_i^2 = \pi(1.5)^2 x10^{-4} = 2.25\pi x10^{-4}$

$$S_o = \pi(r_o^2 - r_i^2) = \pi(4 - 2.25)x10^{-4} = 1.75\pi x10^{-4}$$

$$R = R_i // R_o = \frac{R_i R_o}{R_i + R_o} = \left(\frac{\frac{\rho_i}{S_i} \frac{\rho_o}{S_o}}{\frac{\rho_i}{S_i} + \frac{\rho_o}{S_o}} \right) l = 10 \left(\frac{\frac{1.77x11.8x10^{-16}}{2.25\pi x1.75\pi x10^{-8}}}{\frac{1.77x10^{-8}}{1.75\pi x10^{-4}} + \frac{11.8x10^{-8}}{2.25\pi x10^{-4}}} \right) = \underline{\underline{0.27\text{m}\Omega}}$$

(b) $V = I_i R_i = I_o R_o \quad \longrightarrow \quad \frac{I_i}{I_o} = \frac{R_o}{R_i} = \frac{0.3219}{1.669} = 0.1929$

$$I_i + I_o = 1.1929 I_o = 60 \text{ A}$$

$I_o = \underline{\underline{50.3 \text{ A}}}$ (copper), $\qquad I_i = \underline{\underline{9.7 \text{ A}}}$ (steel)

(c) $R = \frac{10x1.77x10^{-8}}{1.75\pi x10^{-4}} = \underline{\underline{0.322\text{m}\Omega}}$

Prob. 5.10

$$R = \frac{l}{\sigma S} = \frac{h}{\sigma\pi(b^2 - a^2)} = \frac{2}{10^5 \pi(25 - 9)x10^{-4}} = \underline{\underline{4\text{m}\Omega}}$$

Prob. 5.11

$$|P| = n|p| = nQd = 2ned = \chi_e \varepsilon_o E \qquad (Q = 2e)$$

$$\chi_e = \frac{2ned}{\varepsilon_o E} = \frac{2x5x10^{25} x1.602x10^{-19} x10^{-18}}{\frac{10^{-9}}{36\pi} x10^4} = 0.000182$$

$$\varepsilon_r = 1 + \chi_e = \underline{\underline{1.000182}}$$

Prob. 5.12

$$P = \frac{\sum_{i=1}^{N} q_i d_i}{v} = \frac{\sum_{i=1}^{N} p_i}{v}$$

$$|P| = \frac{N}{v}|p| = 2x10^{19} x1.8x10^{-27} = 3.6x10^{-8}$$

$$P = |P|a_x = \underline{\underline{3.6x10^{-18} a_x \text{ C/m}^2}}$$

But $P = \chi_e \varepsilon_o E$ or $\chi_e = \frac{P}{\varepsilon_o E} = \frac{3.6x36\pi x10^9 x10^{-18}}{10^5} = 0.0407$

$$\varepsilon_r = 1 + \chi_e = \underline{\underline{1.0407}}$$

Prob. 5.13 (a) $E = -\nabla V = -\frac{dV}{dz} a_z = 600za_z$

$$D = \varepsilon_o \varepsilon_r E = \frac{10^{-9}}{36\pi}(2.4)600za_z = \underline{\underline{12.73za_z \text{ nC/m}^2}}$$

$$\rho_v = \nabla \bullet D = \frac{\partial D_z}{\partial z} = \underline{\underline{12.73 \text{ nC/m}^3}}$$

(b) $\chi_e = \varepsilon_r - 1 = 1.4$

$$P = \chi_e \varepsilon_o E = \frac{\chi_e D}{\varepsilon_r} = \frac{1.4}{2.4}(12.732)a_z = \underline{\underline{7.427a_z \text{ nC/m}^2}}$$

$$\rho_{pv} = -\nabla \bullet P = \underline{\underline{-7.427 \text{ nC/m}^3}}$$

Prob. 5.14

$$\rho_{pv} = -\nabla \bullet P = \underline{\underline{0}}, \qquad \rho_{ps} = P \bullet a_n = \underline{\underline{5\sin\alpha y}}$$

Prob. 5.15 (a) Applying Coulomb's law.

$$E_r = \begin{cases} \frac{D_r}{\varepsilon_o} = \frac{Q}{4\pi\varepsilon_o r^2}, & b < r < a \\ \frac{D_r}{\varepsilon} = \frac{Q}{4\pi\varepsilon r^2}, & a < r < b \end{cases}$$

$$P = \frac{\varepsilon_r - 1}{\varepsilon_r} D \qquad (= D - \varepsilon E)$$

Hence

$$P_r = \frac{\varepsilon_r - 1}{\varepsilon_r} \cdot \frac{Q}{4\pi r^2}, \qquad a < r < b$$

(b) $$\rho_{pv} = -\nabla \bullet P = -\frac{1}{r^2}\frac{d}{dr}(r^2 P_r) = 0$$

(c)

$$\rho_{ps} = P \bullet (-a_r) = -\frac{Q}{4\pi a^2}\left(\frac{\varepsilon_r - 1}{\varepsilon_r}\right), \qquad r = a$$

$$\rho_{ps} = P \bullet (a_r) = -\frac{Q}{4\pi b^2}\left(\frac{\varepsilon_r - 1}{\varepsilon_r}\right), \qquad r = b$$

Prob. 5.16

$$F_1 = \frac{Q_1 Q_2}{4\pi\varepsilon_o r^2}, \qquad F_2 = \frac{Q_1 Q_2}{4\pi\varepsilon_o \varepsilon_r r^2}$$

$$\frac{F_1}{F_2} = \varepsilon_r = 4.5/2 = 2.25$$

$\varepsilon_r = 2.25$, polystrene

Prob. 5.17

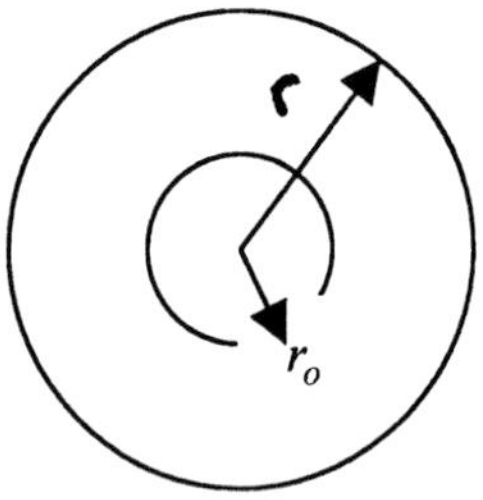

$$Q = 4\pi r_o^2 \rho_s, \qquad r_o = 10 \text{ cm}$$

From Gauss's law,

$$Q = \int D \bullet dS = D_n(4\pi r^2) \longrightarrow D_n = \frac{Q}{4\pi r^2} = \varepsilon E$$

$$E = \frac{Q}{4\pi\varepsilon r^2} a_r$$

At (-3cm, -4cm, 12cm), r = 13 cm

$$E = \frac{4\pi r_o^2 \rho_s}{4\pi\varepsilon r^2} a_r = \frac{(0.1)^2 x 4x10^{-9}}{\frac{10^{-9}}{36\pi} x2.5x(0.13)^2} a_r = \underline{\underline{107.1a_r \text{ V/m}}}$$

Since $\quad a_r = \frac{1}{13}(-3a_x + 4a_y + 12a_x),$

$$\underline{\underline{E = -24.72a_x + 32.95a_y + 98.86a_x \text{ V/m}}}$$

Prob. 5.18

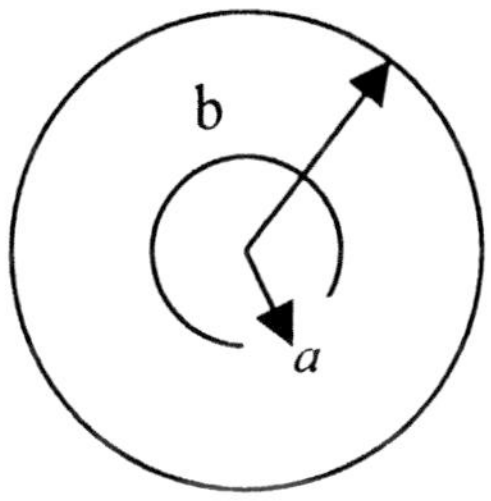

For $0 < r < a$,

$$D = \frac{Q}{4\pi r^2} a_r \quad \longrightarrow \quad E = \frac{Q}{4\pi\varepsilon_o r^2} a_r, \quad P = 0$$

For $a < r < b$,

$$D = \frac{Q}{4\pi r^2} a_r \quad \longrightarrow \quad E = \frac{Q}{4\pi\varepsilon\varepsilon_r r^2} a_r, \quad P = \chi_e \varepsilon_o E = \frac{\varepsilon_r - 1}{\varepsilon_r} \frac{Q}{4\pi\ r^2} a_r$$

For $r > b$,

$$D = \frac{Q}{4\pi r^2} a_r \quad \longrightarrow \quad E = \frac{Q}{4\pi\varepsilon_o r^2} a_r, \quad P = 0$$

Thus,

$$\underline{D = \frac{Q}{4\pi\varepsilon r^2} a_r, \quad r > 0}$$

$$\underline{\underline{E = \begin{cases} \dfrac{Q}{4\pi\varepsilon\varepsilon_r r^2} a_r, & a < r < b \\ \dfrac{Q}{4\pi\varepsilon_o r^2} a_r, & \text{otherwise} \end{cases}}}$$

$$P = \begin{cases} \dfrac{\varepsilon_r - 1}{\varepsilon_r} \dfrac{Q}{4\pi r^2} a_r, & a < r < b \\ 0, \text{ otherwise} \end{cases}$$

Prob. 5.19 (a)

$$\rho_v = \begin{cases} \rho_o, & 0 < r < a \\ 0, & r > a \end{cases}$$

For r < a, $\varepsilon E_r(4\pi r^2) = \rho_o \dfrac{4\pi r^3}{3} \longrightarrow E_r = \dfrac{\rho_o r}{3\varepsilon}$

$$V = -\int E \bullet dl = -\frac{\rho_o r^2}{6\varepsilon} + c_1$$

For r > a, $\varepsilon_o E_r(4\pi r^2) = \rho_o \dfrac{4\pi a^3}{3} \longrightarrow E_r = \dfrac{\rho_o a^3}{3\varepsilon_o r^2}$

$$V = -\int E \bullet dl = \frac{\rho_o a^3}{3\varepsilon_o r} + c_2$$

As $r \longrightarrow \infty$, $V = 0$ and $c_2 = 0$

At r = a, $V(a^+) = V(a^-)$

$$-\frac{\rho_o a^2}{6\varepsilon_o \varepsilon_r} + c_1 = \frac{\rho_o a^2}{3\varepsilon_o} \longrightarrow c_1 = \frac{\rho_o a}{6\varepsilon_o a}(2\varepsilon_r + 1)$$

$$V(r=0) = c_1 = \underline{\underline{\frac{\rho_o(2\varepsilon_r + 1)}{6\varepsilon_o a}}}$$

(b) $V(r = a) = \underline{\underline{\dfrac{\rho_o a^2}{3\varepsilon_o}}}$

Prob. 5.20 Since $\dfrac{\partial \rho_v}{\partial t} = 0$, $\nabla \bullet J = 0$ must hold.

(a) $\nabla \bullet J = 6x^2 y + 0 - 6x^2 y = 0 \longrightarrow$ This is <u>possible</u>.

(b) $\nabla \bullet J = y + (z + 1) \neq 0 \longrightarrow$ This is <u>not possible</u>.

(c) $\nabla \bullet J = \dfrac{1}{\rho}\dfrac{\partial}{\partial \rho}(z^2) + \cos\phi \neq 0 \longrightarrow$ This is <u>not possible.</u>

(d) $\nabla \bullet J = \dfrac{1}{r^2}\dfrac{\partial}{\partial r}(\sin\theta) = 0 \longrightarrow$ This is <u>possible</u>.

Prob. 5.21 (a)

$$\begin{bmatrix} D_x \\ D_y \\ D_z \end{bmatrix} = \varepsilon_o \begin{bmatrix} 4 & 1 & 1 \\ 1 & 4 & 1 \\ 1 & 1 & 4 \end{bmatrix} \begin{bmatrix} 10 \\ 10 \\ 0 \end{bmatrix}$$

$$D_x = 50\varepsilon_o, \qquad D_y = 50\varepsilon_o, \qquad D_z = 20\varepsilon_o$$

$$D = \underline{\underline{0.442a_x + 0.442a_y + 0.1768a_z}} \text{ nC/m}^2$$

(b)

$$\begin{bmatrix} D_x \\ D_y \\ D_z \end{bmatrix} = \varepsilon_o \begin{bmatrix} 4 & 1 & 1 \\ 1 & 4 & 1 \\ 1 & 1 & 4 \end{bmatrix} \begin{bmatrix} 10 \\ 20 \\ -30 \end{bmatrix}$$

$$D_x = 30\varepsilon_o, \qquad D_y = 60\varepsilon_o, \qquad D_z = 90\varepsilon_o$$

$$D = \underline{\underline{0.2653a_x + 0.5305a_y + 0.7958a_z}} \text{ nC/m}^2$$

Prob. 5.22 (a) $\nabla \bullet J = \frac{1}{\rho}\frac{\partial}{\partial \rho}\left(\frac{100}{\rho}\right) = -\frac{100}{\rho^3}$

$$-\frac{\partial \rho_v}{\partial t} = \nabla \bullet J = -\frac{100}{\rho^3} \quad \longrightarrow \quad \frac{\partial \rho_v}{\partial t} = \underline{\underline{\frac{100}{\rho^3} \text{ C/m}^3.s}}$$

(b) $I = \int J \bullet dS = \iint \frac{100}{\rho^3}\rho \, d\phi \, dz\Big|_{\rho=2} = \frac{100}{2^2}\int_0^{2\pi} d\phi \int_0^1 dz = 50\pi = \underline{\underline{157.1}}$ A

Prob. 5.23 (a)

$$I = \int J \bullet dS = \iint \frac{5e^{-10^4 t}}{r} r^2 \sin\theta d\theta d\phi\Big|_{r=2} = (2)(5)e^{-10^4 t}\int_0^{\pi} \sin\theta d\theta \int_0^{2\pi} d\phi = 40\pi e^{-10^4 t}$$

At t=0.1 ms, $I = 40\pi e^{-1} = \underline{\underline{46.23}}$ A

$$-\frac{\partial \rho_v}{\partial t} = \nabla \bullet J \quad \longrightarrow \quad \rho_v = -\int \nabla \bullet J \partial t$$

$$\nabla \bullet J = \frac{1}{r^2}\frac{\partial}{\partial r}(r^2 J_r) = \frac{5}{r^2}e^{-10^4 t}$$

$$\rho_v = -\int \frac{5}{r^2} e^{-10^4 t} dt = \frac{5}{10^4 r^2} e^{-10^4 t}$$

At t=0.1 ms and r = 2m,

$$\rho_v = \frac{5}{10^4 (2)^2} e^{-1} = \underline{\underline{45.98\ \mu\text{C/m}^3}}$$

Prob. 5.24 (a) $$\frac{\varepsilon}{\sigma} = \frac{3.1x\dfrac{10^{-9}}{36\pi}}{10^{-15}} = \underline{\underline{2.741x10^4\ \text{s}}}$$

(b) $$\frac{\varepsilon}{\sigma} = \frac{6x\dfrac{10^{-9}}{36\pi}}{10^{-15}} = \underline{\underline{5.305x10^4\ \text{s}}}$$

(c) $$\frac{\varepsilon}{\sigma} = \frac{80x\dfrac{10^{-9}}{36\pi}}{10^{-4}} = \underline{\underline{7.07\ \mu\text{s}}}$$

Prob. 5.25 (a) $Q = Q_o e^{-t/T_r} \longrightarrow \frac{1}{3} Q_o = Q_o e^{-t_1/T_r} \longrightarrow e^{t_1/T_r} = 3$

$$T_r = \frac{t_1}{\ln 3} = \frac{20\mu s}{\ln 3} = \underline{\underline{18.2\mu s}}$$

(b) But $T_r = \frac{\varepsilon_r \varepsilon_o}{\sigma}$, $$\varepsilon_r = \frac{\sigma T_r}{\varepsilon_o} = \frac{10^{-5} x 18.2x10^{-6}}{\dfrac{10^{-9}}{36\pi}} = \underline{\underline{20.58}}$$

(c) $$\frac{Q}{Q_o} = e^{-t_o/T_r} = e^{-30/18.3} = 0.1923 \quad \text{i.e.} \quad \underline{\underline{19.23\%}}$$

Prob. 5.26

$$T_r = \frac{\varepsilon}{\sigma} = \frac{2.5x10^{-9}}{5x10^{-6} x 36\pi} = 4.42\ \mu\text{s}$$

$$\rho_{vo} = \frac{Q}{V} = \frac{10x10^{-6}}{\dfrac{4\pi}{3} x10^{-6} x8} = \underline{\underline{0.2984\ C/\text{m}^3}}$$

$$\rho_v = \rho_{vo} e^{-t/T_r} = 0.2984 e^{-2/4.42} = \underline{\underline{0.1898\ \text{C/m}^3}}$$

Prob. 5.27 (a) $E_{2t} = E_{1t} = -300a_x + 50a_y, \quad E_{1n} = 70a_z$

$$D_{2n} = D_{1n} \longrightarrow \varepsilon_2 E_{2n} = \varepsilon_1 E_{1n}$$

$$E_{2n} = \frac{\varepsilon_1}{\varepsilon_2} E_{1n} = \frac{2.5}{4}(70a_z) = 43.75a_z$$

$$E_2 = -30a_x + 50a_y + 43.75a_z$$

$$D_2 = \varepsilon_o \varepsilon_r E_2 = 4x\frac{10^{-9}}{36\pi}(-30,50,43.75) = \underline{\underline{-1.061a_x + 1.768a_y + 1.547a_z \text{ nC/m}^2}}$$

(b) $P_2 = \varepsilon_o \chi_{e2} E_2 = 3x\frac{10^{-9}}{36\pi}(-30,50,43.75) = \underline{\underline{0.7958a_x + 1.326a_y + 1.161a_z \text{ nC/m}^2}}$

(c) $E_1 \bullet a_z = E_1 \cos\theta_n$

$$\cos\theta_n = \frac{70}{\sqrt{30^2 + 50^2 + 70^2}} = 0.7683 \longrightarrow \underline{\underline{\theta_n = 39.79^o}}$$

Prob. 5.28

(a) $P_1 = \varepsilon_o \chi_{e1} E_1 = 2x\frac{10^{-9}}{36\pi}(10,-6,12) = \underline{\underline{0.1768a_x - 0.1061a_y + 0.2122a_z \text{ nC/m}^2}}$

(b) $E_{1n} = -6a_x, \quad E_{2t} = E_{1t} = 10a_x + 12a_z$

$$D_{2n} = D_{1n} \longrightarrow \varepsilon_2 E_{2n} = \varepsilon_1 E_{1n}$$

or $$E_{2n} = \frac{\varepsilon_1}{\varepsilon_2} E_{1n} = \frac{3\varepsilon_o}{4.5\varepsilon_o}(-6a_z) = -4a_y$$

$$E_2 = \underline{\underline{10a_x - 4a_y + 12a_z \text{ V/m}}}$$

$$\tan\theta_2 = \frac{E_{2t}}{E_{2n}} = \frac{\sqrt{10^2 + 12^2}}{4} = 3.905 \longrightarrow \underline{\underline{\theta_2 = 75.64^o}}$$

(c) $w_E = \frac{1}{2} D \bullet E = \frac{1}{2}\varepsilon |E|^2$

$$w_{E1} = \frac{1}{2}\varepsilon_1 |E_1|^2 = \frac{1}{2}x3x\frac{10^{-9}}{36\pi}(10^2 + 6^2 + 12^2) = \underline{\underline{0.2219 \text{ nJ/m}^3}}$$

$$w_{E2} = \frac{1}{2}\varepsilon_2 |E_2|^2 = \frac{1}{2}x4.5x\frac{10^{-9}}{36\pi}(10^2 + 4^2 + 12^2) = \underline{\underline{0.3208 \text{ nJ/m}^3}}$$

Prob. 5.29 (a) $D_{2n} = 12a_\rho = D_{1n}, \qquad D_{2t} = -6a_\phi - 9a_z$

$$E_{2t} = E_{2t} \quad \longrightarrow \quad \frac{D_{1t}}{\varepsilon_1} = \frac{D_{2t}}{\varepsilon_2}$$

$$D_{1t} = \frac{\varepsilon_1}{\varepsilon_2} D_{2t} = \frac{3.5\varepsilon_o}{1.5\varepsilon_o}(-6a_\phi + 9a_z) = -14a_\phi + 21a_z$$

$$\underline{\underline{D_1 = 12a_\rho - 14a_\phi + 21a_z \text{ nC/m}^2}}$$

$$E_1 = D_1/\varepsilon_1 = \frac{(12,-14,21)x10^{-9}}{3.5x\dfrac{10^{-9}}{36\pi}} = \underline{\underline{387.8a_\rho - 452.4a_\phi + 678.6a_z \text{ V/m}}}$$

(b) $P_2 = \varepsilon_o \chi_{e2} E_2 = 0.5\varepsilon_o \dfrac{D_2}{\varepsilon_2} = \dfrac{0.5\varepsilon_o}{1.5\varepsilon_o}(12,-6,9) = \underline{\underline{4a_\rho - 2a_\phi + 3a_z \text{ nC/m}^2}}$

$\rho_{v2} = \nabla \bullet P_2 = 0$

(c) $$w_{E1} = \frac{1}{2} D_1 \bullet E_1 = \frac{1}{2}\frac{D_1 \bullet D_1}{\varepsilon_o \varepsilon_{r1}} = \frac{1}{2}\frac{(12^2 + 14^2 + 21^2)x10^{-18}}{3.5x\dfrac{10^{-9}}{36\pi}} = \underline{\underline{12.62 \text{ mJ/m}^2}}$$

$$w_{E1} = \frac{1}{2}\frac{D_2 \bullet D_2}{\varepsilon_o \varepsilon_{r2}} = \frac{1}{2}\frac{(12^2 + 6^2 + 9^2)x10^{-18}}{5x\dfrac{10^{-9}}{36\pi}} = \underline{\underline{9.839 \text{ mJ/m}^2}}$$

Prob. 5.30 (a)

$$P_1 = \varepsilon_o \chi_{e1} E_1 = 1.5x\frac{10^{-9}}{36\pi}(2,5,-4)x10^3 = \underline{\underline{26.53a_\rho + 66.31a_\phi - 53.05a_z \text{ nC/m}^2}}$$

$$\rho_{pv1} = -\nabla \bullet P_1 = -\frac{1}{\rho}\frac{\partial}{\partial\rho}(26.53\rho) = \underline{\underline{-\frac{26.53}{\rho} \text{ nC/m}^3}}$$

(b) $E_{2t} = E_{1t} = 5a_\phi - 4a_z$

$$D_{2n} = D_{1n} \quad \longrightarrow \quad E_{2n} = \frac{\varepsilon_1}{\varepsilon_2}E_{1n} = \frac{2.5}{1.0}(2) = 5$$

$$\underline{\underline{E_2 = 5a_\rho + 5a_\phi - 4a_z \text{ kV/m}}}$$

$$D_2 = \varepsilon_o \varepsilon_r E = 2.5x\frac{10^{-9}}{36\pi}(5,5,-4)x10^3 = \underline{\underline{110.5a_\rho + 110.5a_\phi - 88.42a_z \text{ nC/m}^2}}$$

Prob. 5.31 (a)

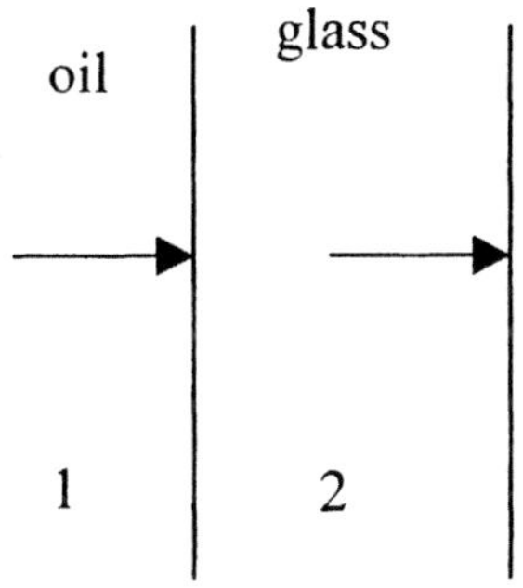

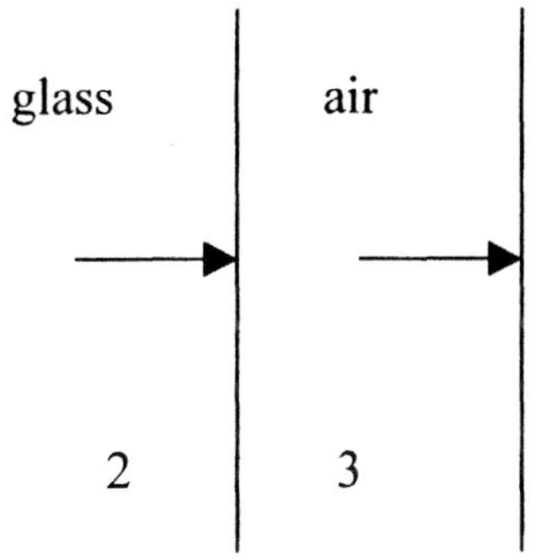

$E_{1n} = 2000, \quad E_{1t} = 0 = E_{2t} = E_{3t}$

$D_{1n} = D_{2n} = D_{3n} \quad \longrightarrow \quad \varepsilon_1 E_{1n} = \varepsilon_2 E_{2n} = \varepsilon_3 E_{3n}$

$$E_{2n} = \frac{\varepsilon_1}{\varepsilon_2} E_{1n} = \frac{3.0}{8.5}(2000) = \underline{\underline{705.9 \text{ V/m}, \ \theta_2 = 0^o}}$$

$$E_{3n} = \frac{\varepsilon_1}{\varepsilon_3} E_{1n} = \frac{3.0}{1.0}(2000) = \underline{\underline{6000 \text{ V/m}, \ \theta_3 = 0^o}}$$

(b)

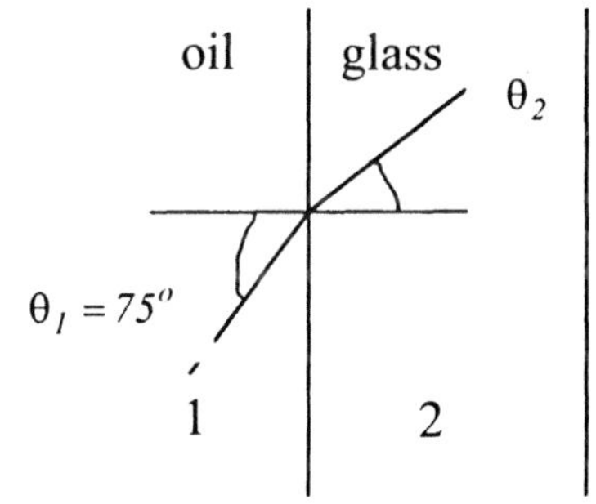

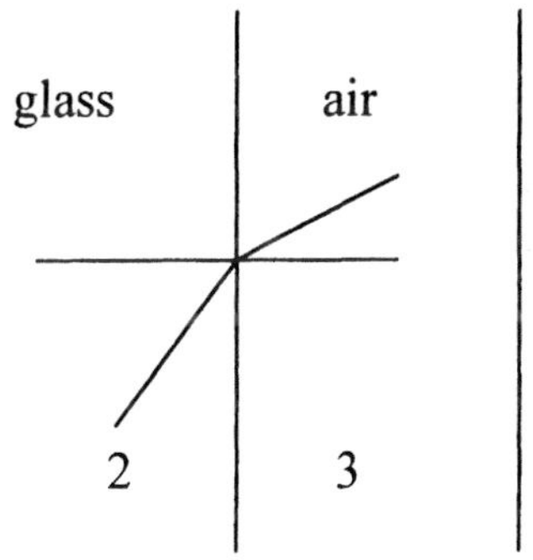

$E_{1n} = 2000\cos 75^o = 517.63, \quad E_{1t} = 2000\sin 75^o = E_{2t} = E_{3t} = 1931.85$

$E_{2n} = \frac{\varepsilon_1}{\varepsilon_2} E_{1n} = \frac{3}{8.5}(517.63) = 182.7, \quad E_{3n} = \frac{\varepsilon_1}{\varepsilon_3} E_{1n} = \frac{3}{1}(517.63) = 1552.89$

$$E_2 = \sqrt{E_{2n}^2 + E_{2t}^2} = 1940.5, \quad \theta_2 = \tan^{-1}\frac{E_{2t}}{E_{2n}} = \underline{\underline{84.6^o}},$$

$$E_3 = \sqrt{E_{3n}^2 + E_{3t}^2} = 2478.6, \quad \theta_3 = \tan^{-1}\frac{E_{3t}}{E_{3n}} = \underline{\underline{51.2^o}}$$

Prob. 5.32 (a) $\rho_s = D_n = \varepsilon_o E_n = \frac{10^{-9}}{36\pi}\sqrt{15^2+8^2} = \underline{\underline{0.1503 \text{ nC/m}^2}}$

(b) $D_n = \rho_s = -20$ nC

$$D = D_n a_n = (-20 \text{ nC})(-a_y) = \underline{\underline{20a_y \text{ nC/m}^2}}$$

Prob. 5.33 (a)

$$D_n = \rho_s = \frac{Q}{4\pi a^2} = \frac{12x10^{-9}}{4\pi x 25x10^{-4}} = \frac{1200}{\pi} \text{ nC/m}^2$$

$\underline{\underline{|D| = 381.97 \text{ nC/m}^2}}$

(b) Using Gauss's law,

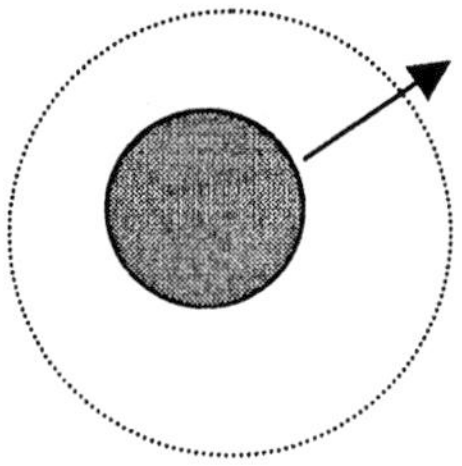

$$D_r 4\pi r^2 = Q \longrightarrow D_r = \frac{Q}{4\pi r^2}$$

$$D = \frac{Q}{4\pi r^2}a_r = \frac{12}{4\pi r^2}a_r \text{ nC/m}^2 = \underline{\underline{\frac{0.955}{r^2}a_r \text{ nC/m}^2}}$$

(c) $W = \frac{1}{2\varepsilon_o}\int |D|^2 dv = \frac{Q^2}{2\varepsilon_o 16\pi^2}\iiint \frac{1}{r^4}r^2 \sin\theta \, d\theta \, d\phi \, dr = \frac{Q^2}{32\pi^2\varepsilon_o}4\pi\int_a^\infty \frac{dr}{r^2}$

$$= \frac{Q^2}{8\pi\varepsilon_o a} = \frac{144x10^{-18}}{8\pi x \frac{10^{-9}}{36\pi} x 5x10^{-2}} = \underline{\underline{12.96 \ \mu\text{J}}}$$

CHAPTER 6

P. E. 6.1

$$\nabla^2 V = -\frac{\rho}{\varepsilon} \longrightarrow \frac{d^2V}{dx^2} = -\frac{\rho_o x}{\varepsilon_o}$$

$$V = -\frac{\rho_o x^3}{6\varepsilon a} + Ax + B$$

$$E = -\frac{dV}{dx}a_x = \left(\frac{\rho_o x^2}{2\varepsilon a} - A\right)a_x$$

If **E** = 0 at x =0, then

$$0 = 0 - A \longrightarrow A = 0$$

If V= 0 at x =a, then

$$0 = -\frac{\rho_o a^3}{6\varepsilon a} + B \longrightarrow B = \frac{\rho_o a^2}{6\varepsilon}$$

Thus

$$\underline{\underline{V = \frac{\rho_o}{6\varepsilon a}(a^3 - x^3)}}, \qquad \underline{\underline{E = \frac{\rho_o x^2}{2\varepsilon a} a_x}}$$

P. E. 6.2 $V_1 = A_1 x + B_1, \quad V_2 = A_2 x + B_2$

$$V_1(x = d) = V_o = A_1 d + B_1 \longrightarrow B_1 = V_o - A_1 d$$

$$V_1(x = 0) = 0 = 0 + B_2 \longrightarrow B_2 = 0$$

$$V_1(x = a) = V_2(x = a) \longrightarrow A_1 + B_1 = A_2 a$$

$$D_{1n} = D_{2n} \longrightarrow \varepsilon_1 A_1 = \varepsilon_2 A_2 \longrightarrow A_2 = \frac{\varepsilon_1}{\varepsilon_2} A_1$$

$$A_1 a + V_o - A_1 d = \frac{\varepsilon_1}{\varepsilon_2} a A_1 \longrightarrow V_o = A_1\left(-a + d + \frac{\varepsilon_1}{\varepsilon_2} a\right)$$

or

$$A_1 = \frac{V_o}{d - a + \varepsilon_1 a / \varepsilon_2}, \qquad A_2 = \frac{\varepsilon_1}{\varepsilon_2} A_1 \frac{\varepsilon_1 V_o}{\varepsilon_2 d - \varepsilon_2 a + \varepsilon_1 a}$$

Hence

$$\underline{\underline{E_1 = -A_1 a_x = \frac{-V_o a_x}{d - a + \varepsilon_1 a / \varepsilon_2}}}, \qquad \underline{\underline{E_2 = -A_2 a_x = \frac{-V_o a_x}{a + \varepsilon_2 d / \varepsilon_1 - \varepsilon_2 a / \varepsilon_1}}}$$

P. E. 6.3 From Example 6.3,

$$E = -\frac{V_o}{\rho\phi_o} a_\phi, \qquad D = \varepsilon_o E$$

$$\rho_s = D_n(\phi = 0) = -\frac{V_o \varepsilon}{\rho \phi_o}$$

The charge on the plate $\phi = 0$ is

$$Q = \int \rho_s dS = -\frac{V_o \varepsilon}{\phi_o} \int_{z=0}^{L} \int_{\rho=a}^{b} \frac{1}{\rho} dz d\rho = -\frac{V_o \varepsilon}{\phi_o} L \ln(b/a)$$

$$C = \frac{|Q|}{V_o} = \frac{\varepsilon L}{\phi_o} \ln\frac{b}{a}$$

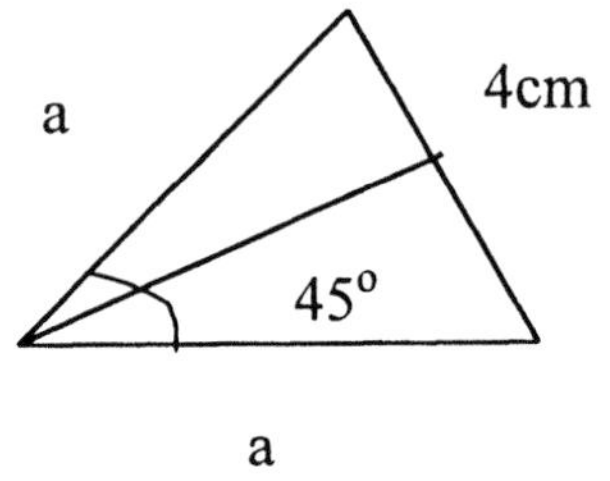

$$a \sin\frac{45^o}{2} = 2 \quad \longrightarrow \quad a = \frac{2}{\sin 22.5^o} = 5.226 \text{ mm}$$

$$C = \frac{1.5x10^{-9}}{36\pi x \frac{\pi}{4}} 5 \ln\frac{1000}{5.226} = 444 \text{ pF}$$

$$Q = CV_o = 444x10^{-12} \; x50 = \underline{\underline{22.2 \text{ nC}}}$$

P. E. 6.4 From Example 6.4,

$$V_o = 50, \quad \theta_1 = \pi/2, \quad \theta_2 = 45^o, \quad r = \sqrt{3^2 + 4^2 + 2^2} = \sqrt{29}, \quad \theta = \tan^{-1}\frac{\rho}{z} = \frac{5}{2} \quad \longrightarrow \quad \theta = 68.2^o$$

$$V = \frac{50 \ln(\tan 34.1^o)}{\tan(22.5^o)} = \underline{\underline{22.13 \text{ V}}}, \qquad E = \frac{-50 a_\theta}{\sqrt{29} \sin 68.2^o \ln\tan(22.5^o)} = \underline{\underline{11.36 a_\theta \text{ V/m}}}$$

P. E. 6.5

$$E = -\nabla V = -\frac{\partial V}{\partial x} a_x - \frac{\partial V}{\partial y} a_y$$

$$= -\frac{4V_o}{b} \sum_{n=odd}^{\infty} \frac{1}{\sinh n\pi a/b} \left[\cos(n\pi x/b) \sinh(n\pi y/b)\, a_x + \sin(n\pi x/b) \cosh(n\pi y/b) a_y \right]$$

(a) At (x,y) = (a, a/2),

$$V = \frac{400}{\pi}(0.3775 - 0.0313 + 0.00394 - 0.000584 + \ldots) = \underline{\underline{44.51 \text{ V}}}$$

$$E = 0a_x + (-115.12 + 19.127 - 3.9431 + 0.8192 - 0.1703 + 0.035 - 0.0094 + \ldots)a_y$$
$$= \underline{\underline{-99.25a_y \text{ V/m}}}$$

(b) At (x,y) = (3a/2, a/4),

$$V = \frac{400}{\pi}(0.1238 + 0.00626 - 0.00383 + 0.000264 + \ldots) = \underline{\underline{16.50 \text{ V}}}$$

$$E = (24.757 - 3.7358 - 0.3834 - 0.0369 + 0.00351 - 0.00033 + \ldots)a_x$$
$$+(-66.25 - 4.518 + 0.3988 + 0.03722 - 0.00352 - 0.000333 + \ldots)a_y$$
$$= \underline{\underline{20.68a_x - 70.34a_y \text{ V/m}}}$$

P. E. 6.6

$$V(y=a) = V_o \sin(7\pi x/b) = \sum_{n=1}^{\infty} c_n \sin(n\pi x/b)\sinh(n\pi a/b)$$

By equating coefficients, we notice that $c_n = 0$ for $n \neq 7$. For n=7,

$$V_o \sin(7\pi x/b) = c_7 \sin(7\pi x/b)\sinh(7\pi a/b) \longrightarrow c_7 = \frac{V_o}{\sinh(7\pi a/b)}$$

Hence

$$\underline{\underline{V(x,y) = \frac{V_o}{\sinh(7\pi a/b)}\sin(7\pi x/b)\sinh(7\pi y/b)}}$$

P. E. 6.7 Let $V(r,\theta,\phi) = R(r)F(\theta)\Phi(\phi)$.
Substituting this in Laplace's equation gives

$$\frac{\Phi F}{r^2}\frac{d}{dr}\left(r^2\frac{dR}{dr}\right) + \frac{R\Phi}{r^2\sin\theta}\frac{d}{d\theta}\left(\sin\theta\frac{dF}{d\theta}\right) + \frac{RF}{r^2\sin^2\theta}\frac{d^2\Phi}{d\phi^2} = 0$$

Dividing by $RF\Phi / r^2\sin^2\theta$ gives

$$\frac{\sin^2\theta}{R}\frac{d}{dr}\left(r^2R'\right) + \frac{\sin\theta}{F}\frac{d}{d\theta}(\sin\theta F') = -\frac{1}{\Phi}\frac{d^2\Phi}{d\phi^2} = \lambda^2$$

$$\underline{\underline{\Phi'' + \lambda^2\Phi = 0}}$$

$$\frac{1}{R}\frac{d}{dr}\left(r^2 R'\right) = \frac{\lambda^2}{\sin^2\theta} - \frac{1}{F\sin\theta}\frac{d}{d\theta}(\sin\theta F') = \mu^2$$

$$2R' + r^2 R'' = \mu^2 R$$

or

$$\underline{\underline{R'' + \frac{2}{r}R' - \frac{\mu^2}{r^2}R = 0}}$$

$$\frac{\sin\theta}{F}\frac{d}{d\theta}(\sin\theta F') - \lambda^2 + \mu^2\sin^2\theta = 0$$

or

$$\underline{\underline{F'' + \cot\theta F' + (\mu^2 - \lambda^2\,\mathrm{cosec}^2\theta)F = 0}}$$

P. E. 6.8 (a) This is similar to Example 6.8(a) except that here $0 < \phi < 2\pi$ instead of $0 < \phi < \pi/2$. Hence

$$I = \frac{2\pi t V_o \sigma}{\ln(b/a)} \qquad \text{and} \qquad R = \frac{V_o}{I} = \underline{\underline{\frac{\ln\frac{b}{a}}{2\pi t\sigma}}}$$

(b) This to similar to Example 6.8(b) except that here $0 < \phi < 2\pi$. Hence

$$I = \frac{V_o\sigma}{t}\int_a^b\int_0^{2\pi}\rho\, d\rho\, d\phi = \frac{V_o\sigma\pi(b^2 - a^2)}{t}$$

and $R = \dfrac{V_o}{I} = \underline{\underline{\dfrac{t}{\sigma\pi(b^2 - a^2)}}}$

P. E. 6.9 From Example 6.9,

$$J_1 = \frac{\sigma_1 V_o}{\rho\ln\frac{b}{a}}, \qquad J_2 = \frac{\sigma_2 V_o}{\rho\ln\frac{b}{a}}$$

$$I = \int J\bullet dS = \int_{z=0}^{L}\left[\int_{\phi=0}^{\pi} J_1\rho\, d\phi + \int_{\phi=\pi}^{2\pi} J_2\rho\, d\phi\right]dz = \frac{V_o l}{\ln\frac{b}{a}}\left[\pi\sigma_1 + \pi\sigma_2\right]$$

$$R = \frac{V_o}{I} = \frac{\ln\frac{b}{a}}{\pi l[\sigma_1 + \sigma_2]}$$

P. E. 6.10 (a) $C = \frac{4\pi\varepsilon}{\frac{1}{a} - \frac{1}{b}}$, C_1 and C_2 are in series.

$$C_1 = 4\pi x \frac{10^{-9}}{36\pi}\left(\frac{2.5}{\frac{10^3}{2} - \frac{10^3}{3}}\right) = 5/3 \text{ pF}, \quad C_1 = 4\pi x \frac{10^{-9}}{36\pi}\left(\frac{3.5}{\frac{10^3}{1} - \frac{10^3}{2}}\right) = 7/9 \text{ pF}$$

$$C = \frac{C_1 C_2}{C_1 + C_2} = \frac{(5/3)(7/9)}{(5/3) + (7/9)} = \underline{\underline{0.53 \text{ pF}}}$$

(b) $C = \frac{2\pi\varepsilon}{\frac{1}{a} - \frac{1}{b}}$, C_1 and C_2 are in parallel.

$$C_1 = 2\pi x \frac{10^{-9}}{36\pi}\left(\frac{2.5}{\frac{10^3}{1} - \frac{10^3}{3}}\right) = 5/24 \text{ pF}, \quad C_2 = 2\pi x \frac{10^{-9}}{36\pi}\left(\frac{3.5}{\frac{10^3}{1} - \frac{10^3}{3}}\right) = 7/24 \text{ pF}$$

$$C = C_1 + C_2 = \underline{\underline{0.5 \text{ pF}}}$$

P. E. 6.11 As in Example 6.8, assuming $V(\rho = a) = 0$, $V(\rho = b) = V_o$,

$$V = V_o \frac{\ln\rho/a}{\ln b/a}, \quad E = -\nabla V = -\frac{V_o}{\rho \ln b/a} a_\rho$$

$$Q = \int \varepsilon E \bullet dS = \frac{V_o \varepsilon}{\ln b/a} \int_{z=0}^{L} \int_{\phi=0}^{2\pi} \frac{1}{\rho} dz \rho d\phi = \frac{V_o 2\pi\varepsilon L}{\ln b/a}$$

$$C = \frac{Q}{V_o} = \underline{\underline{\frac{2\pi\varepsilon L}{\ln b/a}}}$$

P. E. 6.12 (a) C_1 and C_2 are in series.

$$C_1 = \frac{2\pi\varepsilon_{r1}\varepsilon_o}{\ln c/a} = \frac{2\pi x 2.5}{\ln 2/1} \frac{10^{-9}}{36\pi} = 200 \text{ pF/m}, \quad C_2 = \frac{2\pi\varepsilon_{r2}\varepsilon_o}{\ln b/c} = \frac{2\pi x 3.5}{\ln 3/2} \frac{10^{-9}}{36\pi} = 480 \text{ pF/m}$$

$$C = \frac{C_1 C_2}{C_1 + C_2} = \frac{200x480}{200 + 480} = 141.12 \text{ pF}$$

$$C_T = Cl = \underline{\underline{1.41}} \text{ nF}$$

(b) C_1 and C_2 are in parallel.

$$C = C_1 + C_2 = \frac{\pi\varepsilon_{r1}\varepsilon_o}{\ln b/a} + \frac{\pi\varepsilon_{r2}\varepsilon_o}{\ln b/a} = \frac{\pi(\varepsilon_{r1} + \varepsilon_{r2})\varepsilon_o}{\ln b/a} = \frac{6\pi}{\ln 3/7}\frac{10^{-9}}{36\pi} = 151.7 \text{ pF/m}$$

$$C_T = Cl = \underline{\underline{1.52}} \text{ nF}$$

P. E. 6.13 Instead of Eq. (6.31), we now have

$$V = -\int_b^a \frac{Qdr}{4\pi\varepsilon r^2} = -\int_b^a \frac{Qdr}{4\pi \frac{10\varepsilon_o}{r} r^2} = -\frac{Q}{40\pi\varepsilon_o}\ln b/a$$

$$C = \frac{Q}{|V|} = \frac{40\pi}{\ln 4/1.5}\frac{10^{-9}}{36\pi} = \underline{\underline{1.13 \text{ nF}}}$$

P. E. 6.14 Let

$$F = F_1 + F_2 + F_3 + F_4 + F_5$$

where F_i, $i = 1,2,\ldots,5$ are shown on in the figure below.

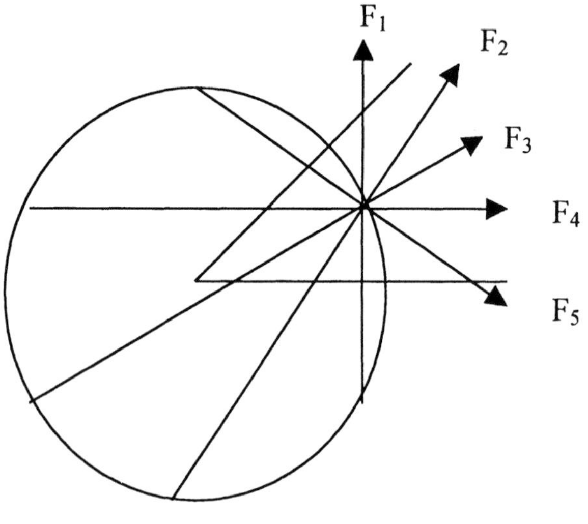

$$F = -\frac{Q^2}{4\pi\varepsilon_o r^2} + \frac{Q^2(a_x \sin 30^o + a_y \cos 30^o)}{4\pi\varepsilon_o (r\cos 30^o)^2} - \frac{Q^2(a_x \cos 30^o + a_y \sin 30^o)}{4\pi\varepsilon_o (2r)^2} + \frac{Q^2 a_x}{4\pi\varepsilon_o (r\cos 30^o)^2}$$
$$- \frac{Q^2(a_x \cos 30^o - a_y \sin 30^o)}{4\pi\varepsilon_o r^2}$$

$$= -\frac{Q^2}{4\pi\varepsilon_o r^2}\left[-a_y + \frac{4}{3}\left(\frac{a_x}{2} + \frac{\sqrt{3}a_y}{2}\right) - \frac{1}{4}\left(\frac{\sqrt{3}a_x}{2} + \frac{a_y}{2}\right) + \frac{4}{3}a_x - \frac{\sqrt{3}a_x}{2} + \frac{a_y}{2}\right]$$

$$= 9x10^{-5}\left[a_x\left(2 - \frac{5\sqrt{3}}{8}\right) + a_y\left(\frac{4\sqrt{3} - 5}{8}\right)\right] = 82.57a_x + 21.69a_y \ \mu\text{N}$$

$$|F| = 85.37 \ \mu\text{N}$$

Prob. 6.1

$$E = -\nabla V = -\frac{\partial V}{\partial x}a_x - \frac{\partial V}{\partial y}a_y - \frac{\partial V}{\partial z}a_z = -6y^2za_x - 12xyza_y - 6xy^2a_z$$

At P(1,2,-5),

$$E = 120a_x + 120a_y - 12a_z \ \text{V/m}$$

$$\nabla^2 V = \frac{\partial^2 V}{\partial x^2} + \frac{\partial^2 V}{\partial y^2} + \frac{\partial^2 V}{\partial z^2} = 0 + 12xz + 0$$

$$\rho_v = -\varepsilon_o \nabla^2 V = -12xz\varepsilon_o$$

At P,

$$\rho_v = 60\varepsilon_o = 60x\frac{10^{-9}}{36\pi} = 530.5 \ \text{pC/m}^3$$

Prob. 6.2

$$\frac{d^2V}{dx^2} = -\frac{\rho_v}{\varepsilon_o} = -\frac{\frac{x}{6\pi}10^{-9}}{10^{-9}/36\pi} = -6x$$

$$\frac{dV}{dx} = -3x^2 + A \longrightarrow V = -x^3 + Ax + B$$

$$-50 = -1 + A + B \longrightarrow A + B = -49$$

$$50 = -64 + 4A + B \longrightarrow 4A + B = 114$$

Thus, A = 54.33 and B = -103.33

$$V = -x^3 + 54.33x - 103.3$$

V(2) = -8 + 108.66 – 103.3 = -2.667

Prob. 6.3 (a)

$$\nabla^2 V = -\frac{\rho_v}{\varepsilon_o} \longrightarrow \frac{d^2V}{dx^2} = -\frac{\rho_o(x-d)}{d\varepsilon_o} = -kx + kd, \qquad k = \frac{\rho_o}{d\varepsilon_o}$$

$$\frac{dV}{dx} = -kx^2/2 + kdx + A \longrightarrow V = -kx^3/6 + kdx^2/2 + Ax + B$$

When x=0, V = 0 ⟶ 0 = B

When x=d, $V = V_o$, ⟶ $V_o = -kd^3/6 + kd^3/2 + Ad$

i.e. $A = V_o/d - kd^2/3$

$$V = -\frac{\rho_o x^3}{6d\varepsilon_o} + \frac{\rho_o x^2}{2\varepsilon_o} + \left(\frac{V_o}{d} - \frac{\rho_o d}{3\varepsilon_o}\right)x$$

$$E = -\nabla V = -\frac{dV}{dx}a_x = \left(\frac{\rho_o x^2}{2d\varepsilon_o} - \frac{\rho_o x}{\varepsilon_o} - \frac{V_o}{d} + \frac{\rho_o d}{3\varepsilon_o}\right)a_x$$

(b) $\rho_s = D_n = \varepsilon_o E_n = \varepsilon_o E \bullet a_n$

At x=0, $a_n = a_x$, $\rho_s = \frac{\rho_o d}{3} - \frac{\varepsilon_o V_o}{d}$

At x=d, $a_n = -a_x$, $\rho_s = -\rho_o d/2 + \rho_o d + \varepsilon_o V_o/d - \rho_o d/3$

$$\rho_s = \frac{\varepsilon_o V_o}{d} + \frac{\rho_o d}{6}$$

Prob. 6.4 If V'' = f,

$$V' = \int_0^x f(x)dx + c_1$$

$$V = \int_0^x \int_0^\lambda f(\mu)d\mu d\lambda + c_1 x + c_2$$

$$V(x=0) = V_1 = c_2 \longrightarrow c_2 = V_1$$

$$V(x=L)=V_2=\int_0^L\int_0^\lambda f(\mu)d\mu d\lambda + c_1 L + c_2$$

$$c_1=\frac{1}{L}\left[V_2 - V_1 - \int_0^L\int_0^\lambda f(\mu)d\mu\, d\lambda\right]$$

Thus,

$$V=\frac{x}{L}\left[V_2 - V_1 - \int_0^L\int_0^\lambda f(\mu)d\mu d\lambda\right]+V_1+\int_0^x\int_0^\lambda f(\mu)d\mu d\lambda$$

Prob. 6.5

$$\nabla^2 V=\frac{d^2V}{dx^2}=-\frac{\rho_v}{\varepsilon}=-\frac{50(1-y^2).10^6}{\varepsilon}=-k(1-y^2)$$

where $k=\dfrac{50x10^{-6}}{3x\dfrac{10^{-9}}{36\pi}}=600\pi x10^3$

$$\frac{dV}{dy}=-k(y-y^3/3)+A$$

$$V=-k\left(\frac{y^2}{2}-\frac{y^4}{12}\right)+Ay+B=50\pi.10^3y^4-300\pi.10^3y^2+Ay+B$$

When y=2cm, $V=30x10^3$,

$$30x10^3=50\pi x10^3x16x10^{-6}-300\pi x10^3x4x10^{-4}+Ay+B$$

or

$$30{,}374.5=0.02A+B \qquad (1)$$

When y=-2cm, $V=30x10^3$,

$$30{,}374.5=-0.02A+B \qquad (2)$$

From (1) and (2), A=0, B=30,374.5. Thus,

$$\underline{\underline{V=157.08y^4-942.5y^2+30.374\text{ kV}}}$$

Prob. 6.6 (a)

$$\nabla^2 V_1=\frac{\partial^2 V_1}{\partial x^2}+\frac{\partial^2 V_1}{\partial y^2}+\frac{\partial^2 V_1}{\partial z^2}=2+2-4=0$$

i.e Yes.

(b) $V_2 = \dfrac{1}{(x^2+y^2+z^2)^{1/2}} = 1/r = r^{-1}$

$$\nabla^2 V_2 = \frac{1}{r^2}\frac{\partial}{\partial r}\left(r^2 \frac{\partial V_2}{\partial r}\right) + 0 = \frac{1}{r^2}\frac{\partial}{\partial r}\left(r^2 \frac{-1}{r^2}\right) = 0$$

i.e Yes.

(c) $\nabla^2 V_3 = \dfrac{1}{\rho}\dfrac{\partial}{\partial \rho}\left(\rho \dfrac{\partial^2 V_3}{\partial \phi^2}\right) + \dfrac{1}{\rho^2}\dfrac{\partial^2 V_3}{\partial \phi^2} + \dfrac{\partial^2 V_3}{\partial z^2}$

$$= \frac{1}{\rho}\frac{\partial}{\partial \rho}(\rho z \sin\phi) - \frac{z}{\rho}\sin\phi + 0 = \frac{z}{\rho}\sin\phi + 4 - \frac{z}{\rho}\sin\phi = 4$$

i.e. No.

(d) $\nabla^2 V_4 = 0 + \dfrac{10\sin\phi}{r^4 \sin\theta}\dfrac{\partial}{\partial \theta}(\sin\theta\cos\theta) - \dfrac{10\sin\theta\sin\phi}{r^4}$

$$= \frac{10\sin\phi(\cos^2\theta - \sin^2\theta)}{r^4\sin\theta} - \frac{10\sin\theta\sin\phi}{r^4} = \frac{10\sin\phi}{r^4\sin\theta} - \frac{30\sin\theta\sin\phi}{r^4} \neq 0$$

i.e. No.

Prob. 6.7 (a)

$$\nabla^2 V = \frac{\partial^2 V}{\partial x^2} + \frac{\partial^2 V}{\partial y^2} + \frac{\partial^2 V}{\partial z^2} = \frac{\partial}{\partial x}\left(-5e^{-5x}\cos 13y \sinh 12z\right) + \ldots = 25V - 169V + 144V = 0$$

(b) $\nabla^2 V = \dfrac{1}{\rho}\dfrac{\partial}{\partial \rho}\left(-\dfrac{z\cos\phi}{\rho}\right) - \dfrac{z\cos\phi}{\rho} + 0 = \dfrac{z\cos\phi}{\rho^3} - \dfrac{z\cos\phi}{\rho^3} = 0$

(c) $V = 30r^{-2}\cos\theta,$

$$\nabla^2 V = \frac{1}{r^2}\frac{\partial}{\partial r}(-60r^{-1}\cos\theta) + \frac{1}{r^2\sin\theta}\frac{\partial}{\partial \theta}(-\sin\theta 30r^{-2}\sin\theta) = \frac{60}{r^2}\cos\theta - \frac{30}{r^4\sin\theta}(2\sin\theta\cos\theta) = 0$$

Prob. 6.8 If

$$\nabla^2 V = \frac{\partial^2 V}{\partial x^2} + \frac{\partial^2 V}{\partial y^2} + \frac{\partial^2 V}{\partial z^2} = 0$$

then

$$0 = -\frac{\partial}{\partial x}\left(\frac{\partial^2 V}{\partial x^2} + \frac{\partial^2 V}{\partial y^2} + \frac{\partial^2 V}{\partial z^2}\right) = \frac{\partial^2}{\partial x^2}\left(-\frac{\partial V}{\partial x}\right) + \frac{\partial^2}{\partial y^2}\left(-\frac{\partial V}{\partial x}\right) + \frac{\partial^2}{\partial z^2}\left(-\frac{\partial V}{\partial x}\right)$$

or

$$0 = \frac{\partial^2}{\partial x^2} E_x + \frac{\partial^2}{\partial y^2} E_x + \frac{\partial^2}{\partial z^2} E_x = \nabla^2 E_x$$

i.e. $\nabla^2 E_x = 0.$

The same holds for E_y and E_z.

Prob. 6.9

$$\frac{\partial V}{\partial x} = (-An\sin nx + Bn\cos nx)(Ce^{ny} + De^{-ny})$$

$$\frac{\partial^2 V}{\partial x^2} = (-An^2 \cos nx - n^2 B\sin nx)(Ce^{ny} + De^{-ny}) = -n^2 V$$

$$\frac{\partial V}{\partial y} = (A\cos nx + B\sin nx)(nCe^{ny} - nDe^{-ny})$$

$$\frac{\partial^2 V}{\partial y^2} = n^2 V$$

Thus

$$\nabla^2 V = \frac{\partial^2 V}{\partial x^2} + \frac{\partial^2 V}{\partial y^2} = -n^2 V + n^2 V = \underline{\underline{0}}$$

Prob. 6.10 (a)

$$\frac{\partial V}{\partial x} = 4xyz, \quad \frac{\partial^2 V}{\partial x^2} = 4yz$$

$$\frac{\partial V}{\partial y} = 2x^2 z - 3y^2 z, \quad \frac{\partial^2 V}{\partial y^2} = -6yz$$

$$\frac{\partial V}{\partial z} = 2x^2 y - y^3, \quad \frac{\partial^2 V}{\partial z^2} = 0$$

$$\nabla^2 V = 4yz - 6yz + 0 = -2yz$$

$\nabla^2 V \neq 0$, V <u>does not</u> satisfy Laplace's equation.

(b)

$$\nabla^2 V = -\frac{\rho_v}{\varepsilon} = -2yz \longrightarrow \rho_v = 2yz\varepsilon$$

$$Q = \int \rho_v dv = \int_0^1\int_0^1\int_0^1 (2yz\varepsilon)dxdydz = 2\varepsilon(1)\frac{y^2}{2}\Big|_0^1 \frac{z^2}{2}\Big|_0^1 = \varepsilon/2 = 2\varepsilon_o/2 = \varepsilon_o$$

Q = 8.854 pC

Prob. 6.11

$$\nabla^2 V = \frac{d^2V}{dz^2} = 0 \quad \longrightarrow \quad V = Az + B$$

When z=0, V = 0 ⟶ B=0

When z=d, V = V_o ⟶ V_o=Ad or A = V_o/d

Hence,

$$V = \frac{V_o z}{d}$$

$$E = -\nabla V = -\frac{dV}{dz}a_z = -\frac{V_o}{d}a_z$$

$$D = \varepsilon E = -\varepsilon_o\varepsilon_r\frac{V_o}{d}a_z$$

Since V_o = 50 V and d = 2mm,

V = 25z kV, E = - 25a_z kV/m

$$D = -\frac{10^{-9}}{36\pi}(1.5)25x10^3 a_z = -332a_z \text{ nC/m}^2$$

$$\rho_s = D_n = \pm 332 \text{ nC/m}^2$$

The surface charge density is positive on the plate at z=d and negative on the plate at z=0.

Prob. 6.12 From Example 6.8, solving $\nabla^2 V = 0$ when $V = V(\rho)$ leads to

$$V = \frac{V_o \ln\rho/a}{\ln b/a}$$

$$E = -\nabla V = -\frac{V_o}{\rho\ln b/a}a_\rho, \qquad D = \varepsilon E = -\frac{\varepsilon_o\varepsilon_r V_o}{\rho\ln b/a}a_\rho$$

$$\rho_s = D_n = \pm\frac{\varepsilon_o\varepsilon_r V_o}{\rho\ln b/a}\Big|_{\rho=a,b}$$

In this case, V_o=100 V, b=5mm, a=15mm, $\varepsilon_r = 2$. Hence at ρ = 10mm,

$$V = \frac{100 \ln 10/5}{\ln 15/5} = \underline{\underline{36.91 \text{ V}}}$$

$$E = -\frac{100}{10x10^{-3} \ln 3} a_\rho = \underline{\underline{-9.102 a_\rho}}$$

$$D = -9.102x10^3 x \frac{10^{-9}}{36\pi} 2a_\rho = \underline{\underline{-161 a_\rho \text{ nC/m}^2}}$$

$$\rho_s(\rho = 5\text{mm}) = \frac{10^{-9}}{36\pi}(2)\frac{10^5}{5 \ln 3} = \underline{\underline{322 \text{ nC/m}^2}}$$

$$\rho_s(\rho = 15\text{mm}) = -\frac{10^{-9}}{36\pi}(2)\frac{10^5}{15 \ln 3} = \underline{\underline{-107.3 \text{ nC/m}^2}}$$

Prob. 6.13

$$\nabla^2 V = \frac{1}{\rho}\frac{\partial}{\partial \rho}\left(\rho \frac{\partial V}{\partial \rho}\right) = 0 \quad \longrightarrow \quad V = A \ln \rho + B$$

Let ρ be in cm.

$$V(\rho = 2) = 60 \quad \longrightarrow \quad 60 = A \ln 2 + B$$

$$V(\rho = 6) = -20 \quad \longrightarrow \quad -20 = A \ln 6 + B$$

Thus, A = -72.82, B = 110.47, and

$$V = 110.47 - 72.82 \ln \rho$$

$$E = -\frac{dV}{d\rho} a_\rho = -\frac{A}{\rho} a_\rho = \frac{72.82}{\rho} a_\rho, \qquad D = \varepsilon_o E$$

At ρ=4, $\underline{\underline{V = 9.52 \text{ V}}}$, $\underline{\underline{E = 18.21\ a_\rho \text{ V/m}}}$

$$D = \varepsilon_o E = \frac{10^{-9}}{36\pi} x 18.21 a_\rho = 0.161 a_\rho \text{ nC/m}^2$$

Prob. 6.14

$$\nabla^2 V = 0 \quad \longrightarrow \quad V = -A/r + B$$

At r=0.5, V=-50 $\longrightarrow$ -50 = -A/0.5 + B

Or

$$-50 = -2A + B \qquad (1)$$

At r = 1, V = 50 ⟶ 50 = -A + B (2)

From (1) and (2), A = 100, B = 150, and

$$V = -\frac{100}{r} + 150$$

$$E = -\nabla V = -\frac{A}{r^2}a_r = -\frac{100}{r^2}a_r \text{ V/m}$$

Prob. 6.15 From Example 6.4,

$$V = \frac{V_o \ln\left(\dfrac{\tan\theta/2}{\tan\theta_1/2}\right)}{\ln\left(\dfrac{\tan\theta_2/2}{\tan\theta_1/2}\right)}$$

$V_o = 100, \quad \theta_1 = 30^o, \quad \theta_2 = 120^o, \quad r = \sqrt{3^2 + 0^2 + 4^2} = 5, \quad \theta = \tan^{-1}\rho/z = \tan^{-1} 3/4 = 36.87^o$

$$V = 100\frac{\ln\left(\dfrac{\tan 18.435^o}{\tan 15^o}\right)}{\ln\left(\dfrac{\tan 60^o}{\tan 15^o}\right)} = 11.7 \text{ V}$$

$$E = \frac{-V_o a_\theta}{r\sin\theta \ln\left(\dfrac{\tan\theta_2/2}{\tan\theta_1/2}\right)} = \frac{-100a_\theta}{5\sin 36.87^o \ln 6.464} = -17.86a_\theta \text{ V/m}$$

Prob. 6.16 (a)

$$\nabla^2 V = \frac{1}{\rho}\frac{\partial}{\partial\rho}\left(\rho\frac{\partial V}{\partial\rho}\right) = 0 \quad \longrightarrow \quad V = A\ln\rho + B$$

$$V(\rho = b) = 0 \quad \longrightarrow \quad 0 = A\ln b + B \quad \longrightarrow \quad B = -A\ln b$$

$$V(\rho = b) = V_o \quad \longrightarrow \quad V_o = A\ln a/b \quad \longrightarrow \quad A = -\frac{V_o}{\ln b/a}$$

$$V = -\frac{V_o}{\ln b/a}\ln\rho/b = \frac{V_o \ln b/\rho}{\ln b/a}$$

$$V(\rho = 15\text{mm}) = 70\frac{\ln 2}{\ln 50} = 12.4 \text{ V}$$

(b) As the electron decelerates, potential energy gained = K.E. loss

$$e[70-12.4]=\frac{1}{2}m[(10^7)^2-u^2] \longrightarrow 10^{14}-u^2=\frac{2e}{m}x57.6$$

$$u^2=10^{14}-\frac{2x1.6x10^{-19}}{9.1x10^{-31}}x57.6=10^{12}(100-20.25)$$

$$\underline{u=8.93x10^6 \text{ m/s}}$$

Prob. 6.17 (a) For the parallel-plate capacitor,

$$E=-\frac{V_o}{d}a_x$$

From Example 6.11,

$$C=\frac{1}{V_o^2}\int\varepsilon|E|^2dv=\frac{1}{V_o^2}\int\varepsilon\frac{V_o^2}{d^2}dv=\frac{\varepsilon}{d^2}Sd=\frac{\varepsilon S}{d}$$

(b) For the cylindrical capacitor,

$$E=-\frac{V_o}{\rho\ln b/a}a_\rho$$

From Example 6.8,

$$C=\frac{1}{V_o^2}\iiint\frac{\varepsilon V_o^2}{(\rho\ln b/a)^2}\rho d\rho d\phi dz=\frac{2\pi\varepsilon L}{(\ln b/a)^2}\int_a^b\frac{d\rho}{\rho}=\frac{2\pi\varepsilon L}{\ln b/a}$$

(c)For the spherical capacitor,

$$E=\frac{V_o}{r^2(1/a-1/b)}a_r$$

From Example 6.10,

$$C=\frac{1}{V_o^2}\iiint\frac{\varepsilon V_o^2}{r^4(1/a-1/b)^2}r^2\sin\theta d\theta dr d\phi=\frac{\varepsilon}{(1/a-1/b)^2}4\pi\int_a^b\frac{dr}{r^2}=\frac{4\pi\varepsilon}{\frac{1}{a}-\frac{1}{b}}$$

Prob. 6.18 This is similar to case 1 of Example 6.5.

$$X=c_1x+c_2,\qquad Y=c_3y+c_4$$

But $X(0)=0 \longrightarrow 0=c_2$, $Y(0)=0 \longrightarrow 0=c_4$

Hence,

$$V(x,y) = XY = a_o xy, \qquad a_o = c_1 c_3$$

Also, $V(xy = 4) = 20 \longrightarrow 20 = 4a_o \longrightarrow a_o = 5$

Thus,

$$V(x,y) = 5xy \text{ and } E = -\nabla V = -5ya_x - 5xa_y$$

At (x,y) = (1,2),

$$V = 10 \text{ V}, \quad E = -10a_x - 5a_y \text{ V/m}$$

Prob. 6.19 (a) As in Example 6.5, $X(x) = A\sin(n\pi x/b)$

For Y,

$$Y(y) = c_1 \cosh(n\pi y/b) + c_2 \sinh(n\pi y/b)$$

$$Y(a) = 0 \longrightarrow 0 = c_1 \cosh(n\pi a/b) + c_2 \sinh(n\pi a/b) \longrightarrow c_1 = -c_2 \tanh(n\pi a/b)$$

$$V = \sum_{n=1}^{\infty} a_n \sin(n\pi x/b)\left[\sinh(n\pi y/b) - \tanh(n\pi a/b)\cosh(n\pi y/b)\right]$$

$$V(x, y=0) = V_o = -\sum_{n=1}^{\infty} a_n \tanh(n\pi a/b)\sinh(n\pi x/b)$$

$$-a_n \tanh(n\pi a/b) = \frac{2}{b}\int_a^b V_o \sin(n\pi y/b)dy = \begin{cases} \dfrac{4V_o}{n\pi}, & n = \text{odd} \\ 0, & n = \text{even} \end{cases}$$

Hence,

$$V = -\frac{4V_o}{\pi}\sum_{n=\text{odd}}^{\infty} \sin(n\pi x/b)\left[\frac{\sin(n\pi y/b)}{n\tanh(n\pi a/b)} - \frac{\cosh(n\pi y/b)}{n}\right]$$

$$= -\frac{4V_o}{\pi}\sum_{n=\text{odd}}^{\infty} \frac{\sin(n\pi x/b)}{n\sinh(n\pi a/b)}\left[\sin(n\pi y/b)\cosh(n\pi a/b) - \cosh(n\pi y/b)\sinh(n\pi a/b)\right]$$

$$= -\frac{4V_o}{\pi}\sum_{n=\text{odd}}^{\infty} \frac{\sin(n\pi x/b)\sinh[n\pi(a-y)/b]}{n\sinh(n\pi a/b)}$$

Alternatively, for Y

$$Y(y) = c_1 \sinh n\pi(y - c_2)/b$$

$Y(a) = 0 \longrightarrow 0 = c_1 \sinh[n\pi(a - c_2)/b] \longrightarrow c_2 = a$

$$V = \sum_{n=1}^{\infty} b_n \sin(n\pi x/b) \sinh[n\pi(y-a)/b]$$

where

$$b_n = \begin{cases} -\dfrac{4V_o}{n\pi \sinh(n\pi a/b)}, & n = \text{odd} \\ 0, & n = \text{even} \end{cases}$$

(b) This is the same as Example 6.5 except that we exchange y and x. Hence

$$V(x,y) = \frac{4V_o}{\pi} \sum_{n=\text{odd}}^{\infty} \frac{\sin(n\pi y/a) \sinh n\pi x/a]}{n \sinh(n\pi b/a)}$$

(c) This is the same as part (a) except that we must exchange x and y. Hence

$$V(x,y) = \frac{4V_o}{\pi} \sum_{n=\text{odd}}^{\infty} \frac{\sin(n\pi y/b) \sinh[n\pi(a-x)/b]}{n \sinh(n\pi a/b)}$$

Prob. 6.20 (a) X(x) is the same as in Example 6.5. Hence

$$V(x,y) = \sum_{n=1}^{\infty} \sin(n\pi x/b)\left[a_n \sinh(n\pi y/b) + b_n \cosh(n\pi y/b)\right]$$

At y=0, $V = V_1$

$$V_1 = \sum_{n=1}^{\infty} b_n \sin(n\pi x/b) \longrightarrow b_n = \begin{cases} \dfrac{4V_1}{n\pi}, & n = \text{odd} \\ 0, & \text{n = even} \end{cases}$$

At y=a, $V = V_2$

$$V_2 = \sum_{n=1}^{\infty} \sin(n\pi x/b)\left[a_n \sinh(n\pi a/b) + b_n \cosh(n\pi a/b)\right]$$

$$a_n \sinh(n\pi a/b) + b_n \cosh(n\pi a/b) = \begin{cases} \dfrac{4V_2}{n\pi}, & n = \text{odd} \\ 0, & \text{n = even} \end{cases}$$

or

$$a_n = \begin{cases} \dfrac{4V_2}{n\pi \sinh(n\pi a/b)}\left(V_2 - V_1 \cosh(n\pi a/b)\right), & n = \text{odd} \\ 0, & \text{n = even} \end{cases}$$

Alternatively, we may apply superposition principle.

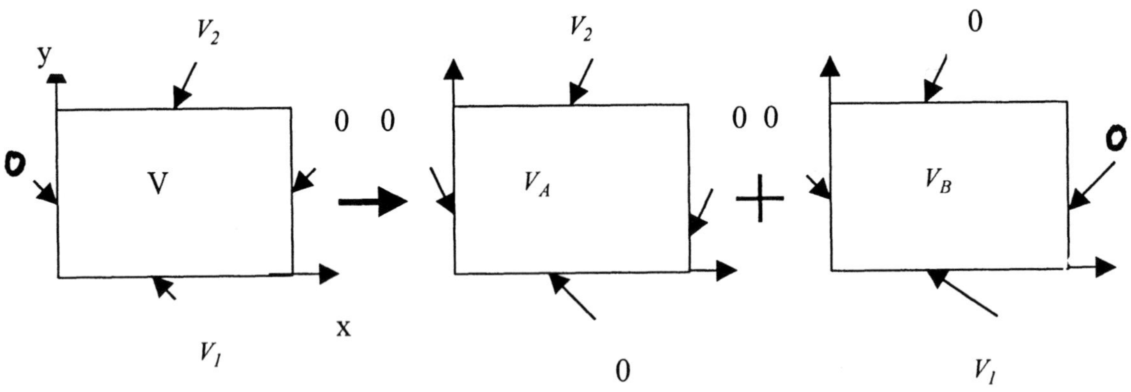

i.e. $V = V_A + V_B$

V_A is exactly the same as Example 6.5 with $V_o = V_2$, while V_B is exactly the same as Prob. 6.19(a). Hence

$$V = \frac{4}{\pi}\sum_{n=\text{odd}}^{\infty} \frac{\sin(n\pi x/b)}{n\sinh(n\pi a/b)}\left[V_1 \sinh[n\pi(a-y)/b] + V_2 \sinh(n\pi y/b)\right]$$

(b)

$$V(x,y) = (a_1 e^{-\alpha x} + a_2 e^{+\alpha x})(a_3 \sin\alpha y + a_4 \cos\alpha y)$$

$$\lim_{x\to\infty} V(x,y) = 0 \longrightarrow a_2 = 0$$

$$V(x, y=0) = 0 \longrightarrow a_4 = 0$$

$$V(x, y=a) = 0 \longrightarrow \alpha = n\pi/a, \quad n = 1,2,3,\ldots$$

Hence,

$$V(x,y) = \sum_{n=1}^{\infty} a_n e^{-n\pi x/a} \sin(n\pi y/a)$$

$$V(x=0,y) = V_o = \sum_{n=1}^{\infty} a_n \sin(n\pi y/a) \longrightarrow a_n = \begin{cases} \frac{4V_o}{n\pi}, & n = \text{odd} \\ 0, & n = \text{even} \end{cases}$$

$$V(x,y) = \frac{4V_o}{\pi}\sum_{n=\text{odd}}^{\infty} \frac{\sin(n\pi y/a)}{n}\exp(-n\pi x/a)$$

(d) The problem is easily solved using superposition theorem, as illustrated below.

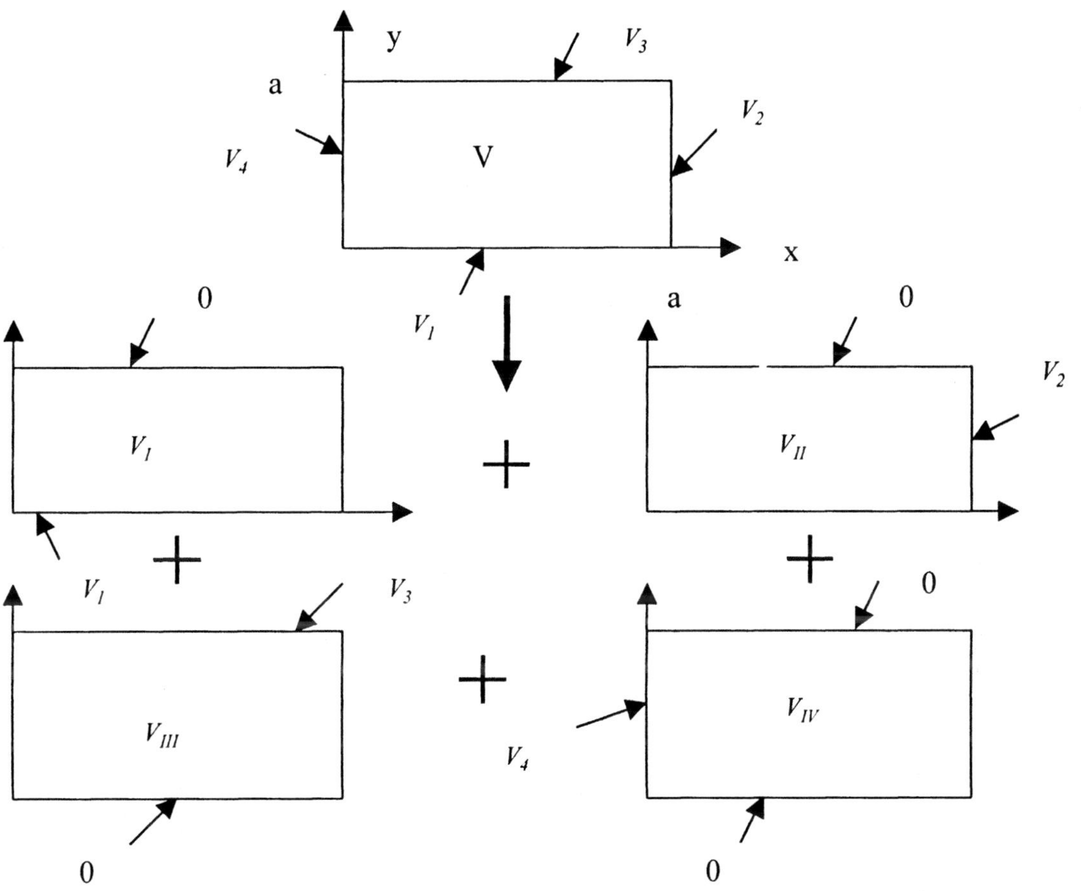

Therefore,

$$V = V_I + V_{II} + V_{III} + V_{IV}$$

where

$$V_I = \frac{4V_1}{\pi} \sum_{n=\text{odd}}^{\infty} \frac{\sin(n\pi x/b)\sinh[n\pi(a-y)/b]}{n\sinh(n\pi a/b)}$$

$$V_{II} = \frac{4V_2}{\pi} \sum_{n=\text{odd}}^{\infty} \frac{\sin(n\pi x/a)\sinh(n\pi y/a)}{n\sinh(n\pi b/a)}$$

$$V_{III} = \frac{4V_3}{\pi} \sum_{n=\text{odd}}^{\infty} \frac{\sin(n\pi x/b)\sinh(n\pi y/b)}{n\sinh(n\pi a/b)}$$

$$V_{IV} = \frac{4V_4}{\pi} \sum_{n=\text{odd}}^{\infty} \frac{\sin(n\pi y/a)\sinh[n\pi(b-x)/a]}{n\sinh(n\pi b/a)}$$

Prob. 6.21

$$\nabla^2 V = \frac{1}{\rho}\frac{\partial}{\partial \rho}\left(\rho \frac{\partial V}{\partial \rho}\right) + \frac{1}{\rho^2}\frac{\partial^2 V}{\partial \phi^2} = 0$$

If we let $V(\rho,\phi) = R(\rho)\Phi(\phi)$,

$$\frac{\Phi}{\rho}\frac{\partial}{\partial \rho}(\rho R') + \frac{1}{\rho^2} R\Phi'' = 0$$

or

$$\frac{\rho}{R}\frac{\partial}{\partial \rho}(\rho R') = -\frac{\Phi''}{\Phi} = \lambda$$

Hence

$$\underline{\underline{\Phi'' + \lambda\Phi = 0}}$$

and

$$\frac{\partial}{\partial \rho}(\rho R') - \frac{\lambda R}{\rho} = 0$$

or

$$\underline{\underline{R'' + \frac{R'}{\rho} - \frac{\lambda R}{\rho^2} = 0}}$$

Prob. 6.22

$$\nabla^2 V = \frac{1}{r^2}\frac{\partial}{\partial r}\left(r^2 \frac{\partial V}{\partial r}\right) + \frac{1}{r^2 \sin\theta}\frac{\partial}{\partial \theta}(\sin\theta \frac{\partial V}{\partial \theta}) = 0$$

If $V(r,\theta) = R(r)F(\theta)$, $r \neq 0$,

$$F\frac{d}{dr}(r^2 R') + \frac{R}{\sin\theta}\frac{d}{d\theta}(\sin\theta F') = 0$$

Dividing through by RF gives

$$\frac{1}{R}\frac{d}{dr}(r^2 R') = -\frac{1}{F\sin\theta}\frac{d}{d\theta}(\sin\theta F') = \lambda$$

Hence,

$$\sin\theta F'' + \cos\theta F' + \lambda F \sin\theta = 0$$

or

$$\underline{\underline{F'' + \cot\theta F' + \lambda F = 0}}$$

Also,

$$\frac{d}{dr}(r^2 R') - \lambda R = 0$$

or

$$R'' + \frac{2R'}{r} - \frac{\lambda}{r^2} R = 0$$

Prob. 6.23 If the centers at $\phi = 0$ and $\phi = \pi/2$ are maintained at a potential difference of V_o, from Example 6.3,

$$E_\phi = \frac{2V_o}{\pi\rho}, \qquad J = \sigma E$$

Hence,

$$I = \int J \bullet dS = \frac{2V_o\sigma}{\pi} \int_{\rho=a}^{b} \int_{z=0}^{t} \frac{1}{\rho} d\rho dz = \frac{2V_o \sigma t}{\pi} \ln(b/a)$$

and

$$R = \frac{V_o}{I} = \frac{\pi}{2\sigma t \ln(b/a)}$$

Prob. 6.24 If $V(r = a) = 0, \quad V(r = b) = V_o$, from Example 6.9,

$$E = \frac{V_o}{r^2(1/a - 1/b)}, \qquad J = \sigma E$$

Hence,

$$I = \int J \bullet dS = \frac{V_o \sigma}{1/a - 1/b} \int_{\theta=0}^{\alpha} \int_{\phi=0}^{2\pi} \frac{1}{r^2} r^2 \sin\theta d\theta d\phi = \frac{2\pi V_o \sigma}{1/a - 1/b}(-\cos\theta)|_0^\alpha$$

$$R = \frac{V_o}{I} = \frac{\frac{1}{a} - \frac{1}{b}}{2\pi\sigma(1 - \cos\alpha)}$$

Prob. 6.25 For a spherical capacitor, from Eq. (6.38),

$$R = \frac{\frac{1}{a} - \frac{1}{b}}{4\pi\sigma}$$

For the hemisphere, $R' = 2R$ since the sphere consists of two hemispheres in parallel. As $b \longrightarrow \infty$,

$$R' = \lim_{b \to \infty} \frac{\frac{1}{a} - \frac{1}{b}}{4\pi\sigma} = \frac{1}{2\pi a\sigma}$$

$$G = 1/R' = 2\pi a\sigma$$

Alternatively, for an isolated sphere, $C = 4\pi\varepsilon a$. But

$$RC = \frac{\varepsilon}{\sigma} \longrightarrow R = \frac{1}{4\pi a\sigma}$$

$$R' = 2R = \frac{1}{2\pi a\sigma} \quad \text{or} \quad G = 2\pi a\sigma$$

Prob. 6.26 l =1.5mm, S=3x4 + 1x4 + 3x4 = 28 cm^2

$$R = \frac{l}{\sigma S} = \frac{1.5x10^{-3}}{5.8x10^{7}x28x10^{-4}} = \underline{\underline{9.236\ \text{n}\Omega}}$$

Prob. 6.27

$$C = \frac{\varepsilon S}{d} \longrightarrow S = \frac{Cd}{\varepsilon_o \varepsilon_r} = \frac{2x10^{-9}x10^{-6}}{4x10^{-9}/36\pi}\ \text{m}^2 = \underline{\underline{0.5655\ \text{cm}^2}}$$

Prob. 6.28

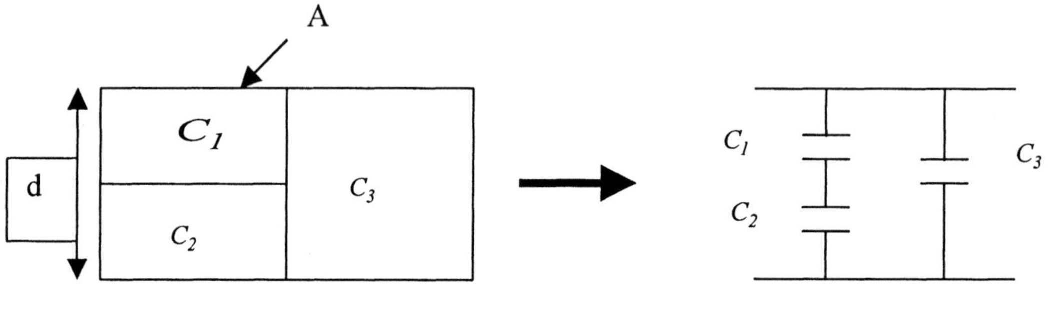

From the figure above,

$$C = \frac{C_1 C_2}{C_1 + C_2} + C_3$$

where

$$C_1 = \frac{\varepsilon_o A/2}{d/2} = \frac{\varepsilon_o A}{d}, \quad C_2 = \frac{\varepsilon_o \varepsilon_r A}{d}, \quad C_3 = \frac{\varepsilon_o A}{2d}$$

$$C = \frac{\varepsilon_o^2\varepsilon_r A^2/d^2}{\varepsilon_o(\varepsilon_r+1)A/d} + \frac{\varepsilon_o A}{2d} = \frac{\varepsilon_o A}{d}\left(\frac{1}{2}+\frac{\varepsilon_r}{\varepsilon_r+1}\right) = \frac{10^{-9}}{36\pi}\frac{10x10^{-4}}{2x10^{-3}}\left(\frac{1}{2}+\frac{6}{7}\right) \cong \underline{\underline{6 \text{ pF}}}$$

Prob. 6.29

$$Fdx = dW_E \longrightarrow F = \frac{dW_E}{dx}$$

$$W_E = \int \frac{1}{2}\varepsilon|E|^2 dv = \frac{1}{2}\varepsilon_o\varepsilon_r E^2 xad + \frac{1}{2}\varepsilon_o E^2 da(1-x)$$

where $E = V_o / d$.

$$\frac{dW_E}{dx} = \frac{1}{2}\varepsilon_o\frac{V_o^2}{d^2}(\varepsilon_r - 1)da \longrightarrow F = \frac{\varepsilon_o(\varepsilon_r-1)V_o^2 a}{2d}$$

Alternatively, $W_E = \frac{1}{2}CV_o^2$, where

$$C = C_1 + C_2 = \frac{\varepsilon_o\varepsilon_r ax}{d} + \frac{\varepsilon_o\varepsilon_r(1-x)}{d}$$

$$\frac{dW_E}{dx} = \frac{1}{2}\varepsilon_o\frac{V_o^2 a}{d}(\varepsilon_r - 1)$$

$$\underline{\underline{F = \frac{\varepsilon_o(\varepsilon_r-1)V_o^2 a}{2d}}}$$

Prob. 6.30 (a)

$$C = \frac{\varepsilon_o S}{d} = \frac{10^{-9}}{36\pi}\frac{200x10^{-4}}{3x10^{-3}} = 59 \text{ pF}$$

(b) $\rho_s = D_n = 10^{-6}$ nC / m². But

$$D_n = \varepsilon E_n = \frac{\varepsilon_o V_o}{d} = \rho_s$$

or

$$V_o = \frac{\rho_s d}{\varepsilon_o} = 10^{-6}x3x10^{-3}x36\pi x10^{9} = 339.3 \text{ V}$$

(c)

$$F = \frac{Q^2}{2S\varepsilon_o} = \frac{\rho_s^2 S}{2\varepsilon_o} = \frac{10^{-12}x200x10^{-4}x36\pi x10^{9}}{2} = 1.131 \text{ mN}$$

Prob. 6.31

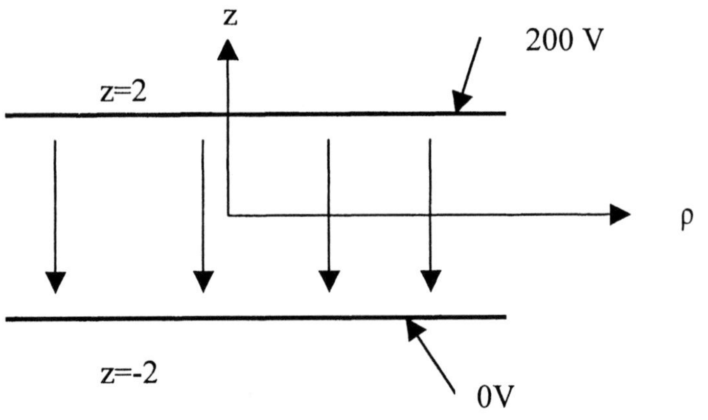

Let z be in cm

$$\frac{d^2V}{dz^2} = 0 \longrightarrow V = Az + B$$

When z= -2, V = 0 ⟶ 0 = -2A + B or B= 2A

When z = 2, V = 200 ⟶ 200 = 2A + 2A ⟶ A = 50

V = 50z + 100

(a) V(z=0) = 100 V

(b) $E = -\nabla V = -Aa_z = -50a_z$ V/cm $= -5a_z$ kV/m

$$\rho_s = D_n = \varepsilon E_n = \varepsilon E \bullet a_n$$

At the upper plate (z=2), $a_n = -a_z$

$$\rho_s = 5000\varepsilon_o\varepsilon_r = 5000x2.25x\frac{10^{-9}}{36\pi}$$

$$= 99.5 \text{ nC/m}^2$$

At the lower plate (z= -2), $a_n = +a_z$

$$\rho_s = -99.5 \text{ nC/m}^2$$

Prob. 6.32 (a)

$$C = \frac{Q}{V_o} = \frac{\varepsilon_o \varepsilon_r S}{d} = 5.6x\frac{10^{-9}}{36\pi}x\frac{80x10^{-4}}{6.4x10^{-4}} = \underline{\underline{619 \text{ pF}}}$$

(b)

$$C = \frac{Q}{V_o} \longrightarrow V_o = Q/C$$

$$E = -\nabla V = -3a_x - 4a_y + 12a_z \text{ kV/m} \longrightarrow |E| = \sqrt{3^2 + 4^2 + 12^2} = 13 \text{ kV/m}$$

$$\rho_s = D_n = \varepsilon_o |E|$$

Since the entire E is normal to each conducting plate.

$$Q = \rho_s S = \varepsilon_o |E| S$$

$$V_o = Q/C = \varepsilon_o |E| S \frac{d}{\varepsilon_o \varepsilon_r S} = \frac{|E| d}{\varepsilon_r} = \frac{13x10^3 x0.64x10^{-3}}{5.6} = \underline{\underline{14.86 \text{ V}}}$$

Prob. 6.33 (a)

$$C = \frac{4\pi\varepsilon}{\frac{1}{a} - \frac{1}{b}} = \frac{4\pi x 2.25x\frac{10^{-9}}{36\pi}}{\frac{1}{5x10^{-2}} - \frac{1}{10x10^{-2}}} = \underline{\underline{25 \text{ pF}}}$$

(b) $Q = C\, V_o = 25x80$ pC

$$\rho_s = \frac{Q}{4\pi r^2} = \frac{25x80}{4\pi x25x10^{-4}} \text{ pC/m}^2 = \underline{\underline{63.66 \text{ nC/m}^2}}$$

Prob. 6.34 (a)

$$\nabla^2 V = 0 \longrightarrow V = -\frac{A}{r} + B$$

When r=20cm, V=0 ⟶ 0 = - A/0.2 + B or B = 5A

When r=30cm, V=50 ⟶ 50 = - A/0.3 + 5A or A = 30, B = 150

$$\underline{\underline{V = -\frac{30}{r} + 150 \text{ V}}}$$

$$E = -\nabla V = -\frac{A}{r^2} a_r = -\frac{30}{r^2} a_r \text{ V/m}$$

$$D = \varepsilon_o \varepsilon_o E = -\frac{30x3.1}{r^2} x \frac{10^{-9}}{36\pi} a_r = -\frac{0.8223}{r^2} a_r \text{ nC/m}^2$$

(b) $\rho_s = D_n = D \bullet a_n$

On r = 30cm, $a_n = -a_r$

$$\rho_s = \frac{0.8223}{0.3^2} \text{ nC/m}^2 = 9.137 \text{ nC/m}^2$$

On r = 20cm, $a_n = +a_r$

$$\rho_s = -\frac{0.8223}{0.2^2} \text{ nC/m}^2 = -20.56 \text{ nC/m}^2$$

(c)

$$R = \frac{\frac{1}{a} - \frac{1}{b}}{4\pi\sigma} = \frac{\frac{1}{0.2} - \frac{1}{0.3}}{4\pi x 10^{-12}} = 132.6 \text{ G}\Omega$$

Prob. 6.35

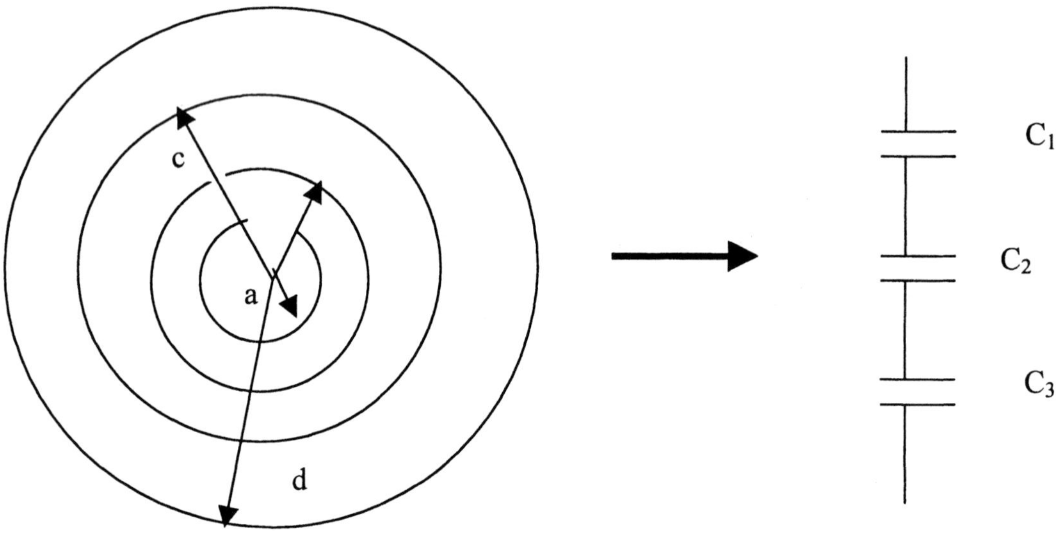

$$\frac{1}{C} = \frac{1}{C_1} + \frac{1}{C_2} + \frac{1}{C_3}$$

where $C_1 = \frac{4\pi\varepsilon_1}{\frac{1}{c} - \frac{1}{d}}$, $\quad C_2 = \frac{4\pi\varepsilon_2}{\frac{1}{b} - \frac{1}{c}}$, $\quad C_3 = \frac{4\pi\varepsilon_3}{\frac{1}{a} - \frac{1}{b}}$,

$$\frac{4\pi}{C} = \frac{1/c - 1/d}{\varepsilon_1} + \frac{1/b - 1/c}{\varepsilon_2} + \frac{1/a - 1/b}{\varepsilon_3}$$

$$C = \frac{4\pi}{\dfrac{\varepsilon_1}{\frac{1}{c} - \frac{1}{d}} + \dfrac{\varepsilon_2}{\frac{1}{b} - \frac{1}{c}} + \dfrac{\varepsilon_3}{\frac{1}{a} - \frac{1}{b}}}$$

Prob. 6.36

$$C = \frac{4\pi\varepsilon}{\frac{1}{a} - \frac{1}{b}}$$

Since b $\rightarrow \infty$,

$$C = 4\pi a \varepsilon_o \varepsilon_r = 4\pi x 5x10^{-2} x 80x \frac{10^{-9}}{36\pi} = \underline{\underline{444 \text{ pF}}}$$

Prob. 6.37

$$C = \frac{4\pi\varepsilon}{\frac{1}{a} - \frac{1}{b}} = \frac{4\pi x 5.9x10^{-9}/36\pi}{\left(\frac{1}{2} - \frac{1}{5}\right)x10^{-2}} = \underline{\underline{21.85 \text{ pF}}}$$

Prob. 6.38

$$C = \frac{2\pi\varepsilon_o L}{\ln(b/a)} = \frac{2\pi x \frac{10^{-9}}{36\pi} x100x10^{-6}}{\ln(600/20)} = 1.633x10^{-15}$$

$$V = Q/C = \frac{50x10^{-15}}{1.633x10^{-15}} = \underline{\underline{30.62 \text{ V}}}$$

Prob. 6.39 $V = V_o e^{-t/T_r}$, where $T_r = RC = 10 \times 10^{-6} \times 100 \times 10^{6} = 1000$

$$50 = 100e^{-t/T_r} \longrightarrow 2 = e^{t/T_r}$$

$$t = 1000 \ln 2 = \underline{\underline{693.1 \text{ s}}}$$

Prob. 6.40

$$RC = C/G = \varepsilon/\sigma \quad \longrightarrow \quad G = \frac{C\sigma}{\varepsilon}$$

$$G = \frac{\pi\sigma}{\cosh^{-1}(d/2a)}$$

Prob. 6.41 $E = \dfrac{Q}{4\pi\varepsilon r^2} a_r$

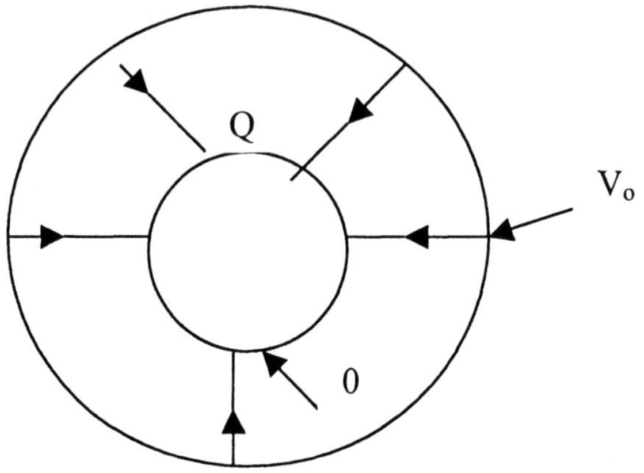

$$W = \frac{1}{2}\int \varepsilon |E|^2 dv = \iiint \frac{Q^2}{32\pi^2\varepsilon^2 r^2} \varepsilon r^2 \sin\theta\, d\theta\, d\phi\, dr$$

$$= \frac{Q^2}{32\pi^2\varepsilon}(2\pi)(2)\int_b^c \frac{dr}{r^2} = \frac{Q^2}{8\pi\varepsilon}\left(\frac{1}{c} - \frac{1}{b}\right)$$

$$W = \frac{Q^2(b-c)}{8\pi\varepsilon bc}$$

Prob. 6.42 (a) Method 1: $E = \dfrac{\rho_s}{\varepsilon}(-a_x)$, where ρ_s is to be determined.

$$V_o = -\int E \bullet dl = -\int \frac{-\rho_s}{\varepsilon} dx = \rho_s \int_0^d \frac{1}{\varepsilon_o} \frac{d}{d+x} dx = \frac{\rho_s}{\varepsilon} d \ln(x+d)|^d_0$$

$$V_o = \rho_s d \ln\frac{2d}{d} \quad \longrightarrow \quad \rho_s = \frac{V_o \varepsilon_o}{d \ln 2}$$

$$E = -\frac{\rho_s}{\varepsilon} a_x = -\frac{V_o}{(x+d)\ln 2} a_x$$

Method 2: We solve Laplace's equation

$$\nabla \bullet (\varepsilon \nabla V) = \frac{d}{dx}(\varepsilon \frac{dV}{dx}) = 0 \quad \longrightarrow \quad \varepsilon \frac{dV}{dx} = A$$

$$\frac{dV}{dx} = \frac{A}{\varepsilon} = \frac{Ad}{\varepsilon_o (x+d)} = \frac{c_1}{x+d}$$

$$V = c_1 \ln(x+d) + c_2$$

$$V(x=0) = 0 \quad \longrightarrow \quad 0 = c_1 \ln d + c_2 \quad \longrightarrow \quad c_2 = -c_1 \ln d$$

$$V(x=d) = V_o \quad \longrightarrow \quad V_o = c_1 \ln 2d - c_1 \ln d = c_1 \ln 2$$

$$c_1 = \frac{V_o}{\ln 2}$$

$$V = c_1 \ln \frac{x+d}{d} = \underline{\underline{\frac{V_o}{\ln 2} \ln \frac{x+d}{d}}}$$

$$E = -\frac{dV}{dx} a_x = \underline{\underline{-\frac{V_o}{(x+d)\ln 2} a_x}}$$

(b) $$P = (\varepsilon_r - 1)\varepsilon_o E = -\left(\frac{x+d}{d} - 1\right) \frac{\varepsilon_o V_o}{(x+d)\ln 2} a_x = \underline{\underline{-\frac{\varepsilon_o x V_o}{d(x+d)\ln 2} a_x}}$$

(c)

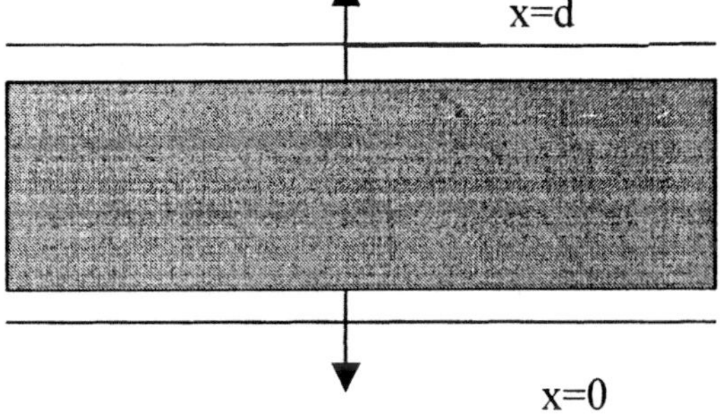

$$\rho_{ps}|_{x=0} = P \bullet (-a_x)|_{x=0} = 0$$

$$\rho_{ps}|_{x=d} = P \bullet a_x|_{x=d} = -\frac{\varepsilon_o V_o}{2d \ln 2}$$

(d) $Q = \int \rho_s dS = \rho_s S = \frac{\varepsilon_o S V_o}{d \ln 2}$

$$C = \frac{Q}{V_o} = \frac{\varepsilon_o S}{d \ln 2} = \frac{10^{-9}}{36\pi} \frac{200x10^{-4}}{2.5x10^{-3} \ln 2} = \underline{\underline{102 \text{ pF}}}$$

Prob. 6.43 Method 1: Using Gauss's law,

$$Q = \int D \bullet dS = 4\pi r D_r \quad \longrightarrow \quad D = \frac{Q}{4\pi r^2} a_r$$

$$E = D/\varepsilon = \frac{Q}{4\pi \varepsilon_o k} a_r$$

$$V = -\int E \bullet dl = -\frac{Q}{4\pi \varepsilon_o k} \int_a^b dr = \frac{Q}{4\pi \varepsilon_o k}(b-a)$$

$$C = \frac{Q}{|V|} = \underline{\underline{\frac{4\pi \varepsilon_o k}{b-a}}}$$

Method 2: Using the inhomogeneous Laplace's equation,

$$\nabla \bullet (\varepsilon \nabla V) = 0 \quad \longrightarrow \quad \frac{1}{r^2} \frac{d}{dr}\left(\frac{\varepsilon_o k}{r^2} r^2 \frac{dV}{dr}\right) = 0$$

$$\varepsilon_o k \frac{dV}{dr} = A' \quad \longrightarrow \quad \frac{dV}{dr} = A \text{ or } V = Ar + B$$

$$V(r = a) = 0 \quad \longrightarrow \quad 0 = Aa + B \quad \longrightarrow \quad B = -Aa$$

$$V(r = b) = V_o \quad \longrightarrow \quad V_o = Ab + B = A(b-a) \quad \longrightarrow \quad A = \frac{V_o}{b-a}$$

$$E = -\frac{dV}{dr} a_r = -A a_r = -\frac{V_o}{b-a} a_r$$

$$\rho_s = D_n = -\frac{V_o}{b-a} \frac{\varepsilon_o k}{r^2} \Big|_{r=a,b}$$

$$Q = \int \rho_s dS = -\frac{V_o \varepsilon_o k}{b-a} \iint \frac{1}{r^2} r^2 \sin\theta \, d\theta \, d\phi = -\frac{V_o \varepsilon_o k}{b-a} 4\pi$$

$$C = \frac{|Q|}{V_o} = \underline{\underline{\frac{4\pi \varepsilon_o k}{b-a}}}$$

Prob. 6.44 Method 1: We use Laplace's equation for inhomogeneous medium.

$$\nabla \bullet \nabla V = 0 = \frac{1}{\rho}\frac{d}{d\rho}\left(\rho\varepsilon\frac{dV}{d\rho}\right) = 0$$

$$\frac{d}{d\rho}\left(\rho\frac{\varepsilon_o k}{\rho}\frac{dV}{d\rho}\right) = 0$$

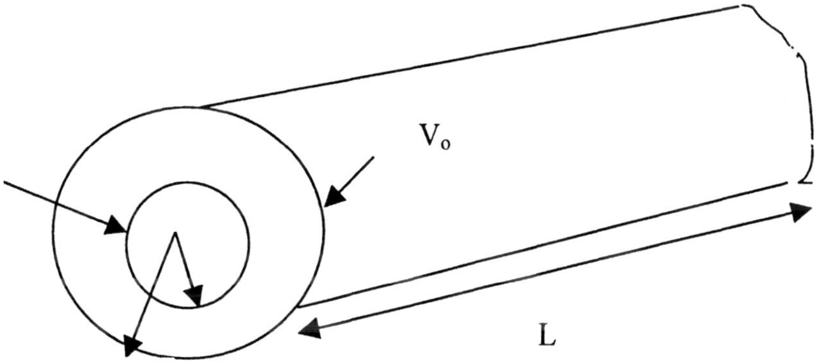

$$\varepsilon_o k\frac{dV}{d\rho} = A' \quad \longrightarrow \quad \frac{dV}{d\rho} = A \quad \text{or} \quad V = A\rho + B$$

$$V(\rho = a) = 0 \quad \longrightarrow \quad 0 = Aa + B \quad \longrightarrow \quad B = -Aa$$

$$V(\rho = b) = V_o \quad \longrightarrow \quad V_o = Ab + B = A(b-a) \quad \longrightarrow \quad A = \frac{V_o}{b-a}$$

$$E = -\frac{dV}{dr}a_\rho = -Aa_\rho = -\frac{V_o}{b-a}a_\rho$$

$$\rho_s = D_n = \varepsilon E_n$$

On $\rho = b, \quad a_n = -a_\rho$

$$\rho_s = \frac{V_o}{b-a}\frac{\varepsilon_o k}{\rho}, \qquad dS = \rho\, d\phi\, dz$$

$$Q = \int \rho_s ds = \iint \frac{V_o}{b-a}\frac{\varepsilon_o k}{\rho}\quad \rho\, d\phi\, dz = 2\pi L\frac{V_o}{b-a}\varepsilon_o k$$

$$C = \frac{Q}{V_o} = \frac{2\pi\varepsilon_o kL}{b-a}$$

$$C' = \frac{C}{L} = \underline{\underline{\frac{2\pi\varepsilon_o k}{b-a}}}$$

Method 2: We use Gauss's law. Assume Q is on the inner conductor and –Q on the outer conductor.

$$D = \frac{Q}{2\pi L} a_\rho$$

$$E = D/\varepsilon = \frac{Q}{2\pi\varepsilon_o kL} a_\rho$$

$$V_o = -\int E \bullet dl = -\frac{Q}{2\pi\varepsilon_o kL}\int d\rho = -\frac{Q(b-a)}{2\pi\varepsilon_o kL}$$

$$C = \frac{Q}{V} = \frac{2\pi\varepsilon_o kL}{b-a}$$

$$C' = \frac{C}{L} = \underline{\underline{\frac{2\pi\varepsilon_o k}{b-a}}}$$

Prob. 6.45

$$C = 4\pi\varepsilon_o a = 4\pi x \frac{10^{-9}}{36\pi} x 6.37x10^6 = \underline{\underline{0.708 \text{ mF}}}$$

Prob. 6.46 (a)

$$V = \frac{Q}{4\pi\varepsilon_o}\left[\frac{1}{|(6,3,2)|} - \frac{1}{|(6,3,8)|}\right] = \frac{10x10^{-9}}{4\pi x10^{-9}/36\pi}\left[\frac{1}{7} - \frac{1}{\sqrt{109}}\right] = 4.237 \text{ V}$$

$$E = \frac{10x10^{-9}}{4\pi x10^{-9}/36\pi}\left[\frac{(6,3,2)}{7^3} - \frac{(6,3,8)}{109^{3/2}}\right] = \underline{\underline{1.1a_x + 0.55a_y - 0.108a_z \text{ V/m}}}$$

(b)

$$F = \frac{Q_1Q_2}{4\pi\varepsilon_o r^2} a_r = \frac{-10x10x10^{-18}[(0,0,-3)-(0,0,3)]}{4\pi x\frac{10^{-9}}{36\pi}|(0,0,-3)-(0,0,3)|^3} = -900x10^{-9}\frac{(0,0,-6)}{6^3} = \underline{\underline{-25a_z \text{ N}}}$$

Prob. 6.47

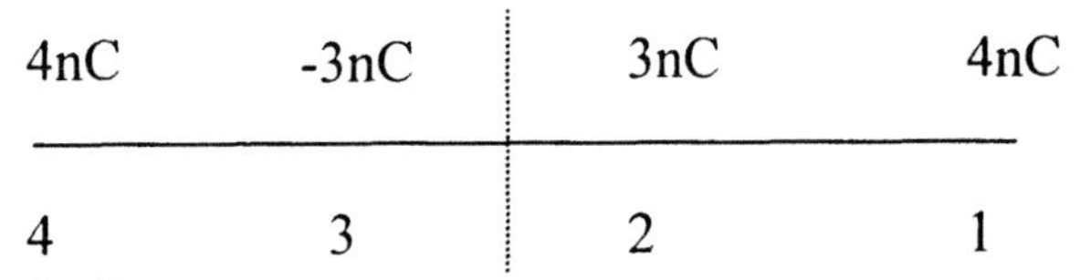

(a) Q_i = -(3nC – 4nC) = <u>1nC</u>

(b) The force of attraction between the charges and the plates is

$$F = F_{13} + F_{14} + F_{23} + F_{24}$$

$$|F| = \frac{10^{-18}}{4\pi x10^{-9}/36\pi}\left[\frac{9}{2^2} - \frac{2(12)}{3^2} + \frac{16}{4^2}\right] = \underline{\underline{5.25 \text{ nN}}}$$

Prob. 6.48

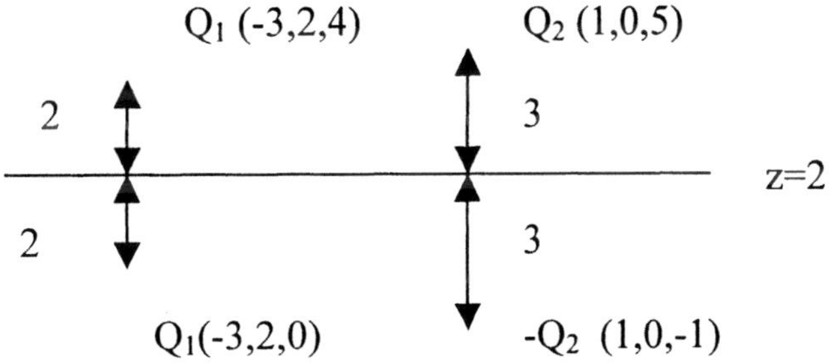

$$D(x,y,z) = \frac{Q_1}{4\pi}\left[\frac{(x,y,z)-(-3,2,4)}{|(x,y,z)-(-3,2,4)|^3} - \frac{(x,y,z)-(-3,2,0)}{|(x,y,z)-(-3,2,0)|^3}\right]$$

$$+\frac{Q_2}{4\pi}\left[\frac{(x,y,z)-(1,0,5)}{|(x,y,z)-(1,0,5)|^3} - \frac{(x,y,z)-(1,0,1)}{|(x,y,z)-(1,0,1)|^3}\right]$$

$$= \frac{50}{4\pi}\left[\frac{(x+3, y-2, z-4)}{|(x+3)^2 + (y-2)^2 + (z-4)^2|^{3/2}} - \frac{(x+3, y-2, z)}{|(x+3)^2 + (y-2)^2 + z^2|^{3/2}}\right]$$

$$-\frac{20}{4\pi}\left[\frac{(x-1, y, z-5)}{|(x-1)^2 + y^2 + (z-5)^2|^{3/2}} - \frac{(x-1, y, z+1)}{|(x-1)^2 + y^2 + (z+1)^2|^{3/2}}\right]$$

(a) At (x,y,z) = (7,-2,2),

$$\rho_s = D_z|_{z=2} = \frac{50}{4\pi}\left[\frac{2-4}{(10^2+4^2+2^2)^{3/2}} - \frac{2}{(10^2+4^2+2^2)^{3/2}}\right]$$

$$-\frac{20}{4\pi}\left[\frac{-3}{(6^2+4^2+3^2)^{3/2}} - \frac{3}{(6^2+4^2+3^2)^{3/2}}\right] \text{nC/m}^2$$

$$\rho_s = 7.934 \text{ pC/m}^2$$

(b) At (3,4,8)

$$D = \frac{50}{4\pi}\left[\frac{(6,2,4)}{(6^2+2^2+4^2)^{3/2}} - \frac{(6,2,8)}{(6^2+2^2+8^2)^{3/2}}\right]$$

$$-\frac{20}{4\pi}\left[\frac{(2,4,3)}{(2^2+4^2+3^2)^{3/2}} - \frac{(2,4,9)}{(2^2+4^2+9^2)^{3/2}}\right] \text{nC/m}^2$$

$$D = 17.21a_x - 16.29a_y - 8.486a_y \text{ pC/m}^2$$

(c) Since (1,1,1) is below the ground plane, $\mathbf{D} = 0$

Prob. 6.49 We have 7 images as follows: -Q at (-1,1,1), -Q at (1,-1,1), -Q at (1,1,-1), -Q at (-1,-1,-1), Q at (1,-1,-1), Q at (-1,-1,1), and Q at (-1,1,-1). Hence,

$$F = \frac{Q}{4\pi\varepsilon_o}\left[-\frac{2}{2^3}a_x - \frac{2}{2^3}a_y - \frac{2}{2^3}a_z - \frac{(2a_x+2a_y+2a_z)}{12^{3/2}} + \frac{(2a_y+2a_z)}{8^{3/2}} + \frac{(2a_x+2a_z)}{8^{3/2}} + \frac{(2a_x+2a_z)}{8^{3/2}}\right]$$

$$= 0.9(a_x + a_y + a_z)\left(-\frac{1}{4} - \frac{1}{12\sqrt{3}} + \frac{1}{4\sqrt{2}}\right) = -0.1891(a_x + a_y + a_z) \text{ N}$$

Prob. 6.50

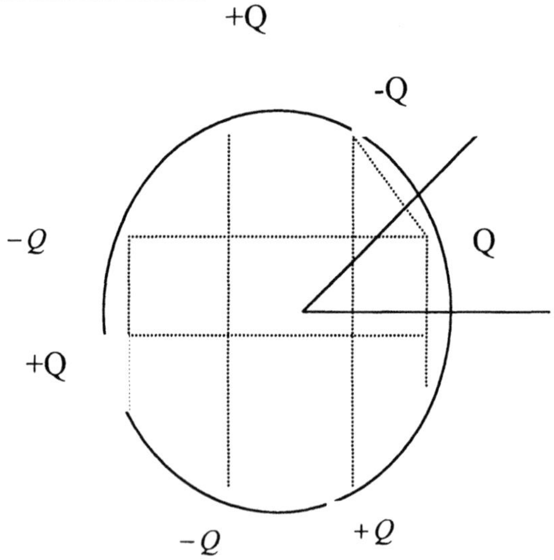

$$N = \left(\frac{360^o}{45^o} - 1\right) = 7$$

Prob. 6.51 (a)

$$E = E_+ + E_- = \frac{\rho_L}{2\pi\varepsilon_o}\left(\frac{a_{\rho 1}}{\rho_1} - \frac{a_{\rho 2}}{\rho_2}\right) = \frac{16x10^{-9}}{2\pi x10^{-9}/36\pi}\left[\frac{(2,-2,3)-(3,-2,4)}{|(2,-2,3)-(3,-2,4)|^2} - \frac{(2,-2,3)-(3,-2,-4)}{|(2,-2,3)-(3,-2,-4)|^2}\right]$$

$$= 18x16\left[\frac{(-1,0,1)}{2} - \frac{(-1,0,7)}{50}\right] = \underline{\underline{-138.2a_x - 184.3a_y \text{ V/m}}}$$

(b) $\rho_s = D_n$

$$D = D_+ + D_- = \frac{\rho_L}{2\pi}\left(\frac{a_{\rho 1}}{\rho_1} - \frac{a_{\rho 2}}{\rho_2}\right) = \frac{16x10^{-9}}{2\pi}\left[\frac{(5,-2,0)-(3,-2,4)}{|(5,-2,0)-(3,-2,4)|^2} - \frac{(5,-6,0)-(3,-2,-4)}{|(5,-6,0)-(3,-2,-4)|^2}\right]$$

$$= \frac{8}{\pi}\left[\frac{(2,0,-4)}{20} - \frac{(2,0,4)}{20}\right] \text{nC/m}^2 = -1.018a_z \text{ nC/m}^2$$

$$\underline{\underline{\rho_s = -1.018 \text{ nC/m}^2}}$$

Prob. 6.52

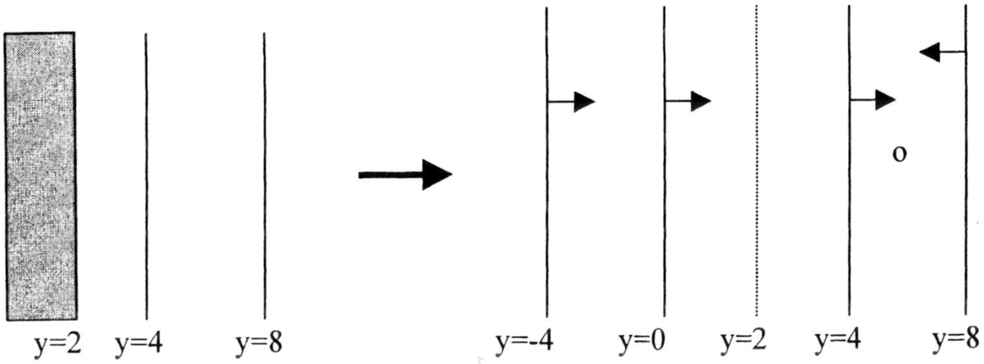

At P(0,0,0), $\underline{\underline{\mathbf{E}=0}}$ since **E** does not exist for y<2.

At Q(-4,6,2), y=6 and

$$E = \sum \frac{\rho_s}{2\varepsilon_o} a_n = \frac{10^{-9}}{2x10^{-9}/36\pi}(-30a_y + 20a_y + 20a_y + 30a_y) = \underline{\underline{2.262a_y \text{ kV/m}}}$$

CHAPTER 7

P.E. 7.1

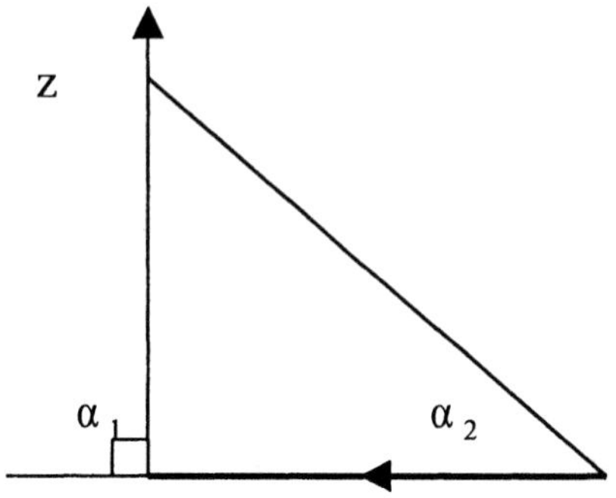

$$\rho = 5, \cos\alpha_1 = 0, \cos\alpha_2 = \sqrt{\frac{2}{27}}$$

$$a_\phi = a_l \times a_\rho = \left(\frac{-a_x - a_y}{\sqrt{2}}\right) \times a_z = \frac{-a_x - a_y}{\sqrt{2}}$$

$$H_3 = \frac{10}{4\pi(5)}\left(\sqrt{\frac{2}{27}} - 0\right)\left(\frac{-a_x + a_y}{2}\right) = \underline{\underline{-30.03a_x + 30.6a_y}} \text{ mA/m}$$

P.E. 7.2

(a) $H = \frac{2}{4\pi(2)}\left(1 + \frac{3}{\sqrt{13}}\right)a_z = \underline{\underline{0.1458}}$ A/m

(b) $\rho = \sqrt{3^2 + 4^2} = 5, \alpha_2 = 0, \cos\alpha_1 = -\frac{12}{13},$

$$a_\phi = a_y x\left(\frac{3a_x - 4a_z}{5}\right) = \frac{4a_x + 3a_z}{5}$$

$$H = \frac{2}{4\pi(5)}\left(1 + \frac{12}{13}\right)\left(\frac{4a_x + 3a_z}{5}\right) = \frac{1}{26\pi}(4a_x + 3a_z)$$

$$= \underline{\underline{48.97a_x + 36.73a_z}} \text{ mA/m}$$

P.E. 7.3

(a) From Example 7.3,

$$H = \frac{Ia^2}{2(a^2 + z^2)^{3/2}} a_z$$

At (0,0,1), z = 2cm,

$$H = \frac{50 \times 10^{-3} \times 25 \times 10^{-4}}{2(5^2 + 2^2)^{3/2} \times 10^{-6}} a_z \text{ A/m}$$

$$= \underline{\underline{400.2a_z \text{ A/m}}}$$

(b) At (0,0,10cm), z = 9cm,

$$H = \frac{50\times 10^{-3} \times 25\times 10^{-4}}{2(5^2+9^2)^{3/2}\times 10^{-6}} a_z$$

$$= \underline{\underline{57.3a_z \quad mA/m}}$$

P.E. 7.4

$$H = \frac{NI}{2L}\left(\cos\theta_2 - \cos\theta_1\right)a_z = \frac{2\times 10^3 \times 50\times 10^{-3}(\cos\theta_2 - \cos\theta_1)a_z}{2\times 0.75}$$

$$= \frac{100}{1.5}\left(\cos\theta_2 - \cos\theta_1\right)\boldsymbol{a}_z$$

(a) At (0,0,0), $\theta = 90^o$, $\cos\theta_2 = \dfrac{0.75}{\sqrt{0.75^2 + 0.05^2}}$

$= 0.9978$

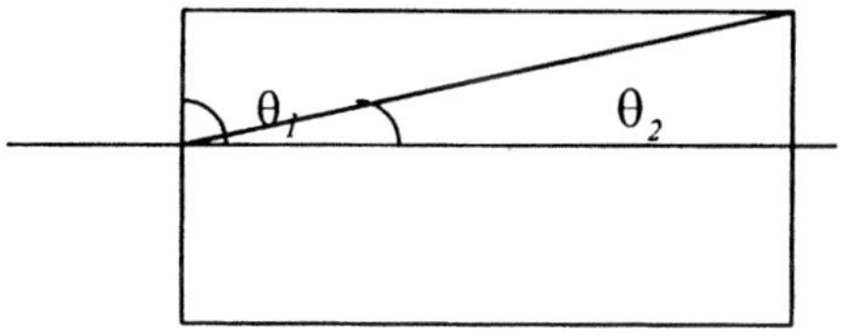

$$\boldsymbol{H} = \frac{100}{1.5}(0.9978 - 1)\boldsymbol{a}_z$$

$$= \underline{\underline{66.52\,\boldsymbol{a}_z \text{ A/m}}}$$

(b) At (0,0,0.75), $\theta_2 = 90^o, \cos\theta_1 = -0.9978$

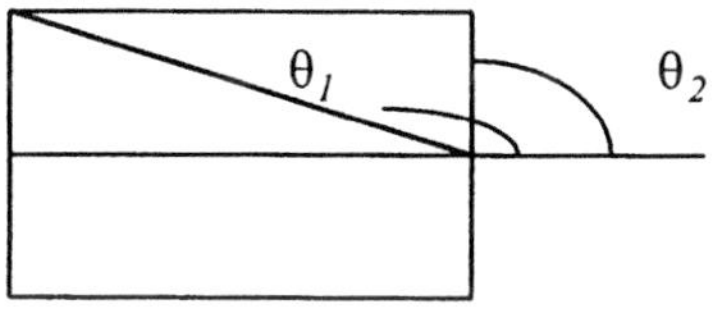

$$\boldsymbol{H} = \frac{100}{1.5}(0 + 0.9978)\boldsymbol{a}_z$$

$$= \underline{\underline{66.52\,\boldsymbol{a}_z \text{ A/m}}}$$

(c) At (0,0,0.5), $\cos\theta_1 = \dfrac{-0.5}{\sqrt{0.5^2 + 0.05^2}} = -0.995$

$$\cos\theta_1 = \frac{0.25}{\sqrt{0.25^2 + 0.05^2}} = 0.9806$$

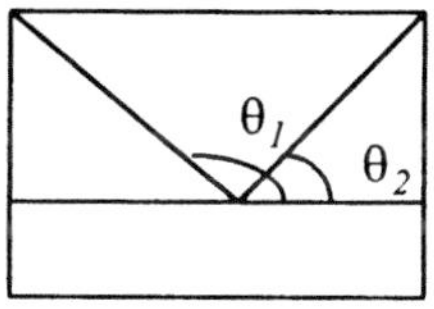

$$\boldsymbol{H} = \frac{100}{1.5}(0.9806 + 0.995)\boldsymbol{a}_z$$

$$= \underline{\underline{131.7\,\boldsymbol{a}_z \text{ A/m}}}$$

P.E. 7.5

$$\boldsymbol{H} = \frac{1}{2}\boldsymbol{k}\times\boldsymbol{a}_n$$

(a) $H(0,0,0) = \frac{1}{2}50a_z \times (-a_y) = \underline{\underline{25a_x}}$ mA/m

(b) $H(1,5,-3) = \frac{1}{2}50a_z \times a_y = \underline{\underline{-25a_x}}$ mA/m

P.E. 7.6

$$|H| = \begin{cases} \dfrac{NI}{2\pi\rho}, \rho - a \langle \rho \langle \rho + a = a \langle \rho \langle 11 \\ 0, \qquad \text{otherwise} \end{cases}$$

(a) At (3,-4,0), $\rho = \sqrt{3^2 + 4^2}$ =5cm ‹ 9cm

$$|H| = \underline{\underline{0}}$$

(b) At (6,9,0), $\rho = \sqrt{6^2 + 9^2} = \sqrt{117}$ ‹ 11

$$|H| = \frac{10^3 \times 100 \times 10^{-3}}{2\pi\sqrt{117} \times 10^2} = \underline{\underline{147.1}} \text{ A/m}$$

P.E. 7.7

(a) $\boldsymbol{B} = \nabla \times \boldsymbol{A} = (-4xz - 0)\boldsymbol{a}_x + (0 + 4yz)\boldsymbol{a}_y + (y^2 - x^2)\boldsymbol{a}_z$

$$B(-1,2,5) = \underline{\underline{20a_x + 40a_y + 3a_z}} \text{ Wb/m}^2$$

(b) $$\psi = \int \mathbf{B}.\partial\mathbf{s} = \int_{y=1}^{4}\int_{x=0}^{1} (\mathbf{y}^2 - \mathbf{x}^2)\partial\mathbf{x}\partial\mathbf{y} = \int_{-1}^{4} \mathbf{y}^2\partial\mathbf{y} - 5\int_{0}^{1} \mathbf{x}^2\partial\mathbf{x}$$

$$= \frac{1}{3}(64 + 1) - \frac{5}{3} = \underline{\underline{20}} \text{ Wb}$$

Alternatively,

$$\psi = \int \mathbf{A}.\partial\mathbf{l} = \int_{0}^{1} \mathbf{x}^2(-1)\partial\mathbf{x} + \int_{-1}^{4} \mathbf{y}^2(1)\partial\mathbf{y} + \int_{1}^{0} \mathbf{x}^2(4)\partial\mathbf{x} + 0$$

$$= -\frac{5}{3} + \frac{65}{3} = \underline{\underline{20}} \text{ Wb}$$

P.E. 7.8

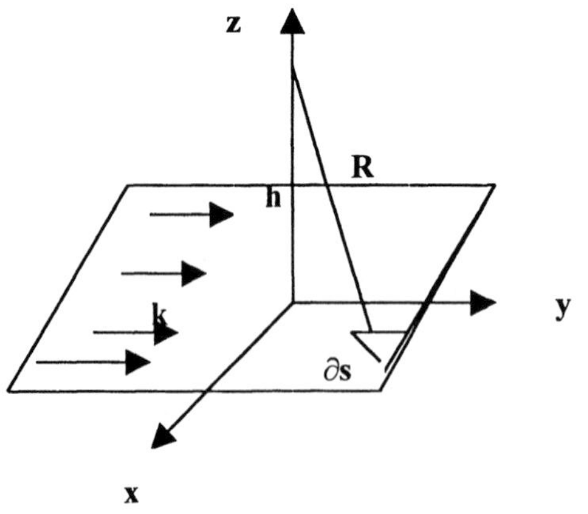

$$\mathbf{H} = \int \frac{\mathbf{k}\partial\mathbf{s} \times \mathbf{R}}{4\pi\mathbf{R}^3},$$

$$\partial s = \partial x \partial y, \mathbf{k} = \mathbf{k_y a_y},$$
$$\mathbf{R} = (-\mathbf{x}, -\mathbf{y}, \mathbf{h}),$$
$$\mathbf{k} \times \mathbf{R} = (\mathbf{h a_x} + \mathbf{x a_z})\mathbf{k_y},$$

$$\mathbf{H} = \int \frac{\mathbf{k_y}(\mathbf{h a_x} + \mathbf{x a_z})\partial \mathbf{x} \partial \mathbf{y}}{4\pi(\mathbf{x}^2 + \mathbf{y}^2 + \mathbf{h}^2)^{3/2}} = \frac{\mathbf{k_y h a_x}}{4\pi} \int_{-\infty}^{\infty}\int_{-\infty}^{\infty} \frac{\partial \mathbf{x} \partial \mathbf{y}}{(\mathbf{x}^2 + \mathbf{y}^2 + \mathbf{h}^2)^{3/2}} + \frac{\mathbf{k_y a_z}}{4\pi} \int_{-\infty}^{\infty}\int_{-\infty}^{\infty} \frac{\mathbf{x}\partial \mathbf{x} \partial \mathbf{y}}{(\mathbf{x}^2 + \mathbf{y}^2 + \mathbf{h}^2)^{3/2}}$$

The integrand in the last term is zero because it is an odd function of x.

$$H = \frac{k_y h a_x}{4\pi} \int_{\phi=0}^{2\pi} \int_{\rho=0}^{\infty} \frac{\rho \partial \phi \partial \rho}{(\rho^2 + h^2)^{3/2}} = \frac{k_y h 2\pi a_z}{4\pi} \int_0^{\infty} (\rho^2 + h^2)^{-3/2} \frac{\partial(\rho^2)}{2}$$

$$= \frac{\mathbf{k_y h}}{2} \mathbf{a_z} \left(\frac{-1}{(\rho^2 + \mathbf{h}^2)^{1/2}} \right) \Big|_0^{\infty} = \frac{\mathbf{k_y}}{2} \mathbf{a_z}$$

Similarly, for point (0,0,-h), $\mathbf{H} = -\frac{1}{2} \mathbf{k_y a_x}$

Hence,

$$H = \begin{bmatrix} \frac{1}{2} k_y a_x, & z \rangle 0 \\ \frac{1}{2} k_y a_x, & z \langle 0 \end{bmatrix}$$

Prob. 7.1

(a) See text

(b) Let $\mathbf{H} = H_y + H_z$

For $\mathbf{H_z} = \frac{\mathbf{I}}{2\pi\rho} \mathbf{a_\phi} \quad \rho = \sqrt{(-3)^2 + 4^2} = 5$

$$\mathbf{a_\phi} = -\mathbf{a_z} \times \frac{(-3\mathbf{a_x} + 4\mathbf{a_y})}{5} = \frac{(3\mathbf{a_y} - 4\mathbf{a_x})}{5}$$

$$\mathbf{H_z} = \frac{20}{2\pi(25)} (4\mathbf{a_x} + 3\mathbf{a_y}) = 0.5093\mathbf{a_x} + 0.382\mathbf{a_y}$$

For $\mathbf{H_y} = \frac{\mathbf{I}}{2\pi\rho} \mathbf{a_\phi}, \ \rho = \sqrt{(-3)^2 + 5^2} = \sqrt{34}$

$$\mathbf{a_\phi} = \mathbf{a_y} \times \frac{(-3\mathbf{a_x} + 5\mathbf{a_z})}{\sqrt{34}} = \frac{3\mathbf{a_z} - 5\mathbf{a_x}}{\sqrt{34}}$$

$$\mathbf{H_y} = \frac{10}{2\pi(34)} (-5\mathbf{a_x} + 3\mathbf{a_z}) = -0234\mathbf{a_x} + 0.1404\mathbf{a_z}$$

$$\mathbf{H} = H_y + H_z$$

$$= 0.2753\mathbf{a}_x + 0.382\mathbf{a}_y + 0.1404\mathbf{a}_z \text{ A/m}$$

Prob. 7.2

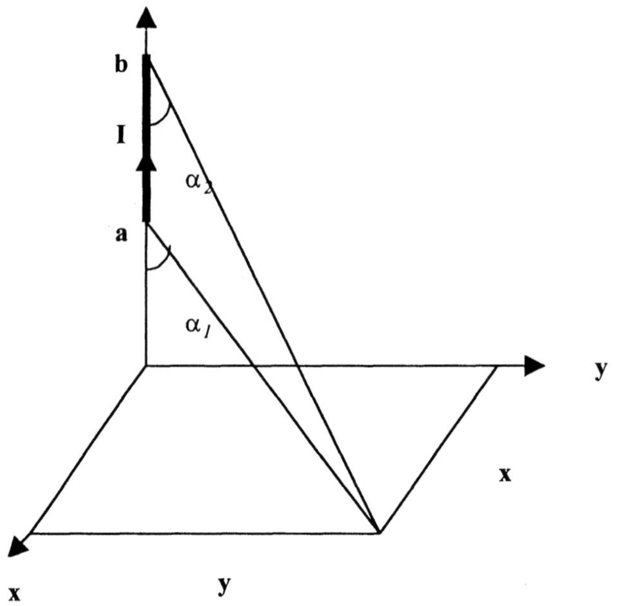

$$H = \frac{I}{4\pi\rho}(\cos\alpha_2 - \cos\alpha_1)a_\phi$$

$$\rho = \sqrt{\mathbf{x}^2 + \mathbf{y}^2}, \cos\alpha_1 = \frac{\mathbf{a}}{\sqrt{\mathbf{a}^2 + \rho^2}}, \cos\alpha_2 = \frac{\mathbf{b}}{\sqrt{\mathbf{b}^2 + \rho^2}}$$

$\mathbf{a}_\rho = \mathbf{a}_l \times \mathbf{a}_\rho = \mathbf{a}_z \times \mathbf{a}_\rho$ i.e $\mathbf{a}_\rho$ is regular $\mathbf{a}_\phi$. Hence,

$$H = \left[\frac{I}{4\pi\sqrt{x^2 + y^2 + b^2}} - \frac{a}{\sqrt{x^2 + y^2 + a^2}}\right]a_\phi$$

Prob. 7.3

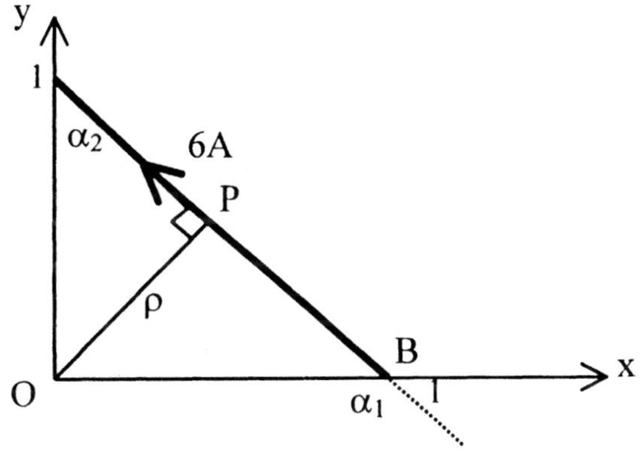

$$\bar{H} = \frac{I}{4\pi\rho}\left(\cos\alpha_2 - \cos\alpha_1\right)\bar{a}_\phi$$

$$\alpha_1 = 135^\circ,\ \alpha_2 = 45^\circ,\ \rho = \frac{1}{2}\sqrt{2} = \frac{\sqrt{2}}{2}$$

$$\bar{a}_p = \bar{a}_l \times \bar{a}_p = \left(\frac{-\bar{a}_x + \bar{a}_y}{\sqrt{2}}\right) \times \left(\frac{-\bar{a}_x - \bar{a}_y}{\sqrt{2}}\right) = \frac{1}{2}\begin{vmatrix} -1 & 1 & 0 \\ -1 & -1 & 0 \end{vmatrix} = \bar{a}_z$$

$$\bar{H} = \frac{6}{4\pi\frac{\sqrt{2}}{2}}\left(\cos 45^\circ - \cos 135^\circ\right)\bar{a}_z = \frac{3}{\pi}\bar{a}_z$$

$$\bar{H}(0,0,0) = \underline{\underline{0.954\bar{a}_z \text{ A/m}}}$$

Prob.7.4

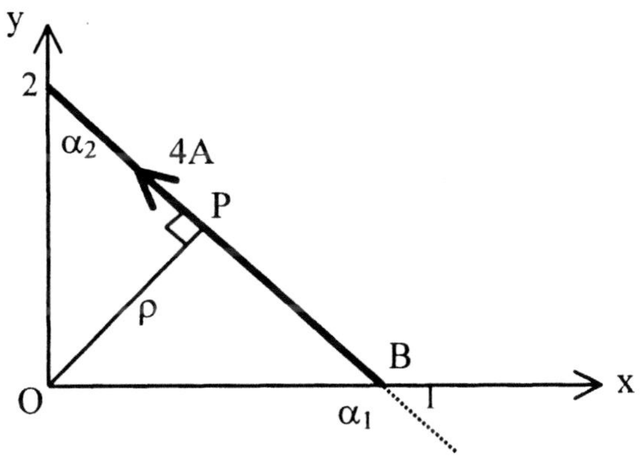

$$\overline{H} = \frac{I}{4\pi\rho}(\cos\alpha_2 - \cos\alpha_1)\overline{a}_\phi$$

$$\cos\alpha_2 = \frac{2}{\sqrt{5}}, \quad \cos(180 - \alpha_1) = \frac{1}{\sqrt{5}} = \cos 180^\circ \cos\alpha_1 + \sin 180^\circ \sin\alpha_1$$

$$= -\cos\alpha_1, \qquad \cos\alpha_1 = -\frac{1}{\sqrt{5}}$$

$$OP = (x-0, y-0) = x\overline{a}_x + y\overline{a}_y$$

$$AB = -\overline{a}_x + 2\overline{a}_y$$

But on AB, $y = 2(1-x)$

$$OP \cdot AB = 0 = -x + 2y = -x + 4(1-x) = 4 - 5x$$

$$x = 0.8, \quad y = 0.4, \qquad \rho = |OP| = 0.4\sqrt{5}$$

$$\bar{a}_p = \bar{a}_l \times \bar{a}_p = \left(\frac{-\bar{a}_x + 2\bar{a}_y}{5}\right) \times \left(\frac{-0.8\bar{a}_x - 0.4\bar{a}_y}{0.4\sqrt{5}}\right) = \bar{a}_z$$

$$\overline{H} = \frac{4}{4\pi(0.4\sqrt{5})}\left[\frac{2}{\sqrt{5}} + \frac{1}{\sqrt{5}}\right]\bar{a}_z = \frac{3}{2\pi}\bar{a}_z = \underline{\underline{0.4775\,\bar{a}_z \text{ A/m}}}$$

Prob. 7.5

(*a*) $$\overline{H} = \frac{I}{4\pi\rho}(\cos\alpha_2 - \cos\alpha_1)\overline{a}_\phi = \frac{2}{4\pi(5)}\left(\frac{10}{5\sqrt{2}} - 0\right)\overline{a}_y$$

$$= \underline{\underline{28.47\ \overline{a}_y \text{ mA/m}}}$$

(b) $$\overline{H} = \frac{2}{4\pi(5\sqrt{2})}\left(\frac{10}{5\sqrt{6}} - 0\right)\overline{a}_\phi, \text{ where } \quad \overline{a}_\phi = \overline{a}_z \times \left(\frac{\overline{a}_x + \overline{a}_y}{\sqrt{2}}\right)$$

$$= \frac{1}{5\pi\sqrt{2}}\left(\frac{-\overline{a}_x + \overline{a}_y}{\sqrt{2}}\right) = \underline{\underline{-13\overline{a}_x + 13\overline{a}_y \text{ mA/m}}}$$

(c) $$\overline{H} = \frac{2}{4\pi(5)\sqrt{10}}\left(\frac{10}{5\sqrt{14}} - 0\right)\overline{a}_\phi, \quad \overline{a}_\phi = \overline{a}_z \times \left(\frac{\overline{a}_x + 3\overline{a}_y}{\sqrt{10}}\right)$$

$$= \frac{1}{50\pi\sqrt{14}}(-3\overline{a}_x + \overline{a}_y) = -5.1\overline{a}_x + 1.7\overline{a}_y \text{ mA/m}$$

$$= \underline{\underline{28.47\ \overline{a}_y \text{ mA/m}}}$$

$$\underline{\underline{5.1\overline{\ \ } \quad 1.7\overline{\ \ }_y \quad 2}}$$

(d) $H = \underline{\underline{5.1a_x + 1.7a_y \text{ mA/m}^2}}$

Prob. 7.6

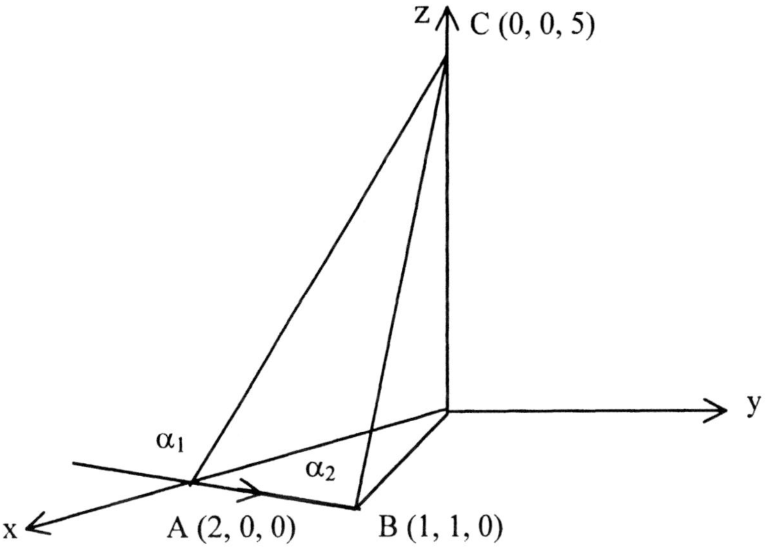

(*a*) Consider the figure above.

$AB = (1,1,0)-(2,0,0) = (-1,1,0)$

$AC = (0,0,5)-(2,0,0) = (-2,0,5)$

$AB \cdot AC = 2$, i.e AB and AC are not perpendicular.

$$\cos(180^\circ - \alpha_1) = \frac{AB \cdot AC}{|AB||AC|} = \frac{2}{\sqrt{2}\sqrt{29}} \rightarrow \cos\alpha_1 = -\sqrt{\frac{2}{29}}$$

$BC = (0,0,5)-(-1,-1,5) = (-1,-1,5)$

$BA = (1,-1,0)$

$$\cos\alpha_2 = \frac{\overline{BC} \cdot \overline{BA}}{|BC||BA|} = \frac{-1+1}{|BC||BA|} = 0$$

i.e. $BC = \bar{\rho} = (-1,-1,5), \quad \rho = \sqrt{27}$

$$\bar{a}_\phi = \bar{a}_l \times \bar{a}_\rho = \frac{(-1,1,0)}{\sqrt{2}} \times \frac{(-1,-1,5)}{\sqrt{27}} = \frac{(5,5,2)}{\sqrt{54}}$$

$$\overline{H}_2 = \frac{10}{4\pi\sqrt{27}}\left(0+\sqrt{\frac{2}{29}}\right)\frac{(5,5,2)}{\sqrt{27}} = \frac{5}{2\pi\sqrt{29}} \cdot \frac{(5,5,2)}{27} \text{ A/m}$$

$$= \underline{\underline{27.37\,\bar{a}_x + 27.37\,\bar{a}_y + 10.95\,\bar{a}_z \text{ mA/m}}}$$

(b) $\overline{H} = \overline{H}_1 + \overline{H}_2 + \overline{H}_3 = (0,-59.1,0) + (27.37, 27.37, 10.95) + (-30.63, 30.63, 0)$

$$= \underline{\underline{-3.26\,\bar{a}_x - 1.1\,\bar{a}_y + 10.95\,\bar{a}_z \text{ mA/m}}}$$

Prob. 7.7

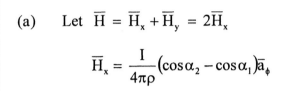

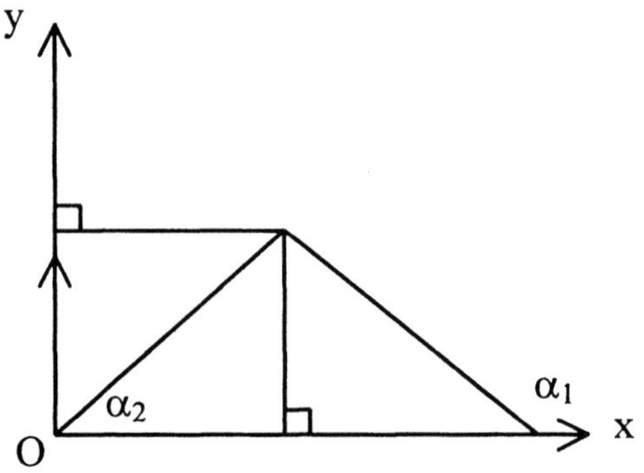

(a) Let $\overline{H} = \overline{H}_x + \overline{H}_y = 2\overline{H}_x$

$$\overline{H}_x = \frac{I}{4\pi\rho}(\cos\alpha_2 - \cos\alpha_1)\overline{a}_\phi$$

where $\bar{a}_\phi = -\bar{a}_x \times \bar{a}_y = -\bar{a}_z$, $\alpha_1 = 180^\circ$, $\alpha_2 = 45^\circ$

$$\overline{H}_x = \frac{5}{4\pi(2)}\left(\cos 45^\circ - \cos 180^\circ\right)\left(-\bar{a}_z\right)$$

$$= \underline{\underline{-0.6792\,\bar{a}_z \text{ A/m}}}$$

(b) $\overline{H} = \overline{H}_x + \overline{H}_y$

where $\overline{H}_x = \dfrac{5}{4\pi(2)}(1-0)\bar{a}_\phi$, $\bar{a}_\phi = -\bar{a}_x \times -\bar{a}_y = \bar{a}_z$

$$= 198.9\,\bar{a}_z \text{ MA/m}$$

$\overline{H}_x = 0$ since $\alpha_1 = \alpha_2 = 0$

$\overline{H} = \underline{\underline{0.1989\,\bar{a}_z \text{ A/m}}}$

(c) $\overline{H} = \overline{H}_x + \overline{H}_y$

where $\overline{H}_x = \dfrac{5}{4\pi(2)}(1-0)\left(-\bar{a}_x \times \bar{a}_z\right) = 198.9\,\bar{a}_y$ MA/m

$$\overline{H}_y = \frac{5}{4\pi(2)}(1-0)\left(\bar{a}_y \times \bar{a}_z\right) = 198.9\,\bar{a}_x \text{ MA/m}$$

$$\overline{H} = \underline{\underline{0.1989\,\bar{a}_x + 0.1989\,\bar{a}_y \text{ A/m.}}}$$

Prob. 7.8

For the side of the loop along y - axis,

$$\overline{H}_1 = \frac{I}{4\pi\rho}\left(\cos\alpha_2 - \cos\alpha_1\right)\bar{a}_\phi$$

where $\bar{a}_\phi = -\bar{a}_x$, $\rho = 2\tan 30^\circ = \dfrac{2}{\sqrt{3}}$, $\alpha_2 = 30^\circ$, $\alpha_1 = 150^\circ$

$$\overline{H}_1 = \frac{5}{4\pi}\frac{2}{\sqrt{3}}\left(\cos 30^\circ - \cos 150^\circ\right)\left(-\bar{a}_x\right) = -\frac{15}{8\pi}\bar{a}_x$$

$$\overline{H} = 3\overline{H}_1 = -1.79\,\bar{a}_x \text{ A/m}$$

Prob. 7.9

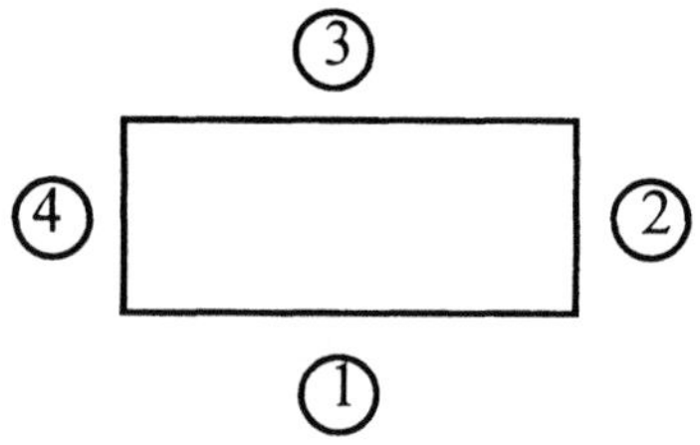

Let $\overline{H} = \overline{H}_1 + \overline{H}_2 + \overline{H}_3 + \overline{H}_4$

where $\overline{H}_n$ is the contribution by side n.

(a) $\overline{H} = 2\overline{H}_1 + \overline{H}_2 + \overline{H}_4$ since $\overline{H}_1 = \overline{H}_3$

$$\overline{H}_1 = \frac{I}{4\pi\rho}(\cos\alpha_2 - \cos\alpha_1)\overline{a}_\phi = \frac{10}{4\pi(2)}\left(\frac{6}{\sqrt{40}} + \frac{1}{\sqrt{2}}\right)\overline{a}_z$$

$$\overline{H}_2 = \frac{10}{4\pi(6)}\left(2 \times \frac{2}{\sqrt{40}}\right)\overline{a}_z, \quad \overline{H}_4 = \frac{10}{4\pi(2)}\left(2 \cdot \frac{1}{\sqrt{2}}\right)\overline{a}_z$$

$$\overline{H} = \left[\frac{5}{2\pi}\left(\frac{3}{\sqrt{10}} + \frac{1}{\sqrt{2}}\right) + \frac{5}{6\pi\sqrt{10}} + \frac{5}{2\pi\sqrt{2}}\right]\overline{a}_z = \underline{\underline{1.964\,\overline{a}_z \text{ A/m}}}$$

(b) At $(4, 2, 0)$, $\overline{H} = 2(\overline{H}_1 + \overline{H}_4)$

$$\overline{H}_1 = \frac{10}{4\pi(2)}\frac{8}{\sqrt{20}}\overline{a}_z, \quad \overline{H}_4 = \frac{10}{4\pi(4)}\frac{4}{\sqrt{20}}\overline{a}_z$$

$$\overline{H} = \frac{2\sqrt{5}}{\pi}\left(1 + \frac{1}{4}\right)\overline{a}_z = \underline{\underline{1.78\,\overline{a}_z \text{ A/m}}}$$

(c) At $(4, 8, 0)$, $\overline{H} = \overline{H}_1 + 2\overline{H}_2 + \overline{H}_3$

$$\overline{H}_1 = \frac{10}{4\pi(8)}\left(2 \cdot \frac{4}{4\sqrt{5}}\right)\overline{a}_z, \quad \overline{H}_2 = \frac{10}{4\pi(4)}\left(\frac{8}{4\sqrt{5}} - \frac{1}{\sqrt{2}}\right)\overline{a}_z$$

$$\overline{H}_3 = \frac{10}{4\pi(4)}\left(\frac{2}{\sqrt{2}}\right)(-\overline{a}_z)$$

$$\overline{H} = \frac{5}{8\pi}(\overline{a}_z)\left(\frac{1}{\sqrt{5}} + \frac{4}{\sqrt{5}} - \frac{4}{\sqrt{2}}\right) = \underline{\underline{-0.1178\,\overline{a}_z \text{ A/m}}}$$

(d) At $(0, 0, 2)$,

$$\overline{H}_1 = \frac{10}{4\pi(22)}\left(\frac{8}{\sqrt{64}} - 0\right)(\overline{a}_x \times \overline{a}_z) = -\frac{10}{\pi\sqrt{64}}\overline{a}_y$$

$$\overline{H}_2 = \frac{10}{4\pi\sqrt{64}}\left(\frac{4}{\sqrt{84}} - 0\right)\overline{a}_y \times \left(\frac{2\overline{a}_x - 8\overline{a}_x}{\sqrt{68}}\right) = \frac{5(\overline{a}_x + 4\overline{a}_y)}{17\pi\sqrt{84}}$$

$$\overline{H}_3 = \frac{10}{4\pi\sqrt{20}}\left(-\frac{8}{\sqrt{84}} - 0\right)\overline{a}_x \times \left(\frac{2\overline{a}_x - 8\overline{a}_y}{\sqrt{20}}\right) = \frac{\overline{a}_y + 2\overline{a}_z}{\pi\sqrt{21}}$$

$$\overline{H}_4 = \frac{10}{4\pi\sqrt{2}}\left(0 + \frac{4}{\sqrt{20}}\right)\left(-\overline{a}_y \times \overline{a}_z\right) = \frac{-5\overline{a}_x}{\pi\sqrt{20}}$$

$$\overline{H} = \left(\frac{1}{34\pi\sqrt{21}} - \frac{5}{\pi\sqrt{20}}\right)\overline{a}_x + \left(\frac{1}{\pi\sqrt{21}} - \frac{10}{\pi\sqrt{68}}\right)\overline{a}_x + \left(\frac{20}{34\pi\sqrt{21}} - \frac{2}{\pi\sqrt{21}}\right)\overline{a}_z$$

$$= -0.3457\,\overline{a}_x - 0.3165\,\overline{a}_y + 0.1798\,\overline{a}_z \text{ A/m}$$

Prob. 7.10

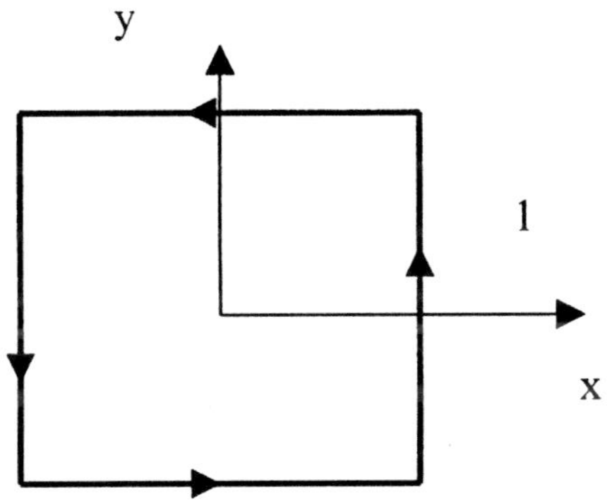

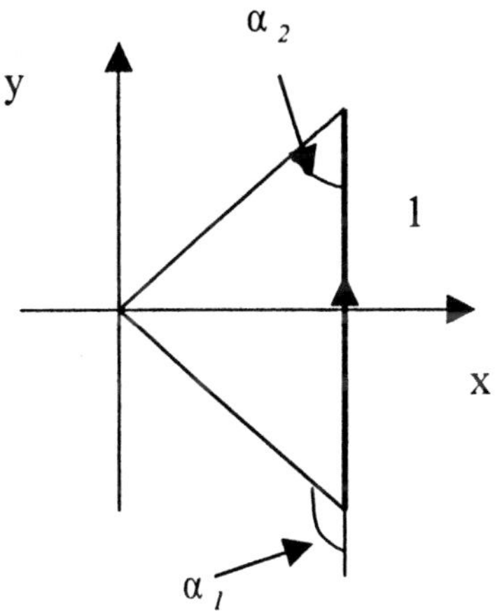

H = 4H₁, where H₁ is due to side 1.

$$H_1 = \frac{I}{4\pi\rho}\left(\cos\alpha_2 - \cos\alpha_1\right)a_\phi$$

$$\rho = a, \quad \alpha_2 = 45^o, \quad \alpha_1 = 135^o, \quad a_\phi = a_y x - a_x = a_z$$

$$H_1 = \frac{I}{4\pi\rho}\left(\frac{1}{\sqrt{2}} + \frac{1}{\sqrt{2}}\right)a_z = \frac{2I}{4\pi a\sqrt{2}}a_z$$

Prob. 7.11

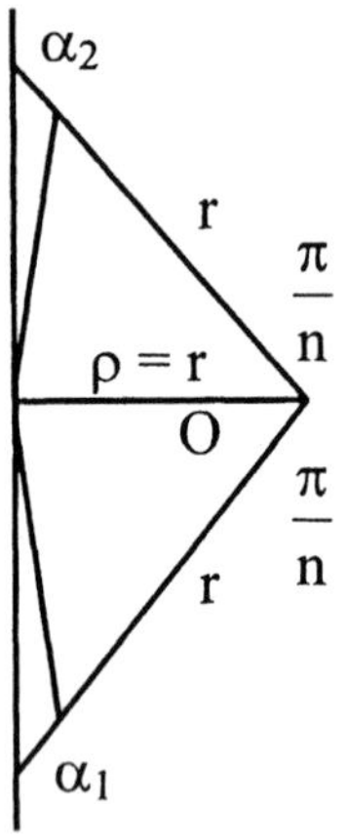

(a) Consider one side of the polygon as shown. The angle subtended by the Side At the center of the circle is

$$\frac{360^\circ}{n} = \frac{2\pi}{n}$$

The field due to this side is

$$H_1 = \frac{I}{4\pi\rho}(\cos\alpha_2 - \cos\alpha_1)$$

where $\rho = r$, $\cos\alpha_2 = \left(\cos 90 - \frac{\pi}{n}\right) = \sin\frac{\pi}{n}$

$$\cos\alpha_1 = -\sin\frac{\pi}{n}$$

$$H_1 = \frac{I}{4\pi r} 2\sin\frac{\pi}{n}$$

$$\overline{H} = n\overline{H}_1 = \frac{nI}{2\pi r}\sin\frac{\pi}{n}$$

(b) For $n = 3$, $H = \frac{3I}{2\pi r}\sin\frac{\pi}{3}$

$$r\cot 30^\circ = 2 \rightarrow r = \frac{2}{\sqrt{3}}$$

$$H = \frac{3\times 5}{2\pi\, {}^{2}\!/\!\sqrt{3}}\cdot\frac{\sqrt{3}}{2} = \frac{48}{8\pi} = 1.78 \text{ A/m.}$$

For $n = 4$, $H = \frac{4I}{2\pi r}\sin\frac{\pi}{4} = \frac{4\times 5}{2\pi(2)}\cdot\frac{1}{\sqrt{2}}$

$$= 1.128 \text{ A/m.}$$

(c) As $n \rightarrow \infty$,

$$H = \lim_{n\to\infty}\frac{nI}{2\pi r}\sin\frac{\pi}{n} = \frac{nI}{2\pi r}\cdot\frac{\pi}{n} = \frac{I}{2r}$$

From Example 7.3, when $h = 0$,

$$H = \frac{I}{2r}$$

which agrees.

Prob. 7.12

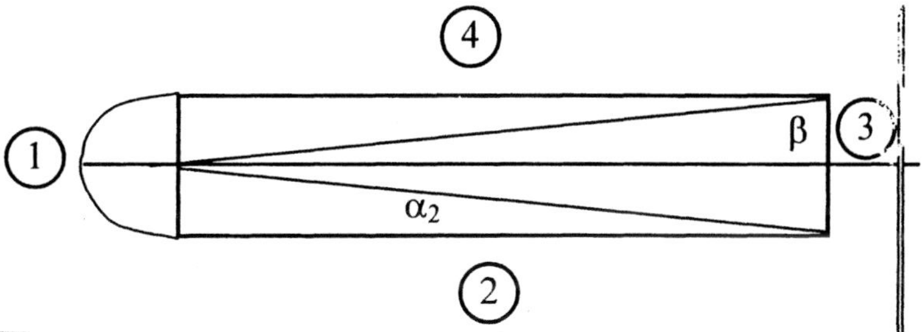

Let $\overline{H} = \overline{H}_1 + \overline{H}_2 + \overline{H}_3 + \overline{H}_4$

$$\overline{H}_1 = \frac{I}{4a}\bar{a}_z = \frac{10}{4\times 4\times 10^{-2}}\bar{a}_z = 62.5\,\bar{a}_z$$

$$\overline{H}_2 = \overline{H}_4 = \frac{I}{4\times 4\times 10^{-2}}\left(\cos\alpha_2 - \cos 90^\circ\right)\bar{a}_z, \quad \alpha_2 = \tan^{-1}\frac{4}{100} = 2.29^\circ$$

$$= 19.99\,\bar{a}_z$$

$$\overline{H}_3 = \frac{I}{4\pi(1)}2\cos\beta\,\bar{a}_z, \quad \beta = \tan^{-1}\frac{100}{4} = 87.7^\circ$$

$$= \frac{10}{4\pi}2\cos 87.7^\circ\,\bar{a}_z = 0.06361\,\bar{a}_z$$

$$\overline{H} = \left(62.5 + 2\times 19.88 + 0.06361\right)\bar{a}_z$$

$$= 120.32\,\bar{a}_z \text{ A/m.}$$

Prob. 7.13

From Example 7.3, $\overline{H}$ due to circular loop is

$$\overline{H}_1 = \frac{I\rho^2}{2\left(\rho^2 + z^2\right)}\bar{a}_z$$

(a) $$\overline{H}(0,0,0) = \frac{5\times 2^2}{2\left(2^2+0^2\right)^{3/2}}\bar{a}_z + \frac{5\times 2^2}{2\left(2^2+4^2\right)^{3/2}}\bar{a}_z$$

$$= 1.36\,\bar{a}_z \text{ A/m}$$

(b) $$\overline{H}(0,0,2) = 2\frac{5\times 2^2}{2\left(2^2+2^2\right)^{3/2}}\bar{a}_z$$

$$= 0.884\,\bar{a}_z \text{ A/m}$$

Prob. 7.14

$$\overline{B} = \mu_o \overline{H} = \frac{\mu_o NI}{L}$$

$$N = \frac{Bl}{\mu \cdot I} = \frac{5\times10^{-3}\times3\times10^{-2}}{4\pi\times10^{-7}\times400\times10^{-3}} = 29.84$$

$N \approx 30$ turns.

Prob. 7.15

(a)

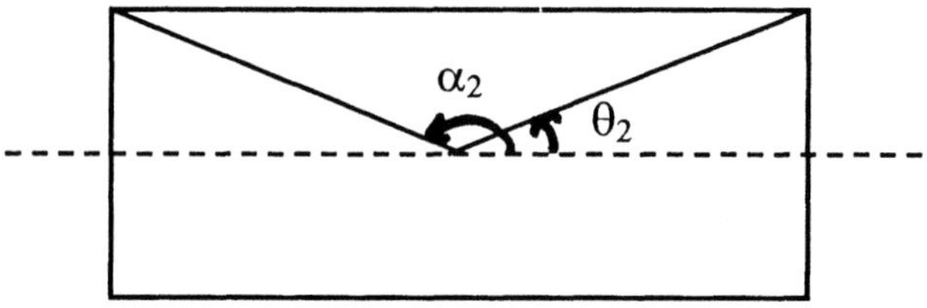

$$|\vec{H}| = \frac{nl}{2}(\cos\theta_2 - \cos\theta_1)$$

$$\cos\theta_2 = -\cos\theta_1 = \frac{l/2}{\left(a^2 + l^2/4\right)^{1/2}}$$

$$|\vec{H}| = \frac{lnl}{2\left(a^2 + l^2/4\right)^{1/2}} = \frac{0.5\times150\times2\times10^{-2}}{2\times10^{-3}\times\sqrt{4^2+10^2}} = \underline{\underline{69.63 \text{ A/m}}}$$

(b)

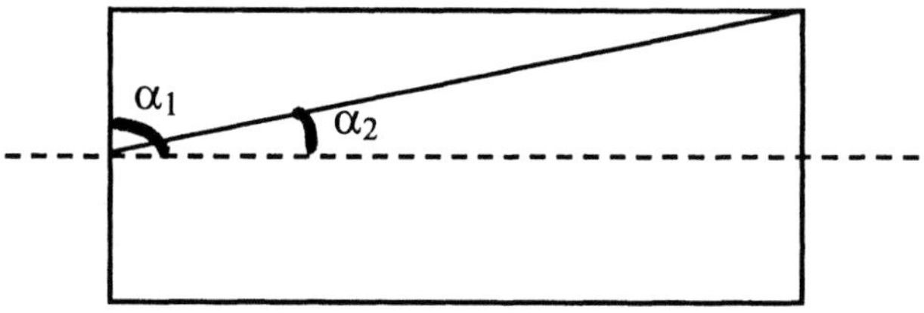

$$\alpha_1 = 90^\circ, \quad \tan\theta_2 = \frac{a}{b} = \frac{4}{20} = 0.2 \rightarrow \quad \theta_2 = 11.31^\circ$$

$$|\vec{H}| = \frac{nl}{2}\cos\theta_2 = \frac{150\times0.5}{2}\cos 11.31^\circ = \underline{\underline{36.77 \text{ A/m.}}}$$

Prob. 7.16

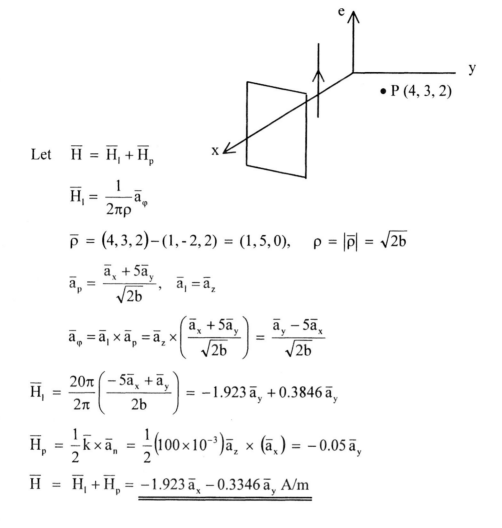

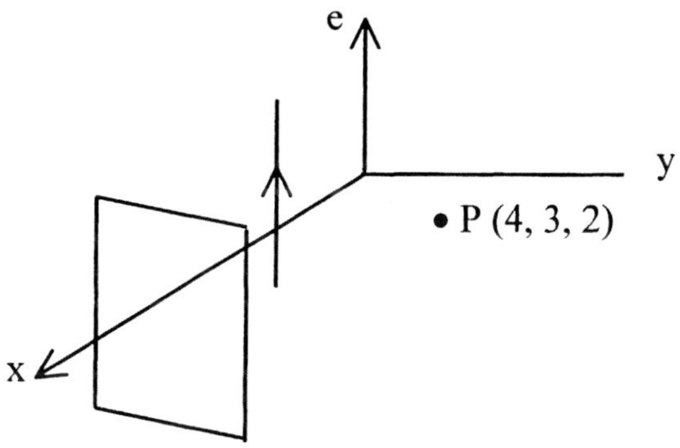

Let $\overline{H} = \overline{H}_l + \overline{H}_p$

$$\overline{H}_l = \frac{1}{2\pi\rho}\bar{a}_\varphi$$

$$\bar{\rho} = (4,3,2)-(1,-2,2) = (1,5,0), \qquad \rho = |\bar{\rho}| = \sqrt{2b}$$

$$\bar{a}_p = \frac{\bar{a}_x + 5\bar{a}_y}{\sqrt{2b}}, \quad \bar{a}_l = \bar{a}_z$$

$$\bar{a}_\varphi = \bar{a}_l \times \bar{a}_p = \bar{a}_z \times \left(\frac{\bar{a}_x + 5\bar{a}_y}{\sqrt{2b}}\right) = \frac{\bar{a}_y - 5\bar{a}_x}{\sqrt{2b}}$$

$$\overline{H}_l = \frac{20\pi}{2\pi}\left(\frac{-5\bar{a}_x + \bar{a}_y}{2b}\right) = -1.923\,\bar{a}_y + 0.3846\,\bar{a}_y$$

$$\overline{H}_p = \frac{1}{2}\bar{k} \times \bar{a}_n = \frac{1}{2}(100\times10^{-3})\bar{a}_z \times (\bar{a}_x) = -0.05\,\bar{a}_y$$

$$\overline{H} = \overline{H}_l + \overline{H}_p = \underline{\underline{-1.923\,\bar{a}_x - 0.3346\,\bar{a}_y \text{ A/m}}}$$

Prob. 7.17 (a) See text.

(b)

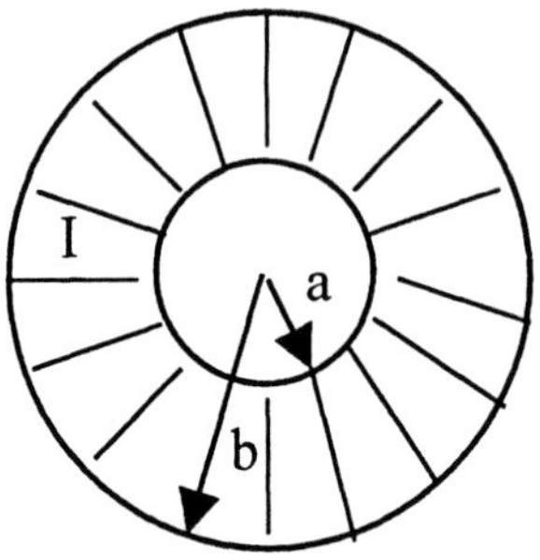

For $\rho < a$, $\oint \overline{H} \cdot dl = I_{enc} = 0 \rightarrow \overline{H} = 0$

For $0 < \rho < b$, $H_\phi \cdot 2\pi\rho = \dfrac{I\pi(\rho^2 - a^2)}{\pi(b^2 - a^2)}$

$$H_\phi = \frac{I}{2\pi\rho}\left(\frac{\rho^2 - a^2}{b^2 - a^2}\right)$$

For $\rho < b$, $H_\phi \cdot 2\pi\rho = I \rightarrow \overline{H}_\varphi = \dfrac{I}{2\pi\rho}$

Thus,

$$H_\phi = \begin{cases} 0, & \rho < a \\ \dfrac{I}{2\pi\rho}\left(\dfrac{\rho^2 - a^2}{b^2 - a^2}\right), & a < \rho < 1 \\ \dfrac{I}{2\pi\rho}, & \rho > b \end{cases}$$

Prob. 7.18

(*a*) Applying Ampere's law,

$$H_\phi \cdot 2\pi\rho = I \cdot \frac{\pi\rho^2}{\pi a^2} \rightarrow H_\phi = I \cdot \frac{I\rho^2}{2\pi a^2}$$

i.e $\overline{H} = \dfrac{I\rho}{2\pi a^2}\overline{a}_\phi$

$$\overline{J} = \overline{\nabla} \cdot \overline{H} = -\frac{\partial H_\phi}{\partial z}\overline{a}_\rho + \frac{I}{\rho}\frac{\partial}{\partial \rho}(\rho H_\phi)\overline{a}_z$$

$$= \frac{I}{\rho}\frac{1}{2\pi a^2} \cdot 2\rho\overline{a}_z = \frac{I}{\pi a^2}\overline{a}_z$$

(*b*) From Prob. 7.15,

$$H_\phi = \begin{cases} \dfrac{I\rho}{2\pi a^2}, & \rho < a \\ \dfrac{I}{2\pi\rho}, & \rho > a \end{cases}$$

At $(0, 1\,\text{cm}, 0)$,

$$H_\phi = \frac{3\times 1\times 10^{-2}}{2\pi\times 4\times 10^{-4}} = \frac{300}{8\pi}$$

$$\overline{H} = \underline{\underline{11.94\,\overline{a}_\phi \text{ A/m}}}$$

At $(0, 4\,\text{cm}, 0)$,

$$H_\phi = \frac{3}{2\pi\times 4\times 10^{-2}} = \frac{300}{8\pi}$$

$$\overline{H} = \underline{\underline{11.94\,\overline{a}_\phi \text{ A/m}}}$$

Prob. 7.19

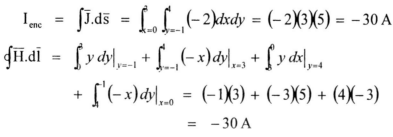

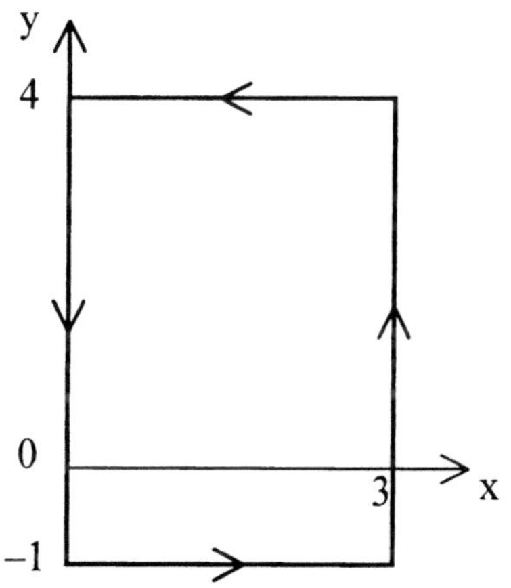

(*a*) $$\overline{J} = \nabla\cdot\overline{H} = \begin{vmatrix} \dfrac{\partial}{\partial x} & \dfrac{\partial}{\partial y} & \dfrac{\partial}{\partial z} \\ y & -x & 0 \end{vmatrix}$$

$$\overline{J} = \underline{\underline{-2\,\overline{a}_z \text{ A/m}^2}}$$

(*b*) $$\oint \overline{H}.d\overline{l} = I_{enc}$$

$$I_{enc} = \int \overline{J}.d\overline{s} = \int_{x=0}^{3}\int_{y=-1}^{4}(-2)\,dx\,dy = (-2)(3)(5) = -30\text{ A}$$

$$\oint \overline{H}.d\overline{l} = \int_0^3 y\,dy\Big|_{y=-1} + \int_{y=-1}^{4}(-x)\,dy\Big|_{x=3} + \int_3^0 y\,dx\Big|_{y=4}$$

$$+ \int_4^{-1}(-x)\,dy\Big|_{x=0} = (-1)(3) + (-3)(5) + (4)(-3)$$

$$= -30\text{ A}$$

Thus, $$\oint \overline{H}.d\overline{l} = I_{enc} = \underline{\underline{-30\text{ A}}}$$

Prob. 7.20

(a) $$\bar{J} = \nabla \times \overline{H} = \begin{vmatrix} \frac{\partial}{\partial x} & \frac{\partial}{\partial y} & \frac{\partial}{\partial z} \\ yz(x^2+y^2) & -y^2 \times z & -4x^2y^2 \end{vmatrix}$$

$$= \left(8x^2y + xy^2\right)\bar{a}_x + \left[y(x^2+y^2) - 4xy^2\right]\bar{a}_y + \left[-y^2z - z(x^2+y^2)\right]\bar{a}_z$$

At $(5,2,-3)$, $x=5$, $y=2$, $z=-3$

$\bar{J} = 420\,\bar{a}_z - 22\,\bar{a}_y + 99\,\bar{a}_z$ A/m^2

(b) $$I = \int J \cdot dS = \iint\left(8x^2y + xy^2\right)dydz\Big|_{x=-1}$$

$$= \int_0^2 dz \int_0^2 \left(8y - y^2\right)dz = 2\left(4y^2 - \frac{y^3}{3}\right)\Big|_0^2$$

$$= 4\left(16 - \frac{8}{3}\right) = 53.33\text{ A}$$

(c) $\overline{B} = \mu\overline{H}$, $\nabla\cdot\overline{B} = 0 \rightarrow \nabla\cdot\overline{H} = 0$

$$\nabla\cdot\overline{H} = \frac{\partial}{\partial x}H_x + \frac{\partial}{\partial y}H_y + \frac{\partial}{\partial z}H_z = 2xy - 2yxz = 0$$

Hence $\nabla\cdot\overline{B} = 0$

Prob. 7.21

(a) $\overline{B} = \dfrac{\mu_o I}{2\pi\ell}\bar{a}_\phi$. At $(-3,4,5)$, $\rho = 5$

$$\overline{B} = \frac{4\pi\times10^{-7}\times2}{2\pi(5)}\bar{a}_\phi = 80\,\bar{a}_\phi\text{ nWb/m}^2$$

(b) $$\phi = \int\overline{B}.dS = \frac{\mu_o I}{2\pi}\iint\frac{d\rho\,dz}{\rho}$$

$$= \frac{4\pi\times10^{-7}\times2}{2\pi}\ln\rho\Big|_2^6 z\Big|_0^4 = 16\times10^{-7}\ln 3$$

$= 1.756\,\mu$Wb.

Prob. 7.22

$$\psi = \int \bar{B}.d\bar{s} = \mu_o \int_{z=0}^{0.2} \int_{\phi=0}^{50^o} \frac{10^6}{\rho} \sin 2\phi \, \rho \, d\phi \, dz$$

$$\psi = 4\pi \times 10^{-7} \times 10^6 (0.2) \left(-\frac{\cos 2\phi}{2} \right) \Bigg|_0^{50^o}$$

$$= 0.04\pi \left(1 - \cos 100^o\right)$$

$$= \underline{\underline{0.1475 \text{ Wb}}}$$

Prob. 7.23

Let $\bar{H} = \bar{H}_1 + \bar{H}_2$

where $\bar{H}_1$ and $\bar{H}_2$ are due to the wires centered at x = 0 and x = 10cm respectively.

(a) For $\bar{H}_1, \rho = 50\text{ cm}, \ \bar{a}_\phi = \bar{a}_l \times \bar{a}_\rho = \bar{a}_z \times \bar{a}_x = \bar{a}_y$

$$\bar{H}_1 = \frac{5}{2\pi\left(5 \times 10^{-2}\right)} \bar{a}_y = \frac{50}{\pi} \bar{a}_y$$

For $\bar{H}_2, \rho = 5\text{ cm}, \ \bar{a}_\phi = -\bar{a}_z \times -\bar{a}_x = \dot{a}_y, \ \bar{H}_2 = \bar{H}_1$

$$\bar{H} = 2\bar{H}_1 = \frac{100}{\pi} \bar{a}_y$$

$$= 31.43\, \bar{a}_y \text{ A/m}$$

(b) For $\bar{H}_1, \ \bar{a}_\phi = \bar{a}_z \times \left(\frac{2\bar{a}_x + \bar{a}_y}{\sqrt{5}} \right) = \frac{2\bar{a}_y - \bar{a}_x}{\sqrt{5}}$

$$\bar{H}_1 = \frac{5}{2\pi 5\sqrt{5} \times 10^{-2}} \left(\frac{-\bar{a}_x + 2\bar{a}_y}{\sqrt{5}} \right) = -3.183\, \bar{a}_x + 6.366\, \bar{a}_y$$

For $\bar{H}_2, \bar{a}_\rho = -\bar{a}_z \times \bar{a}_y = \dot{a}_x$

$$\bar{H}_2 = \frac{5}{2\pi(5)} \bar{a}_x = 15.924\, \bar{a}_x$$

$$\bar{H} = \bar{H}_1 + \bar{H}_2$$

$$= 12.79\, \bar{a}_x + 6.366\, \bar{a}_y \text{ A/m}$$

Prob. 7.24

$$\overline{B} = \frac{\mu_0 I}{2\pi\rho}\overline{a}_\varphi$$

$$\psi = \overline{B}\cdot d\overline{s} = \int_{\rho=d}^{d+a}\int_{z=0}^{b}\frac{\mu_0 I}{2\pi\rho}d\rho dz$$

$$= \frac{\mu_0 Ib}{2\pi}\ln\frac{d+a}{d}$$

Prob. 7.25

On the slant side of the ring, $z = \frac{h}{6}(\rho - a)$

where $\overline{H}_1$ and $\overline{H}_2$ are due to the wires centered at x = 0 and x = 10cm respectively.

$$\psi = \int\overline{B}.d\overline{s} = \int\frac{\mu_0 I}{2\pi\rho}d\rho\, dz$$

$$= \frac{\mu_0 I}{2\pi\rho}\int_{\rho=a}^{a+b}\int_{z=0}^{\frac{h}{b}}\frac{dz\, d\rho}{\rho} = \frac{\mu_0 Ih}{2\pi b}\int_{\rho=a}^{a+b}\left(1-\frac{a}{\rho}\right)d\rho$$

$$= \frac{\mu_0 Ih}{2\pi b}\left(b - a\ln\frac{a+b}{a}\right) \text{ as required.}$$

If $a = 30$ cm, $b = 10$ cm, $h = 5$ cm, $I = 10$ A,

$$\psi = \frac{2\pi\times10^{-7}\times10\times0.05}{2\pi(5\times10^{-2})}\left(0.1 - 0.3\ln\frac{4}{3}\right)$$

$$= 1.37\times10^{-8}\text{ Wb}$$

Prob. 7.26

(*a*) $\nabla\cdot\overline{A} = -ya\sin ax \neq 0$

$$\nabla\times\overline{H} = \begin{vmatrix}\frac{\partial}{\partial x} & \frac{\partial}{\partial y} & \frac{\partial}{\partial z}\\ y\cos ax & 0 & y+e^{x}\end{vmatrix}$$

$$= \overline{a}_x + e^{-x}\overline{a}_y - \cos a_x\overline{a}_z \neq 0$$

$\overline{A}$ is neither electrostatic nor magnetostatic field (*b*) $\nabla \cdot \overline{B} = \frac{1}{\rho}\frac{\partial}{\partial \rho}(\rho\, B_\rho) = \frac{1}{\rho}\frac{\partial}{\partial \rho}(20) = 0$

$\nabla \times \overline{B} = 0$

$\overline{B}$ can be $\overline{E}$ - field in a charge - free region.

(*c*) $\nabla \cdot \overline{C} = \frac{1}{r^2} 4r^3 \sin\theta \neq 0$

$\nabla \times \overline{C} = \frac{1}{r\sin\theta}\frac{\partial}{\partial r}\left(r^2 \sin^2\theta\right) \neq 0$

$\overline{C}$ is neither or $\overline{E}$ nor $\overline{H}$ field.

Prob. 7.27

(*a*) $\nabla \cdot \overline{D} = 0$

$$\nabla \times \overline{H} = \begin{vmatrix} \frac{\partial}{\partial x} & \frac{\partial}{\partial y} & \frac{\partial}{\partial z} \\ u^2 z & 2(x+1)yz & -(x+1)z^2 \end{vmatrix}$$

$$= 2(x+1)y\overline{a}_x + \ldots \neq 0$$

$\overline{D}$ is a magnetostatic field.

(*b*) $\nabla \cdot \overline{E} = 0$

$\nabla \times \overline{E} = \frac{1}{\rho^2}\cos\theta\,\overline{a}_\rho + \ldots \neq 0$

$\overline{E}$ can be a magnetostatic field.

(*c*) $\nabla \cdot \overline{F} = 0$

$\nabla \times \overline{F} = \frac{1}{r}\left[\frac{\partial}{\partial r}\left(r^{-1}\sin\theta\right) + \frac{2\sin\theta}{r^2}\right]\overline{a}_\theta \neq 0$

$\overline{F}$ can be a magnetostatic field.

Prob. 7.28

(a) $$\overline{B} = \nabla \times \overline{A} = \begin{vmatrix} \frac{\partial}{\partial x} & \frac{\partial}{\partial y} & \frac{\partial}{\partial z} \\ 2x^2y + yz & xy^2 - xz^2 & -6xy + 2z^2y^2 \end{vmatrix}$$

$$\overline{B} = (-6xz + 4z^2y + 2xz^2)\overline{a}_x + (y + 4yz)\overline{a}_y + (y^2 - z^2 - 2x^2 - z)\overline{a}_z \text{ Wb/m}^2$$

(*b*) $\psi = \int \overline{B} \cdot dS, \qquad dS = dy\, dz\, dx$

$$\psi = \int_{z=0}^{2} \int_{y=0}^{2} \left(-6xz + 4zy - 2xy\right) dy\, dz \Big|_{x=1}$$

$$= \int \int_0 \left(-6z\right) dy\, dz + 4 \int \int_0 z^2 y\, dy\, dz + 2 \int \int_0 y\, dy\, dz$$

$$= -8 \int_0^2 z\, dz \int_0^2 dy + 4 \int_0^2 z^2\, dz \int_0^2 y\, dy$$

$$= -8 \frac{z^2}{2}\Big|_0^2 (2) + 4 \frac{z^3}{3}\Big|_0^2 \left(\frac{y^2}{2}\Big|_0^2\right) = -32 + \frac{64}{3}$$

$\psi = -10.67$ Wb

$\overline{E}$ can be a magnetost atic field.

(*c*) $\overline{\nabla} \cdot \overline{A} = \partial A_x + \frac{\partial A_y}{} + \frac{\partial A_z}{xz} = 4xy + 2xy - 6xy = 0$

$$\overline{\nabla} \cdot \overline{B} = -6z + 3z^3 + 1 + 6z - 3z^3 - 1 = 0$$

Prob. 7.29

$$\overline{B} = \overline{\nabla} \times \overline{A} = \frac{1}{\rho}\frac{\partial A_z}{\partial \phi}\overline{a}_\rho - \frac{\partial A_z}{\partial \rho}\overline{a}_\phi$$

$$= \frac{15}{\rho} e^{-\rho} \cos\phi\, \overline{a}_\rho + 15\, e^{-\rho} \sin\phi\, \overline{a}_\phi$$

$$\overline{B}\left(3, \frac{\pi}{4}, -10\right) = 5\, e^{-3} \frac{1}{\sqrt{2}} \overline{a}_\rho + 15\, e^{-3} \frac{1}{\sqrt{2}} \overline{a}_\phi$$

$$\overline{H} = \frac{\overline{B}}{\mu_o} = \frac{10^7}{4\pi} \frac{15}{\sqrt{2}} e^{-3} \left(\frac{1}{3}\overline{a}_\rho + \overline{a}_\phi\right)$$

$$\overline{H} = \left(14\, \overline{a}_\rho + 42\, \overline{a}_\phi\right) \cdot 10^4 \text{ A/m}$$

$$\psi = \int \overline{B} \cdot d\overline{s} = \iint \frac{15}{\rho} e^{-\rho} \cos\phi\, \rho\, d\phi\, dz$$

$$= 15\, z\Big|_0^{10} \left(-\sin\phi\right)\Big|_0^{\pi/2} e^{-5} = -150\, e^{-5} \qquad \Rightarrow \qquad \psi = -1.011 \text{ Wb}$$

Prob. 7.30

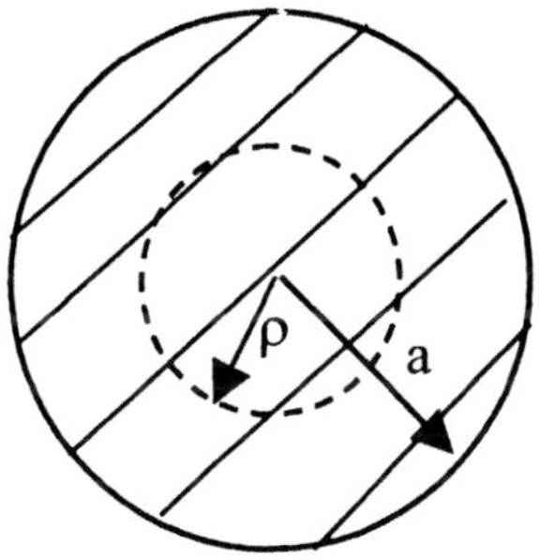

Applying Ampere's law gives

$$H_\varphi \cdot 2\pi\rho = \tau_o \cdot \pi\rho^2$$

$$H_\varphi = \frac{\tau_o}{2}\rho$$

$$B_\varphi = \mu_o H_\varphi = \mu_o \frac{\tau_o \rho}{2}$$

But $B_\varphi = \nabla \times \overline{A} = -\frac{\partial A_Z}{\partial \rho}\bar{a}_\varphi + \ldots$

$$-\frac{\partial A_Z}{\partial \rho} = \frac{1}{2}\mu\tau_o\rho$$

$$A_Z = -\mu_o \frac{\tau_o \rho^2}{4}$$

or $\overline{A} = \frac{1}{4}\mu_o\tau_o\rho^2\,\bar{a}_z$

Prob. 7.31

$$\overline{A} = \frac{I_o\mu_o}{4\pi a^2}\left(x^2 + y^2\right)\bar{a}_z = -\frac{I_o\mu_o\rho^2}{4\pi a^2}\bar{a}_z$$

$$\overline{B} = \nabla \times \overline{A} = \frac{I_o\mu_o\rho}{4\pi a^2}\bar{a}_\phi = \mu_o\overline{H}$$

i.e. $\overline{H} = \frac{I_o\rho}{2\pi a^2}\bar{a}_\phi = \frac{I_o\sqrt{x^2+y^2}}{2\pi a^2}\bar{a}_\phi$

By Ampere's law, $\oint \overline{H}.d\bar{l} = I_{enc}$

$$H_\phi \cdot 2\pi\rho = I_o \cdot \frac{\rho^2}{a^2}$$

or $\overline{H} = \frac{I_o\rho}{2\pi a^2}\,\bar{a}_\phi$

Prob. 7.32

$$\overline{A} = \frac{\mu I}{2\pi}\left[In(d-\rho) - In\,\rho\right]\overline{a}_z$$

$$\overline{B} = \overline{\nabla}\cdot\overline{A} = \frac{-\partial A_Z}{\partial\rho}\overline{a}_\varphi = -\frac{\mu_o I}{2\pi}\left[-\frac{1}{d-\rho} - \frac{1}{\rho}\right]\overline{a}_\varphi$$

$$= \frac{\mu_o Ld}{2\pi\rho(d-\rho)}\overline{a}_\varphi$$

Prob. 7.33

$$\overline{J} = \overline{\nabla}\times\overline{H} = \overline{\nabla}\times\frac{\overline{\nabla}\times\overline{A}}{\mu_o} = \frac{1}{\mu_o}\overline{\nabla}\times\overline{\nabla}\times\overline{A}$$

$$\overline{\nabla}\times\overline{A} = \frac{\partial A_z}{\partial\rho}\overline{a}_\psi = \frac{+20}{\rho^3}\overline{a}_\psi$$

$$\overline{\nabla}\times\overline{\nabla}\times\overline{A} = \frac{1}{\rho}\frac{\partial}{\partial d}(\rho A_\phi)\overline{a}_z = -\frac{40}{\rho^4}\overline{a}_z$$

$$\overline{J} = -\frac{40}{\mu_o\rho^4}\overline{a}_z \text{ A/m}^2$$

or $\quad \overline{\nabla}^2\overline{A} = -\mu_o\overline{J}$

or $\quad \overline{J} = -\frac{1}{\mu_o}\overline{\nabla}^2\overline{A} = -\frac{1}{\mu_o}\overline{\nabla}^2 A_z\overline{a}_z$

$$= -\frac{1}{\mu_o}\overline{a}_z\left[\frac{1}{\rho}\frac{\partial}{\partial\rho}\left(\rho\frac{\partial A_z}{\partial\rho}\right) + \frac{1}{\rho^2}\frac{\partial^2 A_z}{\partial\phi^2} + \frac{\partial^2 A_z}{\partial z^2}\right]$$

$$= \frac{1}{\mu_o}\overline{a}_z\frac{1}{\rho}\frac{\partial}{\partial\rho}\left(\frac{20}{\rho^2}\right) = -\frac{40}{\mu_o\rho^4}\overline{a}_z \text{ A/m}^2$$

Prob. 7.34

$$\overline{H} = -\overline{\nabla}V_m \rightarrow V_m = -\int \overline{H}\cdot d\overline{l}$$

From Example 7.3, $\overline{H} = \dfrac{Ia^2}{2(z^2+a^2)^{3/2}}\,\overline{a}_z$

$$V_m = -\frac{Ia^2}{2}\int 2(z^2+a^2)^{-3/2}dz = \frac{-Ia^2}{2(z^2+a^2)^{1/2}} + c$$

As $z\rightarrow\infty$, $V_m = 0$, i.e.

$$0 = -\frac{I}{2} + c \rightarrow c = \frac{I}{2}$$

Hence,

$$V_m = \frac{I}{2}\left[1 - \frac{z}{\sqrt{z^2+a^2}}\right]$$

Prob. 7.35

For the outer conductor,

$$J_z = -\frac{I}{\pi(c^2-b^2)} = -\frac{I}{\pi(16-9)a^2} = -\frac{I}{7\pi a^2}$$

Let $\overline{A} = A_z\,\overline{a}_z$. Using Poisson's equation,

$$\overline{\nabla}^2 A_z = -\mu_o J_z$$

$$\frac{1}{\rho}\frac{\partial}{\partial\rho}\left(\rho\frac{\partial A_z}{\partial\rho}\right) = \frac{\mu_o I}{7a^2\pi}$$

or
$$\frac{\partial}{\partial\rho}\left(\rho\frac{\partial A_z}{\partial\rho}\right) = \frac{\mu_o I\rho}{7\pi a^2}$$

Integrating once,

$$\rho\frac{\partial A_z}{\partial\rho} = \frac{\mu_o I\rho^2}{14\pi a^2} + c_1$$

or
$$\frac{\partial A_z}{\partial\rho} = \frac{\mu_o I\rho}{14\pi a^2} + \frac{c_1}{\rho}$$

Integrating again,

$$A_z = \frac{\mu_o I\rho^2}{28\pi a^2} + c_1 \, In\,\rho + c_2$$

But $A_z = 0$ when $\rho = 3a$.

$$0 = \frac{9}{28\pi}\mu_o I + c_1 \, In\,3a + c_2$$

$$c_2 = -c_2 \,\text{In}\,3a - \frac{9}{28\pi}\mu_o I$$

i.e. $$A_z = \frac{\mu_o I}{28\pi}\left(\frac{\rho^2}{a^2} - a\right) + c_1 \, In\frac{\rho}{3a}$$

But $\nabla \times \overline{A} = \overline{B} = \mu_o \overline{H}$

$$\nabla \times \overline{A} = \frac{\partial A_z}{\partial \rho}\overline{a}_\phi = -\left(\frac{\mu_o I\rho}{14\pi a^2} + \frac{c_1}{\rho}\right)\overline{a}_\phi$$

At $\rho = 3a$, $\int \overline{H}.d\overline{l} = I \rightarrow 2\pi(3a)H_\phi = 1$

or $$H_\phi = \frac{1}{6\pi a}$$

Thus $\nabla \times \overline{A}\big|_{\rho=3a} = \mu_o \overline{H}\,(\rho = 3a)$ implies that

$$-\left(\frac{3\mu_o I}{14\pi a} + \frac{c_1}{3a}\right) = \frac{\mu_o I}{6\pi a}$$

or $$c_1 = -\frac{I\mu_o}{2\pi} - \frac{9\mu_o I}{14\pi} = -\frac{16\mu_o I}{14\pi}$$

Thus,

$$A_z = \frac{\mu_o I}{28\pi}\left(\frac{\rho^2}{a^2} - a\right) - \frac{8\mu_o I}{7\pi}\,\text{In}\,\frac{\rho}{3a}$$

Prob. 7.36

$$\overline{H} = \frac{I}{2\pi\rho}\overline{a}_\varphi$$

But $$\overline{H} = -\nabla V_m \quad (\overline{T} = 0)$$

$$\frac{I}{2\pi\rho}\bar{a}_\varphi = -\frac{1}{\rho}\frac{\partial V_m}{\partial \phi}\bar{a}_\varphi \rightarrow V_m = -\frac{I}{2\pi}\phi + C$$

At $(10, 60°, 7)$, $\phi = 60° = \frac{\pi}{3}$, $V_m = 0 \rightarrow 0 = -\frac{I}{2\pi}\cdot\frac{\pi}{3} + C$

or $\quad C = \frac{I}{6}$

$$V_m = -\frac{I}{2\pi}\phi + \frac{I}{6}$$

At $(4, 30°, -2)$, $\phi = 30° = \frac{\pi}{6}$,

$$V_m = -\frac{I}{2\pi}\cdot\frac{n}{6} + \frac{I}{6} = \frac{I}{12} = \frac{12}{12}$$

$$V_m = 1\,A$$

Prob. 7.37

For an infinite current sheet,

$$\bar{H} = \frac{1}{2}\bar{K}\times\bar{a}_n = \frac{1}{2}50\bar{a}_y\times\bar{a}_n = 25\bar{a}_x$$

But $\bar{H} = -\nabla V_m \; (\bar{J} = 0)$

$$25\,\bar{a}_x = -\frac{\partial V_m}{\partial x}\bar{a}_n \rightarrow V_m = -25x + c$$

At the origin, $x = 0$, $V_m = 0$, $c = 0$, i.e.

$$V_m = -25x$$

(*a*) At $(-2, 0, 5)$, $V_m = 50A$.

(*b*) At $(10, 3, 1)$, $V_m = -250A$.

Prob. 7.38

(a)
$$\nabla x \nabla V = \nabla \times \left(\frac{\partial V}{\partial \rho} \bar{a}_\rho + \frac{1}{\rho} \frac{\partial V}{\partial \phi} \bar{a}_\phi + \frac{\partial V}{\partial z} \bar{a}_z \right)$$

$$= \left(\frac{1}{\rho} \frac{\partial^2 V}{\partial \phi \partial z} - \frac{1}{\rho} \frac{\partial^2 V}{dz \partial \phi} \right) \bar{a}_\rho + \left(\frac{\partial^2 V}{\partial z \partial \rho} - \frac{\partial^2 V}{\partial \rho \partial z} \right) \bar{a}_\phi$$

$$+ \frac{1}{\rho} \left(\frac{\partial^2 V}{\partial \rho \partial \phi} - \frac{\partial^2 V}{\partial \phi d \rho} \right) \bar{a}_z = 0$$

(b)
$$\nabla \cdot \left(\nabla \times \bar{A} \right) = \nabla \bullet \left[\left(\frac{1}{\rho} \frac{\partial A_m}{\partial \phi} - \frac{\partial A_\rho}{\partial z} \right) \bar{a}_\rho \right.$$

$$\left. + \left(\frac{\partial A_m}{\partial z} - \frac{\partial A_\rho}{\partial \rho} \right) \bar{a}_\phi + \frac{1}{\rho} \left(\frac{\partial}{\partial \rho} (\rho A_\phi) - \frac{\partial A_\rho}{\partial \phi} \right) \bar{a}_z \right]$$

$$= \frac{1}{\rho} \frac{\partial^2 A_m}{\partial \rho \partial \phi} - \frac{1}{\rho} \frac{\partial}{\partial \rho} \left(\rho \frac{\partial A_\phi}{\partial z} \right) + \frac{1}{\rho} \frac{\partial^2 A_\rho}{\partial \phi \partial z} - \frac{1}{\rho} \frac{\partial^2 A_z}{\partial \phi \partial \rho} + \frac{\partial}{\partial z} \left(\frac{1}{\rho} \frac{\partial}{\partial \rho} (\rho A_\phi) \right)$$

$$- \frac{\partial}{\partial z} \left(\frac{1}{\rho} \frac{\partial A_\rho}{\partial \phi} \right)$$

$$= - \frac{\partial^2 A_\phi}{\partial \rho \partial z} - \frac{1}{\rho} \frac{\partial A_\phi}{\partial z} + \frac{\partial^2 A_\phi}{\partial z \partial \rho} + \frac{1}{\rho} \frac{\partial A_\phi}{\partial z} = 0$$

Prob. 7.39

$$R = | \bar{r} - \bar{r} | = \left[(x - x^1)^2 + (y - y^1)^2 + (z - z^1)^2 \right]^{1/2}$$

$$\nabla \frac{1}{R} = \left(\frac{\partial}{\partial x} \bar{a}_x + \frac{\partial}{\partial y} \bar{a}_y + \frac{\partial}{\partial z} \bar{a}_z \right) \left[(x - x^1)^2 + (y - y^1)^2 + (z - z^1)^2 \right]^{1/2}$$

$$= -\frac{1}{2} 2 (x - x^1) \bar{a}_x \left[(x - x^1)^2 + (y - y^1)^2 + (z - z^1)^2 \right]^{-3/2}$$

$$= -\left[(x - x^1) \bar{a}_x + (y - y^1) \bar{a}_y + (z - z^1) \bar{a}_z \right] / R^3 = -\frac{\bar{R}}{R^3}$$

$$\nabla' \frac{1}{R} = \left(\frac{\partial}{\partial x^1} \bar{a}_x + \frac{\partial}{\partial y^1} \bar{a}_y + \frac{\partial}{\partial z^1} \bar{a}_z \right) \left[(x - x^1)^2 + (y - y^1)^2 + (z - z^1)^2 \right]^{-1/2}$$

$$= \left(-\frac{1}{2} \right) (-2) (x - x^1) \bar{a}_x \left[(x - x^1)^2 + (y - y^1)^2 + (z - z^1)^2 \right]^{-3/2} = \frac{\bar{R}}{R^3}$$

CHAPTER 8

P.E. 8.1

(a) $F = m\frac{\partial \bar{u}}{\partial t} = Q\bar{E} = \underline{\underline{6\bar{a}_z N}}$

(b) $\frac{\partial \bar{u}}{\partial t} = 6\bar{a}_z = \frac{\partial}{\partial t}(u_x, u_y, u_z) \Rightarrow$

$$\frac{\partial u_x}{\partial t} = 0 \rightarrow u_x = A$$

$$\frac{\partial u_y}{\partial t} = 0 \rightarrow u_y = B$$

$$\frac{\partial u_z}{\partial t} = 6 \rightarrow u_z = 6t + C$$

Since $\bar{u}(t=0) = 0$, $\quad$ A = B = C = 0

$u_x = 0 = u_y, \quad u_z = 6t$

$$u_x \frac{\partial x}{\partial t} = 0 \rightarrow x = A$$

$$u_y \frac{\partial y}{\partial t} = 0 \rightarrow y = B$$

$$u_z \frac{\partial z}{\partial t} = 6t \rightarrow z = 3t^2 + C_1$$

At t = 0, (x,y,z) = (0,0,0) $\rightarrow A_1 = 0 = B_1 = C_1$

Hence , (x,y,z) = $(0,0,3t^2)$,

$u = 6ta_z$ at any time. At P(0,0,12), $z = 12 = 3t^2 \rightarrow t = 2s$

$\underline{\underline{t = 2s}}$

(c) $u = 6t\bar{a}_z = 12a_z m/s$.

$$a = \frac{\partial \bar{U}}{\partial t} = \underline{\underline{6a_z \, m/s^2}}$$

(d) $K.E = \frac{1}{2}m|\bar{U}|^2 = \frac{1}{2}(1)(144) = \underline{\underline{72J}}$

P.E. 8.2

(a) $\quad m\vec{a} = e\vec{u}xB$ = $(eB_o uy, -eB_o ux, 0)$

$$\frac{d^2x}{dt^2} = \frac{eBo}{m}\frac{dy}{dt} = w\frac{dy}{dt} \qquad (1)$$

$$\frac{d^2y}{dt^2} = -\frac{eBo}{m}\frac{dx}{dt} = -w\frac{dx}{dt} \qquad (2)$$

$$\frac{d^2z}{dt^2} = 0; \Rightarrow \frac{dz}{dt} = C_1 \qquad (3)$$

From (1) and (2),

$$\frac{d^3x}{dt^3} = w\frac{d^2y}{dt^2} = -w^2\frac{dx}{dt}$$

$(D^2 + w^2 D)x = 0 \rightarrow Dx = (0, \pm jw)x$

$x = c_2 + c_3 \cos wt + c_4 \sin wt$

$$\frac{dy}{dt} = \frac{1}{w}\frac{d^2x}{dt^2} = -c_3 w \cos twt - c_4 w \sin wt$$

At $t = 0$, $\vec{u} = (\alpha, 0, \beta)$. Hence,

$$c_1 = \beta, c_3 = 0, c_4 = \frac{\alpha}{w}$$

$$\frac{dx}{dt} = \alpha \cos wt, \frac{dy}{dt} = -\alpha \sin wt, \frac{dz}{dt} = \beta$$

(b) Solving these yields

$$x = \frac{a}{w}\sin wt, y = \frac{\alpha}{w}\cos wt, z = \beta t$$

(c) $x^2 + y^2 = \frac{\alpha^2}{w^2}$, $z=\beta t$

showing that the particles move along a helix of radius α/w placed along the z-axis.

P.E. 8.3

(a) From Example 8.3, QuB = QE regardless of the sign of the charge.

$E = uB = 8 \times 10^6 \times 0.5 \times 10^{-3} = 4$ kV/m

(b) Yes, since QuB = QE holds for any Q and m.

P.E. 8.4

By Newton's 3rd law, $\vec{F}_{12} = \vec{F}_{21}$, the force on the infinitely long wire is:

$$\vec{F}_l = -\vec{F} = \frac{\mu_o I_1 I_2 b}{2\pi}\left(\frac{1}{\rho_o} - \frac{1}{\rho_o + a}\right)\vec{a}_\rho$$

$$= \frac{4\pi\times10-7\times50\times3}{2\pi}\left(\frac{1}{2}-\frac{1}{3}\right)\vec{a}_\rho = \underline{\underline{5\vec{a}_\rho\ \mu N}}$$

P.E. 8.5

$$\vec{m} = IS\vec{a}_n = 10\times10^{-4}x50\frac{(2,6,-3)}{7}$$

$$= 7.143 \text{ x } 10^{-3} (2, 6, -3)$$

$$= \underline{\underline{(1.429\,\vec{a}_x + 4.286\,\vec{a}_y - 2.143\,\vec{a}_z) \text{ x } 10^{-2}\text{ A-m}^2}}$$

P.E. 8.6

(a) $$\vec{T} = \vec{m}\times B = \frac{10\times10^{-4}\times50}{7\times10}\begin{vmatrix}2 & 6 & -3\\ 6 & 4 & 5\end{vmatrix}$$

$$= \underline{\underline{0.03\,\vec{a}_x - 0.02\,\vec{a}_y - 0.02\,\vec{a}_z \text{ N-m}}}$$

(b) $$\left|\vec{T}\right| = ISB\sin\theta \rightarrow \left|\vec{T}\right|_{\max} = ISB$$

$$\left|\vec{T}\right|_{\max} = \frac{50\times10^{-2}}{10}\left|6\vec{a}_x+4\vec{a}_y+5\vec{a}_z\right| = 0.4387$$

or $$\left|\vec{T}\right|_{\max} = \left|\vec{m}\times\vec{B}\right| = \left|-0.3055\vec{a}_x + 0.076\vec{a}_y + 0.3055\vec{a}_z\right| = \underline{\underline{0.4387}}\text{ Nm}$$

P.E. 8.7

(a) $$\mu_r = \frac{\mu}{\mu_o} = 4.6,\ \chi_m = \mu_r - 1 = \underline{\underline{3.6}}$$

(b) $$\vec{H} = \frac{\vec{B}}{\mu} = \frac{10\times10^{-3}e^{-y}}{4\pi\times10^{-7}\times4.6}\vec{a}_z\ A/m \quad = \underline{\underline{1730e^{-y}\vec{a}_z}}\text{ A/m}$$

(c) $$M = \chi_m\vec{H} = \underline{\underline{6228e^{-y}}}\text{ A/m}$$

P.E. 8.8

$$\vec{a}_n = \frac{3\vec{a}_x+4\vec{a}_y}{5}$$

$$\vec{B}_{1n} = (\vec{B}_1\bullet\vec{a}_n)\vec{a}_n = \frac{(6+32)(6\vec{a}_x+8\vec{a}_y)}{1000}$$

$$= 0.228\,\vec{a}_x + 0.304\,\vec{a}_y = B_{2n}$$

$$\vec{B}_{1t} = (\vec{B}_1\bullet\vec{B}_{1n}) = -0.128\vec{a}_x + 0.096\vec{a}_y + 0.2\vec{a}_z$$

$$\vec{B}_{2t} = \frac{\mu_2}{\mu_1}\vec{B}_{1t} = 10\vec{B}_{1t} = -1.28\vec{a}_x + 0.96\vec{a}_y + 2\vec{a}_z$$

$$\vec{B}_2 = \vec{B}_{2n} + \vec{B}_{2t} = \underline{\underline{-1.052\vec{a}_x + 1.264\vec{a}_y + 2\vec{a}_z}} \text{ Wb/m}^2$$

P.E. 8.9

(a) $\vec{B}_{1n} = \vec{B}_{2n} \rightarrow \mu_1\vec{H}_{1n} =_z \mu_2\vec{H}_{2n}$

or $\mu_1\vec{H}_1 \bullet \vec{a}_{n21} = \mu_2\vec{H}_2 \bullet \vec{a}_{n21}$

$$\mu_o \frac{(60+2-36)}{7} = 2\mu_o \frac{(6H_{2x}+10-12)}{7}$$

$$35 = 6H_{2x}$$

$$\underline{\underline{H_{2x} = 5.833}}$$

(b) $\vec{K} = (\vec{H}_1 - \vec{H}_2)\times\vec{a}_{n12} = \vec{a}_{n21}\times(\vec{H}_1 - \vec{H}_2)$

$$= \vec{a}_{n21}\times\left[(10,1,12) - (35/6, -5, 4)\right]$$

$$= \frac{1}{7}\begin{vmatrix} 6 & 2 & -3 \\ 25/6 & 6 & 8 \end{vmatrix}$$

$$\vec{K} = \underline{\underline{4.86\vec{a}_x - 8.64\vec{a}_y + 3.95\vec{a}_z}} \text{ A/m}$$

(c) Since $\vec{B} = \mu\vec{H}$, $\vec{B}_1$ and $\vec{H}_1$ are parallel, i.e. they make the same angle with the normal to the interface.

$$\cos\theta_1 = \frac{\vec{H}_1 \bullet \vec{a}_{n21}}{|\vec{H}_1|} = \frac{26}{7\sqrt{100+1+144}} = 0.2373$$

$$\underline{\underline{\theta_1 = 76.27^o}}$$

$$\cos\theta_2 = \frac{\vec{H}_2 \bullet \vec{a}_{n21}}{|\vec{H}_2|} = \frac{13}{7\sqrt{(5.833)^2+25+16}} = 0.2144$$

$$\underline{\underline{\theta_2 = 77.62^o}}$$

P.E. 8.10

(a) $L' = \mu_o\mu_r n^2 S = 4\pi\times10^{-7}\times1000\times16\times10^6\times4\times10^{-4}$

$= \underline{\underline{8.042 \text{ H/m}}}$

(b) $W_m' = \frac{1}{2}L'I^2 = \frac{1}{2}(8.042)(0.5^2) = \underline{\underline{1.005}}$ J/m

P.E. 8.11 From Example 8.11,

$$L_{in} = \frac{8I}{8\pi}$$

$$L_{ext} = \frac{2w_m}{I^2} = \frac{1}{I^2}\iiint \frac{\mu I^2}{4\pi^2\rho^2}\rho d\rho d\phi dz$$

$$= \frac{1}{4\pi^2}\int_0^l dz \int_0^{2\pi} d\phi \int_a^b \frac{2\mu_o}{(1+\rho)\rho} d\rho$$

$$= \frac{\mu_o l}{\pi} \bullet 2\pi l \int_a^b \left[\frac{1}{\rho} - \frac{1}{(1+\rho)}\right] d\rho$$

$$= \frac{\mu_o l}{\pi}\left[\ln\frac{b}{a} - \ln\frac{1+b}{1+a}\right]$$

$$L = L_{in} + L_{ext} = \underline{\underline{\frac{\mu_o l}{8\pi} + \frac{\mu_o l}{\pi}\left[\ln\frac{b}{a} - \ln\frac{1+b}{1+a}\right]}}$$

P.E. 8.12

(a) $L'_{in} = \frac{\mu_o}{8\pi} = \frac{4\pi\times10^{-7}}{8\pi} = \underline{\underline{0.05\ \mu H/m}}$

$L'_{ext} = L' - L'_{in} = 1.2 - 0.05 = \underline{\underline{1.15\ \mu H/m}}$

(b) $L' = \frac{\mu_o}{2\pi}\left[\frac{1}{4} + \ln\frac{d-a}{a}\right]$

$$\ln\frac{d-a}{a} = \frac{2\pi l'}{\mu_o} - 0.25 = \frac{2\pi\times1.2\times10^{-6}}{4\pi x10^{-7}} - 0.25$$

$$= 6 - 0.25 = 5.75$$

$$\frac{d-a}{a} = e^{5.75} = 314.19$$

$$d - a = 314.19a = 314.19\times\frac{2.588\times10^{-3}}{2} = 406.6mm$$

$$d = 407.9mm = \underline{\underline{40.79cm}}$$

P.E. 8.13

This is similar to Example 8.13. In this case, however, h=0 so that

$$\vec{A}_1 = \frac{\mu_o I_1 a^2 b}{4b^3}\vec{a}_\phi$$

$$\phi_{12} = \frac{\mu_o I_1 a^2}{4b^2} \bullet 2\pi b = \frac{\mu_o \pi I_1 a^2}{2b}$$

$$m_{12} = \frac{\phi_{12}}{I_1} = \frac{\mu_o \pi a^2}{b} = \frac{4\pi \times 10^{-7} \times \pi \times 4}{2 \times 3}$$

$= \underline{\underline{2.632\ \mu\text{H}}}$

P.E. 8.14

$$\text{L}_{\text{in}} = \frac{\mu_o}{8\pi} l = \frac{\mu_o 2\pi \rho_o}{8\pi} = \frac{4\pi \times 10^{-7} \times 10 \times 10^{-7}}{4}$$

$= \underline{\underline{31.42\ \text{nH}}}$

P.E. 8.15

(a) From Example 7.6,

$$B_{ave} = \frac{\mu_o NI}{L} = \frac{\mu_o NI}{2\pi \rho_o}$$

$$\phi = B_{ave} \bullet S = \frac{\mu_o NI}{2\pi \rho_o} \bullet \pi a^2$$

$$\text{or } I = \frac{2\rho_o \phi}{\mu a^2 N} = \frac{2 \times 10 \times 10^{-2} \times 0.5 \times 10^{-3}}{4\pi \times 10^{-7} \times 10^{-4} \times 10^3}$$

$= \underline{\underline{795.77\text{A}}}$

Alternatively, using circuit approach

$$R = \frac{l}{\mu S} = \frac{2\pi \rho_o}{\mu_o S} = \frac{2\pi \rho_o}{\mu_o \pi a^2}$$

$$\Im = NI = \frac{\phi \Re}{N} = \frac{2\rho_o \phi}{\mu a^2 N}, \quad \text{as obtained before.}$$

$$\Re = \frac{2\rho_o}{\mu a^2} = \frac{2 \times 10 \times 10^{-2}}{4\pi \times 10^{-7} \times 10^{-4}} = 1.591 \times 10^9$$

$$\Im = \phi \Re = 0.5\text{x}10^{-3}\text{x}1.591\text{x}10^9 = 7.955\text{x}10^5$$

$$I = \frac{\Im}{N} = 795A \text{ as obtained before.}$$

(b) If $\mu = 500\mu_o$,

$$I = \frac{795.77}{500} = \underline{\underline{1.592}}\text{ A}$$

P.E. 8.16

$$\Im = \frac{B^2{}_a S}{2\mu_o} = \frac{(1.5)^2 \times 10 \times 10^{-4}}{2 \times 4\pi \times 10^{-7}} = \frac{22500}{8\pi} = \underline{\underline{895.25\text{N}}}$$

Prob. 8.1

$$\overline{F} = q(\overline{E}+u\times\overline{B})$$

If $\overline{F}=0, \quad \overline{E} = -u\times\overline{B} = \overline{B}\times u$

$$= \begin{vmatrix} 10 & 20 & 30 \\ 3 & 12 & -4 \end{vmatrix} \times 10^5 \times 10^{-3}$$

$$\overline{E} = -4.4\bar{a}_x + 1.3\bar{a}_x + 11.4\bar{a}_x \text{ kV/m}$$

Prob.8.2

$$\overline{F} = ma = q\,u\times B$$

$$\bar{a} = \frac{q}{m} u\times \overline{B}$$

$$\frac{d}{dt}(u_x, u_y, u_z) = 0 \rightarrow \frac{2}{1}\begin{vmatrix} u_x & u_y & u_z \\ 1 & 0 & 0 \end{vmatrix} = 2(0, u_z, -u_y)$$

$$\frac{du_x}{dt} = 0 \rightarrow u_x = C_o \qquad \ldots \quad (1)$$

$$\frac{du_y}{dt} = 2u_z, \quad \frac{du_z}{dt} = -2u_y$$

$$\frac{d^2u_y}{dt^2} = 2, \quad \frac{du_z}{dt} = -4u_y$$

$$\ddot{u}_y + 4u_y = 0$$

$$u_y = C_1 \cos 2t + C_2 \sin 2t \qquad \ldots \quad (2)$$

$$u_z = \frac{1}{2}\frac{du_y}{dt} = -C_1 \sin 2t + C_2 \cos 2t \qquad \ldots \quad (3)$$

At $t = 0$, $u_x = 0 \rightarrow c_0 = 0$

$u_y = 0 \rightarrow c_1 = 0$

$u_z = 10 \rightarrow c_2 = 10$

Hence,

$$\bar{u} = (0, 10\sin 2t, 10\cos 2t)$$

$$u_x = \frac{dx}{dt} = 0 \qquad \rightarrow x = c_4$$

$$u_y = \frac{dy}{dt} = 10\sin 2t \rightarrow y = -5\cos 2t + c_5$$

$$u_z = 10\cos 2t \qquad \rightarrow z = 5\sin 2t + c_6$$

At $t = 0$,

$$x = 0 \rightarrow c_4 = 0$$
$$y = 0 \rightarrow c_5 = 5$$
$$z = 0 \rightarrow c_6 = 0$$

Hence,

$$(x, y, z) = (0, 5 - 5\cos 2t, 5\sin 2t)$$

At $t = 0$,

$$(x, y, z) = (0, 5 - 5\cos 4, 5\sin 4)$$
$$= (0, 8.268, -3.724)$$

$$\bar{u} = (0, 10\sin 4, 10\cos 4) = (0, -7.568, -6.536)$$

$$K.E = \frac{1}{2}m|\bar{u}|^2 = \frac{1}{2}\left(100\sin^2 4 + 100\cos^2 4\right)$$
$$= \underline{\underline{50\ J}}$$

Prob. 8.3

(a) $F = m\vec{a} = Q(\vec{E} + \vec{u} \times \vec{B})$

$$\frac{d}{dt}(u_x, u_y, u_z) = 2\left[-4\vec{a}_y + \begin{vmatrix} u_x & u_y & u_z \\ 5 & 0 & 0 \end{vmatrix}\right] = -8\vec{a}_y + 10u_z\vec{a}_y - 10u_y\vec{a}_z$$

i.e. $\frac{du_x}{dt} = 0 \rightarrow u_x = A_1$ (1)

$$\frac{du_y}{dt} = -8 + 10u_z \quad (2)$$

$$\frac{du_z}{dt} = -10u_y \quad (3)$$

$$\frac{d^2u_y}{dt^2} = 0 + 10\frac{du_z}{dt} = -100u_y$$

$$\ddot{u}_y + 100u_y = 0 \rightarrow u_y = B_1\cos 10t + B_2\sin 10t$$

From (2),

$$10u_z = 8 + \dot{u}_y = 8 - 10B_1\sin 10t + 10B_2\cos 10t$$

$$u_z = 0.8 - B_1\sin 10t + B_2\cos 10t$$

At t=0, $\bar{u} = 0 \rightarrow A_1 = 0, B_1 = 0, B_2 = -0.8$

Hence,

$$\bar{u} = (0,\ 0.8\sin 10t,\ 0.8 - 0.8\cos 10t) \qquad (4)$$

$$u_x = \frac{dx}{dt} = 0 \rightarrow x = c_1$$

$$u_y = \frac{dy}{dt} = -0.8\sin 10t \rightarrow y = 0.08\cos 10t + c_2$$

$$u_z = \frac{dz}{dt} = 0.8 - 0.8\cos 10t \rightarrow z = 0.8t + c_3 - 0.08\sin 10t$$

At t=0, (x, y, z) = (2, 3, -4) $\Rightarrow$ $c_1=2$, $c_2=2.92$, $c_3=-4$

Hence (x, y, z) = (2, 2 + 0.08cos10t, 0.8t – 0.08sin10t – 4)

At t=1,

(x, y, z) = (2, 1.933, -3.156)

(b) From (4), at t=1, $\vec{u} = (0, 0.435, 1.471)$ m/s

$$\text{K.E.} = \frac{1}{2}m|\vec{u}|^2 = \frac{1}{2}(1)(0.435^2 + 1.471^2) = 1.177\text{J}$$

Prob. 8.4

$$m\vec{a} = Q\vec{u} \times \vec{B}$$

$$10^{-3}\vec{a} = -2\times 10^{-3}\begin{vmatrix} u_x & u_y & u_z \\ 0 & 6 & 0 \end{vmatrix}$$

$$\frac{d}{dt}(u_x, u_y, u_z) = (12u_z, 0, -12u_x)$$

i.e. $$\frac{du_x}{dt} = -12u_z \qquad (1)$$

$$\frac{du_y}{dt} = 0 \rightarrow u_y = A_1 \qquad (2)$$

$$\frac{du_z}{dt} = -12u_x \qquad (3)$$

From (1) and (2),

$$\ddot{u}_x = -12\dot{u}_z = -144u_x$$

or

$$\ddot{u}_x + 144u_x = 0 \rightarrow u_x = c_1\cos 12t + c_2\sin 12t$$

From (1), $u_z = -c_1\sin 12t + c_2\cos 12t$

At t=0,

u_x=2, u_y=0, u_z=0 → A_1=0=c_2, c_1=5

Hence,

$$\vec{u} = (5\cos 12t, 0, -5\sin 12t)$$

$$\vec{u}(t = 10s) = (5\cos 120, 0, -5\sin 120) = \underline{\underline{4.071\vec{a}_x - 2.903\vec{a}_z \text{ m/s}}}$$

$$u_x = \frac{dx}{dt} = 5\cos 12t \to x = \frac{5}{12}\sin 12t + B_1$$

$$u_y = \frac{dy}{dt} = 0 \to y = B_2$$

$$u_z = \frac{dz}{dt} = -5\sin 12t \to z = \frac{5}{12}\cos 12t + B_3$$

At t=0, (x, y, z) = (0, 1, 2) → B_1=0, B_2=1, $B_3=\frac{19}{12}$

$$(x, y, z) = \left(\frac{5}{12}\sin 12t, 1, \frac{5}{12}\cos 12t + \frac{19}{12}\right) \qquad (4)$$

At t=10s,

$$(x, y, z) = \left(\frac{5}{12}\sin 120, 1, \frac{5}{12}\cos 120 + \frac{19}{12}\right) = \underline{(0.2419,\ 1,\ 1.923)}$$

By eliminating t from (4),

$$x^2 + (z = 19/12) = (5/12)^2,\ y = 1$$ which is a helix with axis on line y=1, $z=\frac{19}{12}$

Prob. 8.5

(a) $m\vec{a} = e(\vec{u} \times \vec{B})$

$$\frac{m}{e}\frac{d}{dt}(u_x, u_y, u_z) = \begin{vmatrix} u_x & u_y & u_z \\ 0 & 0 & B_z \end{vmatrix} = u_y B_o \vec{a}_x - B_o u_x \vec{a}_y$$

$$\frac{du_z}{dt} = 0 \to u_z = c = 0$$

$$\frac{du_x}{dt} = u_y \frac{B_o e}{m} = u_y w, \text{ where } w = \frac{B_o e}{m}$$

$$\frac{du_z}{dt} = -u_x w$$

Hence,

$$\ddot{u}_x = w\dot{u}_y = -w^2 u_x$$

or $\ddot{u}_x + w^2 u_x = 0 \to u_x = A\cos wt + B\sin wt$

$$u_y = \frac{\dot{u}_x}{w} = -A\sin wt + B\cos wt$$

At t=0, $u_x = u_o$, $u_y = 0 \rightarrow A = u_o$, B=0

Hence,

$$u_x = u_o \cos wt = \frac{dx}{dt} \rightarrow x = -\frac{u_o}{w}\sin wt + c_1$$

$$u_y = -u_o \sin wt = \frac{dy}{dt} \rightarrow y = -\frac{-u_o}{w}\cos wt + c_2$$

At t=0, x = 0 = y → c_1=0, $c_2=\frac{u_o}{w}$. Hence,

$$x = -\frac{u_o}{w}\sin wt,\ y = \frac{u_o}{w}(1-\cos wt)$$

$$\frac{u^2_o}{w^2}(\cos^2 wt + \sin^2 wt) = \left(\frac{u_o}{w}\right)^2 = x^2 + (y - \frac{u_o}{w})^2$$

showing that the electron would move in a circle centered at (0, $\frac{u_o}{w}$). But since the field does not exist throughout the circular region, the electron passes through a semi-circle and leaves the field horizontally.

(b) d = twice the radius of the semi-circle

$$= \frac{2u_o}{w} = \frac{2u_o m}{B_o e}$$

Prob.8.6
$$\bar{F} = \int I d\bar{l} \times \bar{R}$$

$$= I\int_{x=1}^{3} dx\, \bar{a}_x \times \bar{B} + I\int_{y=1}^{3} dy\, \bar{a}_y \times \bar{B} + I\int_{x=3}^{1} dx\, \bar{a}_x \times \bar{B} + I\int_{y=2}^{1} dy\, \bar{a}_y \times \bar{B}$$

$$\bar{a}_x \times \bar{B} = \begin{vmatrix} 1 & 0 & 0 \\ 6x & -9x & 3z \end{vmatrix} = -3z\bar{a}_y - 9y\bar{a}_z$$

$$\bar{a}_y \times \bar{B} = \begin{vmatrix} 0 & 1 & 0 \\ 6x & -9x & 3z \end{vmatrix} = 3z\bar{a}_x - 6x\bar{a}_z$$

$$\bar{F} = I\int_1^3 dx\left(-3z\bar{a}_y - 9y\bar{a}_z\right)\Big|_{z=0,\ y=1} + I\int_1^2 dy\left(3z\bar{a}_x - 6x\bar{a}_z\right)\Big|_{x=3,\ z=0}$$

$$+ I\int_3^1 dx\left(-3z\bar{a}_y - 9y\bar{a}_z\right)\Big|_{z=0,\ y=2} + I\int_2^1 dy\left(3z\bar{a}_x - 6x\bar{a}_z\right)\Big|_{z=1,\ x=1}$$

$$= I(-18 - 18 + 36 + 6)\bar{a}_z = 6I\bar{a}_z$$

$$= 6 \times 5\bar{a}_z = \underline{\underline{30\bar{a}_z\ N}}$$

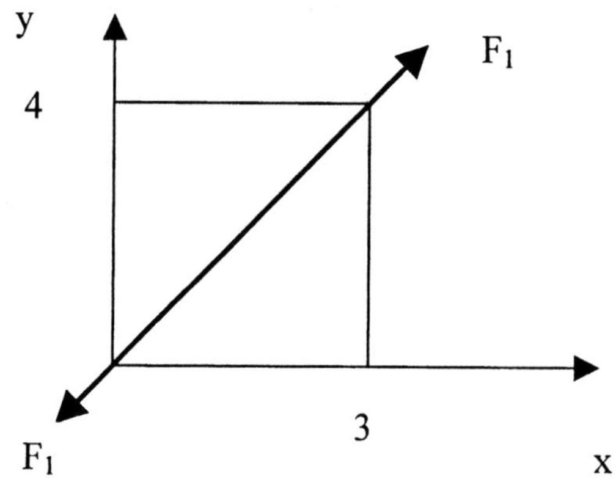

Prob. 8.7

$$\overline{B}_1 = \frac{\mu I_1}{2\pi\rho}\bar{a}_\phi, \quad \rho = 3$$

$$\bar{a}_y = \bar{a}_z \times \left(\frac{-3\bar{a}_y - 4\bar{a}_y}{5}\right) = -\frac{3\bar{a}_y + 4\bar{a}_y}{5}$$

$$\overline{B}_1 = \frac{4\pi\times10^{-7}\times15}{r-1}\left(\frac{4}{5}\bar{a}_x - \frac{3}{5}\bar{a}_y\right) = \frac{6\times10^{-7}}{5}\left(4a_x - 3\bar{a}_y\right)$$

$$\overline{F}_2 = d\overline{F} = Id\bar{l}\times\overline{B} = 2\times10^{-2}\times12\times10^{-3}\,\bar{a}_x \times \frac{6\times10^{-7}}{5}\left(4a_x - 3\bar{a}_y\right)$$

$$= \underline{\underline{-86.4\,\bar{a}_x \text{ pN}}}$$

Prob. 8.8

$$\vec{\Im} = L\vec{L}\times\vec{B} \rightarrow \vec{\Im} = \frac{\vec{F}}{L} = I_1\vec{a}_l\times\vec{B}_2 = \frac{\mu o I_1 I_2 a_l \times \vec{a}_\phi}{2\pi\rho}$$

(a) $$\vec{F}_{21} = \frac{\vec{a}_z\times(-\vec{a}_x)}{2\pi}\frac{4\times10^{-7}(-2\times10^4)}{4}$$

$$= \underline{\underline{\vec{a}_y}} \text{ mN/m (repulsive)}$$

(b) $$\vec{F}_{12} = -\vec{F}_{21} = \underline{\underline{-\vec{a}_y}} \text{ mN/m (repulsive)}$$

(c) $$\vec{a}_l\times\vec{a}_\phi = \vec{a}_z\times\left(-\frac{4}{5}\vec{a}_x + \frac{3}{5}\vec{a}_y\right) = -\frac{3}{5}\vec{a}_x - \frac{4}{5}\vec{a}_y, \rho = 5$$

$$\vec{F}_{31} = \frac{4\pi\times10^{-7}(-3\times10^4)}{2\pi(5)}\left(-\frac{3}{5}\vec{a}_x - \frac{4}{5}\vec{a}_y\right)$$

$$= \underline{\underline{0.72\vec{a}_x + 0.96\vec{a}_y}} \text{ mN/m (attractive)}$$

(d) $$\vec{F}_3 = \vec{F}_{31} + \vec{F}_{32}$$

$$\vec{F}_{32} = \frac{4\pi\times10^{-7}\times6\times10^4)}{2\pi(3)}\left(\vec{a}_z\times\vec{a}_y\right) = -4\vec{a}_x \text{ mN/m(attractive)}$$

$\vec{F}_3 = -3.28\vec{a}_x + 0.96\vec{a}_y$ mN/m

(attractive due to L_2 and repulsive due to L_1)

Prob. 8.9

$$W = -\int \vec{F} \bullet d\vec{l}, \vec{F} = \int L d\vec{l} \times \vec{B} = 3(2\vec{a}_z) \times \cos{}^{d}\!/_{3}\, \vec{a}_\phi$$

$$= 6\cos{}^{d}\!/_{3}\, \vec{a}_\phi mN$$

$$W = -\int_0^{2\pi} 6\cos{}^{d}\!/_{3}\, \rho_o dd = -6 \times 3\sin{}^{d}\!/_{3}\Big|_0^{2\pi} \text{ mJ}$$

$$= -18\sin\frac{2\pi}{3} = \underline{-15.59\text{mJ}}$$

Prob. 8.10

(a) $$\vec{F}_1 = \int_{\rho=2}^{4} \frac{\mu_o I_1 I_2}{2\pi\rho} d\rho \vec{a}_\rho \times \vec{a}_\phi = \frac{4\pi \times 10^{-7}}{2\pi}(2)(5)\ln{}^{4}\!/_{2}\, \vec{a}_z$$

$= 2 \ln 2\, \vec{a}_z$ μN = $\underline{1.3863\, \vec{a}_z}$ μN

(b) $$\vec{F}_2 = \int I_2 d\vec{l}_2 \times \vec{B}_1$$

$$= \frac{\mu_o I_1 I_2}{2\pi} \int \frac{1}{\rho}\left[d\rho \vec{a}_\rho + dz\vec{a}_z\right] \times \vec{a}_\phi$$

$$= \frac{\mu_o I_1 I_2}{2\pi} \int \frac{1}{\rho}\left[d\rho \vec{a}_z - dz\vec{a}_\rho\right]$$

But ρ = z+2, dz=dρ

$$\vec{F}_2 = \frac{4\pi \times 10^{-7}}{2\pi}(5)(2) \int_{\rho=4}^{2} \frac{1}{\rho}\left[d\rho \vec{a}_z - dz\vec{a}_\rho\right]$$

$$2\ln{}^{2}\!/_{4}(\vec{a}_z - \vec{a}_\rho)\mu N = 1.386\vec{a}_\rho - 1.386\vec{a}_z \mu N$$

$$\vec{F}_3 = \frac{\mu_o I_1 I_2}{2\pi} \int \frac{1}{\rho}\left[d\rho \vec{a}_z - dz\vec{a}_\rho\right]$$

But z = -ρ + 6, dz = -dρ

$$\vec{F}_3 = \frac{4\pi \times 10^{-7}}{2\pi}(5)(2) \int_{\rho=6}^{4} \frac{1}{\rho}\left[d\rho \vec{a}_z - dz\vec{a}_\rho\right]$$

$$2\ln{}^{4}\!/_{6}(\vec{a}_z + \vec{a}_\rho)\mu N = -0.8109\vec{a}_\rho - 0.8109\vec{a}_z \mu N$$

$$\vec{F} = \vec{F}_1 + \vec{F}_2 + \vec{F}_3$$

$$= 1.3863\vec{a}_z + 1.386\vec{a}_\rho - 1.3863\vec{a}_z - 0.8109\vec{a}_\rho - 0.8109\vec{a}_z$$

$$= 0.5751\vec{a}_\rho - 0.8109\vec{a}_z\ \mu N$$

Prob. 8.11

From Prob. 8.7,

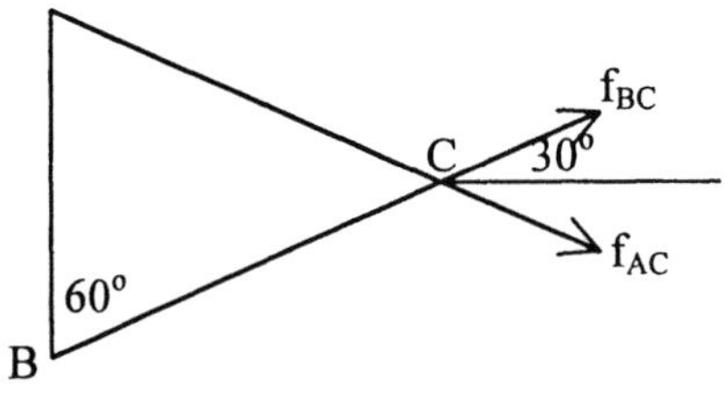

$$f = \frac{\mu_o I_1 I_2}{2\pi\rho}\vec{a}_\rho$$

$$\vec{f} = \vec{f}_{AC} + \vec{f}_{BC}$$

$$\vec{f}_{AC} = \vec{f}_{BC} = \frac{4\pi\times10^{-7}\times75\times150}{2\pi\times2} = 1.125\times10^{-3}$$

$$\vec{f} = 2\times1.125\cos30^o\,\vec{a}_x \text{ mN/m}$$

$$= \underline{\underline{1.949\vec{a}_x}} \text{ mN/m}$$

Prob. 8.12

$$\vec{F} = \int L d\vec{l}\times\vec{B} = \int \vec{J}dv\times\vec{B}$$

$$\vec{J} = \frac{I}{\pi(b^2-a^2)}\vec{a}_z,\ \vec{B} = B_o\vec{a}_\rho$$

$$\vec{F} = \frac{I}{\pi(b^2-a^2)}\int \vec{a}_z dv\times B_o\vec{a}_\rho = \frac{IB_o\vec{a}_\rho}{\pi(b^2-a^2)}\int dv$$

$$= \frac{IB_o}{\pi(b^2-a^2)}\pi(a^2-b^2)l$$

$$\vec{f} = \frac{\vec{F}}{l} = \underline{\underline{IB_o\vec{a}_\phi}}$$

Prob. 8.13

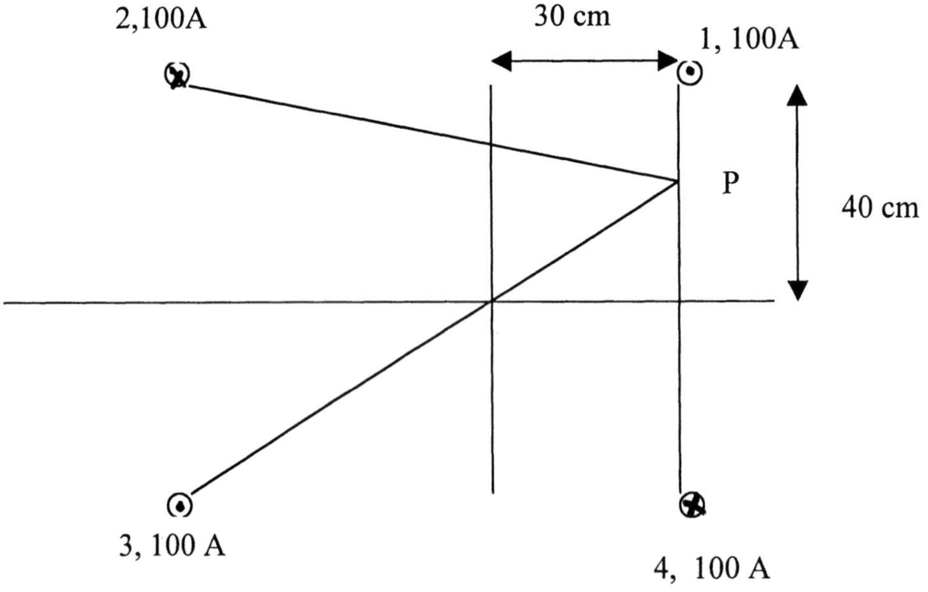

Let $\vec{B} = \vec{B}_1 + \vec{B}_2 + \vec{B}_3 + \vec{B}_4$

where $\vec{B}_n = \dfrac{\mu_o \mu_r I}{2\pi\rho}\vec{a}_\phi$

For (1), $\vec{a}_\phi = \vec{a}_l \times \vec{a}_\rho)\vec{a}_z \times (-\vec{a}_y) = \vec{a}_x$,

$$\vec{B}_1 = \frac{4\pi \times 10^{-7} \times 2000 \times 100}{2\pi \times 20 \times 10^{-3}}\vec{a}_x = 2\vec{a}_x$$

For (2), $\vec{\rho} = 6\vec{a}_x - 2\vec{a}_y$,

$$\vec{a}_\phi = -\vec{a}_z \times \frac{(6\vec{a}_x - 2\vec{a}_y)}{\sqrt{40}} = \frac{(-2\vec{a}_x - 6\vec{a}_y)}{\sqrt{40}}$$

$$\vec{B}_2 = \frac{4\pi \times 10^{-7} \times 2000 \times 100}{2\pi \times 400 \times 10^{-3}}(-2\vec{a}_x - 6\vec{a}_y)$$

$$= -0.2\vec{a}_x - 0.6\vec{a}_y$$

For (3), $\vec{\rho} = 6\vec{a}_x + 6\vec{a}_y$,

$$\vec{a}_\phi = \vec{a}_z \times \frac{(6\vec{a}_x + 6\vec{a}_y)}{\sqrt{72}} = \frac{(-6\vec{a}_x + 6\vec{a}_y)}{\sqrt{72}}$$

$$\vec{B}_3 = \frac{4\pi \times 10^{-7} \times 2000 \times 100}{2\pi \times 720 \times 10^{-3}}(-6\vec{a}_x + 6\vec{a}_y)$$

$$= -0.3333\vec{a}_x + 0.3333\vec{a}_y$$

For (4), $\vec{a}_\phi = -\vec{a}_z \times \vec{a}_y = \vec{a}_x$,

$$\vec{B}_4 = \frac{4\pi \times 10^{-7} \times 2000 \times 100}{2\pi \times 60 \times 10^{-3}}\vec{a}_x = 0.6667\vec{a}_x$$

$$\vec{B} = (2 + \tfrac{2}{3} - \tfrac{1}{5} - \tfrac{1}{3})\vec{a}_x + (-\tfrac{3}{5} + \tfrac{1}{3})\vec{a}_y$$

$$= \underline{\underline{2.1333\vec{a}_x - 0.2667\vec{a}_y}} \text{ Wb/m}^2$$

Prob. 8.14

T = mB = NISB = 1000 x 2 x 10^{-3} x 300 x 10^{-6} x 0.4

= $\underline{\underline{240\mu\text{Nm}}}$

Prob. 8.15

$$\vec{B} = \frac{k}{r^3}(2\cos\theta\vec{a}_r + \sin\theta\vec{a}_\theta)$$

At (10, 0, 0), r = 10; $\theta = \pi/2, \vec{a}_r = \vec{a}_x, \vec{a}_\theta = -\vec{a}_z$

$$-0.5\times 10^{-3}\vec{a}_z = \frac{k}{10^3}(0-\vec{a}_x) \rightarrow k = 0.5$$

Thus,

$$\vec{B} = \frac{0.5}{r^3}(2\cos\theta\vec{a}_r + \sin\theta\vec{a}_\theta)$$

(a) At (0, 3, 0), r=3, $\theta = \pi/2, \vec{a}_r = \vec{a}_y, \vec{a}_\theta = -\vec{a}_z$

$$\vec{B} = \frac{0.5}{27}(0-\vec{a}_z) = \underline{\underline{-18.52\vec{a}_z}} \text{ mWb/m}^2$$

(b) At (3, 4, 0), r=5, $\theta = \pi/2, \vec{a}_\theta = -\vec{a}_z$

$$\vec{B} = \frac{0.5}{125}(0-\vec{a}_z) = \underline{\underline{-4\vec{a}_z}} \text{ mWb/m}^2$$

(c) At (1, -1, 1), r=$\sqrt{3}$, $\tan\theta = \rho/z = \sqrt{2}/-1$, i.e.

$$\sin\theta = \sqrt{2}/3, \cos\theta = -1/3$$

$$\vec{B} = \frac{0.5}{3\sqrt{3}}(-2/3\,\vec{a}_r + \sqrt{2}/3\,\vec{a}_\theta) = \underline{\underline{-111\vec{a}_r + 78.6\vec{a}_\theta}} \text{ mWb/m}^2$$

Prob. 8.16 (a)

$$\overline{M} = x_m H = x_m \frac{B}{\mu_o \mu}$$

$$= \frac{4999}{5000}\times\frac{1.5}{4\pi\times 10^{-7}} = \underline{\underline{1.194\times 10^6 \text{ A/m}}}$$

(b)

$$\overline{M} = \frac{\sum_{k=1}^{N} m_k}{\Delta v}$$

If we assume that all $\overline{m}_k$ align with the applied $\overline{B}$ field,

$$M = \frac{Nm_k}{\Delta v} \rightarrow m_k = \frac{Nm_k}{N/\Delta v} = \frac{1.194\times 10^6}{8.5\times 10^{28}}$$

$$m_k = \underline{\underline{1.047\times 10^{-23} \text{ A}\cdot\text{m}^2}}$$

Prob. 8.17

(a) $\psi_m = \mu_r - 1 = \underline{\underline{5.5}}$

(b)

$$\overline{B} = \mu_0\mu_r\overline{H} = 4\pi\times 10^{-7}\times 6.5(10, 25, -40)$$

$$= \underline{\underline{81.68\,\overline{a}_x + 204.2\,\overline{a}_y - 326.7\,\overline{a}_z \ \mu\text{Wb/m}^2}}$$

(c)

$$\overline{M} = \psi_m\overline{H} = \underline{\underline{55\,\overline{a}_x + 137.5\,\overline{a}_y - 220\,\overline{a}_z \text{ A/m}}}$$

(d)

$$W_m = \frac{1}{2}\mu\overline{H}\cdot\overline{H} = \frac{1}{2}(6.5)4\pi\times 10^{-7}\times 6.5(100+625+1600)$$

$$= \underline{\underline{9.5 \text{ mJ/m}^2}}$$

Prob. 8.18 (a) $\psi_m = \mu_r - 1 = \underline{\underline{3.5}}$

(b) $$\overline{H} = \frac{\overline{B}}{\mu} = \frac{4y\,\bar{a}_z \times 10^{-3}}{4\pi \times 10^{-7} \times 4.5} = \underline{\underline{707.3y\,\bar{a}_z\ A/m}}$$

(c) $$\overline{M} = \psi_m \overline{H} = \underline{\underline{2.476y\,\bar{a}_z \text{ kA/m}}}$$

(d) $$\bar{\tau}_b = \overline{\nabla} \times \overline{M} = \begin{vmatrix} \frac{\partial}{\partial x} & \frac{\partial}{\partial y} & \frac{\partial}{\partial z} \\ 0 & 0 & M_z(y) \end{vmatrix} = \frac{dM_z}{dy}\bar{a}_x$$

$$= \underline{\underline{2.476\,\bar{a}_x \text{ kA/m}^2}}$$

Prob. 8.19

For case 1,

$$\mu = \frac{B_1}{H_1} = \frac{2}{1200}$$

$$\mu_r = \frac{\mu}{\mu_o} = \frac{1}{600} \times \frac{1}{4\pi \times 10^{-7}} = 1326.3$$

$$\psi_m = \mu_r - 1 = 1325.3$$

$$\overline{M}_1 = \psi_m H_1 = 1{,}590{,}366$$

For case 2,

$$\mu = \frac{B_2}{H_2} = \frac{1.4}{400}$$

$$\mu_r = \frac{\mu}{\mu_o} = \frac{1.4}{400} \times \frac{1}{4\pi \times 10^{-7}} = 2785.2$$

$$\psi_m = \mu_r - 1 = 2784.2$$

$$M = \psi_m H = 1{,}113{,}630$$

$$\Delta M = M_1 - M_2 = 476{,}680$$

$$= \underline{\underline{476.7 \text{ kA/m}}}$$

Prob. 8.20 $\oint \overline{H} \cdot d\overline{l} = I_{enc}$

$$H_\varphi \cdot 2\pi\rho = \frac{\pi\rho^2}{\pi a^2} \cdot I \rightarrow H_\varphi = \frac{I\rho}{2\pi a^2}$$

$$\overline{M} = \psi_m \overline{H} = \underline{\underline{(\mu_r - 1)\frac{I\rho}{2\pi a^2}\overline{a}_\varphi}}$$

$$J_b = \nabla \times \overline{M} = \frac{1}{\rho}\frac{\partial}{\partial\rho}(\rho M_\phi) = \underline{\underline{(\mu_r - 1)\frac{I}{\pi a^2}\overline{a}_z}}$$

Prob. 8.21 $J_b = \nabla \times \overline{M} = \dfrac{k_o}{a}\begin{vmatrix} \dfrac{\partial}{\partial x} & \dfrac{\partial}{\partial y} & \dfrac{\partial}{\partial z} \\ -y & x & 0 \end{vmatrix} = \underline{\underline{2\dfrac{k_o}{a}\overline{a}_z}}$

Prob. 8.22

(a) From $H_{1t} - H_{2t} = k$ and $M = \chi_m H$, we obtain:

$$\underline{\underline{\frac{M_{1t}}{\chi_{m1}} - \frac{M_{2t}}{\chi_{m2}} = k}}$$

Also from $B_{1n} - B_{2n} = k$ and $B = \mu H = (\mu/\chi_m)M$, we get:

$$\underline{\underline{\frac{\mu_1 M_{1n}}{\chi_{m1}} = \frac{\mu_2 M_{2n}}{\chi_{m2}}}}$$

(b) From $B_1\cos\theta_1 - B_{1n} = B_{2n} = B_2\cos\theta_2$ (1)

and $\dfrac{B_1 \sin\theta_1}{\mu_1} = H_{2t} = k + H_{2t} = k + \dfrac{B_2 \sin\theta_2}{\mu_2}$ (2)

Dividing (2) by (1) gives

$$\frac{\tan\theta_1}{\mu_1} = \frac{k}{B_2\cos\theta_2} + \frac{\tan\theta_2}{\mu_2} = \frac{\tan\theta_2}{\mu_2}\left(1 + \frac{k\mu_2}{B_2\sin\theta_2}\right)$$

i.e. $$\underline{\underline{\frac{\tan\theta_1}{\tan\theta_2} = \frac{\mu_1}{\mu_2}\left(1 + \frac{k\mu_2}{B_2\sin\theta_2}\right)}}$$

Prob. 8.23 (a) $\overline{B}_{1n} = \overline{B}_{2n} = 1.5\,\overline{a}_\phi$

$$\overline{H}_{1t} = \overline{H}_{2t} \rightarrow \frac{\overline{B}_{1t}}{\mu_1} = \frac{\overline{B}_{2t}}{\mu_2}$$

$$\overline{B}_{1t} = \frac{\mu_1}{\mu_2}\overline{B}_{2t} = \frac{5\mu_1}{2\mu_2}\left(10\,\overline{a}_\rho - 20\,\overline{a}_z\right) = 25\,\overline{a}_\rho - 50\,\overline{a}_z$$

Hence,

$$\overline{B}_{1t} = \underline{\underline{25\,\overline{a}_\rho + 15\,\overline{a}_\phi - 50\,\overline{a}_z \text{ mWb/m}^2}}$$

(b) $$W_{m1} = \frac{1}{2}\overline{B}_1 \cdot \overline{H}_1 = \frac{B_1^2}{2\mu_1} = \frac{\left(25^2 + 15^2 + 50^2\right)\times 10^{-6}}{2\times 2\times 4\pi\times 10^{-7}}$$

$$W_1 = \underline{\underline{666.5\ \ J/m^3}}$$

$$W_2 = \frac{B_2^2}{2\mu_2} = \frac{\left(10^2 + 15^2 + 20^2\right)\times 10^{-6}}{2\times 5\times 4\pi\times 10^{-7}} = \underline{\underline{57.7\ \ J/m^3}}$$

Prob. 8.24 (a) $$W_{m1} = \frac{1}{2}\overline{B}_1 \cdot \overline{H}_1 = \frac{1}{2}\mu_0\mu_{r1}\overline{H}_1 \cdot \overline{H}_1\,,\quad \mu_r = 1$$

$$W_{m1} = \frac{1}{2}\times 4\pi\times 10^{-7}\times 1\left(16+9+1\right)$$

$$= \underline{\underline{16.34\,\mu J/m^3}}$$

(b)

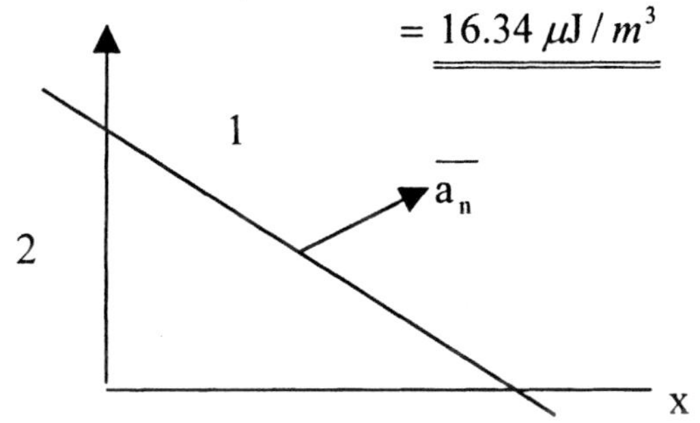

$$f(x, y) = 2x + y - 8 = 0$$

$$\nabla f = 2\bar{a}_x + \bar{a}_y, \quad \bar{a}_n = \frac{\nabla f}{|\nabla f|} = \frac{2\bar{a}_x + \bar{a}_y}{\sqrt{5}}$$

$$\overline{H}_{1n} = (\overline{H}_1 \cdot \bar{a}_n)\bar{a}_n = \left(\frac{-8+3}{5}\right)(2\bar{a}_x + \bar{a}_y) = -2\bar{a}_x - \bar{a}_y$$

$$\overline{H}_{1t} = \overline{H}_1 - \overline{H}_{1n} = -2\bar{a}_x + 4\bar{a}_y - \bar{a}_z = \overline{H}_{2t}$$

$$\overline{B}_{2n} = \overline{B}_{1n} \rightarrow \mu_2\overline{H}_{2n} = \mu_1\overline{H}_{1n}$$

$$\overline{H}_{2n} = \frac{\mu_1}{\mu_2}\overline{H}_{1n} = \frac{1}{10}(-2\bar{a}_x - \bar{a}_y)$$

$$= -0.2\bar{a}_x - 0.1\bar{a}_y$$

$$\overline{H}_2 = \overline{H}_{2t} + \overline{H}_{2n} = -2.2\bar{a}_x + 3.9\bar{a}_y - \bar{a}_z$$

$$\overline{M}_2 = \psi_{m2}\overline{H}_2 = 9H_2 = -19.8\bar{a}_x + 35.1\bar{a}_y - 9\bar{a}_z \text{ A/m}$$

$$\overline{B}_2 = \mu_2\overline{H}_2 = 10\mu_0\overline{H}_2$$

$$= 4\pi(-2.2, 2.9, -1)\mu\text{Wb/m}^2$$

$$\overline{B}_2 = -27.65\,\bar{a}_x + 49\,\bar{a}_y - 12.56\,\bar{a}_5\ \mu\text{Wb/m}^3$$

$$H_2 = H_{2t} + H_{2n} = -2.2a_x + 3.9a_y - a_z$$

$$H_2 = H_{2t} + H_{2n} = -2.2a_x + 3.9a_y - a_z$$

$$M_2 = \chi_{m2}H_2 = 9H_2 = \underline{\underline{-19.8a_x + 35.1a_y - 9a_z \text{ A/m}}}$$

$$B_2 = \mu_2 H_2 = 10\mu_o H_2 = 4\pi x(-2.2, 2.9, -1)\ \mu\text{Wb/m}^2$$

$$= \underline{\underline{-27.75a_x + 49a_y - 12.56a_z\ \mu\text{Wb/m}^2}}$$

(c) $H_1 \bullet a_n = H_1 \cos\theta_1$

$$\cos\theta_1 = \frac{H_1 \bullet a_n}{H_1} = \frac{(-8+3)/\sqrt{9}}{\sqrt{16+9+1}} = -0.4389 \longrightarrow \underline{\underline{\theta_1 = 116^o}}$$

$$\cos\theta_2 = \frac{H_2 \bullet a_n}{H_2} = \frac{(-4.4 + 3.9)/\sqrt{5}}{\sqrt{4.588}} = -0.1044 \longrightarrow \underline{\underline{\theta_1 = 96^o}}$$

Prob. 8.25

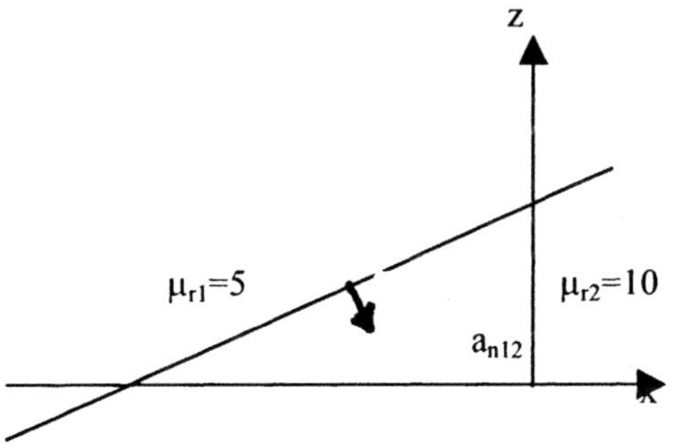

Let $\vec{H}_2 = (H_x, H_y, H_z)$

$$(\vec{H}_1 - \vec{H}_2) \times \vec{a}_{n12} = \vec{k}$$

where f(x, z)= 5z – 4x = 0 and

$$a_{n12} = -\frac{\nabla f}{|\nabla f|} = \frac{4\vec{a}_x - 5\vec{a}_z}{\sqrt{41}}$$

$$(\vec{H}_1 - \vec{H}_2) \times \vec{a}_{n12} = \frac{1}{\sqrt{41}} \begin{vmatrix} 25 - H_x & -30 - H_y & 45 - H_z \\ 4 & 0 & -5 \end{vmatrix}$$

$$= \frac{1}{\sqrt{41}} \left[150 + 5H_y, 180 - 4H_z, 120 + 4H_y \right] = \vec{k} = 35\vec{a}_y$$

Equating components,

$\vec{a}_x$: $150 + 5H_y = 0 \rightarrow H_y = -30$

$\vec{a}_y$: $300 - 4H_z - 5H_x = 35 \rightarrow 4H_z + 5H_x = 270$

$\vec{a}_z$: $120 + 4H_y = 0 \rightarrow H_y = -30$

Also, $\vec{B}_{1n} = \vec{B}_{2n} \rightarrow \mu_1 \vec{H}_{1n} = \mu_2 \vec{H}_{2n}$

$$5\mu_o(2, -30, 45)\frac{(4,0,5)}{\sqrt{41}} = 10\mu_o(H_x, H_y, H_z)\frac{(4,0,5)}{\sqrt{41}}$$

100 – 225 = 68H$_x$ – 10H$_z$

or 125 = 10H$_z$ – 8H$_x$
= 10H$_z$ – 8(54-0.8H$_z$) $\longrightarrow$ H$_z$ = 33.96

and H$_x$ = 54 – 0.8 H$_z$ = 26.83

Thus,

$$\vec{H}_z = \underline{\underline{26.83\vec{a}_x - 30\vec{a}_y + 33.96\vec{a}_z}} \text{ A/m}$$

Prob. 8.26

$$\vec{H}_{1n} = -3\vec{a}_z, = \vec{H}_{1t} = 10\vec{a}_x + 15\vec{a}_y$$

$$\vec{H}_{2t} = \vec{H}_{1t} = 10\vec{a}_x + 15\vec{a}_y$$

$$\vec{H}_{2n} = \frac{\mu_1}{\mu_2}\vec{H}_{1n} = \frac{1}{200}(-3\vec{a}_z) = -0.015\vec{a}_z$$

$$\vec{H}_2 = 10\vec{a}_x + 15\vec{a}_y - 0.015\vec{a}_z$$

$$\vec{B}_2 = \mu_2\vec{H}_2 = 200 \times 4\pi \times 10^{-7}(10,15,-0.015)$$

$$\vec{B}_2 = \underline{\underline{2.51\vec{a}_x + 3.77\vec{a}_y - 0.0037\vec{a}_z}} \text{ mWb/m}^2$$

$$\tan\alpha = \frac{B_{2n}}{B_{2t}}$$

or $\alpha = \tan^{-1}\dfrac{0.0037}{\sqrt{2.51^2 + 3.77^2}} = \underline{\underline{0.047^\circ}}$

Prob. 8.27

(a) $\vec{H} = \frac{1}{2}\bar{k} \times \vec{a}_n = \frac{1}{2}(30-40)\vec{a}_x \times (-\vec{a}_r) = \underline{\underline{-5\vec{a}_y}}$ A/m

$\vec{B} = \mu_o\vec{H} = 4\pi \times 10^{-7}(-5\vec{a}_y) = \underline{\underline{-6.28\vec{a}_y}}\mu$ Wb/m^2

(b) $\vec{H} = \frac{1}{2}(-30-40)\vec{a}_y = \underline{\underline{-35\vec{a}_y}}$ A/m

$\vec{B} = \mu_o\mu_r\vec{H} = 4\pi \times 10^{-7}(-35\vec{a}_y) = \underline{\underline{-110\vec{a}_y}}\mu$ Wb/m^2

(c) $\vec{H} = \frac{1}{2}(-30+40)\vec{a}_y = \underline{\underline{5\vec{a}_y}}$

$\vec{B} = \mu_o \vec{H} = \underline{\underline{6.283\vec{a}_y}}\mu \text{ Wb/m}^2$

Prob. 8.28 $\mu_r = \psi_m + 1 = 20$

$$W_m = \frac{1}{2}\overline{B}_1 \cdot \overline{H}_1 = \frac{1}{2}\mu\overline{H}\cdot\overline{H}$$

$$= \frac{1}{2}\mu\left(25x^4y^2z^2 + 100x^2y^4z^2 + 225x^2y^2z^4\right)$$

$$W_m = \int W_m dv$$

$$= \frac{1}{2}\mu\left[25\int_0^1 x^4dx \int_0^2 y^2dy \int_{-1}^2 z^2dz + 100\int_0^1 x^2dx\int_0^2 y^4dy\int_{-1}^2 z^2dz\right.$$

$$\left. = +\ 225\int_0^1 x^2dx \int_0^2 y^2dy \int_{-1}^2 zdz\right]$$

$$= \frac{25\mu}{2}\left[\left.\frac{x^5}{5}\right|_0^1 \left.\frac{y^3}{3}\right|_0^2 \left.\frac{z^3}{3}\right|_{-1}^2 + 4\left.\frac{x^3}{3}\right|_0^1 \left.\frac{y^5}{5}\right|_0^2 \left.\frac{z^3}{3}\right|_{-1}^2\right.$$

$$\left. = +\ 9\left.\frac{x^3}{3}\right|_0^1 \left.\frac{y^3}{3}\right|_0^2 \left.\frac{z^5}{5}\right|_{-1}^2\right]$$

$$= \frac{25\mu}{2}\left(\frac{1}{5}\cdot\frac{8}{3}\cdot\frac{9}{3} + \frac{4}{3}\cdot\frac{32}{3}\cdot\frac{9}{3} + \frac{9}{3}\cdot\frac{8}{3}\cdot\frac{33}{5}\right)$$

$$= \frac{25}{2}\times 4\pi\times 10^{-7}\times 20\times\frac{3600}{45}$$

$W_m = \underline{\underline{25.13 \text{ mJ}}}$

Prob. 8.29

(a) $B = 70 + (210)^2 = 44.17 \text{Wb/m}^2$

$$\mu_r = \frac{B}{\mu_o H} = \frac{44.17\times 10^3}{4\pi\times 10^{-7}\times 210} = \underline{\underline{167.4}}$$

(b) $W_m = \int_0^{H_o} HdB = \int_0^{H_o} H(\frac{1}{3} + 2H)dH$

$= \frac{H_o^2}{6} + \frac{2}{3}H_o^3 = 7350 + 6174000$

$= \underline{\underline{6181.35}}$ kJ/m^3

Prob. 8.30

(a) $L = \frac{\lambda}{I} = \frac{N\psi}{I} = \frac{\mu_o N^2 I_a}{2\pi}\ln\left(\frac{2\rho_o + a}{2\rho_o - a}\right)$

(b) $L = \frac{N\psi}{I} = \mu_o N^2[\rho_o - (\rho^2{}_o - a^2)^{\frac{1}{2}}]$

when ρ_o>>a, binomial series expansion gives:

$$L = \frac{\mu_o N^2 a^2}{2\rho_o}$$

Or from Example 8.10,

$$L = L'l = \frac{\mu_o N^2 lS}{l^2} = \frac{\mu_o N^2 \pi a^2}{2\pi\rho_o} = \frac{\mu_o N^2 a^2}{2\rho_o}$$

Prob. 8.31

For d >> a,

$$L' = \frac{L}{l} = \frac{\mu_o}{\pi}\ln\frac{d}{a} = \frac{4\pi\times10^{-7}}{\pi}\ln\frac{d}{a} = 2.5\times10^{-6}$$

or $\ln\frac{d}{a} = 6.25 \rightarrow \frac{d}{a} = e^{6.25} = 518.01$

$a = \frac{3}{518.01} = 5.78mm$

$D = 2a = 11.58mm$

Prob. 8.32

$$L = \frac{\mu_o N^2 S}{l} = \frac{4\pi x10^{-7} x(450)^2 x\pi(10^{-2})^2}{0.1} = \underline{\underline{80\mu\text{H}}}$$

Prob. 8.33

$$L = \frac{\mu N^2 S}{l} \rightarrow N^2 = \frac{Ll}{\mu S} = \frac{L2\pi\rho_o}{\mu_o \mu_r S}$$

$$= \frac{2.5 \times 2\pi \times 0.5}{4\pi \times 10^{-7} \times 200 \times 12 \times 10^{-4}} = \frac{25}{96} \times 10^8$$

N = <u>5103 turns</u>

Prob. 8.34

$$\psi_{12} = \int \vec{B}_1 \bullet d\vec{S} = \int_{z=0}^{b} \frac{\mu_o I}{2\pi\rho} dz d\rho = \frac{\mu_o I b}{2\pi} \ln \frac{a+\rho_o}{\rho_o}$$

For N = 1,

$$M_{12} = \frac{N\psi_{12}}{I_1} = \frac{\mu_o b}{2\pi} \ln \frac{a+\rho_o}{\rho_o}$$

$$= \frac{4\pi \times 10^{-7}}{2\pi} (1) \ln 2 = \underline{0.1386 \mu \text{H}}$$

Prob. 8.35

We may approximate the longer solenoid as infinite so that $B_1 = \frac{\mu_o N_1 I_1}{l_1}$. The flux linking the second solenoid is:

$$\psi_2 = N_2 B_1 S_1 = \frac{\mu_o N_1 I_1}{l_1} \bullet \pi r_1^2$$

$$M = \frac{\psi_2}{I_1} = \frac{\mu_o N_1 N_2}{l_1} \bullet \pi r_1^2$$

Prob. 8.36

$$NI = Hl = \frac{Bl}{\eta}$$

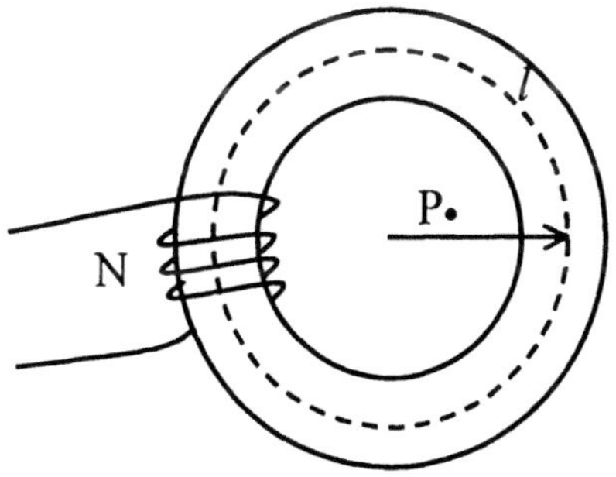

$$N = \frac{Bl}{\lambda_o \eta_r I} = \frac{1.5 \times 0.6\pi}{4\pi \times 10^{-7} \times 600 \times 12}$$

$$\underline{= 312.5}$$

Prob. 8.37

F = NI = 400 x 0.5 = 200 A.t

$$F_a = \frac{R_a}{R_a + R_3 + R_1 // R_2} = \frac{796 \times 10^3 \times 200}{(796 + 383) \times 10^3} = \underline{190.8} \text{ A.t}$$

$$H_a = \frac{F_a}{l_a} = \frac{190.8}{1 \times 10^{-2}} = \underline{19080} \text{ A/m}$$

Prob. 8.38

Total F = NI = 2000 x 10 = 20,000 A.t

$$R_c = \frac{l_c}{\mu_o \mu_r S} = \frac{(24+20-0.6)\times 10^{-2}}{4\pi \times 10^{-7} \times 1500 \times 2 \times 10^{-4}} = \underline{\underline{0.115 \text{ x } 10^7 \text{ A.t/m}}}$$

$$R_a = \frac{l_a}{\mu_o \mu_r S} = \frac{0.6\times 10^{-2}}{4\pi \times 10^{-7}(1) \times 2 \times 10^{-4}} = \underline{\underline{2.387 \text{ x } 10^7 \text{ A.t/m}}}$$

$R = R_a + R_c = 2.502 \text{ x } 10^7$ A.t/m

$$\psi = \frac{\Im}{R} = \psi_a = \psi_c = \frac{20{,}000}{2.502 \times 10^7} = \underline{\underline{8 \text{ x } 10^{-4} \text{ Wb/m}^2}}$$

$$\Im_a = \frac{R_a}{R_a + R_c}\Im = \frac{2.387 \times 20{,}000}{2.502} = \underline{\underline{19{,}081}} \text{ A.t}$$

$$\Im_c = \frac{R_c}{R_a + R_c}\Im = \frac{0.115 \times 20{,}000}{2.502} = \underline{\underline{919}} \text{ A.t}$$

Prob. 8.39

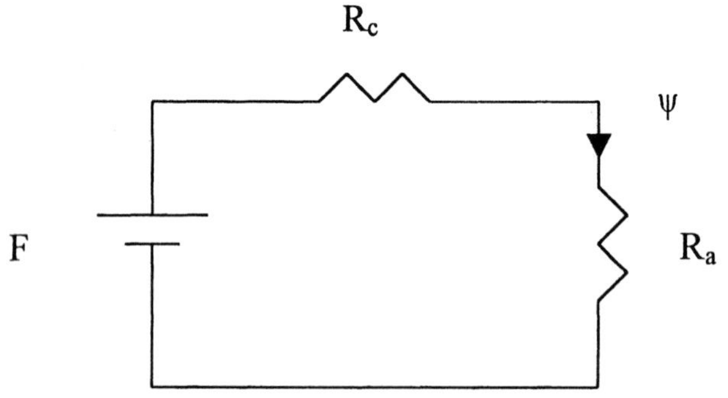

F = NI = 500 x 0.2 = 100 A.t

$$R_c = \frac{l_c}{\mu S} = \frac{42x10^{-2}}{4\pi x 10^{-7} x 10^3 x 4 x 10^{-4}} = \frac{42x10^6}{16\pi}$$

$$R_a = \frac{l_a}{\mu_o S} = \frac{10^{-3}}{4\pi x 10^{-7} x 4 x 10^{-4}} = \frac{10^8}{16\pi}$$

$$R_a + R_c = \frac{1.42x10^8}{16\pi}$$

$$\psi = \frac{F}{R_a + R_c} = \frac{16\pi x 100}{1.42x10^8} = \frac{16\pi}{1.42} \ \mu\text{Wb}$$

$$B_a = \frac{\psi}{S} = \frac{16\pi x10^{-6}}{1.42x4x10^{-4}} = \underline{\underline{88.5 \text{ mWb/m}^2}}$$

Prob. 8.40

$$F = \frac{B_2 S}{2\mu_o} = \frac{\psi^2}{2\mu_o S} = \frac{4\times10^{-6}}{2\times4\pi\times10^{-7}\times0.3\times10^{-4}} = \underline{\underline{53.05}} \text{ kN}$$

Prob. 8.41

(a) F = NI = 200 x 10^{-3} x 750 = 150 A.t.

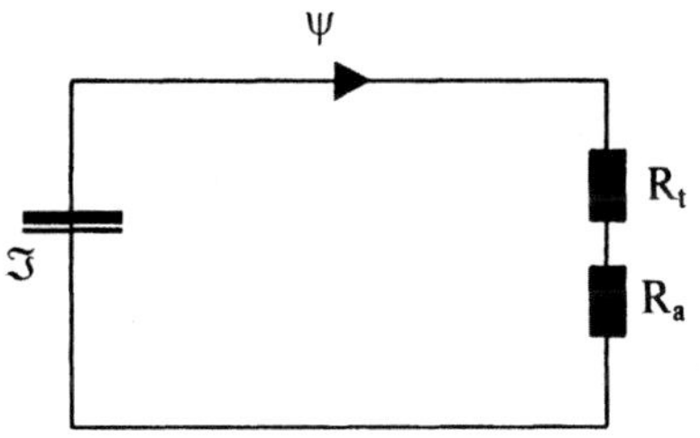

$$R_a = \frac{l_a}{\mu_o S} = \frac{10^{-3}}{25\times10^{-6}\mu_o} = 3.183\times10^7$$

$$R_t = \frac{l_t}{\mu_o\mu_r S} = \frac{2\pi\times0.1}{\mu_o\times300\times25\times10^{-6}} = 20 \text{ x } 10^7$$

$$\psi = \frac{\Im}{R_a + R_t} = \frac{150}{10^7(3.183+20)} = 20\times10^7$$

$$F = \frac{B^2 S}{2\mu_o} = \frac{\psi^2}{2\mu_o S} = \frac{41.861\times10^{-14}}{2\times4\pi\times10^{-7}\times25\times10^{-6}}$$

$= \underline{\underline{6.66 \text{ mN}}}$

(c) If $\mu_t \to \infty$, $R_t = 0$, $\psi = \dfrac{\Im}{R_a} = \dfrac{150}{3.183\times10^7}$

$$F_2 = I_2 dl_2 \bullet B_1 = I_2 dl_2 \frac{\psi_1}{S} = \frac{2\times10^{-3}\times5\times10^{-3}\times150}{3.183\times10^7\times25\times10^{-6}}$$

F_2 = $\underline{\underline{1.885 \text{ nN}}}$

Prob. 8.42

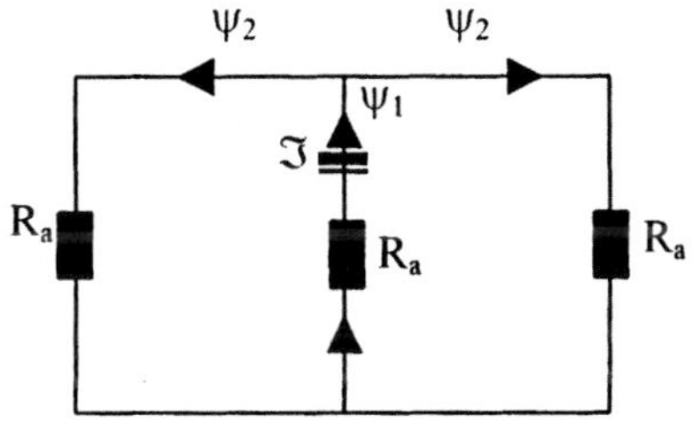

⇒

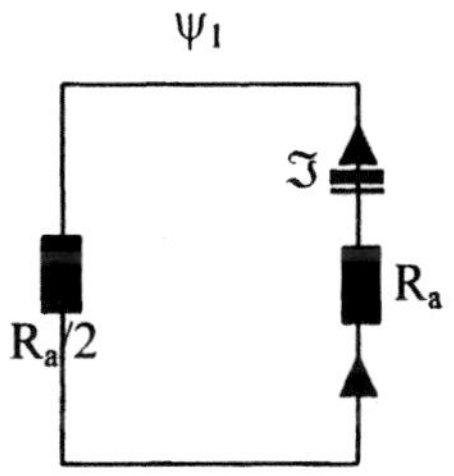

$$\psi_1 = 2\psi_2, \psi_1 = \frac{\Im}{\frac{3}{2}R_a} = \frac{2\Im}{3R_a} \to \psi_2 = \frac{\Im}{3R_a}$$

$$\Im = 2\left(\frac{\psi_2^{\,2}}{2\mu_o S}\right) + \frac{\psi_1}{2\mu_o S} = \frac{3\psi_1^{\,2}}{\mu_o S} = \frac{\Im^2}{3R_a^{\,2}\mu_o^{\,2}}$$

$$= \frac{\mu_o S \Im^2}{3 l_a^{\,2}} = \frac{4\pi \times 10^{-7} \times 210 \times 10^{-4} \times 9 \times 10^{6}}{3 \times 10^{6}}$$

$$= 24\pi \times 10^{3} = mg \rightarrow m = \frac{24\pi \times 10^{3}}{9.8} = \underline{\underline{7694}} \text{ kg}$$

Prob. 8.43

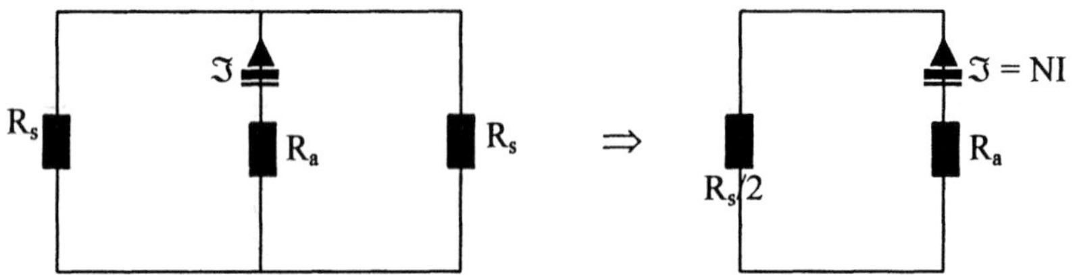

Since $\mu \rightarrow \infty$ for the cure, $R_c = 0$.

$$\Im = NI = \psi\left(R_a + \frac{R_s}{2}\right) = \frac{\psi(a/2 + x)}{\mu_o S}$$

$$= \frac{\psi(2x + a)}{2\mu_o S}$$

$$\Im = \frac{B^2 S}{2\mu_o} = \psi \frac{4}{2\mu_o S} = \frac{1}{2\mu_o S} \bullet \frac{N^2 I^2 4\mu_o S^2}{(a + 2x)^2}$$

$$= \frac{2N^2 I^2 \mu_o S}{(a + 2x)^2}$$

$\vec{F} = -F\vec{a}_x$ since the force is attractive, i.e.

$$\vec{F} = \frac{-2N^2 I^2 \mu_o S \vec{a}_x}{(a + 2x)^2}$$

CHAPTER 9

P.E. 9.1

(a) $V_{emf} = \int (\overline{u} \times \overline{B}) \cdot \partial \overline{l} = uBl = 8(0.5)(0.1) = \underline{\underline{0.4}}$ V

(b) $I = \frac{V_{emf}}{R} = \frac{0.4}{20} = \underline{\underline{20}}$ mA

(c) $\overline{F_m} = I\overline{l} \times \overline{B} = 0.2(0.1\overline{a_y} \times -0.5\overline{a_z}) = \underline{\underline{-\overline{a_x}}}$ mN

(d) $P = FU = I^2 R = 8$ mW

or $P = \frac{V_{emf}}{R} = \frac{(0.4)^2}{20} = \underline{\underline{8}}$ mW

P.E. 9.2

(a) $V_{emf} = \int (\overline{u} \times \overline{B}) \cdot \partial \overline{l}$

where $\overline{B} = B_o \overline{a_y} = B_o(\sin\phi \overline{a_o} + \cos\phi \overline{a_o})$, $B_o = 0.05$

$(\overline{u} \times \overline{B}) \cdot \overline{\partial l} = -\rho w B_o \sin\phi \partial z = -0.2\pi \sin(wt + \pi/2)\partial z$

$$V_{emf} = \int_0^{0.03} (\overline{u} \times \overline{B}) \cdot \partial \overline{l} = -6\pi \cos(100\pi t) \text{ mV}$$

At t = 1ms,

$V_{emf} = -6\pi \cos 0.1\pi = \underline{\underline{-17.93}}$ mV

$i = \frac{V_{emf}}{R} = -60\pi \cos(100\pi.t)$ mA

At t = 3ms, $i = -60\pi \cos 0.3\pi = \underline{\underline{-0.1108}}$ A

(b) <u>Method 1</u>:

$$\Psi = \int \overline{B} \cdot \partial \overline{l} = \int B_o t(\cos\phi \overline{a_\phi} - \sin\phi \overline{a_\phi}) \cdot \partial\rho \partial z \overline{a_\phi} = -\int_0^{\rho^o} \int_0^{Z_o} Bt \sin\phi \partial\rho \partial z = -B_o \rho_o z_o t \sin\phi$$

where $B_o = 0.02$, $\rho_o = 0.04$, $z_o = 0.03$

$\phi = wt + \pi/2$

$\Psi = -B_o \rho_o z_o t \cos wt$

$$V_{emf} = -\frac{\partial \Psi}{\partial t} = B_o \rho_o z_o \cos wt - B_o \rho_o z_o t \sin wt$$

$$= (0.02)(0.04)(0.03)[\cos wt - wt\sin wt]$$

$$= 24[\cos wt - wt\sin wt]\mu V$$

Method 2:

$$V_{emf} = -\int \frac{\partial \vec{B}}{\partial t} \bullet d\vec{s} + \int (\vec{u} \times \vec{B}).d\vec{l}$$

$$\vec{B} = B_o t\vec{a}_x = B_o t(\cos\phi\vec{a}_p - \sin\phi\vec{a}_p), \phi = wt + \pi/2$$

$$\frac{\partial \vec{B}}{\partial t} = B_o(\cos\phi\vec{a}_p - \sin\phi\vec{a}_p)$$

Note that only explicit dependence of $\vec{B}$ on time is accounted for, i.e. we make ϕ = constant because it is transformer (stationary) emf. Thus,

$$V_{emf} = -B_o \int_0^{\rho_o}\int_0^{z_o}(\cos\phi\vec{a}_p - \sin\phi\vec{a}_p)dpdz\vec{a}_\phi + \int_{z_o}^{0} -\rho_o w B_o t\cos\phi dz$$

$$= B_o\rho_o z_o(\sin\phi + wt\cos\phi), \phi = wt + \pi/2$$

$$= B_o\rho_o z_o(\cos wt + wt\sin wt)$$ as obtained earlier.

At t = 1ms,

$$V_{emf} = 24[\cos 18^o - 100\pi \times 10^{-3}\sin 18^o]\mu V$$

$$= \underline{\underline{20.5\mu V}}$$

At t = 3ms,

$$i = 240[\cos 54^o - .03\pi\sin 54^o]mA$$

$$= \underline{\underline{-41.92mA}}$$

P.E. 9.3

$$V_1 = -N_1\frac{d\psi}{dt}, V_2 = -N_2\frac{d\psi}{dt}$$

$$\frac{V_2}{V_1} = \frac{N_2}{N_1} \rightarrow V_2 = \frac{N_2}{N_1}V_1 = \frac{300 \times 120}{500} = \underline{\underline{72V}}$$

P.E. 9.4

(a) $\vec{J}_a = \frac{\partial \vec{D}}{\partial t} = \underline{\underline{-20w\varepsilon_o\sin(wt - 50x)\vec{a}_y A/m^2}}$

(b) $\nabla \times \vec{H} = \vec{J}_a \rightarrow -\frac{\partial \vec{H}_z}{\partial x}\vec{a}_y = -20w\varepsilon_o\sin(wt - 50x)\vec{a}_y$

or $\vec{H} = \frac{20w\varepsilon_o}{50}\cos(wt - 50x)\vec{a}_z$

$= \underline{0.4w\varepsilon_o \cos(wt - 50x)\vec{a}_z}$ A/m

(c) $\nabla \times \vec{E} = -\mu_o \frac{\partial \vec{H}}{\partial t} \rightarrow -\frac{\partial \vec{E}_z}{\partial x}\vec{a}_z = 0.4\mu_o w\varepsilon_o \sin(wt - 50x)\vec{a}_z$

$1000 = 0.4\mu_o\varepsilon_o w^2 = 0.4\frac{u^2}{c_2}$

or w = $\underline{1.5 \text{ x } 10^{10} \text{ rad/s}}$

P.E. 9.5

(a) $j^3\left(\frac{1+j}{2-j}\right)^2 = -j\left[\frac{\sqrt{2}\angle 45^o}{\sqrt{5}\angle -26.56^o}\right]^2 = -j\left(2/5\angle 143.13^o\right)$

$= \underline{0.24 + j0.32}$

(b) $6\angle 30^o + j5 - 3 + ej^{45^o} = 5.196 + j3 + j5 - 3 + 0.7071(1+j)$

$= \underline{2.903 + j8.707}$

P.E. 9.6

$\vec{P} = 2\sin(10t + x - \pi/4)\vec{a}_y = 2\cos(10t + x - \pi/4 - \pi/2)\vec{a}_y$, $w = 10$

$= R_e\left(2e^{j(x - 3\pi/4)}\vec{a}_y e^{jwt}\right) = R_e\left(\vec{P}_s e^{jwt}\right)$

i.e. $P_s = \underline{2e^{j(x-3\pi/4)}\vec{a}_y}$

$\vec{Q} = R_e\left(\vec{Q}_s e^{jwt}\right) = R_e\left(e^{j(x+wt)}(\vec{a}_x - \vec{a}_z)\right)\sin \pi y$

$= \underline{\sin \pi y \cos(wt + x)(\vec{a}_x - \vec{a}_z)}$

P.E. 9.7

$-\mu\frac{\partial \vec{H}}{\partial t} = \nabla \times \vec{E} = \frac{1}{r\sin\theta}\frac{\partial}{\partial \theta}(E_\phi \sin\theta)\vec{a}_r - \frac{1}{r}\frac{\partial}{\partial r}(rE_\phi)\vec{a}_\theta$

$= \frac{2\cos\theta}{r^2}\cos(wt - \beta r)\vec{a}_r - \frac{\beta}{r}\sin\theta\sin(wt - \beta r)\vec{a}_\theta$

$\vec{H} = \frac{2\cos\theta}{wr^2}\sin(wt - \beta r)\vec{a}_r + \frac{\beta}{wr}\sin\theta\cos(wt - \beta r)\vec{a}_\theta$

$$\beta = \frac{w}{c} = \frac{6\times10^7}{3\times10^8} = \underline{\underline{0.2}} \text{ rad/m}$$

$$\vec{H} = \underline{\underline{\frac{10^{-7}}{3r^2}\cos\theta\sin(6\times10^7 - 0.2r)\vec{a}_r + \frac{10^{-8}}{3r}\sin\theta\cos(6\times10^7 - 0.2r)\vec{a}_\theta}}$$

P.E. 9.8

$$\omega = \frac{3}{\sqrt{\mu\varepsilon}} = \frac{3c}{\sqrt{\mu_r\varepsilon_r}} = \frac{9\times10^8}{\sqrt{10}} = \underline{\underline{2.846\times10^8}} \text{ rad/s}$$

$$\vec{E} = \frac{1}{\varepsilon}\int\nabla\times\vec{H}\partial t = -\frac{6}{w\varepsilon}\cos(wt-3y)\vec{a}_x$$

$$= \frac{-6}{\frac{9\times10^8}{\sqrt{10}}\bullet\frac{10^{-9}}{36}(5)}\cos(wt-3y)\vec{a}_x$$

$$\vec{E} = \underline{\underline{-476.8\cos(2.846\times10^8 t - 3y)\vec{a}_x}} \text{ V/m}$$

Prob. 9.1

$$V = -\frac{\partial\psi}{\partial t} = -\frac{\partial}{\partial t}\int\vec{B}\bullet dS = -\frac{\partial\vec{B}}{\partial t}\bullet S$$

$= 3770 \sin377t \times \pi(0.2)^2 \times 10^{-3}$

$= \underline{\underline{0.4738 \sin377t \text{ V}}}$

Prob. 9.2 $V_{emf} = \int(u\times\overline{B})\cdot d\bar{l}, \quad d\bar{l} = d\rho\bar{a}_\rho, \quad u = \rho\frac{d\phi}{dt} = \rho w\bar{a}_\phi$

$$u\times\overline{B} = \rho w\bar{a}_\phi\times B_o\bar{a}_z = B_o\rho w\bar{a}_\rho$$

$$V_{emf} = \int_{\rho=0}^{l} B_o\rho w\bar{a}_\rho\cdot d\rho\bar{a}_\rho = B_o w\left.\frac{\rho^2}{2}\right|_0^l = \frac{1}{2}B_o wl^2$$

$$V_{emf} = \underline{\underline{\frac{1}{2}B_o wl^2}}$$

Prob. 9.3

$$V_{emf} = -\frac{\partial\lambda}{\partial t} = -W\frac{\partial}{\partial t}\int\vec{B}\bullet dS = -NBS\frac{d\phi}{dt}$$

$= -NBSW = -50 \times 0.06 \times 0.3 \times 0.4 \quad = \underline{\underline{-54V}}$

Prob. 9.4

$$\psi = \int \overline{B} \cdot d\overline{S} = BS$$

$$V_{emf} = -\frac{d\psi}{dt} = -\frac{dB}{dt}S = +40\times10^4 \sin\left(10^4\right)\cdot10^{-3}\times20\times10^{-4}$$

$$= 0.8\sin 10^4 t$$

$$I = \frac{V_{emf}}{R} = \underline{\underline{0.2\sin 10^4 t \text{ A}}}$$

I flows $\underline{\underline{\text{clockwise}}}$ for increasing $\overline{B}$ field.

Prob. 9.5 (a)

$$v = \int\left(u\times\overline{B}\right)\cdot d\bar{l}, \quad d\bar{l} = dy\bar{a}_y$$

$$u\times\overline{B} = 2\bar{a}_x\times0.1\bar{a}_z = -0.2\bar{a}_y$$

$y = x$ since the angle of the v - shaped conductor is 45°. Hence $y = x = ut$. At $t = 0, x = 0 = y$

$$v = -\int 0.2\, du = -0.2y, \quad y = ut = 2t$$

$$\underline{\underline{v = -0.4t \text{ V}}}$$

(b)

$$v = \int\left(u\times\overline{B}\right)\cdot d\bar{l}, \quad d\bar{l} = dy\bar{a}_y$$

$$u\times\overline{B} = 2\bar{a}_x\times0.5x\bar{a}_z = -x\bar{a}_y$$

But $y = x$ and $x = ut$. When $t = 0$, $x = 0 = y$

$$v = -\int x\, dy = -\int y\, dy = -y^2/2$$

But $x = y = ut = 2t$

$$\underline{\underline{v = -2t^2 \text{ V}}}$$

Prob. 9.6

$$B = \frac{\mu_o I}{2\pi y}(-a_x)$$

$$\psi = \int \vec{B}\bullet d\vec{S} = \frac{\mu_o I}{2\pi}\int_{z=0}^{a}\int_{y=\rho}^{\rho+a}\frac{dzdy}{y} = \frac{\mu_o Ia}{2\pi}\ln\frac{\rho+a}{\rho}$$

$$V_{emf} = -\frac{\partial\psi}{\partial t} = -\frac{\partial\psi}{\partial\rho}\bullet\frac{\partial\rho}{\partial t} = -\frac{\mu_o Ia}{2\pi}u_o\frac{d}{d\rho}[\ln(\rho+a)-\ln\rho]$$

$$= -\frac{\mu_o Ia}{2\pi}u_o\left[\frac{1}{\rho+a}-\frac{1}{\rho}\right] = \underline{\underline{\frac{\mu_o a^2 Iu_o}{2\pi\rho(\rho+a)}}}$$

Prob. 9.7 This is similar to Prob. 9.6. Assume loop is of width z.

$$\psi = \frac{\mu_o I z}{2\pi} \ln \frac{\rho + a}{\rho}$$

$$V_{emf} = -\frac{\partial \psi}{\partial t} = -\frac{\partial \psi}{\partial z} \bullet \frac{\partial z}{\partial t} = -\frac{\mu_o I}{2\pi} \ln \frac{\rho + a}{\rho} \bullet u$$

$$= -\frac{4\pi \times 10^{-7}}{2\pi} \times 15 \times 3 \ln \frac{60}{20} = -9.888 \mu V$$

Thus the induced emf = 9.888μV, point A at higher potential.

Prob. 9.8

$$V_{emf} = -\int \frac{\partial \vec{B}}{\partial t} \bullet dS + \int (\vec{u} \times \vec{B}) \bullet d\vec{l}$$

where $\vec{B} = B_o \cos wt \vec{a}_x, \vec{u} = u_o \cos wt \vec{a}_y, d\vec{l} = dz \vec{a}_z$

$$V_{emf} = \int_{z=0}^{l} \int_{y=-a}^{y} B_o w \sin wt dy dz - \int_{0}^{l} B_o u_o \cos^2 wt dz$$

$= B_o wl(y+a)\sin wt - B_o u_o l \cos^2 wt$

Alternatively,

$$\psi = \int \vec{B} \bullet d\vec{s} = \int_{z=0}^{l} \int_{y=-a}^{y} Bo \cos wt \vec{a}_x \bullet dy dz \vec{a}_x = B_o(y+a)l \cos wt$$

$$V_{emf} = -\frac{\partial \psi}{\partial t} = B_o (y + a) l w \sin wt - B_o \frac{dy}{dt} l \cos wt$$

But $\frac{dy}{dt} = u = u_o \cos wt \rightarrow y = \frac{u_o}{w} \sin wt$

$V_{emf} = B_o wl(y+a)\sin wt - B_o u_o l \cos^2 wt$

$= B_o u_o l \sin^2 wt + B_o wal \sin wt - B_o u_o l \cos^2 wt$

$= -B_o u_o l \cos 2wt + B_o wal \sin wt$

$= 6 \times 10^{-3} \times 5[10 \times 10 \sin 10t - 2\cos 20t]$

$V_{emf} = 3\sin 10t - 0.06\cos 20t$ V

Prob. 9.9

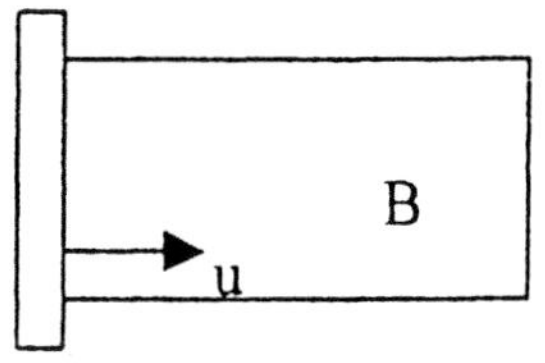

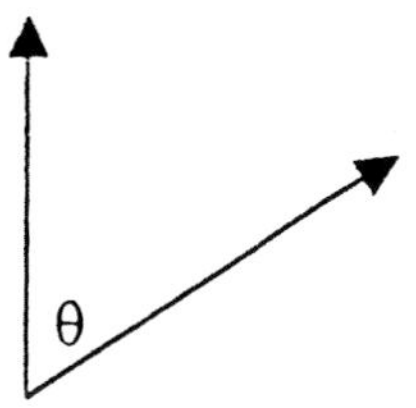

$$V_{emf} = \int(\bar{u} \times \bar{B}) \cdot \partial l = uBlCos\theta$$

$$= \left(\frac{120 \times 10^3}{3600} \text{ mls}\right)(4.3 \times 10^{-5})(1.6)Cos65^\circ$$

$$= 2.293Cos65^o = \underline{\underline{0.97}} \text{ mV}$$

Prob. 9.10

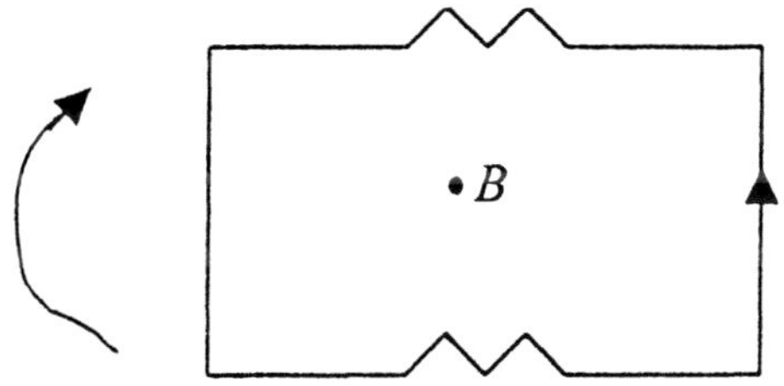

$$\oint \vec{E} \bullet d\vec{l} = -\frac{d}{dt}\int \vec{B} \bullet dS$$

$$= I(R_1 + R_2)$$

$$\frac{dB}{dt} \bullet S = I(R_1 + R_2) \qquad (1)$$

Also, $\oint \vec{E} \bullet d\vec{l} = V_1 - V_2 = -\frac{dB}{dt} \bullet S \qquad (2)$

Hence, $V_1 = IR_1 = -\frac{SR_1}{R_1 + R_2}\frac{dB}{dt}$

$$V_2 = -IR_2 = \frac{SR_2}{R_1 + R_2}\frac{dB}{dt}$$

$$V_1 = \frac{10 \times 10^{-4} \times 10}{15} \times 0.2 \times 150\pi \sin 150\pi t = \underline{\underline{0.0628 \sin 150\pi t}}$$

$$V_2 = \frac{-10 \times 10^{-4} \times 5}{15} \times 0.2 \times 150\pi \sin 150\pi t = \underline{\underline{-0.0314 \sin 150\pi t}}$$

Prob. 9.11

$d\psi = 0.63 - 0.45 = 0.18$, $dt = 0.02$

$$V_{emf} = N\frac{d\psi}{dt} = 10\left(\frac{0.18}{0.02}\right) = 90V$$

$$I = \frac{V_{emf}}{R} = \left(\frac{90}{15}\right) = \underline{\underline{6}}\text{ A}$$

Using Lenz's law, the direction of the induced current is counterclockwise.

Prob. 9.12

$$V = \int(\vec{u} \times \vec{B}) \bullet d\vec{l}\text{ , where } \vec{u} = \rho\omega\vec{a}_\phi,\ \vec{B} = B_o\vec{a}_z$$

$$V = \int_{\rho_2}^{\rho_1} \rho\omega B_o d\rho = \frac{wB_o}{2}(\rho^2{}_2 - \rho^2{}_1)$$

$$V = \frac{60 \times 5}{2} \bullet 10^{-3}(100 - 4) \bullet 10^{-4} = \underline{\underline{4.32}}\text{ mV}$$

Prob. 9.13

$$J_{ds} = j\omega D_s \rightarrow |J_{ds}|_{\max} = \omega\varepsilon E_s = \omega\varepsilon\frac{V_s}{d}$$

$$= \frac{10^{-9}}{36\pi} \times \frac{2\pi \times 20 \times 10^6 \times 50}{0.2 \times 10^{-3}}$$

$$= \underline{\underline{277.8}}\text{ A/m}^2$$

$$I_{ds} = J_{ds} \bullet S = \frac{1000}{3.6} \times 2.8 \times 10^{-4} = \underline{\underline{77.78}}\text{ mA}$$

Prob. 9.14 $\quad \dfrac{J_c}{J_d} = \dfrac{\sigma E}{\omega\varepsilon E} = \dfrac{\sigma}{\omega\varepsilon}$

(a) $$\frac{\sigma}{\omega\varepsilon} = \frac{2x10^{-3}}{2\pi x10^9 x81x\frac{10^{-9}}{36\pi}} = \underline{\underline{0.444x10^{-3}}}$$

(b) $$\frac{\sigma}{\omega\varepsilon} = \frac{25}{2\pi x10^9 x81x\frac{10^{-9}}{36\pi}} = \underline{\underline{5.555}}$$

(c) $\frac{\sigma}{\omega\varepsilon} = \frac{2x10^{-4}}{2\pi x10^9 x5x\frac{10^{-9}}{36\pi}} = \underline{\underline{7.2x10^{-4}}}$

Prob. 9.15 $\frac{J}{J_d} = \frac{\sigma E}{\omega\varepsilon E} = \frac{\sigma}{\omega\varepsilon} = 10$

$\omega = \frac{\sigma}{10\varepsilon} = 2\pi f \longrightarrow f = \frac{\sigma}{20\pi\varepsilon} = \frac{20}{20\pi x\frac{10^{-9}}{36\pi}}$

f = $\underline{\underline{36 \text{ GHz}}}$

Prob. 9.16

$$J_c = \frac{I_c}{S} = \sigma E \rightarrow E = \frac{I_c}{\sigma S}$$

$$J_a = j\omega\varepsilon E \rightarrow |J_a| = \omega\varepsilon E = \frac{\omega\varepsilon I_s}{\sigma S}$$

$$|J_d| = \frac{10^9 \times 4.6\times 10^{-9}/36\pi \times 0.2\times 10^{-3}}{25\times 10^6 \times 10\times 10^{-4}} A/m = \underline{\underline{3.254}} \text{ nA/m}^2$$

Prob. 9.17

(a) $\nabla \bullet \vec{E}_s = \rho_s/\varepsilon, \nabla \bullet \vec{H}_s = 0$

$$\underline{\underline{\nabla\times\vec{E}_s = j\omega\mu\vec{H}_s, \nabla\times\vec{H}_s = (\sigma - j\omega\varepsilon)\vec{E}_s}}$$

(b) $\nabla \bullet \vec{D} = \rho_v \rightarrow \frac{\partial D_x}{\partial x} + \frac{\partial D_y}{\partial y} + \frac{\partial D_z}{\partial z} = \rho_v$ (1)

$\nabla \bullet \vec{B} = 0 \rightarrow \frac{\partial B_x}{\partial x} + \frac{\partial B_y}{\partial y} + \frac{\partial B_z}{\partial z} = 0$ (2)

$\nabla\times\vec{E} = -\frac{\partial\vec{B}}{\partial t} \rightarrow \frac{\partial E_z}{\partial y} - \frac{\partial E_y}{\partial z} = -\frac{\partial B_x}{\partial t}$ (3)

$\frac{\partial E_x}{\partial z} - \frac{\partial E_z}{\partial x} = -\frac{\partial B_y}{\partial t}$ (4)

$\frac{\partial E_y}{\partial x} - \frac{\partial E_x}{\partial y} = -\frac{\partial B_z}{\partial t}$ (5)

$\nabla\times\vec{H} = \vec{J} + \frac{\partial\vec{D}}{\partial t} \rightarrow \frac{\partial H_z}{\partial y} - \frac{\partial H_y}{\partial z} = J_x + \frac{\partial D_x}{\partial t}$ (6)

$$\frac{\partial H_x}{\partial z} - \frac{\partial H_z}{\partial x} = J_y + \frac{\partial D_y}{\partial t} \quad (7)$$

$$\frac{\partial H_y}{\partial x} - \frac{\partial H_x}{\partial y} = J_z + \frac{\partial D_z}{\partial t} \quad (8)$$

Prob. 9.18

If $\vec{J} = 0 = \rho_v$, then

$$\nabla \bullet \vec{B} = 0 \quad (1)$$

$$\nabla \bullet \vec{D} = \rho_v \quad (2)$$

$$\nabla \times \vec{E} = -\frac{\partial \vec{B}}{\partial t} \quad (3)$$

$$\nabla \times \vec{H} = \vec{J} + \frac{\partial \vec{D}}{\partial t} \quad (4)$$

Since $\nabla \bullet \nabla \times \vec{A} = 0$ for any vector field $\vec{A}$,

$$\nabla \bullet \nabla \times \vec{E} = -\frac{\partial}{\partial t} \nabla \bullet \vec{B} = 0$$

$$\nabla \bullet \nabla \times \vec{H} = -\frac{\partial}{\partial t} \nabla \bullet \vec{D} = 0$$

showing that (1) and (2) are incorporated in (3) and (4). Thus Maxwell's equations can be reduced to (3) and (4), i.e.

$$\nabla \times \vec{E} = -\frac{\partial \vec{B}}{\partial t}, \nabla \times \vec{H} = -\frac{\partial \vec{D}}{\partial t}$$

Prob. 9.19

$$-\frac{\partial \rho_v}{\partial t} = \nabla \bullet J = \nabla \bullet \sigma E = \sigma \nabla \bullet \frac{D}{\varepsilon} = \frac{\sigma}{\varepsilon} \rho_v$$

Hence,

$$\frac{\partial \rho_v}{\partial t} + \frac{\sigma}{\varepsilon} \rho_v = 0$$

Prob. 9.20

$$\nabla x E = -\frac{\partial B}{\partial t}$$

$$\nabla x \nabla x E = -\frac{\partial}{\partial t} \nabla x B = -\mu \frac{\partial}{\partial t} \nabla x H = -\mu \frac{\partial J}{\partial t}$$

But

$$\nabla x \nabla x E = \nabla(\nabla \bullet E) - \nabla^2 E$$

$$\nabla(\nabla \bullet E) - \nabla^2 E = -\mu \frac{\partial J}{\partial t}, \qquad J = \sigma E$$

In a source-free region, $\nabla \bullet E = \rho_v / \varepsilon = 0$. Thus,

$$\underline{\underline{\nabla^2 E = \mu\sigma \frac{\partial E}{\partial t}}}$$

Prob. 9.21

$$\nabla \bullet J = (0 + 0 + 3z^2)\sin 10^4 t = -\frac{\partial \rho_v}{\partial t}$$

$$\rho_v = \int \nabla \bullet J dt = \int 3z^2 \sin 10^4 t dt = -\frac{3z^2}{10^4}\sin 10^4 t + C_o$$

If $\rho_v|_{z=0} = 0$, then $C_o = 0$ and

$$\underline{\underline{\rho_v = -0.3z^2 \sin 10^4 t \ \text{mC/m}^3}}$$

Prob. 9.22 (a)

$$J_d = \nabla x H = \begin{vmatrix} \frac{\partial}{\partial x} & \frac{\partial}{\partial y} & \frac{\partial}{\partial z} \\ 0 & 0 & H_z(y,t) \end{vmatrix} = \frac{\partial H_z}{\partial y} a_x = \underline{\underline{20\sin(10^9 t - 4y) a_x \ \text{A/m}}}$$

But $J_d = \frac{\partial D}{\partial t}$.

$$D = \int J_d dt = -\frac{20}{10^9}\cos(10^9 t - 4y) a_x = \underline{\underline{-20\cos(10^9 t - 4y) a_x \ \text{nC/m}^2}}$$

(b) $\nabla x E = -\mu \frac{\partial H}{\partial t} = \nabla x \frac{D}{\varepsilon}$

$$\frac{1}{\varepsilon}\begin{vmatrix} \frac{\partial}{\partial x} & \frac{\partial}{\partial y} & \frac{\partial}{\partial z} \\ D_x(y,t) & 0 & 0 \end{vmatrix} = -\frac{1}{\varepsilon}\left(\frac{-20}{10^9}\right)(-4)\sin(10^9 t - 4y) a_z$$

$$-\frac{1}{\varepsilon}\left(\frac{80}{10^9}\right)\sin(10^9 t - 4)a_z = -5\mu x 10^9 \sin(10^9 t - 4y)a_z$$

$$\frac{80}{10^9\varepsilon_o\varepsilon_r} = 5\mu x 10^9 \quad \longrightarrow \quad \varepsilon_r = \frac{80}{5x4\pi x10^{-7}x10^{18}x\frac{10^{-9}}{36\pi}} = \underline{\underline{1.44}}$$

Prob. 9.23

$$\varepsilon\frac{\partial\vec{E}}{\partial t} = \nabla\times\vec{H} = \begin{vmatrix} \frac{\partial}{\partial x} & \frac{\partial}{\partial y} & \frac{\partial}{\partial z} \\ 0 & 0 & H_z \end{vmatrix} = -\frac{\partial H_z}{\partial x}\vec{a}_y$$

$$= 0.6\beta\sin\beta x\cos wt\vec{a}_y$$

$$E = \frac{1}{\varepsilon}\int\nabla\times\vec{H}\partial t = \frac{0.6\beta}{w\varepsilon}\sin\beta x\sin wt\vec{a}_y$$

$$\nabla\times\vec{E} = -\mu\frac{\partial\vec{H}}{\partial t} = \begin{vmatrix} \frac{\partial}{\partial x} & \frac{\partial}{\partial y} & \frac{\partial}{\partial z} \\ 0 & E_y & 0 \end{vmatrix} = -\frac{\partial E_y}{\partial x}\vec{a}_z$$

$$= \frac{0.6\beta^2}{w\varepsilon}\cos\beta x\sin wt\vec{a}_z$$

$$\vec{H} = -\frac{1}{\mu}\int\nabla\times\vec{E}\partial t = \frac{0.6\beta^2}{w^2\mu\varepsilon}\cos\beta x\cos wt\vec{a}_z$$

Thus $\beta = w\sqrt{\mu\varepsilon} = \frac{w}{c}\sqrt{\mu_r\varepsilon_r} = \frac{10^8(2.25)}{3\times10^8}$

$= \underline{\underline{0.8333 \text{ rad/m}}}$

$$E_o = \frac{0.6\beta}{w\varepsilon} = \frac{0.6w\sqrt{\mu\varepsilon}}{w\varepsilon} = 0.6\sqrt{\mu/\varepsilon} = 0.6(377)\sqrt{\mu_r/\varepsilon_r}$$

$$= \frac{0.6\times337}{2.25} = 100.5$$

$$\underline{\underline{\vec{E} = 100.5\sin\beta x\sin wt\vec{a}_y}} \text{ V/m}$$

Prob. 9.24

$$\nabla xE = -\frac{\partial B}{\partial t} = -\mu_o \frac{\partial H}{\partial t}$$

$$\nabla xE = \begin{vmatrix} \frac{\partial}{\partial x} & \frac{\partial}{\partial y} & \frac{\partial}{\partial z} \\ 0 & E_y(x) & E_z(x) \end{vmatrix} = -\frac{\partial E_z}{\partial x}a_y + \frac{\partial E_y}{\partial x}a_z \quad = 40x8\cos(10^9 t - 8x)a_y + 50x8\sin(10^9 t - 8x)a_z$$

$$H = -\frac{1}{\mu_o}\int \nabla xE dt = -\frac{10^{-9}}{\mu_o}\left[40x8\sin(10^9 t - 8x)a_y - 50x8\cos(10^9 t - 8x)a_z\right]$$

$$= -\frac{10^{-2}}{4\pi}\left[320\sin(10^9 t - 8x)a_y - 400\cos(10^9 t - 8x)a_z\right]$$

$$H = -0.2546\sin(10^9 t - 8x)a_y + 0.3184\cos(10^9 t - 8x)a_z \text{ A/m}$$

$$\beta = \omega\sqrt{\mu\varepsilon} = \frac{\omega}{c}\sqrt{\mu_r\varepsilon_r}, \quad (\mu_r = 1) \quad \longrightarrow \quad \sqrt{\varepsilon_r} = \frac{\beta c}{\omega} = \frac{8x3x10^8}{10^9} = 2.4$$

$$\varepsilon_r = 5.76$$

Prob. 9.25 (a) $\nabla \bullet A = 0$

$$\nabla xA = \begin{vmatrix} \frac{\partial}{\partial x} & \frac{\partial}{\partial y} & \frac{\partial}{\partial z} \\ 0 & 0 & E_z(x,t) \end{vmatrix} = -\frac{\partial E_z(x,t)}{\partial x}a_y \neq 0$$

Yes, A is a possible EM field.

(b) $\nabla \bullet B = 0$

$$\nabla xB = \frac{1}{\rho}\frac{\partial}{\partial \rho}\left[10\cos(\omega t - 2\rho)\right]a_z \neq 0$$

Yes, B is a possible EM field.

(c) $\nabla \bullet C = \frac{1}{\rho}\frac{\partial}{\partial \rho}\left(3\rho^3 \cot\phi\right) - \frac{\sin\phi}{\rho^2} \neq 0$

$$\nabla xC = \frac{1}{\rho}\frac{\partial}{\partial \rho}(\cos\phi \sin\omega t) - 3\rho^2\frac{\partial}{\partial \phi}(\cot\phi) \neq 0$$

No, C is not an EM field.

(d) $\nabla \bullet D = \frac{1}{r^2 \sin\theta} \sin(\omega t - 5r) \frac{\partial}{\partial\theta}(\sin^2\theta) \neq 0$

$$\nabla x D = -\frac{\partial D_\theta}{\partial\phi} a_r + \frac{1}{r}\frac{\partial}{\partial r}(r D_\theta) a_\phi = \frac{1}{r}\sin\theta(-5)\sin(\omega t - 5r) a_\phi \neq 0$$

No, D is not an EM field.

Prob. 9.26 From Maxwell's equations,

$$\nabla \times \vec{E} = -\frac{\partial \vec{B}}{\partial t} \quad (1)$$

$$\nabla \times \vec{H} = \vec{J} + \frac{\partial \vec{D}}{\partial t} \quad (2)$$

Dotting both sides of (2) with $\vec{E}$ gives:

$$\vec{E} \bullet (\nabla \times \vec{H}) = \vec{E} \bullet \vec{J} + \vec{E} \bullet \frac{\partial \vec{D}}{\partial t} \quad (3)$$

But for any arbitrary vectors $\vec{A}$ and $\vec{B}$,

$$\nabla \bullet (\vec{A} \times \vec{B}) = \vec{B} \bullet (\nabla \times \vec{A}) - \vec{A} \bullet (\nabla \times \vec{B})$$

Applying this on the left-hand side of (3) by letting $\vec{A} \equiv \vec{B}$ and $\vec{B} \equiv \vec{E}$, we get

$$\vec{H} \bullet (\nabla \times \vec{E}) + \nabla \bullet (\vec{H} \times \vec{E}) = \vec{E} \bullet \vec{J} + \tfrac{1}{2}\frac{\partial}{\partial t}(\vec{D} \bullet \vec{E}) \quad (4)$$

From (1),

$$\vec{H} \bullet (\nabla \times \vec{E}) = \vec{H} \bullet \left(-\frac{\partial \vec{B}}{\partial t}\right) = \tfrac{1}{2}\frac{\partial}{\partial t}(\vec{B} \bullet \vec{H})$$

Substituting this in (4) gives:

$$-\tfrac{1}{2}\frac{\partial}{\partial t}(\vec{B} \bullet \vec{H}) - \nabla \bullet (\vec{E} \times \vec{H}) = \vec{J} \bullet \vec{E} + \tfrac{1}{2}\frac{\partial}{\partial t}(\vec{D} \bullet \vec{E})$$

Rearranging terms and then taking the volume integral of both sides:

$$\int_v \nabla \bullet (\vec{E} \times \vec{H}) dv = -\frac{\partial}{\partial t}\tfrac{1}{2}\int_v (\vec{E} \bullet \vec{D} + \vec{H} \bullet \vec{B}) dv - \int_v \vec{J} \bullet \vec{E} dv$$

$$\oint_s (\vec{E} \times \vec{H}) \bullet dS = -\frac{\partial w}{\partial t} - \int_v \vec{J} \bullet \vec{E} dv$$

or $\frac{\partial w}{\partial t} = -\oint_s (\vec{E} \times \vec{H}) \bullet dS - \int_v \vec{E} \bullet \vec{J} dv$ as required.

Prob. 9.27 $\nabla x H = J + J_d$

$J = \sigma E = 0$ in free space.

$$J_d = \nabla x H = \left[\frac{1}{\rho}\frac{\partial H_z}{\partial\phi} - \frac{\partial H_\phi}{\partial z}\right] a_\rho + \left[\frac{\partial H_\rho}{\partial z} - \frac{\partial H_z}{\partial\rho}\right] a_\phi + \frac{1}{\rho}\left[\frac{\partial}{\partial\rho}(\rho H_\phi) - \frac{\partial H_\rho}{\partial\phi}\right] a_z$$

$$= 0 + \frac{1}{\rho}\left[\frac{\partial}{\partial\rho}(2\rho^2\cos\phi) - \rho\cos\phi\right]\cos 4x10^6 t a_z = \frac{a_z}{\rho}(4\cos\phi - \rho\cos\phi)\cos 4x10^6 t$$

$$\underline{\underline{J_d = 3\cos\phi\cos 4x10^6 t a_z}}$$

$$J_d = \frac{\partial D}{\partial t} = \varepsilon_o\frac{\partial E}{\partial t} \quad \longrightarrow \quad E = \frac{1}{\varepsilon_o}\int J_d dt$$

$$E = \frac{3}{\varepsilon_o}\frac{\cos\phi}{4x10^6}\sin 4x10^6 t a_z = \frac{3}{4x10^6 x\frac{10^{-9}}{36\pi}}\cos\phi\sin 4x10^6 t a_z$$

$$E = \underline{\underline{84.82\cos\phi\sin 4x10^6 t a_z}}\text{ kV/m}$$

Prob. 9.28 Using Maxwell's equations,

$$\nabla xH = \sigma E + \varepsilon\frac{\partial E}{\partial t} \qquad (\sigma = 0) \qquad \longrightarrow \qquad E = \frac{1}{\varepsilon}\int\nabla xH dt$$

But

$$\nabla xH = -\frac{1}{r\sin\theta}\frac{\partial H_\theta}{\partial\phi}a_r + \frac{1}{r}\frac{\partial}{\partial r}(rH_\theta)a_\phi = \frac{12\sin\theta}{r}\beta\sin(2\pi x10^8 t - \beta r)a_\phi$$

$$E = \frac{12\sin\theta}{\varepsilon_o}\beta\int\sin(2\pi x10^8 t - \beta r)dt a_\phi$$

$$\underline{\underline{= -\frac{12\sin\theta}{\omega\varepsilon_o r}\beta\sin(\omega t - \beta r)a_\phi}}, \qquad \omega = 2\pi x10^8$$

Prob. 9.29

$$\nabla\times\vec{E} = \frac{1}{\rho}\frac{\partial}{\partial\rho}(\rho E_\phi)\vec{a}_z = \frac{1}{\rho}\frac{\partial}{\partial\rho}(\rho^2 + e^{-\rho-t})\vec{a}_z$$

$$= (2-\rho)te^{-\rho-t}\vec{a}_z$$

$$\frac{\partial\vec{B}}{\partial t} = \nabla\times\vec{E} \rightarrow \vec{B} = -\int\nabla\times\vec{E}dt = \int\frac{(\rho-2)t}{V}\frac{e^{-\rho-t}dt}{du}\vec{a}_z$$

Integrating by parts yields

$$\vec{B} = [-(\rho-2)te^{-\rho-t} + \int(\rho-2)e^{-\rho-t}dt]\vec{a}_z$$

$$= \underline{\underline{(2-\rho)(1+t)e^{-\rho-t}\vec{a}_z}} \text{ Wb/m}^2$$

$$\vec{J} = \nabla \times \vec{H} = \nabla \times \frac{\vec{B}}{\mu_o} = -\frac{1}{\mu_o}\frac{\partial B_z}{\partial \rho}\vec{a}_\phi$$

$$= -\frac{1}{\mu_o}(1+t)(-1-2t+\rho)e^{-\rho-t}\vec{a}_\phi$$

$$\underline{\underline{\vec{J} = \frac{(1+t)(3-\rho)e^{-\rho-t}}{4\pi}\vec{a}_\phi}} \text{ A/m}^2$$

Prob. 9.30 For time factor e^{-jwt}, replace every j by –j and obtain:

$$\vec{B}_s = \nabla \times \vec{A}_s$$

$$\vec{E}_s = -\nabla V_s - jw\vec{A}_s$$

$$\nabla \times \vec{A}_s = -jw\mu\varepsilon V_s$$

$$\nabla^2 V_s + w^2\mu\varepsilon V_s = -\rho_s/\varepsilon$$

$$\nabla^2 \vec{A}_s + w^2\mu\varepsilon\vec{A}_s = -\mu\vec{J}_s$$

Prob. 9.31

(a)

$$z = 4\angle 30^o - 10\angle 50^o = 3.464 + j - 6.427 - j7.66$$

$$= -2.296 - 5.60 = \underline{\underline{6.39\angle 242.37^o}}$$

(b)

$$\frac{1+j2}{6-j8-7\angle 15^o} = \frac{2.236\angle 63.43^o}{6-j8-7.761-j1.812} = \frac{2.236\angle 63.43^o}{9.841\angle 265.57^o}$$

$$= \underline{\underline{0.2272\angle -202.1^o}}$$

(c) $$z = \frac{(5\angle 53.13^o)^2}{12-j7-6-j10} = \frac{25\angle 106.26^o}{18.028\angle -70.56^o}$$

$$= \underline{\underline{1.387\angle 176.8^o}}$$

(d)

$$\frac{1.897\angle -100^o}{(5.76\angle 90^o)(9.434\angle -122^o)} = \underline{\underline{0.0349\angle -68^o}}$$

Prob. 9.32 (a) $\sin\theta = \cos(\theta - 90^o)$

$$E = 4\cos(\omega t - 3x - 10^o)a_y - 5\cos(\omega t + 3x - 70^o)a_z$$

$$= \text{Re}\left[4e^{j(-3x-10^o)}e^{j\omega t}a_y - 5e^{j(3x-70^o)}e^{j\omega t}a_z\right] = \text{Re}\left[E_s e^{j\omega t}\right]$$

$$E_s = \underline{\underline{4e^{-j(3x+10^o)}a_y - 5e^{j(3x-70^o)}a_z}}$$

(b) $$H = \text{Re}\left[\frac{\sin\theta}{r}e^{j\omega t}e^{-j5r}a_\theta\right] = \text{Re}\left[H_s e^{j\omega t}\right]$$

$$H_s = \underline{\underline{\frac{\sin\theta}{r}e^{-j5r}a_\theta}}$$

(c) $$J = \text{Re}\left[6e^{-3x}e^{-j2x}e^{-j90^o}e^{j\omega t}a_y + \ldots\right] = \text{Re}\left[J_s e^{j\omega t}\right]$$

$$J_s = \underline{\underline{-j6e^{-(3+j2)x}a_y + 10e^{-(1+j5)x}a_z}}$$

Prob. 9.33 (a) $(4 - j3) = 5e^{-j36.87^o}$

$$A_s = 5e^{-j(\beta x + 36.37^o)}a_y$$

$$A = \text{Re}\left[A_s e^{j\omega t}\right] = \underline{\underline{5\cos(\omega t - \beta x - 36.37^o)a_y}}$$

(b)

$$B = \text{Re}\left[B_s e^{j\omega t}\right] = \text{Re}\left[\frac{20}{\rho}e^{j(\omega t - 2z)}a_\rho\right]$$

$$= \underline{\underline{\frac{20}{\rho}\cos(\omega t - 2z)a_\rho}}$$

(c) $1 + j2 = 2.23e^{j63.43^o}$

$$C_s = \frac{10}{r^2}(2.236)e^{j63.43^o}e^{-j\phi}\sin\theta\, a_\phi$$

$$C = \text{Re}\left[C_s e^{j\omega t}\right] = \text{Re}\left[\frac{22.36}{r^2} e^{j(\omega t - \phi + 63.43°)} \sin\theta \, a_\phi\right]$$

$$= \underline{\underline{\frac{22.36}{r^2} \cos(\omega t - \phi + 63.43°) \sin\theta \, a_\phi}}$$

Prob. 9.34

$$A = 4\cos(\omega t - 90°)a_x + 3\cos\omega t a_y = \text{Re}\left[4e^{j(\omega t - 90°)}a_x + 3e^{j\omega t}a_y\right] = \text{Re}\left[A_s e^{j\omega t}\right]$$

$$A_s = 4e^{-j90°}a_x + 3a_y = \underline{\underline{-j4a_x + 3a_y}}$$

$$B_s = 10ze^{j90°}e^{-jz}a_x$$

$$B = \text{Re}\left[B_s e^{j\omega t}\right] = 10z\cos(\omega t - z + 90°)a_x = \underline{\underline{-10z\sin(\omega t - z)a_z}}$$

Prob. 9.35 We begin with Maxwell's equations:

$$\nabla \bullet D = \rho_v / \varepsilon = 0, \qquad \nabla \bullet B = 0$$

$$\nabla x E = -\frac{\partial B}{\partial t}, \qquad \nabla x H = J + \frac{\partial D}{\partial t}$$

We write these in phasor form and in terms of E_s and H_s only.

$$\nabla \bullet E_s = 0 \qquad (1)$$

$$\nabla \bullet H_s = 0 \qquad (2)$$

$$\nabla x E_s = -j\omega\mu H_s \qquad (3)$$

$$\nabla x H_s = (\sigma + j\omega\varepsilon)E_s \qquad (4)$$

Taking the curl of (3),

$$\nabla x \nabla x E_s = -j\omega\mu \nabla x H_s$$

$$\nabla(\nabla \bullet E_s) - \nabla^2 E_s = -j\omega\mu(\sigma + j\omega\varepsilon)E_s$$

$$\nabla^2 E_s + (\omega^2\mu\varepsilon - j\omega\mu\sigma)E_s = 0 \quad \longrightarrow \quad \underline{\underline{\nabla^2 E_s + \gamma^2 E_s = 0}}$$

Similarly, by taking the curl of (4),

$$\nabla x \nabla x H_s = (\sigma + j\omega\varepsilon)\nabla x E_s$$

$$\nabla(\nabla \bullet H_s) - \nabla^2 H_s = -j\omega\mu(\sigma + j\omega\varepsilon)H_s$$

$$\nabla^2 H_s + (\omega^2\mu\varepsilon - j\omega\mu\sigma)H_s = 0 \quad \longrightarrow \quad \underline{\underline{\nabla^2 H_s + \gamma^2 H_s = 0}}$$

CHAPTER 10

P. E. 10.1 (a)

$$T = \frac{2\pi}{\omega} = \frac{2\pi}{2x10^8} = \underline{\underline{31.42 \text{ ns}}},$$

$$\lambda = uT = 3x10^8 x 31.42x10^{-9} = \underline{\underline{9.425 \text{ m}}}$$

$$k = \beta = 2\pi / \lambda = \underline{\underline{0.677 \text{ rad/m}}}$$

(b) $t_1 = T/8 = \underline{\underline{3.927 \text{ ns}}}$

(c)

$$H(t = t_1) = 0.1\cos(2x10^8 \frac{\pi}{8x10^8} - 2x/3)a_y = 0.1\cos(2x/3 - \pi/4)a_y$$

as sketched below.

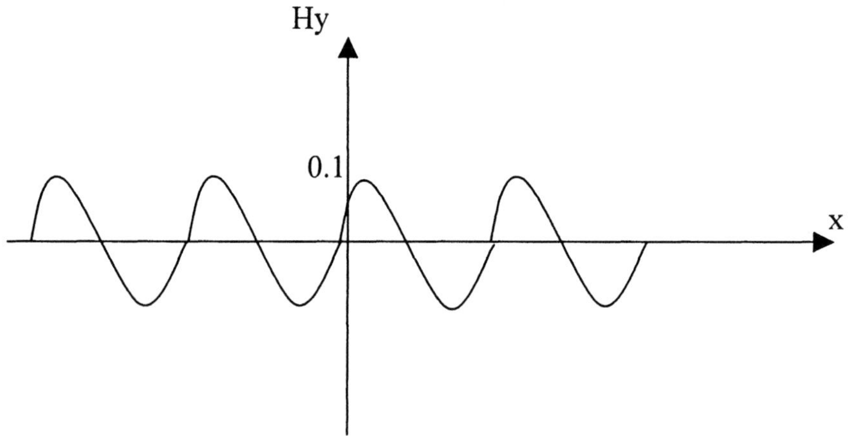

P. E. 10.2 Let $x_o = \sqrt{1 + (\sigma / \omega\varepsilon)^2}$, then

$$\alpha = \omega\sqrt{\frac{\mu_o\varepsilon_o}{2}\mu_r\varepsilon_r(x_o - 1)} = \frac{\omega}{c}\sqrt{\frac{16}{2}}\sqrt{x_o - 1}$$

or $$\sqrt{x_o - 1} = \frac{\alpha c}{\omega\sqrt{8}} = \frac{1/3x3x10^8}{10^8\sqrt{8}} = \frac{1}{\sqrt{8}} \longrightarrow x_o = 9/8$$

$$x_o^2 = \frac{81}{64} = 1 + (\sigma / \omega\varepsilon)^2 \longrightarrow \frac{\sigma}{\omega\varepsilon} = 0.5154$$

$$\tan 2\theta_\eta = 0.5154 \longrightarrow \theta_\eta = 13.63^o$$

$$\frac{\beta}{\alpha} = \sqrt{\frac{x_o + 1}{x_o - 1}} = \sqrt{17}$$

(a) $\beta = \alpha\sqrt{17} = \frac{\sqrt{17}}{3} = \underline{1.374 \text{ rad/m}}$

(b) $\frac{\sigma}{\omega\varepsilon} = \underline{0.5154}$

(c) $|\eta| = \frac{\sqrt{\mu/\varepsilon}}{\sqrt{x_o}} = \frac{120\pi\sqrt{2/8}}{\sqrt{9/8}} = 177.72$

$\eta = \underline{177.72\angle 13.63^o\, \Omega}$

(d) $u = \frac{\omega}{\beta} = \frac{10^8}{1.374} = \underline{7.278x10^7} \text{ m/s}$

(e) $a_H = a_k x a_E \longrightarrow a_x x a_H = a_z \longrightarrow a_H = a_y$

$$H = \frac{0.5}{177.5} e^{-z/3} \sin(10^8 t - \beta z - 13.63^o) a_y = \underline{2.817 e^{-z/3} \sin(10^8 t - \beta z - 13.63^o) a_y \text{ mA/m}}$$

P. E. 10.3 (a) Along <u>-z direction</u>

(b) $\lambda = \frac{2\pi}{\beta} = 2\pi/2 = \underline{3.142 \text{ m}}$

$f = \frac{\omega}{2\pi} = \frac{10^8}{2\pi} = \underline{15.92 \text{ MHz}}$

$\beta = \omega\sqrt{\mu\varepsilon} = \omega\sqrt{\mu_o\varepsilon_o}\sqrt{\mu_r\varepsilon_r} = \frac{\omega}{c}\sqrt{(1)\varepsilon_r}$

or $\sqrt{\varepsilon_r} = \beta c/\omega = \frac{3x10^8 x2}{2x10^8} = 6 \longrightarrow \underline{\varepsilon_r = 3.6}$

(c) $\theta_\eta = 0, |\eta| = \sqrt{\mu/\varepsilon} = \sqrt{\mu_o/\varepsilon_o}\sqrt{1/\varepsilon_r} = \frac{120\pi}{6} = 20\pi$

$a_k = a_E x a_H \longrightarrow -a_z = a_y x a_H \longrightarrow a_H = a_x$

$$H = \frac{50}{20\pi}\sin(\omega t + \beta z)a_x = \underline{\underline{795.8\sin(10^8 t + 2z)a_x}} \quad \text{mA/m}$$

P. E. 10.4 (a)

$$\frac{\sigma}{\omega\varepsilon} = \frac{10^{-2}}{10^9 \pi x 4x \dfrac{10^{-9}}{36\pi}} = 0.09$$

$$\alpha \cong \omega\sqrt{\frac{\mu\varepsilon}{2}\left[1 + \frac{1}{2}\left(\frac{\sigma}{\omega\varepsilon}\right)^2 - 1\right]} = \frac{\omega}{2c}\sqrt{\mu_r\varepsilon_r}\,\frac{\sigma}{\omega\varepsilon} = \frac{10^9\pi}{2x3x10^8}(2)(0.09) = 0.9425$$

$$\beta \cong \omega\sqrt{\frac{\mu\varepsilon}{2}\left[1 + \frac{1}{2}\left(\frac{\sigma}{\omega\varepsilon}\right)^2 + 1\right]} = \frac{10^9\pi}{3x10^8}\sqrt{2[2 + 0.5(0.09)^2]} = 20.965$$

$$E = 30e^{-0.9425y}\cos(10^9\pi t - 20.96y + \pi/4)a_z$$

At t = 2ns, y = 1m,

$$E = 30e^{-0.9425}\cos(2\pi - 20.96 + \pi/4)a_z = \underline{\underline{2.787a_z}} \quad \text{V/m}$$

(b) $\beta y = 10^o = \dfrac{10\pi}{180}$ rad

or

$$y = \frac{\pi}{18}\frac{1}{\beta} = \frac{\pi}{18x20.905} = \underline{\underline{8.325\ \text{mm}}}$$

(c) $30(0.6) = 30\ e^{-\alpha y}$

$$y = \frac{1}{\alpha}\ln(1/0.6) = \frac{1}{0.9425}\ln\frac{1}{0.6} = \underline{\underline{542\ \text{mm}}}$$

(d)

$$|\eta| \cong \frac{\sqrt{\mu/\varepsilon}}{[1 + \frac{1}{4}(0.09)^2]} = \frac{60\pi}{1.002} = 188.11$$

$$2\theta_\eta = \tan^{-1} 0.09 \longrightarrow \theta_\eta = 2.571^o$$

$a_H = a_k x a_E = a_y x a_z = a_x$

$$H = \frac{30}{188.11} e^{-0.9425y} \cos(10^9 \pi t - 20.96y + \pi/4 - 2.571^o) a_x$$

At y = 2m, t = 5ns,

$$H = (0.1595)(0.1518) \cos(-4.5165 rad) a_x = \underline{\underline{-4.71 a_x}} \text{ mA/m}$$

P. E. 10.5

$$I_s = \int_0^w \int_0^\infty J_{xs} dy dz = J_{xs}(0) \int_0^w dy \int_0^\infty e^{-z(1+j)\delta} dz = \frac{J_{xs}(0) w\delta}{1+j}$$

$$\underline{\underline{|I_s| = \frac{J_{xs}(0) w\delta}{\sqrt{2}}}}$$

P. E. 10.6 (a)

$$\frac{R_{ac}}{R_{dc}} = \frac{a}{2\delta} = \frac{a}{2}\sqrt{\pi f \mu \sigma} = \frac{1.3x10^{-3}}{2}\sqrt{\pi x 10^7 x 4\pi x 10^{-7} x 3.5x10^7} = \underline{\underline{24.16}}$$

(b)

$$\frac{R_{ac}}{R_{dc}} = \frac{1.3x10^{-3}}{2}\sqrt{\pi x 2x10^9 x 4\pi x 10^{-7} x 3.5x10^7} = \underline{\underline{1080.54}}$$

P. E. 10.7

$$\mathcal{P}_{ave} = \frac{1}{2}\eta H_o^2 a_x$$

(a) Let f(x,z) = x + z – 1 = 0

$$a_n = \frac{\nabla f}{|\nabla f|} = \frac{a_x + a_z}{\sqrt{2}}, \qquad d\mathbf{S} = dSa_n$$

$$P_t = \int \mathcal{P}.d\mathbf{S} = \mathcal{P}.\mathbf{S}\mathbf{a_n} = \frac{1}{2}\eta H_o^2 a_x . \frac{a_x + a_z}{\sqrt{2}}$$

$$= \frac{1}{2\sqrt{2}}(120\pi)(0.2)^2(0.1)^2 = \underline{\underline{53.31 \text{ mW}}}$$

(d) $d\mathbf{S} = dydz\mathbf{a}_x$, $P_t = \int \mathcal{P}.d\mathbf{S} = \frac{1}{2}\eta H_o^2 S$

$$P_t = \frac{1}{2}(120\pi)(0.2)^2 \pi (0.05)^2 = \underline{\underline{59.22 \text{ mW}}}$$

P. E. 10.8 $\eta_1 = \eta_o = 120\pi, \eta_2 = \sqrt{\frac{\eta}{\varepsilon}} = \frac{\eta_o}{2}$

$$\tau = \frac{2\eta_2}{\eta_2 + \eta_1} = 2/3, \Gamma = \frac{\eta_2 - \eta_1}{\eta_2 + \eta_1} = -1/3$$

$$E_{ro} = \Gamma E_{io} = -\frac{10}{3}$$

$$\underline{\underline{E_{rs} = -\frac{10}{3} e^{j\beta_1 z} a_x \text{ V/m}}}$$

where $\beta_1 = \omega / c = 100\pi / 3$.

$$E_{to} = \tau E_{io} = \frac{20}{3}$$

$$\underline{\underline{E_{ts} = \frac{20}{3} e^{-j\beta_2 z} a_x \text{ V/m}}}$$

where $\beta_2 = \omega \sqrt{\varepsilon_r} / c = 2\beta_1 = 200\pi / 3$.

P. E. 10.9

$$\alpha_1 = 0, \ \beta_1 = \frac{\omega}{c}\sqrt{\mu_r \varepsilon_r} = \frac{2\omega}{c} = 5 \longrightarrow \omega = 5c/2 = 7.5x10^8$$

$$\frac{\sigma_2}{\omega \varepsilon_2} = \frac{0.1}{7.5x10^8 x 4 x \frac{10^{-9}}{36\pi}} = 1.2\pi$$

$$\alpha_2 = \frac{\omega}{c}\sqrt{\frac{4}{2}\left[\sqrt{1 + 1.44\pi^2} - 1\right]} = 6.021$$

$$\beta_2 = \frac{\omega}{c}\sqrt{\frac{4}{2}\left[\sqrt{1 + 1.44\pi^2} + 1\right]} = 7.826$$

$$|\eta_2| = \frac{60\pi}{\sqrt[4]{1+1.44\pi^2}} = 95.445, \eta_1 = 120\pi\sqrt{\varepsilon_{r1}} = 754$$

$$\tan 2\theta_{\eta_2} = 1.2\pi \longrightarrow \theta_{\eta_2} = 37.57^o$$

$$\eta_2 = 95.445\angle 37.57^o$$

(a)

$$\Gamma = \frac{\eta_2 - \eta_1}{\eta_2 + \eta_1} = \frac{95.445\angle 37.57^o - 754}{95.445\angle 37.57^o + 754} = \underline{\underline{0.8186\angle 171.08^o}}$$

$$\tau = 1 + \Gamma = \underline{\underline{0.2295\angle 33.56^o}}$$

$$s = \frac{1+|\Gamma|}{1-|\Gamma|} = \frac{1+0.8186}{1-0.8186} = \underline{\underline{10.025}}$$

(b) $E_i = 50\sin(\omega t - 5x)a_y = \text{Im}(E_{is}e^{j\omega t})$, where $E_{is} = 50e^{-j5x}a_y$.

$$E_{ro} = \Gamma E_{io} = 0.8186e^{j171.08^o}(50) = 40.93e^{j171.08^o}$$

$$E_{rs} = 40.93e^{j5x+j171.08^o}a_y$$

$$E_r = \text{Im}(E_{rs}e^{j\omega t}) = \underline{\underline{40.93\sin(\omega t + 5x + 171.1^o)a_y}} \text{ V/m}$$

$$a_H = a_k x a_E = -a_x x a_y = -a_z$$

$$H_r = -\frac{40.93}{754}\sin(\omega t + 5x + 171.1^o)a_z = \underline{\underline{-0.0543\sin(\omega t + 5x + 171.1^o)a_z}} \text{ A/m}$$

(c)

$$E_{to} = \tau E_{io} = 0.229e^{j33.56^o}(50) = 11.475e^{j33.56^o}$$

$$E_{ts} = 11.475e^{-j\beta_2 x + j33.56^o}e^{-\alpha_2 x}a_y$$

$$E_t = \text{Im}(E_{ts}e^{j\omega t}) = \underline{\underline{11.475e^{-6.021x}\sin(\omega t - 7.826x + 33.56^o)a_y}} \text{ V/m}$$

$$a_H = a_k x a_E = a_x x a_y = a_z$$

$$H_t = \frac{11.495}{95.445} e^{-6.021x} \sin(\omega t - 7.826x + 33.56^o - 37.57^o) a_z$$

$$= \underline{\underline{0.1202 e^{-6.021x} \sin(\omega t - 7.826x - 4.01^o) a_z}} \text{ A/m}$$

(d)

$$\mathcal{P}_{1ave} = \frac{E_{io}^{\ 2}}{2\eta_1} a_x + \frac{E_{ro}^{\ 2}}{2\eta_1}(-a_x) = \frac{1}{2(240\pi)}[50^2 a_x - 40.93^2 a_x] = \underline{\underline{0.5469 a_x}} \text{ W/m}^2$$

$$\mathcal{P}_{2ave} = \frac{E_{to}^{\ 2}}{2|\eta_2|} e^{-2\alpha_2 x} \cos\theta_{\eta_2} a_x = \frac{(11.475)^2}{2(95.445)} \cos 37.57^o e^{-2(6.021)x} a_x = \underline{\underline{0.5469 e^{-12.04} a_x}} \text{ W/m}^2$$

P. E. 10.10 (a)

$$k = -2a_y + 4a_z \longrightarrow k = \sqrt{2^2 + 4^2} = \sqrt{20}$$

$$\omega = kc = 3x10^8 \sqrt{20} = \underline{\underline{1.342x10^9 \text{ rad/s}}},$$

$$\lambda = 2\pi k = \underline{\underline{28.1 \text{m}}}$$

$$\text{(b) } H = \frac{a_k x E}{\eta_o} = \frac{(-2a_y + 4a_z)}{\sqrt{20}(120\pi)} x (10a_y + 5a_z)\cos(\omega t - k.r)$$

$$= \underline{\underline{-29.66 \cos(1.342x10^9 t + 2y - 4z) a_x}} \text{ mA/m}$$

$$\text{(c) } \mathcal{P}_{ave} = \frac{|E_o|^2}{2\eta_o} a_k = \frac{125}{2(120\pi)} \frac{(-2a_y + 4a_z)}{\sqrt{20}} = \underline{\underline{-74.15a_y + 148.9a_z}} \text{ W/m}^2$$

P. E. 10.11 (a)

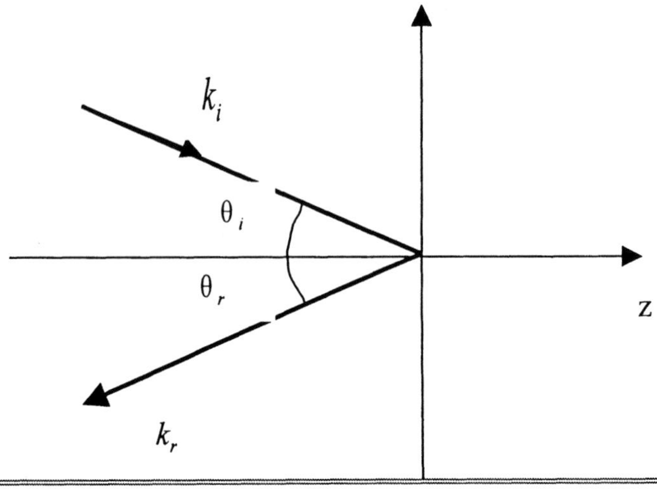

$$\tan\theta_i = \frac{k_{iy}}{k_{iz}} = \frac{2}{4} \longrightarrow \underline{\underline{\theta_i = 26.56 = \theta_r}}$$

$$\sin\theta_t = \sqrt{\frac{\mu_1\varepsilon_1}{\mu_2\varepsilon_2}}\sin\theta_i = \frac{1}{2}\sin 26.56^o \longrightarrow \underline{\underline{\theta_t = 12.92^o}}$$

(b) $\eta_1 = \eta_o, \eta_2 = \eta_o/2$, **E** is parallel to the plane of incidence. Since $\mu_1 = \mu_2 = \mu_o$, we may use the result of Prob. 10.42, i.e.

$$\Gamma_{\backslash\backslash} = \frac{\tan(\theta_t - \theta_i)}{\tan(\theta_t + \theta_i)} = \frac{\tan(-13.64^o)}{\tan(39.48^o)} = \underline{\underline{-0.2946}}$$

$$\tau_{\backslash\backslash} = \frac{2\cos 26.56^o \sin 12.92^o}{\sin 39.48^o \cos(-13.64^o)} = \underline{\underline{0.6474}}$$

(c) $k_r = -\beta_1 \sin\theta_r a_y - \beta_1 \cos\theta_r a_z$. Once k_r is known, E_r is chosen such that

$k_r . E_r = 0$ or $\nabla . E_r = 0$. Let

$E_r = \pm E_{or}(-\cos\theta_r a_y + \sin\theta_r a_z)\cos(\omega t + \beta_1 \sin\theta_r y + \beta_1 \cos\theta_r z)$

Only the positive sign will satisfy the boundary conditions. It is evident that

$E_i = E_{oi}(\cos\theta_i a_y + \sin\theta_i a_z)\cos(\omega t + 2y - 4z)$

Since $\theta_r = \theta_i$,

$E_{or}\cos\theta_r = \Gamma_{//} E_{oi}\cos\theta_i = 10\Gamma_{//} = -2.946$

$E_{or}\sin\theta_r = \Gamma_{//} E_{oi}\sin\theta_i = 5\Gamma_{//} = -1.473$

$\beta_1 \sin\theta_r = 2, \beta_1 \cos\theta_r = 4$

i.e.

$E_r = -(2.946a_y - 1.473a_z)\cos(\omega t + 2y + 4z)$

$E_1 = E_i + E_r = \underline{\underline{(10a_y + 5a_z)\cos(\omega t + 2y - 4z) + (-2.946a_y + 1.473a_z)\cos(\omega t + 2y + 4z)}}$ V/m

(d) $k_t = -\beta_2 \sin\theta_t a_y + \beta_2 \cos\theta_t a_z$. Since $k_r \bullet E_r = 0$, let

$$E_t = E_{ot}(\cos\theta_t a_y + \sin\theta_t a_z)\cos(\omega t + \beta_2 y \sin\theta_t - \beta_2 z \cos\theta_t)$$

$$\beta_2 = \omega\sqrt{\mu_2 \varepsilon_2} = \beta_1 \sqrt{\varepsilon_{r2}} = 2\sqrt{20}$$

$$\sin\theta_t = \frac{1}{2}\sin\theta_i = \frac{1}{2\sqrt{5}}, \qquad \cos\theta_t = \frac{\sqrt{9}}{\sqrt{20}}$$

$$\beta_2 \cos\theta_t = 2\sqrt{20}\sqrt{\frac{19}{20}} = 8.718$$

$$E_{ot}\cos\theta_t = \tau_{//} E_{oi}\cos\theta_t = 0.6474\sqrt{125}\sqrt{\frac{19}{20}} = 7.055$$

$$E_{ot}\sin\theta_t = \tau_{//} E_{oi}\sin\theta_t = 0.6474\sqrt{125}\sqrt{\frac{1}{20}} = 1.6185$$

Hence

$$E_2 = E_t = \underline{\underline{(7.055 a_y + 1.6185 a_z)\cos(\omega t + 2y - 8.718z)}} \ \text{V/m}$$

(d) $\tan\theta_{B//} = \sqrt{\frac{\varepsilon_2}{\varepsilon_1}} = 2 \longrightarrow \underline{\underline{\theta_{B//} = 63.43^o}}$

Prob. 10.1 (a) Wave propagates <u>along $+\mathbf{a}_x$</u>.

(b)

$$T = \frac{2\pi}{\omega} = \frac{2\pi}{2\pi x 10^6} = \underline{\underline{1\mu s}}$$

$$\lambda = \frac{2\pi}{\beta} = \frac{2\pi}{6} = \underline{\underline{1.047\,\text{m}}}$$

$$u = \frac{\omega}{\beta} = \frac{2\pi x 10^6}{6} = \underline{\underline{1.047 x 10^6\,\text{m/s}}}$$

(c) At t=0, $E_z = 25\sin(-6x) = -25\sin 6x$

At t=T/8, $E_z = 25\sin(\frac{2\pi}{T}\frac{T}{8} - 6x) = 25\sin(\frac{\pi}{4} - 6x)$

At t=T/4, $E_z = 25\sin(\frac{2\pi}{T}\frac{T}{4} - 6x) = 25\sin(-6x + 90^o) = 25\cos 6x$

At t=T/2, $E_z = 25\sin(\frac{2\pi}{T}\frac{T}{2} - 6x) = 25\sin(-6x + \pi) = 25\sin 6x$

These are sketched below.

t=0

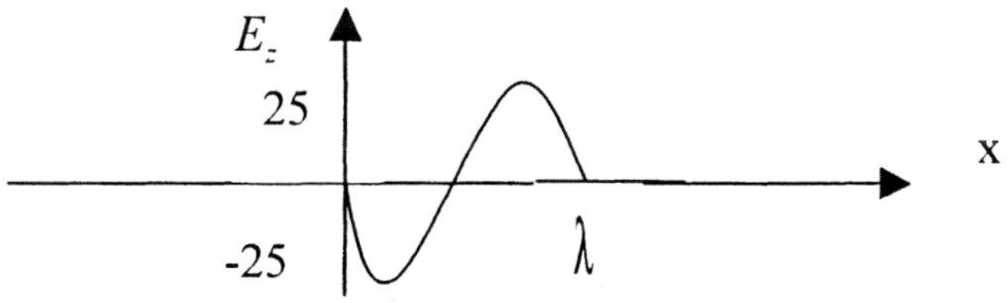

t=T/8

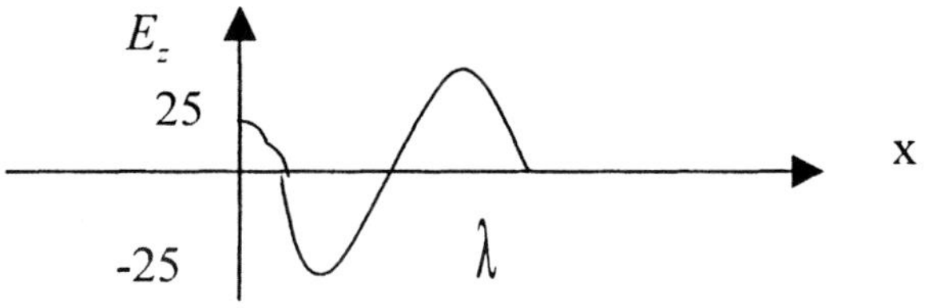

t=T/4

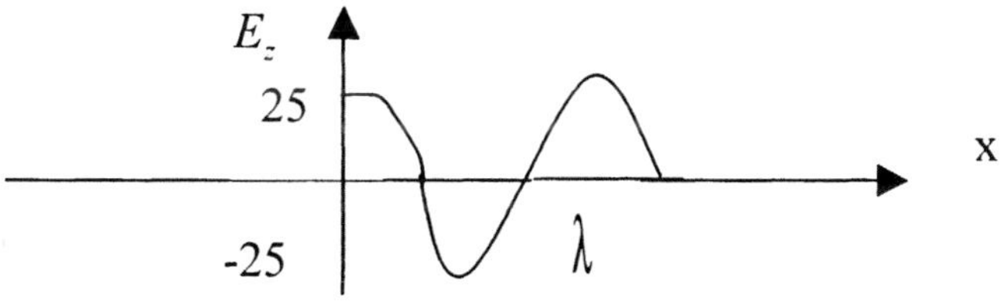

t=T/2

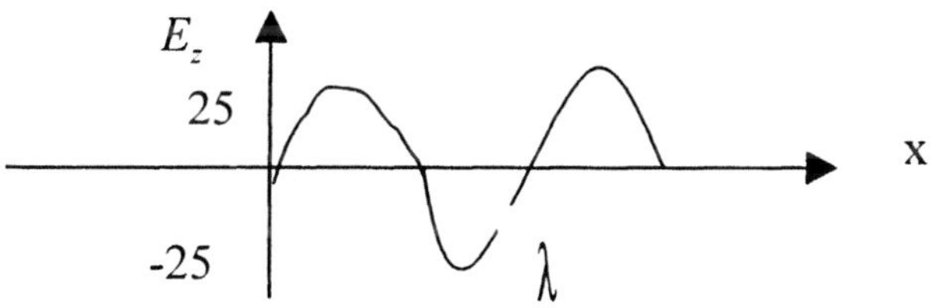

Prob. 10.2 If

$$\gamma^2 = j\omega\mu(\sigma + j\omega\varepsilon) = -\omega^2\mu\varepsilon + j\omega\mu\sigma \quad \text{and} \; \gamma = \alpha + j\beta, \text{ then}$$

$$|\gamma^2| = \sqrt{(\alpha^2 - \beta^2) + 4\alpha^2\beta^2} = \sqrt{(\alpha^2 + \beta^2)^2} = \alpha^2 + \beta^2$$

i.e.

$$\alpha^2 + \beta^2 = \omega\mu\sqrt{(\sigma^2 + \omega^2\varepsilon^2)} \qquad (1)$$

$$\text{Re}(\gamma^2) = \alpha^2 - \beta^2 = -\omega^2\mu\varepsilon$$

$$\beta^2 - \alpha^2 = \omega^2\mu\varepsilon \qquad (2)$$

Subtracting and adding (1) and (2) lead respectively to

$$\alpha = \omega\sqrt{\frac{\mu\varepsilon}{2}\left[\sqrt{1 + \left(\frac{\sigma}{\omega\varepsilon}\right)^2} - 1\right]}$$

$$\beta = \omega\sqrt{\frac{\mu\varepsilon}{2}\left[\sqrt{1 + \left(\frac{\sigma}{\omega\varepsilon}\right)^2} + 1\right]}$$

(b) From eq. (10.25), $E_s(z) = E_o e^{-\gamma z} a_x$.

$$\nabla x E = -j\omega\mu H_s \longrightarrow H_s = \frac{j}{\omega\mu}\nabla x E_s = \frac{j}{\omega\mu}(-\gamma E_o e^{-\gamma z} a_y)$$

But $H_s(z) = H_o e^{-\gamma z} a_y$, hence $H_o = \dfrac{E_o}{\eta} = -\dfrac{j\gamma}{\omega\mu}E_o$

$$\eta = \frac{j\omega\mu}{\gamma}$$

(c) From (b),

$$\eta = \frac{j\omega\mu}{\sqrt{j\omega\mu(\sigma + j\omega\varepsilon)}} = \sqrt{\frac{j\omega\mu}{\sigma + j\omega\varepsilon}} = \frac{\sqrt{\mu/\varepsilon}}{\sqrt{1 - j\dfrac{\sigma}{\omega\varepsilon}}}$$

$$|\eta| = \frac{\sqrt{\mu / \varepsilon}}{\sqrt[4]{1+\left(\frac{\sigma}{\omega\varepsilon}\right)^2}}, \tan 2\theta_\eta = \left(\frac{\omega\varepsilon}{\sigma}\right)^{-1} = \frac{\sigma}{\omega\varepsilon}$$

Prob. 10.3 (a)

$$\frac{\sigma}{\omega\varepsilon} = \frac{8x10^{-2}}{50x10^6 x3.6x\frac{10^{-9}}{36\pi}} = 8$$

$$\alpha = \omega\sqrt{\frac{\mu\varepsilon}{2}\left[\sqrt{1+\left(\frac{\sigma}{\omega\varepsilon}\right)^2} - 1\right]} = \frac{2\pi x50x10^6}{3x10^8}\sqrt{\frac{2.1x3.6}{2}}[\sqrt{65} - 1] = 5.41$$

$$\beta = \omega\sqrt{\frac{\mu\varepsilon}{2}\left[\sqrt{1+\left(\frac{\sigma}{\omega\varepsilon}\right)^2} + 1\right]} = 6.129$$

$$\gamma = \alpha + j\beta = \underline{\underline{5.41 + j6.129}} \text{ /m}$$

(b) $\lambda = \frac{2\pi}{\beta} = \frac{2\pi}{6.129} = \underline{\underline{1.025}}$ m

(c) $u = \frac{\omega}{\beta} = \frac{2\pi x50x10^6}{6.129} = \underline{\underline{5.125x10^7}}$ m/s

(d) $$|\eta| = \frac{\sqrt{\frac{\mu}{\varepsilon}}}{\sqrt[4]{1+\left(\frac{\sigma}{\omega\varepsilon}\right)^2}} = \frac{120\pi\sqrt{\frac{2.1}{3.6}}}{\sqrt[4]{65}} = 101.4$$

$$\tan 2\theta_\eta = \frac{\sigma}{\omega\varepsilon} = 8 \longrightarrow \theta_\eta = 41.44^o$$

$$\eta = \underline{\underline{101.41\angle 41.44^o \,\Omega}}$$

(e) $H_s = a_k x\frac{E_s}{\eta} = a_x x\frac{6}{\eta}e^{-\gamma z}a_z = -\frac{6}{\eta}e^{-\gamma z}a_y = \underline{\underline{-59.16e^{-j41.44^o}e^{-\gamma z}a_y}}$ mA/m

Prob. 10.4 (a) Let $u = \dfrac{\sigma}{\omega\varepsilon}$ = loss tangent

$$\beta = \omega\sqrt{\frac{\mu\varepsilon}{2}\left[\sqrt{1+u^2}+1\right]}$$

$$10 = \frac{\omega}{c}\sqrt{\frac{5x2}{2}\left[\sqrt{1+u^2}+1\right]} = \frac{2\pi x5x10^6\sqrt{5}}{3x10^8}\sqrt{\left[\sqrt{1+u^2}+1\right]}$$

which leads to

$$u = \frac{\sigma}{\omega\varepsilon} = \underline{\underline{1823}}$$

(b) $\sigma = \omega\varepsilon u = 2\pi x5x10^6 x1823x\dfrac{10^{-9}}{36\pi} = \underline{\underline{1.013}}$ S/m

(c) $\varepsilon_c = \varepsilon' - j\varepsilon'' = \varepsilon - j\dfrac{\sigma}{\omega} = 2x\dfrac{10^{-9}}{36\pi} - j\dfrac{1.023}{2\pi x5x10^6} = \underline{\underline{1.768x10^{-11} - j3.224x10^{-8}}}$ F/m

d) $$\frac{\alpha}{\beta} = \frac{\sqrt{\sqrt{1+u^2}-1}}{\sqrt{\sqrt{1+u^2}+1}} = \sqrt{\frac{1822}{1824}}$$

$\alpha = \underline{\underline{9.995}}$ Np/m

(e) $$|\eta| = \frac{\sqrt{\dfrac{\mu}{\varepsilon}}}{\sqrt[4]{1+u^2}} = \frac{120\pi\sqrt{\dfrac{5}{2}}}{\sqrt[4]{1+1823^2}} = 13.96$$

$$\tan 2\theta_\eta = u = 1823 \longrightarrow \theta_\eta = 44.98^o$$

$\eta = \underline{\underline{13.96\angle 44.98^o\,\Omega}}$

Prob. 10.5 (a) $\dfrac{\sigma}{\omega\varepsilon} = \tan 2\theta_\eta = \tan 60^o = \underline{\underline{1.732}}$

(b) $$|\eta| = 240 = \frac{\dfrac{120\pi}{\sqrt{\varepsilon_r}}}{\sqrt[4]{1+3}} = \frac{120\pi}{\sqrt{2\varepsilon_r}} \longrightarrow \varepsilon_r = \frac{\pi^2}{8} = \underline{\underline{1.234}}$$

(c) $\varepsilon_c = \varepsilon(1 - j\dfrac{\sigma}{\omega\varepsilon}) = 1.234x\dfrac{10^{-9}}{36\pi}(1 - j1.732) = \underline{\underline{(1.091 - j1.89)x10^{-11}}}$ F/m

(d)

$$\alpha = \frac{\omega}{c}\sqrt{\frac{\mu_r\varepsilon_r}{2}\left[\sqrt{1+\left(\frac{\sigma}{\omega\varepsilon}\right)^2} - 1\right]} = \frac{2\pi x10^6}{3x10^8}\sqrt{\frac{1}{2}\frac{\pi^2}{8}}\left[\sqrt{1+3} - 1\right] = \underline{\underline{0.0164}} \text{ Np/m}$$

Prob. 10.6 (a) $|E| = E_o e^{-\alpha z}$

$$E_o e^{-\alpha(1)} = (1 - 0.18)E_o \longrightarrow e^{-\alpha} = 0.82$$

$$\alpha = \ln\frac{1}{0.82} = 0.1984$$

$$\theta_\eta = 24^o \longrightarrow \tan 2\theta_\eta = \frac{\sigma}{\omega\varepsilon} = 1.111$$

$$\frac{\alpha}{\beta} = \frac{\sqrt{\sqrt{1+\left(\frac{\sigma}{\omega\varepsilon}\right)^2} - 1}}{\sqrt{\sqrt{1+\left(\frac{\sigma}{\omega\varepsilon}\right)^2} + 1}} = \sqrt{\frac{\sqrt{2.233} - 1}{\sqrt{2.233} + 1}} = 2.247, \quad \beta = 0.4458$$

$$\gamma = \alpha + j\beta = \underline{\underline{0.1984 + j0.4458}} \text{ /m}$$

(b) $\lambda = \dfrac{2\pi}{\beta} = 2\pi / 0.4458 = \underline{\underline{14.09}}$ m

(c) $\delta = 1/\alpha = \underline{\underline{5.04}}$ m

(d) Since

$$\alpha = \omega\sqrt{\frac{\mu\varepsilon}{2}\left[\sqrt{1+\left(\frac{\sigma}{\omega\varepsilon}\right)^2} - 1\right]} = \frac{\omega}{c}\sqrt{\frac{\mu_r\varepsilon_r}{2}}\sqrt{0.494}, \qquad \mu_r = 1$$

$$\sqrt{\frac{\varepsilon_r}{2}} = \frac{\alpha c}{\omega\sqrt{0.494}} = \frac{0.1984x3x10^8}{2\pi x10^7\sqrt{0.494}} = 1.348 \longrightarrow \varepsilon_r = 3.633$$

Since $\frac{\sigma}{\omega \varepsilon} = 1.111$

$$\sigma = \omega \varepsilon_o \varepsilon_r x1.111 = 2\pi x10^7 x \frac{10^{-9}}{36\pi} x3.633x1.111 = \underline{\underline{2.24x10^{-3}}} \text{ S/m}$$

Prob. 10.7

$$\frac{\sigma}{\omega \varepsilon} = \frac{4}{2\pi x10^5 x81x10^{-9} / 36\pi} = \frac{80,000}{9} >> 1$$

$$\alpha = \beta = \sqrt{\frac{\omega \mu \sigma}{2}} = \sqrt{\frac{2\pi x10^5}{2} x4\pi x10^{-7} x4} = 0.4\pi$$

(a) $u = \omega / \beta = \frac{2\pi x10^5}{0.4\pi} = \underline{\underline{5x10^5}}$ m/s

(b) $\lambda = 2\pi / \beta = \frac{2\pi}{0.4\pi} = \underline{\underline{5}}$ m

(c) $\delta = 1/\alpha = \frac{1}{0.4\pi} = \underline{\underline{0.796}}$ m

(d) $\eta = |\eta| \angle \theta_\eta, \theta_\eta = 45^o$

$$|\eta| = \frac{\sqrt{\frac{\mu}{\varepsilon}}}{\sqrt[4]{1 + \left(\frac{\sigma}{\omega \varepsilon}\right)^2}} \cong \sqrt{\frac{\mu}{\varepsilon} \frac{\omega \varepsilon}{\sigma}} = \sqrt{\frac{4\pi x10^{-7} x2\pi x10^8}{4}} = 14.05$$

$\eta = \underline{\underline{14.05\angle 45^o \quad \Omega}}$

Prob. 10.8 (a)

$$T = 1/f = 2\pi / \omega = \frac{2\pi}{\pi x10^8} = \underline{\underline{20 \text{ ns}}}$$

(b) Let $x = \sqrt{1 + \left(\frac{\sigma}{\omega \varepsilon}\right)^2}$

$$\frac{\alpha}{\beta} = \left(\frac{x-1}{x+1}\right)^{1/2}$$

But $\alpha = \frac{\omega}{c}\sqrt{\frac{\mu_r\varepsilon_r}{2}}\sqrt{x-1}$

$$\sqrt{x-1} = \frac{\alpha c}{\omega\sqrt{\frac{\mu_r\varepsilon_r}{2}}} = \frac{0.1x3x10^8}{\pi x10^8\sqrt{2}} = 0.06752 \longrightarrow x = 1.0046$$

$$\beta = \left(\frac{x+1}{x-1}\right)^{1/2}\alpha = \left(\frac{2.0046}{0.0046}\right)^{1/2} 0.1 = 2.088$$

$$\lambda = 2\pi/\beta = \frac{2\pi}{2.088} = \underline{\underline{3}}\text{ m}$$

(c) $|\eta| = \frac{\sqrt{\mu/\varepsilon}}{\sqrt{x}} = \frac{377}{2\sqrt{1.0046}} = 188.1$

$$x = \sqrt{1+\left(\frac{\sigma}{\omega\varepsilon}\right)^2} = 1.0046$$

$$\frac{\sigma}{\omega\varepsilon} = 0.096 = \tan 2\theta_\eta \longrightarrow \theta_\eta = 2.74^o$$

$$\eta = 188.1\angle 2.74^o \quad \Omega$$

$$E_o = \eta H_o = 12x188.1 = 2256.84$$

$$a_E x a_H = a_k \longrightarrow a_E x a_x = a_y \longrightarrow a_E = a_z$$

$$E = \underline{\underline{2.256e^{-0.1y}\sin(\pi x10^8 t - 2.088y + 2.74^o)a_z}} \text{ kV/m}$$

(e) The phase difference is $\underline{\underline{2.74^o}}$.

Prob. 10.9 (a) $\gamma = \alpha + j\beta = \underline{\underline{0.05 + j2}}$ /m

(b) $\lambda = 2\pi/\beta = \pi = \underline{\underline{3.142}}$ m

(c) $u = \omega/\beta = \frac{2x10}{2} = \underline{\underline{10^8}}$ m/s

(d) $\delta = 1/\alpha = \dfrac{1}{0.05} = \underline{\underline{20}}$ m

Prob. 10.10 (a) $\beta = \omega / c = \dfrac{2\pi x10^6}{3x10^8} = \underline{\underline{0.02094}}$ rad/m,

$$\lambda = 2\pi / \beta = \underline{\underline{300}} \text{ m}$$

(b) When $z = 0, \quad E_y = 10\cos\omega t$

$$z = \lambda / 4, \quad E_y = 10\cos(\omega t - \frac{2\pi}{\lambda}\frac{\lambda}{4}) = 10\sin\omega t$$

$$z = \lambda / 2, \quad E_y = 10\cos(\omega t - \pi) = -10\cos\omega t$$

Thus E is sketched below.

z = 0

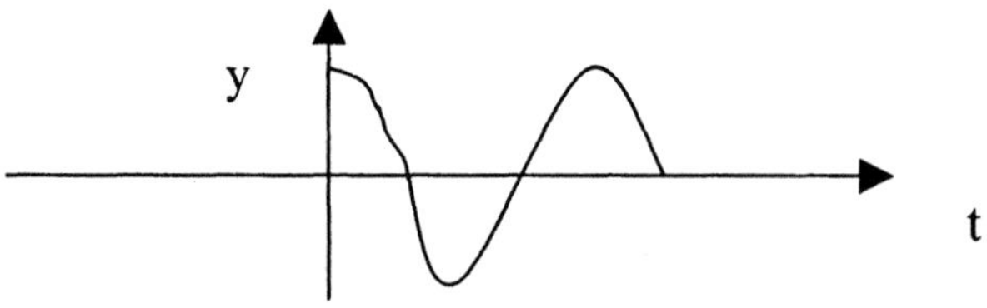

$z = \lambda / 4$

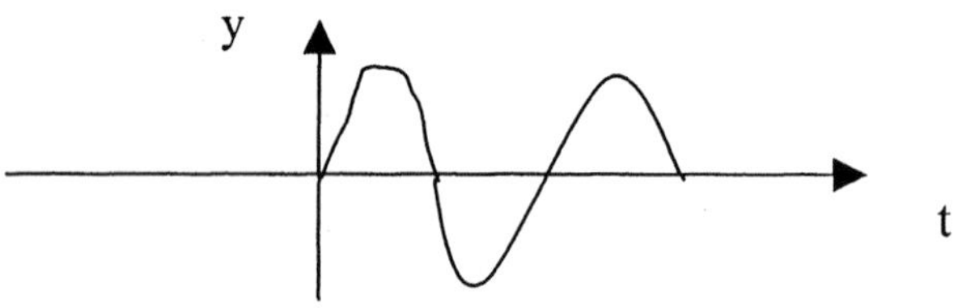

$z = \lambda / 2$

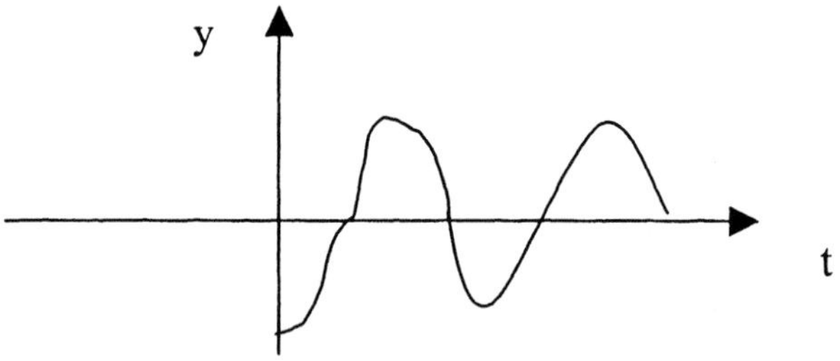

(c)

$H = \dfrac{1}{120\pi}\cos(2\pi x10^6 t - 2\pi z / 300)a_x = \underline{\underline{26.53\cos(2\pi x10^6 t - 0.02094)a_x}}$ A/m

Prob. 10.11 (a) Along $\underline{\underline{\text{-x direction.}}}$

(b) $\beta = 6, \quad \omega = 2x10^8$,

$\beta = \omega\sqrt{\mu\varepsilon} = \dfrac{\omega}{c}\sqrt{\mu_r\varepsilon_r}$

$\sqrt{\varepsilon_r} = \beta c/\omega = \dfrac{6x3x10^8}{2x10^8} = 9 \longrightarrow \varepsilon_r = 81$

$\varepsilon = \varepsilon_o\varepsilon_r = \dfrac{10^{-9}}{36\pi}x81 = \underline{\underline{7.162x10^{-10}}}$ F/m

(c) $\eta = \sqrt{\mu/\varepsilon} = \sqrt{\mu_o/\varepsilon_o}\sqrt{\mu_r/\varepsilon_r} = \dfrac{120\pi}{9}$

$E_o = H_o\eta = 25x10^{-3}x377/9 = 1.047$

$a_E x a_H = a_k \longrightarrow a_E x a_y = -a_x \longrightarrow a_E = a_z$

$E = \underline{\underline{1.047\sin(2x10^8 t + 6x)a_z}}$ V/m

Prob. 10.12 $\beta = 4 \longrightarrow \lambda = 2\pi/\beta = \underline{\underline{1.571}}$ m

Also, $\beta = \omega/u = \omega\sqrt{\mu\varepsilon} = \dfrac{\omega}{c}\sqrt{\mu_r\varepsilon_r}$

$\omega = \dfrac{\beta c}{\sqrt{\mu_r\varepsilon_r}} = \dfrac{4x3x10^8}{\sqrt{4}} = \underline{\underline{6x10^8}}$ rad/s

$$J_d = \nabla x H = \begin{vmatrix} \dfrac{\partial}{\partial x} & \dfrac{\partial}{\partial y} & \dfrac{\partial}{\partial z} \\ H_x(z) & 0 & 0 \end{vmatrix} = \frac{\partial H_x}{\partial z}a_y$$

$J_d = -40\cos(\omega t - 4z)x10^{-3}a_y = \underline{\underline{-40\cos(\omega t - 4z)a_y}}$ mA/m^2

Prob. 10.13 (a) $\dfrac{\sigma}{\omega\varepsilon} = \dfrac{10^{-6}}{2\pi x10^7 x5x\dfrac{10^{-9}}{36\pi}} = 3.6x10^{-4} << 1$

Thus, the material is <u>lossless</u> at this frequency.

(b) $\beta = \omega\sqrt{\mu\varepsilon} = \dfrac{2\pi x10^7}{3x10^8}\sqrt{5x750} = \underline{\underline{12.83}}$ rad/m

$$\lambda = \frac{2\pi}{\beta} = \frac{2\pi}{12.83} = \underline{\underline{0.49}} \text{ m}$$

(c) Phase difference $= \beta l = \underline{\underline{25.66 \text{ rad}}}$

(d) $\eta = \sqrt{\mu/\varepsilon} = 120\pi\sqrt{\dfrac{\mu_r}{\varepsilon_r}} = 120\pi\sqrt{\dfrac{750}{5}} = \underline{\underline{4617\Omega}}$

Prob. 10.14 If **A** is a uniform vector and $\Phi(r)$ is a scalar,

$$\nabla x(\Phi A) = \nabla\Phi xA + \Phi(\nabla xA) = \nabla\Phi xA$$

since $\nabla XA = 0$.

$$\nabla xE = (\frac{\partial}{\partial x}a_x + \frac{\partial}{\partial y}a_y + \frac{\partial}{\partial z}a_z)xE_o e^{j(k_x x + k_y y + k_z z - \omega t)} = j(k_x a_x + k_y a_y + k_z a_z)e^{j(\quad)}xE_o$$

$$= jkxE_o e^{j(\quad)} = jkxE$$

Also, $-\dfrac{\partial B}{\partial t} = j\omega\mu H$. Hence $\nabla xE = -\dfrac{\partial B}{\partial t}$ becomes $kxE = \omega\mu H$

From this, $\underline{\underline{a_k x a_E = a_H}}$

Prob. 10.15

$$\nabla \bullet E = (\frac{\partial}{\partial x}a_x + \frac{\partial}{\partial y}a_y + \frac{\partial}{\partial z}a_z)\bullet E_o e^{j(k_x x + k_y y + k_z z - \omega t)} = j(k_x a_x + k_y a_y + k_z a_z)e^{j(\quad)}\bullet E_o$$

$$= jk\bullet E_o e^{j(\quad)} = jk\bullet E = 0 \quad \longrightarrow \quad k\bullet E = 0$$

Similarly,

$$\nabla\bullet H = jk\bullet H = 0 \quad \longrightarrow \quad k\bullet H = 0$$

It has been shown in Prob. 10.14 that

$$\nabla xE = -\frac{\partial B}{\partial t} \quad \longrightarrow \quad kxE = \omega\mu H$$

Similarly,

$$\nabla x H = \frac{\partial D}{\partial t} \longrightarrow kxH = -\varepsilon\omega E$$

From $kxE = \omega\mu H$, $\quad a_k x a_E = a_H$ and

From $kxH = -\varepsilon\omega E$, $\quad a_k x a_H = -a_E$

Prob. 10.16 (a)

$$\beta = \frac{\omega}{c}\sqrt{\varepsilon_r} \longrightarrow \sqrt{\varepsilon_r} = \frac{\beta c}{\omega} = \frac{5x3x10^8}{2\pi x10^8} = \frac{15}{2\pi}$$

$$\underline{\underline{\varepsilon_r = 5.6993}}$$

(b) $\lambda = 2\pi/\beta = 2\pi/5 = \underline{\underline{1.2566}}$ m

$$u = \frac{c}{\sqrt{\mu_r \varepsilon_r}} = \frac{3x10^8}{\frac{15}{2\pi}} = \underline{\underline{1.257x10^8}} \text{ m/s}$$

(c) $\eta = \eta_o\sqrt{\frac{\mu_r}{\varepsilon_r}} = \frac{120\pi}{\frac{15}{2\pi}} = \underline{\underline{157.91\Omega}}$

(d) $a_E x a_H = a_k \longrightarrow a_E x a_z = a_x \longrightarrow a_E = \underline{\underline{a_y}}$

(e) $E = 30x10^{-3}(157.91)\sin(\omega t - \beta x)a_E = \underline{\underline{4.737\sin(2\pi x10^8 t - 5x)a_y}}$ V/m

(f) $J_d = \frac{\partial D}{\partial t} = \nabla x H = \underline{\underline{0.15\cos(2\pi x10^8 t - 5x)a_y}}$ A/m

Prob. 10.17 $\beta = \omega\sqrt{\mu\varepsilon} = \frac{\omega}{c}\sqrt{\varepsilon_r}\sqrt{\mu_r}$, $\quad \mu_r = 1$

$$\sqrt{\varepsilon_r} = \frac{\beta c}{\omega} = \frac{8x3x10^8}{10^9} = 2.4 \longrightarrow \underline{\underline{\varepsilon_r = 5.76}}$$

Let **E** = $\mathbf{E}_1 + \mathbf{E}_2$,

$$E_1 = 50\cos(10^9 t - 8x)a_y, \qquad E_2 = 40\sin(10^9 t - 8x)a_z$$

$$H_1 = H_{o1}\cos(10^9 t - 8x)a_{H1}, \qquad H_{o1} = \frac{50x2.4}{120\pi} = \frac{1}{\pi}$$

$$a_{E1}xa_{H1} = a_{k1} \longrightarrow a_y x a_{H1} = a_x \longrightarrow a_{H1} = a_z$$

$$H_1 = \frac{1}{\pi}\cos(10^9 t - 8x)a_z,$$

$$H_2 = H_{o2}\sin(10^9 t - 8x)a_{H2}, \qquad H_{o2} = \frac{40x2.4}{120\pi} = \frac{0.8}{\pi}$$

$$a_{E2}xa_{H2} = a_{k2} \longrightarrow a_z x a_{H2} = a_x \longrightarrow a_{H2} = -a_y$$

$$H_2 = -\frac{0.8}{\pi}\sin(10^9 t - 8x)a_y,$$

$$\mathbf{H} = \mathbf{H}_1 + \mathbf{H}_2 = \underline{\underline{-0.2546\sin(10^9 t - 8x)a_y + 0.3183\cos(10^9 t - 8x)a_z}} \quad \text{A/m}$$

Prob. 10.18 $\beta = \omega\sqrt{\mu\varepsilon} = \frac{\omega}{c}\sqrt{\mu_r\varepsilon_r} = \frac{2\pi x10^7}{3x10^8}(10) = \underline{\underline{2.0943}}$ rad/m

$$H = -\frac{1}{\mu}\int \nabla xE dt$$

$$\nabla xE = \begin{vmatrix} \frac{\partial}{\partial x} & \frac{\partial}{\partial y} & \frac{\partial}{\partial z} \\ 0 & E_y(x) & E_z(x) \end{vmatrix} = -\frac{\partial E_z}{\partial x}a_y + \frac{\partial E_y}{\partial x}a_z = -10\beta\sin(\omega t - \beta x)(a_y - a_z)$$

$$H = -\frac{10\beta}{\omega\mu}\cos(\omega t - \beta x)(a_y - a_z) = -\frac{10x2\pi/3}{2\pi x10^7 x50x4\pi x10^{-7}}\cos(\omega t - \beta x)(a_y - a_z)$$

$$H = \underline{\underline{5.305\cos(2\pi x10^7 t - 2.0943x)(-a_y + a_z)}} \quad \text{mA/m}$$

Prob. 10.19 For a good conductor, $\frac{\sigma}{\omega\varepsilon} >> 1$, say $\frac{\sigma}{\omega\varepsilon} > 100$

(a) $\quad \dfrac{\sigma}{\omega\varepsilon} = \dfrac{10^{-2}}{2\pi x 8 x 10^{6} x 15 x \dfrac{10^{-9}}{36\pi}} = 1.5 \quad ----\rightarrow \quad$ lossy

<u>No</u>, not conducting.

(b) $\quad \dfrac{\sigma}{\omega\varepsilon} = \dfrac{0.025}{2\pi x 8 x 10^{6} x 16 x \dfrac{10^{-9}}{36\pi}} = 3.515 \quad ----\rightarrow \quad$ lossy

<u>No</u>, not conducting.

(c) $\quad \dfrac{\sigma}{\omega\varepsilon} = \dfrac{25}{2\pi x 8 x 10^{6} x 81 x \dfrac{10^{-9}}{36\pi}} = 694.4 \quad ----\rightarrow \quad$ conducting

<u>Yes</u>, conducting.

Prob. 10.20

$$\alpha = \omega \sqrt{\dfrac{\mu\varepsilon}{2}\left[\sqrt{1+\left(\dfrac{\sigma}{\omega\varepsilon}\right)^{2}}-1\right]} = \dfrac{2\pi f}{c}\sqrt{\dfrac{\mu_r \varepsilon_r}{2}\left[\sqrt{1.0049}-1\right]} = \dfrac{2\pi x 6 x 10^{6}}{3 x 10^{8}}\sqrt{\dfrac{4}{2}x 2.447 x 10^{-3}}$$

$$\alpha = 8.791 x 10^{-3}$$

$$\delta = 1/\alpha = 113.75 \text{ m}$$

$$\beta = \omega\sqrt{\dfrac{\mu\varepsilon}{2}\left[\sqrt{1+\left(\dfrac{\sigma}{\omega\varepsilon}\right)^{2}}+1\right]} = \dfrac{4\pi}{100}\sqrt{\dfrac{4}{2}\left[\sqrt{1.0049}+1\right]} = 0.2515$$

$$u = \omega/\beta = \dfrac{2\pi x 6 x 10^{6}}{0.2525} = \underline{\underline{1.5 x 10^{8}}} \text{ m/s}$$

Prob. 10.21 $\qquad 0.4E_{o} = E_{o}e^{-\alpha z} \qquad \longrightarrow \qquad \dfrac{1}{0.4} = e^{2\alpha}$

Or $\quad \alpha = \dfrac{1}{2}\ln\dfrac{1}{0.4} = 0.4581 \qquad \longrightarrow \qquad \delta = 1/\alpha = \underline{\underline{2.183}} \text{ m}$

$$\lambda = 2\pi/\beta = 2\pi/1.6$$

$$u = f\lambda = 10^{7} x \dfrac{2\pi}{1.6} = \underline{\underline{3.927 x 10^{7}}} \text{ m/s}$$

Prob. 10.22 (a)

$$R_{dc} = \frac{l}{\sigma S} = \frac{l}{\sigma \pi a^2} = \frac{600}{5.8x10^7 x\pi x(1.2)^2 x10^{-6}} = 2.287\Omega$$

(b) $R_{ac} = \dfrac{l}{\sigma 2\pi a\delta}$. At 100 MHz, $\delta = 6.6x10^{-3}$ mm for copper (see Table 10.2).

$$R_{ac} = \frac{600}{5.8x10^7 x2\pi x(1.2)x6.6x10^{-3} x10^{-6}} = \underline{\underline{207.88\Omega}}$$

(c) $\dfrac{R_{ac}}{R_{dc}} = \dfrac{a}{2\delta} = 1 \longrightarrow \delta = a/2 = \dfrac{66.1x10^{-3}}{\sqrt{f}}$

$$\sqrt{f} = \frac{66.1x2x10^{-3}}{a} = \frac{66.1x2}{1.2} \longrightarrow f = \underline{\underline{12.137}} \text{ kHz}$$

Prob. 10.23

$$\omega = 10^6 \pi = 2\pi f \longrightarrow f = 0.5x10^6$$

$$\delta = \frac{1}{\sqrt{\pi f \sigma \mu}} = \frac{1}{\sqrt{\pi x0.5x10^6 x3.5x10^7 x4\pi x10^{-7}}} = \underline{\underline{0.1203}} \text{ mm}$$

$$R_{ac} = \frac{l}{\sigma \delta w}$$

since δ is very small, $w = 2\pi\rho_{outer}$

$$R_{ac} = \frac{l}{\sigma 2\pi \rho_{outer}\delta} = \frac{40}{3.5x10^7 x0.1203x2\pi x12x10^{-6}} = \underline{\underline{0.126\Omega}}$$

Prob. 10.24 $\alpha = \beta = 1/\delta$

$\lambda = 2\pi/\beta = 2\pi\delta = 6.283\delta \longrightarrow \delta = 0.1591\lambda$

showing that δ is shorter than λ.

Prob. 10.25

$$t = 5\delta = \frac{5}{\sqrt{\pi f \mu \sigma}} = \frac{5}{\sqrt{\pi x12x10^9 x4\pi x10^{-7} x6.1x10^7}} = \underline{\underline{2.94x10^{-6}}} \text{ m}$$

Prob. 10.26 (a)

$$E = \text{Re}[E_s e^{j\omega t}] = (5a_x + 12a_y)e^{-0.2z}\cos(\omega t - 3.4z)$$

At z = 4m, t = T/8, $\omega t = \frac{2\pi}{T}\frac{T}{8} = \frac{\pi}{4}$

$$E = (5a_x + 12a_y)e^{-0.8}\cos(\pi/4 - 13.6)$$

$$|E| = 13e^{-0.8}|\cos(\pi/4 - 13.6)| = \underline{\underline{5.662}}$$

(b) loss $= \alpha\Delta z = 0.2(3) = 0.6$ Np. Since 1 Np = 8.686 dB,

loss = 0.6 x 8.686 = <u>5.212 dB</u>

(c) Let $x = \sqrt{1 + \left(\frac{\sigma}{\omega\varepsilon}\right)^2}$

$$\frac{\alpha}{\beta} = \left(\frac{x-1}{x+1}\right)^{1/2} = 0.2/3.4 = \frac{1}{17}$$

$$\frac{x-1}{x+1} = 1/289 \longrightarrow x = 1.00694$$

$$\alpha = \omega\sqrt{\mu\varepsilon/2}\sqrt{x-1} = \frac{\omega}{c}\sqrt{\varepsilon_r/2}\sqrt{x-1}$$

$$\sqrt{\frac{\varepsilon_r}{2}} = \frac{\alpha c}{\omega\sqrt{x-1}} = \frac{0.2x3x10^3}{10^8\sqrt{0.00694}} = 2.4 \longrightarrow \varepsilon_r = 11.52$$

$$|\eta| = \frac{\sqrt{\frac{\mu_o}{\varepsilon_o}}\cdot\frac{1}{\sqrt{\varepsilon_r}}}{\sqrt{x}} = \frac{120\pi}{\sqrt{11.52x1.00694}} = 32.5$$

$$\tan 2\theta_\eta = \frac{\sigma}{\omega\varepsilon} = \sqrt{x^2-1} = 0.118 \longrightarrow \theta_\eta = 3.365^o$$

$$\eta = 32.5\angle 3.365^o$$

$$H_s = a_k x \frac{E_s}{\eta} = \frac{a_z}{\eta} x(5a_x + 12a_y)e^{-\gamma z} = \frac{(5a_x + 12a_y)}{|\eta|} e^{-j3.365^o} e^{-\gamma z}$$

$$H = (-369.2a_x + 153.8a_y)e^{-0.2z} \cos(\omega t - 3.4z - 3.365^o) \text{ mA}$$

$$P = ExH = \begin{vmatrix} 5 & 12 & 0 \\ -369.2 & 153.8 & 0 \end{vmatrix} x10^{-3} e^{-0.4z} \cos(\omega t - 3.4z)\cos(\omega t - 3.4z - 3.365^o)$$

$$P = 5.2e^{-0.4z} \cos(\omega t - 3.4z)\cos(\omega t - 3.4z - 3.365^o)a_z$$

At z = 4, t = T/4,

$$P = 5.2e^{-1.6} \cos(\pi / 4 - 13.6)\cos(\pi / 4 - 13.6 - 0.0587)a_z = \underline{\underline{0.9702a_z \text{W} / \text{m}^2}}$$

Prob. 10.27 (a) This is a lossless medium,

$$\beta = \omega \sqrt{\mu\varepsilon}, \qquad \eta = \sqrt{\frac{\mu}{\varepsilon}}$$

$$\eta = \frac{\omega \mu_o}{\beta} = \frac{2\pi x10^8 x4\pi x10^{-7}}{6} = \underline{\underline{131.6\Omega}}$$

(b) $E_o = \eta H_o = 131.6x30x10^{-3} = 3.948$

$$a_E x a_H = a_k \longrightarrow a_E x a_y = a_x \longrightarrow a_E = -a_z$$

$$P = ExH = \eta H_o^2 \cos^2(2\pi x10^8 t - 6x)a_x = \underline{\underline{0.1184\cos^2(2\pi x10^8 t - 6x)a_x \text{ W} / \text{m}^2}}$$

(c) $\mathcal{P}_{ave} = \frac{1}{2}\eta H_o^2 = 0.0592a_x \text{ W} / \text{m}^2$

$$P_{ave} = \int \mathcal{P}_{ave} \bullet dS = \mathcal{P}_{ave} \bullet S = 0.0592x3x2 = \underline{\underline{0.3535 \text{ W}}}$$

Prob. 10.28 Let $E_s = E_r + jE_i$ and $H_s = H_r + jH_i$

$$E = \text{Re}(E_s e^{j\omega t}) = E_r \cos\omega t - E_i \sin\omega t$$

Similarly,

$$H = H_r \cos\omega t - H_i \sin\omega t$$

$$\mathcal{P} = ExH = E_r x H_r \cos^2 \omega t + E_i x H_i \sin^2 \omega t - \frac{1}{2}(E_r x H_i + E_i x H_r)\sin\ 2\omega t$$

$$\mathcal{P}_{\text{ave}} = \frac{1}{T}\int_0^T \mathcal{P} dt = \frac{1}{T}\int_0^T \cos^2 \omega\, dt(E_r x H_r) + \frac{1}{T}\int_0^T \sin^2 \omega\, dt(E_i x H_i) - \frac{1}{2T}\int_0^T \sin 2\omega\, dt(E_i x H_i + F_i x H_r)$$

$$= \frac{1}{2}(E_r x H_r + E_i x H_i) = \frac{1}{2}\text{Re}[(E_r + jE_i)x(H_r - jH_i)]$$

$$\mathcal{P}_{\text{ave}} = \frac{1}{2}\ \text{Re}(E_s x H_s^*)$$

as required.

Prob. 10.29 (a)

$$u = \omega / \beta \quad \longrightarrow \quad \omega = u\beta = \frac{\beta}{c}\frac{1}{\sqrt{4.5}} = \frac{2x3x10^8}{\sqrt{4.5}} = \underline{\underline{2.828x10^8}} \text{ rad/s}$$

$$\eta = \frac{120\pi}{\sqrt{4.5}} = 177.7\Omega$$

$$H = a_k x \frac{E}{\eta} = \frac{a_z}{\eta} x \frac{40}{\rho}\sin(\omega t - 2z)a_\rho = \underline{\underline{\frac{0.225}{\rho}\sin(\omega t - 2z)a_\phi}} \text{ A/m}$$

(b) $$\mathcal{P} = ExH = \underline{\underline{\frac{9}{\rho^2}\sin^2(\omega t - 2z)a_z \text{ W/m}^2}}$$

(c) $$\mathcal{P}_{\text{ave}} = \frac{4.5}{\rho^2}a_z, \ \text{dS} = \rho\, d\phi\, d\rho\, a_z$$

$$P_{\text{ave}} = \int \mathcal{P}_{\text{ave}} \bullet \text{dS} = 4.6 \int_{2\text{mm}}^{3\text{mm}} \frac{d\rho}{\rho}\int_0^{2\pi} d\phi = 4.5\ln(3/2)(2\pi) = \underline{\underline{11.46 \text{ W}}}$$

Prob. 10.30 (a) $P_{i,ave} = \frac{E_{io}^2}{2\eta_1}$, $\quad P_{r,ave} = \frac{E_{ro}^2}{2\eta_1}$, $\quad P_{t,ave} = \frac{E_{to}^2}{2\eta_2}$

$$R = \frac{P_{r,ave}}{P_{i,ave}} = \frac{E_{ro}^2}{E_{io}^2} = \Gamma^2 = \underline{\underline{\left(\frac{\eta_2 - \eta_1}{\eta_2 + \eta_1}\right)^2}}$$

$$R = \left(\frac{\sqrt{\frac{\mu_o}{\varepsilon_2}} - \sqrt{\frac{\mu_o}{\varepsilon_1}}}{\sqrt{\frac{\mu_o}{\varepsilon_2}} + \sqrt{\frac{\mu_o}{\varepsilon_1}}} \right)^2 = \left(\frac{\sqrt{\mu_o \varepsilon_1} - \sqrt{\mu_o \varepsilon_2}}{\sqrt{\mu_o \varepsilon_1} + \sqrt{\mu_o \varepsilon_2}} \right)^2$$

Since $n_1 = c\sqrt{\mu_1 \varepsilon_1} = c\sqrt{\mu_o \varepsilon_1}, \qquad n_2 = c\sqrt{\mu_o \varepsilon_2},$

$$R = \left(\frac{n_1 + n_2}{n_1 + n_2} \right)^2$$

$$T = \frac{P_{t,ave}}{P_{i,ave}} = \frac{\eta_1}{\eta_2} \frac{E_{to}^2}{E_{io}^2} = \frac{\eta_1}{\eta_2} \tau^2 = \frac{4 n_1 n_2}{(n_1 + n_2)^2}$$

(b) If $P_{r,ave} = P_{t,ave} \longrightarrow RP_{i,ave} = TP_{i.ave} \longrightarrow R = T$

i.e. $(n_1 - n_2)^2 = 4 n_1 n_2 \qquad \longrightarrow \qquad n_1^2 - 6 n_1 n_2 + n_2^2 = 0$

$$\frac{n_1}{n_2} = 3 \pm \sqrt{8} = \underline{\underline{5.828 \quad \text{or} \quad 0.1716}}$$

Prob. 10.31 (a) $\eta_1 = \eta_o, \qquad \eta_o = \sqrt{\frac{\mu}{\varepsilon}} = \eta_o / 2$

$$\Gamma = \frac{\eta_2 - \eta_1}{\eta_2 + \eta_1} = \frac{\eta_o / 2 - \eta_o}{3\eta_o / 2} = \underline{\underline{-1/3}}, \qquad \tau = \frac{2\eta_2}{\eta_2 + \eta_1} = \frac{\eta_o}{3\eta_o / 2} = \underline{\underline{2/3}}$$

$$s = \frac{1 + |\Gamma|}{1 - |\Gamma|} = \frac{1 + 1/3}{1 - 1/3} = \underline{\underline{2}}$$

(b) $E_{or} = \Gamma E_{oi} = -\frac{1}{3} x(30) = -10$

$$\underline{\underline{E_r = -10 \cos(\omega t + z) a_x}} \text{ V/m}$$

Let $H_r = H_{or} \cos(\omega t + z) a_H$

$a_E x a_H = a_k \longrightarrow -a_k x a_H = -a_z \longrightarrow a_H = a_y$

$$H_r = \frac{10}{120\pi}\cos(\omega t + z)a_y = \underline{\underline{26.53\cos(\omega t + z)a_y}} \text{ mA/m}$$

Prob. 10.32 (a) $\eta_1 = \eta_o$

$$E_i = E_{io}\sin(\omega t - 5x)a_E$$

$$E_{io} = H_{io}\eta_o = 120\pi x 4 = 480\pi$$

$$a_E x a_H = a_k \longrightarrow a_E x a_y = a_x \longrightarrow a_E = -a_z$$

$$E_i = -480\pi\sin(\omega t - 5x)a_z$$

$$\eta_2 = \sqrt{\frac{\mu_o}{\varepsilon_o}} = \frac{120\pi}{\sqrt{4}} = 60\pi$$

$$\Gamma = \frac{\eta_2 - \eta_1}{\eta_2 + \eta_1} = \frac{60\pi - 120\pi}{60\pi + 120\pi} = -1/3, \qquad \tau = 1 + \Gamma = 2/3$$

$$E_{ro} = \Gamma E_{io} = (-1/3)(480\pi) = -160\pi$$

$$E_r = 160\pi\sin(\omega t + 5x)a_z$$

$$E_1 = E_i + E_r = \underline{\underline{-1.508\sin(\omega t - 5x)a_z + 0.503\sin(\omega t + 5x)a_z}} \text{ kV/m}$$

(b) $E_{to} = \tau E_{io} = (2/3)(480\pi) = 320\pi$

$$\mathcal{P} = \frac{E_{to}^2}{2\eta_2}a_x = \frac{(320\pi)^2}{2(60\pi)}a_x = \underline{\underline{2.68a_x \text{ kW/m}^2}}$$

(c) $s = \dfrac{1+|\Gamma|}{1-|\Gamma|} = \dfrac{1+1/3}{1-1/3} = \underline{\underline{2}}$

Prob. 10.33 $\eta_1 = \eta_o = 120\pi, \qquad \eta_2 = \sqrt{\dfrac{\mu_2}{\varepsilon_2}}$

$$\frac{E_{ro}}{E_{io}} = \Gamma = \frac{\eta_2 - \eta_1}{\eta_2 + \eta_1} \qquad (1)$$

But $E_{ro} = \eta_o H_{ro}$ (2)

Combining (1) and (2),

$$E_{ro} = \eta_o H_{ro} = \left(\frac{\eta_2 - \eta_1}{\eta_2 + \eta_1}\right) E_{io} \quad \longrightarrow \quad \eta_o = \left(\frac{\eta_2 - \eta_1}{\eta_2 + \eta_1}\right) \frac{E_{io}}{H_{ro}}$$

But $\frac{E_{io}}{H_{ro}} = \frac{3.6}{1.2x10^{-3}} = 3000$

$$\eta_o = 3000\left(\frac{\eta_2 - \eta_1}{\eta_2 + \eta_1}\right) \quad \longrightarrow \quad 377 = 3000\left(\frac{\eta_2 - 377}{\eta_2 + 377}\right)$$

Thus, $\eta_2 = 485.37$. Since $\eta_2 = \sqrt{\frac{\mu_2}{\varepsilon_2}}$,

$$\mu_2 = \varepsilon_o \varepsilon_r \eta_2^2 = \frac{10^{-9}}{36\pi} x12.5x(485.37)^2 = \underline{\underline{2.604x10^{-5}}} \text{ H/m}$$

Prob. 10.34 $\eta_1 = \sqrt{\frac{\mu_1}{\varepsilon_1}} = \eta_o / 2, \qquad \eta_2 = \eta_o$

$$\Gamma = \frac{\eta_2 - \eta_1}{\eta_2 + \eta_1} = 1/3, \qquad \tau = 1 + \Gamma = 4/3$$

$$E_{or} = \Gamma E_{io} = (1/3)(5) = 5/4, \qquad E_{ot} = \tau E_{io} = 20/3$$

$$\beta = \frac{\omega}{c}\sqrt{\mu_r \varepsilon_r} = \frac{10^8}{3x10^8}\sqrt{4} = 2/3$$

(a) $E_r = \frac{5}{3}\cos(10^8 t - 2y/3)a_z$

$$E_1 = E_i + E_r = \underline{\underline{5\sin(10^8 t + \frac{2}{3}y)a_z + \frac{5}{3}\cos(10^8 t - \frac{2}{3}y)a_z}} \text{ V/m}$$

(b) $\mathcal{P}_{\text{ave1}} = \frac{E_{io}^2}{2\eta_1}(-a_y) + \frac{E_{ro}^2}{2\eta_1}(+a_y) = \frac{25}{2(60\pi)}(1 - \frac{1}{9})(-a_y) = \underline{\underline{-0.0589a_y \text{ W/m}^2}}$

(c) $\mathcal{P}_{\text{ave2}} = \frac{E_{to}^2}{2\eta_2}(-a_y) = \frac{400}{9(2)(120\pi)}(-a_y) = \underline{\underline{-0.0589a_y \text{ W/m}^2}}$

Prob. 10.35 (a) $\beta = 1 = \omega / u = \dfrac{\omega}{c}\sqrt{\mu_r \varepsilon_r}$

$$\omega = \frac{c}{\sqrt{\mu_r \varepsilon_r}} = \frac{3x10^8}{\sqrt{3x12}} = \underline{\underline{0.5x10^8 \text{ rad/s}}}$$

(b) $\eta_1 = \eta_o, \qquad \eta_2 = \eta_o\sqrt{\dfrac{\mu_r}{\varepsilon_r}} = \eta_o\sqrt{\dfrac{3}{12}} = \eta_o / 2$

$$\Gamma = \frac{\eta_2 - \eta_1}{\eta_2 + \eta_1} = -1/3, \qquad \tau = 1 + \Gamma = 2/3$$

$$s = \frac{1+|\Gamma|}{1-|\Gamma|} = \frac{1+1/3}{1-1/3} = \underline{\underline{2}}$$

(c) Let $H_r = H_{or}\cos(\omega t + z)a_H$, where

$$E_r = -\frac{1}{3}(3)\cos(\omega t + z)a_y = -10\cos(\omega t + z)a_y, \qquad H_{or} = \frac{10}{\eta_o} = \frac{10}{120\pi}$$

$$a_E x a_H = a_k \longrightarrow -a_y x a_H = -a_z \longrightarrow a_H = -a_x$$

$$H_r = -\frac{10}{120\pi}\cos(0.5x10^8 t + z)a_x \text{ A/m} = -26.53\cos(0.5x10^8 t + z)a_x \text{ mA/m}$$

Prob. 10.36 (a)

$$a_E x a_H = a_k \longrightarrow a_E x a_z = a_x \longrightarrow a_E = -a_y$$

i.e. polarization is <u>along the y-axis</u>.

(b) $\beta = \omega\sqrt{\mu\varepsilon} = \dfrac{2\pi f}{c}\sqrt{\mu_r \varepsilon_r} = \dfrac{2\pi x 30x10^6}{3x10^8}\sqrt{4x9} = \underline{\underline{3.77}}$ rad/m

(c) $J_d = \nabla x H = \begin{vmatrix} \dfrac{\partial}{\partial x} & \dfrac{\partial}{\partial x} & \dfrac{\partial}{\partial x} \\ 0 & 0 & H_z(x,t) \end{vmatrix} = -\dfrac{\partial H_z}{\partial x}a_y$

$$= -10\beta\cos(\omega t + \beta x)a_y = \underline{\underline{-37.6\cos(\omega t + \beta x)a_y \text{ mA/m}}}$$

(d) $\eta_2 = \eta_o, \qquad \eta_1 = \eta_o\sqrt{\frac{4}{9}} = \frac{2}{3}\eta_o$

$$\Gamma = \frac{\eta_2 - \eta_1}{\eta_2 + \eta_1} = 1/5, \qquad \tau = 1 + \Gamma = 6/5$$

$$E_i = 10\eta_1 \sin(\omega t + \beta x)a_E \text{ mV/m}, \ a_E = -a_y$$

$$E_r = \Gamma 10\eta_1 \sin(\omega t - \beta x)(-a_y) \text{ mV/m}$$

$$a_E x a_H = a_k \longrightarrow -a_y x a_H = a_x \longrightarrow a_H = -a_z$$

$$H_r = \Gamma 10 \sin(\omega t - \beta x)(-a_z) \text{ mA/m} = \underline{\underline{-2\sin(\omega t - \beta x)a_z \text{ mA/m}}}$$

$$E_t = \tau 10\eta_1 \sin(\omega t + \beta x)(-a_y) \text{ mV/m}$$

$$a_E x a_H = a_k \longrightarrow -a_y x a_H = -a_x \longrightarrow a_H = a_z$$

$$H_t = 10(6/5)(\eta_1/\eta_2)\sin(\omega t + \beta x)a_z \text{ mA/m} = \underline{\underline{8\sin(\omega t + \beta x)a_z \text{ mA/m}}}$$

(e) $$\mathcal{P}_{\text{ave1}} = \frac{E_{io}^2}{2\eta_1}(-a_x) + \frac{E_{ro}^2}{2\eta_1}(+a_x) = \frac{-E_{io}^2}{2\eta_1}(1-\Gamma^2)a_x$$

$$= -\frac{\eta_1^2 H_{io}^2}{2\eta_1}(1-\Gamma^2)a_x = -\frac{1}{3}\eta_o 100(1-\frac{1}{25})a_x = \underline{\underline{-0.012064a_x \text{ W/m}^2}}$$

$$E_{ot} = \tau E_{oi} = \tau\eta_1 H_{io}$$

$$\mathcal{P}_{\text{ave2}} = \frac{E_{to}^2}{2\eta_2}(-a_x) = \frac{\tau^2\eta_1^2 H_{io}^2}{2\eta_2}(-a_x) = 32\eta_o(-a_x)\ \mu\text{W/m}^2 = \underline{\underline{-0.012064a_x \text{ W/m}^2}}$$

Prob. 10.37 (a) In air, $\beta_1 = 1, \lambda_1 = 2\pi/\beta_1 = 2\pi = \underline{\underline{6.283 \text{ m}}}$

$$\omega = \beta_1 c = \underline{\underline{3x10^8 \text{ rad/s}}}$$

In the dielectric medium, ω is the same .

$$\omega = \underline{\underline{3x10^8 \text{ rad/s}}}$$

$$\beta_2 = \frac{\omega}{c}\sqrt{\varepsilon_{r2}} = \beta_1\sqrt{\varepsilon_{r2}} = \sqrt{3}$$

$$\lambda_2 = \frac{2\pi}{\beta_2} = \frac{2\pi}{\sqrt{3}} = \underline{\underline{3.6276 \text{ m}}}$$

(b) $H_o = \frac{E_o}{\eta_o} = \frac{10}{120\pi} = 0.0265$

$$a_H = a_k x a_E = a_z x a_y = a_x$$

$$H_i = \underline{\underline{-26.5\cos(\omega t - z)a_x \text{ mA / m}}}$$

(c) $\eta_1 = \eta_o, \quad \eta_1 = \eta_o / \sqrt{3}$

$$\Gamma = \frac{\eta_2 - \eta_1}{\eta_2 + \eta_1} = \frac{(1/\sqrt{3}) - 1}{(1/\sqrt{3}) + 1} = \underline{\underline{-0.268}}, \qquad \tau = 1 + \Gamma = \underline{\underline{0.732}}$$

(d) $E_{to} = \tau E_{io} = 7.32, \qquad E_{ro} = \Gamma E_{io} = -2.68$

$$E_1 = E_i + E_r = \underline{\underline{10\cos(\omega t - z)a_y - 2.68\cos(\omega t + z)a_y \text{ V / m}}}$$

$$E_2 = E_t = \underline{\underline{7.32\cos(\omega t - z)a_y \text{ V / m}}}$$

$$\mathcal{P}_{\text{ave1}} = \frac{1}{2\eta_1}(a_z)[E_{io}^2 - E_{ro}^2] = \frac{1}{2(120\pi)}(a_z)(10^2 - 2.68^2) = \underline{\underline{0.1231a_z \text{ W / m}^2}}$$

$$\mathcal{P}_{\text{ave2}} = \frac{E_{to}^2}{2\eta_2}(a_z) = \frac{\sqrt{3}}{2x120\pi}(7.32)^2(a_z) = \underline{\underline{0.1231a_z \text{ W / m}^2}}$$

Prob. 10.38 (a) $\omega = \beta c = 3x3x10^8 = \underline{\underline{9x10^8 \text{ rad / s}}}$

(b) $\lambda = 2\pi / \beta = 2\pi / 3 = \underline{\underline{2.094}}$

(c) $\frac{\sigma}{\omega\varepsilon} = \frac{4}{9x10^8 x80x10^{-9} / 36\pi} = 2\pi = \underline{\underline{6.288}}$

$$\tan 2\theta_\eta = \frac{\sigma}{\omega\varepsilon} = 6.288 \longrightarrow \theta_\eta = 40.47^o$$

$$|\eta_2| = \frac{\sqrt{\mu_2/\varepsilon_2}}{\sqrt[4]{1+\left(\frac{\sigma_2}{\omega\varepsilon_2}\right)^2}} = \frac{377/\sqrt{80}}{\sqrt[4]{1+4\pi^2}} = 16.71$$

$$\eta_2 = \underline{\underline{16.71\angle 40.47^o\ \Omega}}$$

(d) $$\Gamma = \frac{\eta_2 - \eta_1}{\eta_2 + \eta_1} = \frac{16.71\angle 40.47^o - 377}{16.71\angle 40.47^o + 377} = 0.935\angle 176.7^o$$

$$E_{or} = \Gamma E_{oi} = 9.35\angle 176.7^o$$

$$E_r = \underline{\underline{9.35\sin(\omega t - 3z + 176.7)a_x \text{ V/m}}}$$

$$\alpha_2 = \frac{\omega}{c}\sqrt{\frac{\mu_{r2}\varepsilon_{r2}}{2}\left[\sqrt{1+\left(\frac{\sigma_2}{\omega\varepsilon_2}\right)^2} - 1\right]} = \frac{9x10^9}{3x10^8}\sqrt{\frac{80}{2}\left[\sqrt{1+4\pi^2} - 1\right]} = 43.94 \text{ Np/m}$$

$$\beta_2 = \frac{9x10^9}{3x10^8}\sqrt{\frac{80}{2}\left[\sqrt{1+4\pi^2} + 1\right]} = 51.48 \text{ rad/m}$$

$$\tau = \frac{2\eta_2}{\eta_2 + \eta_1} = \frac{2x16.71\angle 40.47^o}{16.71\angle 40.47^o + 377} = 0.0857\angle 38.89^o$$

$$E_{ot} = \tau E_o = 0.857\angle 38.59^o$$

$$\underline{\underline{E_t = 0.857e^{43.94z}\sin(9x10^8 t + 51.48z + 38.89^o) \text{ V/m}}}$$

Prob. 10.39

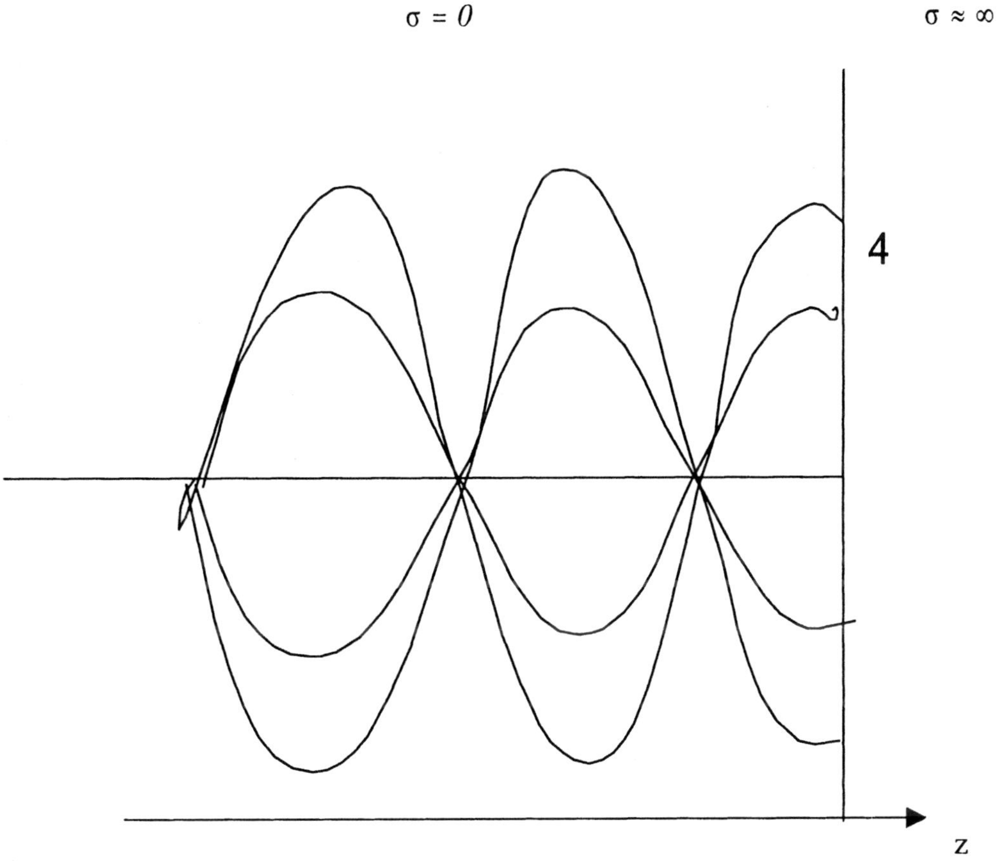

Curve 0 is at t = 0; curve 1 is at t = T/8; curve 2 is at t = T/4; curve 3 is at t = 3T/8, etc.

Prob. 10.40 Since $\mu_o = \mu_1 = \mu_2$,

$$\sin\theta_{t1} = \sin\theta_i \sqrt{\frac{\varepsilon_o}{\varepsilon_1}} = \frac{\sin 45^o}{\sqrt{4.5}} = 0.3333 \quad \longrightarrow \quad \underline{\underline{\theta_{t1} = 19.47^o}}$$

$$\sin\theta_{t2} = \sin\theta_{t1} \sqrt{\frac{\varepsilon_2}{\varepsilon_1}} = \frac{1}{3}\sqrt{\frac{2.25}{4.5}} = 0.2357 \quad \longrightarrow \quad \underline{\underline{\theta_{t2} = 13.63^o}}$$

Prob. 10.41

$$E_s = \frac{20(e^{jk_x x} - e^{-jk_x x})}{2} \frac{(e^{jk_y y} - e^{-jk_y y})}{2} a_z$$

$$= -j5\left[e^{j(k_x x+k_y y)} + e^{j(k_x x-k_y y)} - e^{-j(k_x x-k_y y)} - e^{-j(k_x x+k_y y)}\right]a_z$$

which consists of four plane waves.

$$\nabla x E_s = -j\omega\mu_o H_s \quad \longrightarrow \quad H_s = \frac{j}{\omega\mu_o}\nabla x E_s = \frac{j}{\omega\mu_o}\left(\frac{\partial E_z}{\partial y}a_x - \frac{\partial E_z}{\partial x}a_y\right)$$

$$H_s = -\frac{j20}{\omega\mu_o}\left[k_y \sin(k_x x)\sin(k_y y)a_x + k_x \cos(k_x x)\cos(k_y y)a_y\right]$$

Prob. 10.42 If $\mu_o = \mu_1 = \mu_2$, $\eta_1 = \dfrac{\eta_o}{\sqrt{\varepsilon_{r1}}}, \eta_2 = \dfrac{\eta_o}{\sqrt{\varepsilon_{r2}}}$

$$\Gamma_{\backslash\backslash} = \frac{\frac{1}{\sqrt{\varepsilon_{r2}}}\cos\theta_t - \frac{1}{\sqrt{\varepsilon_{r1}}}\cos\theta_i}{\frac{1}{\sqrt{\varepsilon_{r2}}}\cos\theta_t + \frac{1}{\sqrt{\varepsilon_{r1}}}\cos\theta_i}$$

$$\sqrt{\varepsilon_{r1}}\sin\theta_i = \sqrt{\varepsilon_{r2}}\sin\theta_t \quad \longrightarrow \quad \frac{\sqrt{\varepsilon_{r2}}}{\sqrt{\varepsilon_{r1}}} = \frac{\sin\theta_i}{\sin\theta_t}$$

$$\Gamma_{\backslash\backslash} = \frac{\cos\theta_t - \frac{\sin\theta_i}{\sin\theta_t}\cos\theta_i}{\cos\theta_t + \frac{\sin\theta_i}{\sin\theta_t}\cos\theta_i} = \frac{\sin\theta_t\cos\theta_t - \sin\theta_i\cos\theta_i}{\sin\theta_t\cos\theta_t + \sin\theta_i\cos\theta_i}$$

Dividing both numerator and denominator by $\cos\theta_i\cos\theta_t$ gives

$$\Gamma_{\backslash\backslash} = \frac{\tan\theta_t - \tan\theta_i}{\tan\theta_t + \tan\theta_i} = \frac{\frac{\tan\theta_t - \tan\theta_i}{1+\tan\theta_t\tan\theta_i}}{\frac{\tan\theta_t + \tan\theta_i}{1+\tan\theta_t\tan\theta_i}} = \frac{\tan(\theta_t - \theta_i)}{\tan(\theta_t + \theta_i)}$$

Similarly,

$$\tau_{\backslash\backslash} = \frac{\frac{2}{\sqrt{\varepsilon_{r2}}}\cos\theta_i}{\frac{1}{\sqrt{\varepsilon_{r2}}}\cos\theta_t + \frac{1}{\sqrt{\varepsilon_{r1}}}\cos\theta_i} = \frac{2\cos\theta_i}{\cos\theta_t + \frac{\sin\theta_i}{\sin\theta_t}\cos\theta_i}$$

$$= \frac{2\cos\theta_i\sin\theta_t}{\sin\theta_t\cos\theta_t(\sin^2\theta_i + \cos^2\theta_i) + \sin\theta_i\cos\theta_i(\sin^2\theta_t + \cos^2\theta_t)}$$

$$= \frac{2\cos\theta_i \sin\theta_t}{(\sin\theta_i \cos\theta_t + \sin\,\theta_t \cos\,\theta_i)(\cos\theta_i \cos\theta_t + \sin\,\theta_i \sin\,\theta_t)}$$

$$= \frac{2\cos\theta_i \sin\theta_t}{\sin(\theta_i + \theta_t)\cos(\theta_i - \theta_t)}$$

$$\Gamma_\perp = \frac{\frac{1}{\sqrt{\varepsilon_{r2}}}\cos\theta_i - \frac{1}{\sqrt{\varepsilon_{r1}}}\cos\theta_t}{\frac{1}{\sqrt{\varepsilon_{r2}}}\cos\theta_i + \frac{1}{\sqrt{\varepsilon_{r1}}}\cos\theta_t} = \frac{\cos\theta_i - \frac{\sin\theta_i}{\sin\theta_t}\cos\theta_t}{\cos\theta_i + \frac{\sin\theta_i}{\sin\theta_t}\cos\theta_t} = \frac{\sin(\theta_t - \theta_i)}{\sin(\theta_t + \theta_i)}$$

$$\tau_\perp = \frac{\frac{2}{\sqrt{\varepsilon_{r2}}}\cos\theta_i}{\frac{1}{\sqrt{\varepsilon_{r2}}}\cos\theta_i + \frac{1}{\sqrt{\varepsilon_{r1}}}\cos\theta_t} = \frac{2\cos\theta_i}{\cos\theta_i + \frac{\sin\theta_i}{\sin\theta_t}\cos\theta_t} = \frac{2\cos\theta_i \sin\theta_i}{\sin(\theta_t + \theta_i)}$$

Prob. 10.43 (a) $k_i = 4a_y + 3a_z$

$k_i \bullet a_n = k_i \cos\theta_i \quad \longrightarrow \quad \cos\theta_i = 4/5 \quad \longrightarrow \quad \underline{\underline{\theta_i = 36.87^o}}$

(b)

$$\mathcal{P}_{ave} = \frac{1}{2}\,\mathrm{Re}(E_s x H_s^*) = \frac{E_o^2}{2\eta}a_k = \frac{(\sqrt{8^2+6^2})^2}{2x120\pi}\frac{(3a_y + 4a_z)}{5} = \underline{\underline{79.58a_y + 106.1a_z \text{ mW/m}^2}}$$

(c) $\theta_r = \theta_i = 36.87^o$. Let

$$E_r = (E_{ry}a_x + E_{rz}a_z)\sin(\omega t - k_r \bullet r)$$

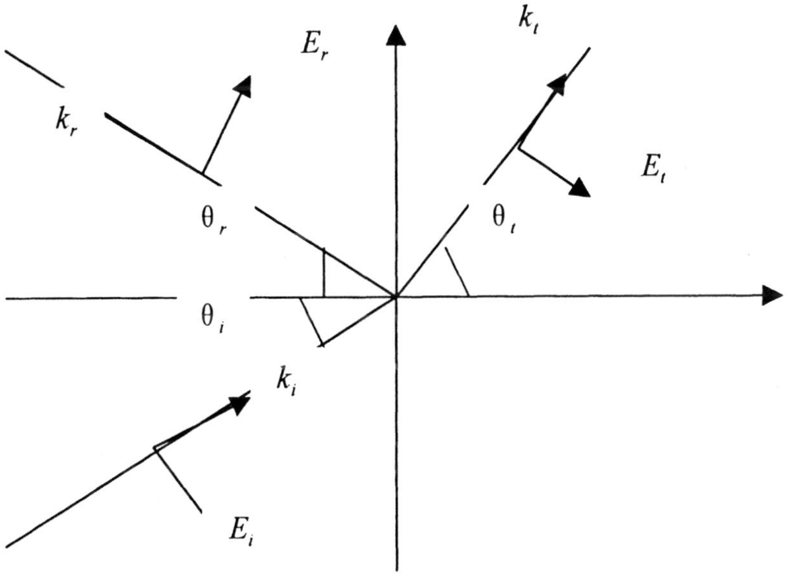

From the figure, $k_r = k_{rz}a_z - k_{ry}a_y$. But $k_r = k_i = 5$

$$k_{rz} = k_r \sin\theta_r = 5(3/5) = 3, \qquad k_{ry} = k_r \cos\theta_r = 5(4/5) = 4,$$

Hence, $k_r = -4a_y + 3a_z$

$$\sin\theta_t = \frac{n_1}{n_2}\sin\theta_i = \frac{c\sqrt{\mu_1\varepsilon_1}}{c\sqrt{\mu_2\varepsilon_2}}\sin\theta_i = \frac{3/5}{\sqrt{4}} = 0.3$$

$$\theta_t = 17.46, \cos\theta_t = 0.9539, \qquad \eta_1 = \eta_o = 120\pi, \eta_2 = \eta_o/2 = 60\pi$$

$$\Gamma_{//} = \frac{E_{ro}}{E_{io}} = \frac{\eta_2\cos\theta_t - \eta_1\cos\theta_i}{\eta_2\cos\theta_t + \eta_1\cos\theta_i} = \frac{\frac{\eta_o}{2}(0.9539) - \eta_o(0.8)}{\frac{\eta_o}{2}(0.9539) + \eta_o(0.8)} = -0.253$$

$$E_{ro} = \Gamma_{//}E_{io} = -0.253(10) = -2.53$$

But $(E_{ry}a_y + E_{rz}a_z) = E_{ro}(\sin\theta_r a_y + \cos\theta_r a_z) = -2.53(\frac{3}{5}a_y + \frac{4}{5}a_z)$

$$\underline{E_r = -(1.518a_y + 2.024a_z)\sin(\omega t + 4y - 3z) \text{ V/m}}$$

Similarly, let

$$E_t = (E_{ty}a_y + E_{tz}a_z)\sin(\omega t - k_t \bullet r)$$

$$k_t = \beta_2 = \omega\sqrt{\mu_2\varepsilon_2} = \omega\sqrt{4\mu_o\varepsilon_o}$$

But $k_i = \beta_1 = \omega\sqrt{\mu_o\varepsilon_o}$

$$\frac{k_t}{k_i} = 2 \longrightarrow k_t = 2k_i = 10$$

$$k_{ty} = k_t\cos\theta_t = 9.539, \qquad k_{tz} = k_t\sin\theta_t = 3,$$

$$k_t = 9.539a_y + 3a_z$$

Note that $k_{iz} = k_{rz} = k_{tz} = 3$

$$\tau_{\backslash\backslash} = \frac{E_{to}}{E_{io}} = \frac{2\eta_2 \cos\theta_i}{\eta_2 \cos\theta_t + \eta_1 \cos\theta_i} = \frac{\eta_o(0.8)}{\frac{\eta_o}{2}(0.9539) + \eta_o(0.8)} = 0.6263$$

$$E_{to} = \tau_{\backslash\backslash} E_{io} = 0.265$$

But

$$(E_{ty}a_y + E_{tz}a_z) = E_{to}(\sin\theta_t a_y - \cos\theta_t a_z) = 0.256(0.3a_y - 0.9539a_z)$$

Hence,

$$\underline{\underline{E_t = (1.877a_y - 5.968a_z)\sin(\omega t - 9.539y - 3z)\ \text{V/m}}}$$

Prob. 10.44 (a)

$$\tan\theta_i = \frac{k_{ix}}{k_{iz}} = \frac{1}{\sqrt{8}} \longrightarrow \underline{\underline{\theta_i = \theta_r = 19.47^o}}$$

$$\sin\theta_t = \sin\theta_i\sqrt{\frac{\varepsilon_{r1}}{\varepsilon_{r2}}} = \frac{1}{3}(3) = 1 \longrightarrow \underline{\underline{\theta_t = 90^o}}$$

(b) $\beta_1 = \frac{\omega}{c}\sqrt{\varepsilon_{r1}} = \frac{10^9}{3x10^8}x3 = 10 = k\sqrt{1+8} = 3k \longrightarrow \underline{\underline{k = 3.333}}$

(c) $\lambda = 2\pi/\beta, \quad \lambda_1 = 2\pi/\beta_1 = 2\pi/10 = \underline{\underline{0.6283}}$ m

$\beta_2 = \omega/c = 10/3, \quad \lambda_2 = 2\pi/\beta_2 = 2\pi x3/10 = \underline{\underline{1.885}}$ m

(d) $E_i = \eta_1 a_k x H_i = 40\pi\frac{(a_x + \sqrt{8}a_z)}{3}x0.2\cos(\omega t - k \bullet r)a_y$

$$\underline{\underline{= (-213.3a_x + 75.4a_z)\cos(10^9 t - kx - k\sqrt{8}z)\ \text{V/m}}}$$

(e) $\tau_{//} = \frac{2\cos\theta_i \sin\theta_t}{\sin(\theta_i + \theta_t)\cos(\theta_t - \theta_i)} = \frac{2\cos 19.47^o \sin 90^o}{\sin 19.47^o \cos 19.47^o} = 6$

$$\Gamma_{//} = -\frac{\cot 19.47^o}{\cot 19.47^o} = -1$$

Let $E_t = -E_{to}(\cos\theta_t a_x - \sin\theta_t a_z)\cos(10^9 t - \beta_2 x\sin\theta_t - \beta_2 z\cos\theta_t)$

where

$$E_i = -E_{io}(\cos\theta_i a_x - \sin\theta_i a_z)\cos(10^9 t - \beta_1 x\sin\theta_i - \beta_1 z\cos\theta_i)$$

$$\sin\theta_t = 1, \quad \cos\theta_t = 0, \quad \beta_2\sin\theta_t = 10/3$$

$$E_{to}\sin\theta_t = \tau_{\parallel}E_{io} = 6(24\pi)(3)(1) = 1357.2$$

Hence,

$$\underline{\underline{E_t = 1357\cos(10^9 t - 3.333x)a_z}} \text{ V/m}$$

Since $\Gamma = -1, \quad \theta_r = \theta_i$

$$\underline{\underline{E_r = (213.3a_x + 75.4a_z)\cos(10^9 t - kx + k\sqrt{8}z)}} \text{ V/m}$$

(f) $\tan\theta_{B//} = \sqrt{\dfrac{\varepsilon_2}{\varepsilon_1}} = \sqrt{\dfrac{\varepsilon_o}{9\varepsilon_o}} = 1/3 \quad \longrightarrow \quad \underline{\underline{\theta_{B//} = 18.43^o}}$

Prob. 10.45

$$\beta_1 = \sqrt{3^2 + 4^2} = 5 = \omega / c \quad \longrightarrow \quad \underline{\underline{\omega = \beta_1 c = 15x10^8 \text{ rad / s}}}$$

Let $E_r = (E_{ox}, E_{oy}, E_{oz})\sin(\omega t + 3x + 4y)$. In order for

$$\nabla \bullet E_r = 0, \qquad 3E_{ox} + 4E_{oy} = 0 \qquad (1)$$

Also, at y=0, $E_{1tan} = E_{2tan} = 0$

$$E_{1\tan} = 0, \qquad 8a_x + 5a_z + E_{ox}a_x + E_{oz}a_z = 0$$

Equating components, $E_{ox} = -8, \quad E_{oz} = -5$

From (1), $4E_{oy} = -3E_{ox} = 24 \qquad E_{oy} = 6$

Hence,

$$\underline{\underline{E_r = (-8a_x + 6a_y - 5a_z)\sin(15x10^8 t + 3x + 4y)}} \text{ V/m}$$

Prob. 10.46 Since both media are nonmagnetic,

$$\tan\theta_{B//} = \sqrt{\frac{\varepsilon_2}{\varepsilon_1}} = \sqrt{\frac{2.6\varepsilon_o}{\varepsilon_o}} = 1.612 \quad \longrightarrow \quad \theta_{B//} = 58.19^o$$

But

$$\cos\theta_t = \frac{\eta_1}{\eta_2}\cos\theta_{B//} = \frac{\eta_o}{\eta_o/\sqrt{2.6}}\cos\theta_{B//} = \sqrt{2.6}\cos 58.19^o \quad \longrightarrow \quad \underline{\underline{\theta_t = 31.8^o}}$$

CHAPTER 11

P.E. 11.1 Since Z_o is real and $\alpha \neq 0$, this is a distortionless line.

$$Z_o = \sqrt{\frac{R}{G}} \qquad (1)$$

or $$\frac{L}{R} = \frac{C}{G} \qquad (2)$$

$$\alpha = \sqrt{RG} \qquad (3)$$

$$\beta = \omega L\sqrt{\frac{G}{R}} = \frac{\omega L}{Z} \qquad (4)$$

$(1) \times (3) \rightarrow R_o = \alpha Z_o = 0.04 \times 80 = \underline{\underline{3.2\Omega/\text{m}}}$,

$(3) \div (1) \rightarrow G = \frac{\alpha}{Z_o} = \frac{0.04}{80} = \underline{\underline{5 \times 10^{-4}\Omega/\text{m}}}$

$L = \frac{\beta Z_o}{\omega} = \frac{1.5 \times 80}{2\pi \times 5 \times 10^8} = \underline{\underline{38.2 \text{ nH/m}}}$

$C = \frac{LG}{R} = \frac{12}{\pi}.10^{-8} \times \frac{0.04}{80} \times \frac{1}{0.04 \times 80} = \underline{\underline{5.97 \text{ pF/m}}}$

P.E. 11.2

(a) $Z_o = \sqrt{\frac{R + j\omega L}{G + j\omega C}} = \sqrt{\frac{0.03 + j2\pi \times 0.1 \times 10^{-3}}{0 + j2\pi \times 0.02 \times 10^{-6}}}$

$= 70.73 - j1.688 = \underline{\underline{70.75\angle -1.367^o\,\Omega}}$

(b) $\gamma = \sqrt{(R + j\omega L)(G + j\omega C)} = \sqrt{(0.03 + j0.2\pi)(j0.4 \times 10^{-4}\pi)}$

$\underline{\underline{= 2.121 \times 10^{-4} + j8.888 \times 10^{-3}\text{/m}}}$

(c) $u = \frac{w}{\beta} = \frac{2\pi \times 10^3}{8.888 \times 10^{-3}} = \underline{\underline{7.069 \times 10^5 \text{ m/s}}}$

P.E. 11.3

(a) $Z_o = Z_l \rightarrow Z_{in} = Z_o = \underline{\underline{30 + j60\Omega}}$

(b) $V_{in} = V_o = \frac{Z_{in}}{Z_{in} + Z_o} V_g = \frac{V_g}{2} = \underline{\underline{7.5\angle 0^o \text{ V}_{rms}}}$

$$I_{in} = I_o = \frac{V_g}{Z_g + Z_{in}} = \frac{V_g}{2Z_o} = \frac{15\angle 0^o}{2(30 + j60^o)}$$

$$\underline{\underline{= 0.05\angle -63.43^o\, A}}$$

(c) Since $Z_o = Z_r$, $\Gamma = 0 \rightarrow V_o^- = 0, V_o^+ = V_o$

The load voltage is $V_L = V_s(z = l) = V_o^+ e^{-\gamma l}$

$$e^{-\gamma l} = \frac{V_o^+}{V_L} = \frac{7.5\angle 0^o}{5\angle -48^o} 1.5\angle 48^o$$

$$e^{\alpha l} e^{j\beta l} = 1.5\angle 48^o$$

$$e^{\alpha l} = 1.5 \rightarrow \alpha = \frac{1}{l}\ln(1.5) = \frac{1}{40}\ln(1.5) = 0.0101$$

$$e^{j\beta l} = e^{j48^o} \rightarrow \beta = \frac{1}{l}\frac{48^o}{180^o}\pi rad = 0.02094$$

$$\underline{\underline{\gamma = 0.0101 + j0.2094 \text{ /m}}}$$

P.E. 11.4

(a) Using the Smith chart, locate S at s = 1.6. Draw a circle of radius OS. Locate P where $\theta_\Gamma = 300^o$. At P,

$$|\Gamma| = \frac{OP}{OQ} = \frac{2.1cm}{9.2cm} = 0.228$$

$$\underline{\underline{\Gamma = 0.228\angle 300^o}}$$

Also at P, $\underline{z_L = 1.15 - j0.48,}$

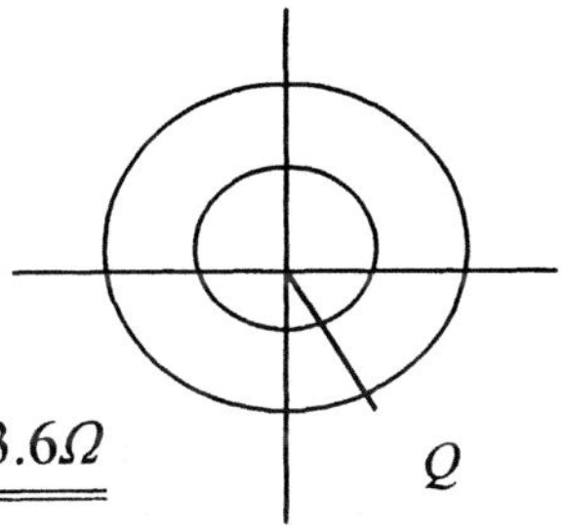

$Z_L = Z_o z_L = 70(1.15 - j0.48) = \underline{\underline{80.5 - j33.6\Omega}}$

$l = 0.6\lambda \rightarrow 0.6 \times 720^o = 432^o = \underline{\underline{360^o + 73^o}}$

From P, move 432° to R. At R, $z_{in} = 0.68 - j025$

$$Z_{in} = Z_o Z_{in} = 70(0.68 - j0.25) = \underline{\underline{47.6 - j17.5\Omega}}$$

(b) The maximum voltage (the only one) occurs at $\theta_\Gamma = 180^o$; its distance from the load is $\dfrac{180-60}{720}\lambda = \dfrac{\lambda}{6} = \underline{\underline{0.1667\lambda}}$

P.E. 11.5

(a) $\Gamma = \dfrac{Z_L - Z_o}{Z_L + Z_o} = \dfrac{60 + j60 - 60}{60 + j60 + 60} = \dfrac{j}{2+j} = \underline{\underline{0.4472\angle 63.43^o}}$

$$s = \frac{1+|\Gamma|}{1-|\Gamma|} = \frac{1+0.4472}{1-0.4472} = \underline{\underline{2.618}}$$

Let $x = \tan(\beta l) = \tan\dfrac{2\pi l}{\lambda}$

$$Z_{in} = Z_o\left[\frac{Z_L + jZ_o\tan(\beta l)}{Z_o + jZ_L\tan(\beta l)}\right]$$

$$120 - j60 = 60\left[\frac{60 + j60 + j60x}{60 + j(60 + j60)x}\right]$$

Or $2 - j = \dfrac{1 + j(1+x)}{1 - x + jx} \rightarrow 1 - x + j(2x - 2) = 0$

Or $x = 1 = \tan(\beta l)$

$$\frac{\pi}{4} + n\pi = \frac{2\pi l}{\lambda}$$

i.e $\underline{\underline{l = \dfrac{\lambda}{8}(1 + 4n), n = 0,1,2,3\ldots}}$

(b) $Z_L = \dfrac{Z_L}{Z_o}\dfrac{60 + j60}{60} = 1 + j$

Locate the load point P on the Smith chart.

$$|\Gamma| = \frac{OP}{OQ} = \frac{4.1cm}{9.2cm} = 0.4457, \theta_\Gamma = 62^o$$

$\underline{\underline{\Gamma = 0.4457\angle 62^o}}$

Locate the point S on the Smith chart. At S, r = $\underline{\underline{s = 2.6}}$

$z_{in} = \frac{Z_{in}}{Z_o} = \frac{120 + j60}{60} = 2 - j$, which is located at R on the chart. The angle between CP and OR is 64°-(-25°) = 90° which is equivalent to $\frac{90\lambda}{720} = \frac{\lambda}{8}$.

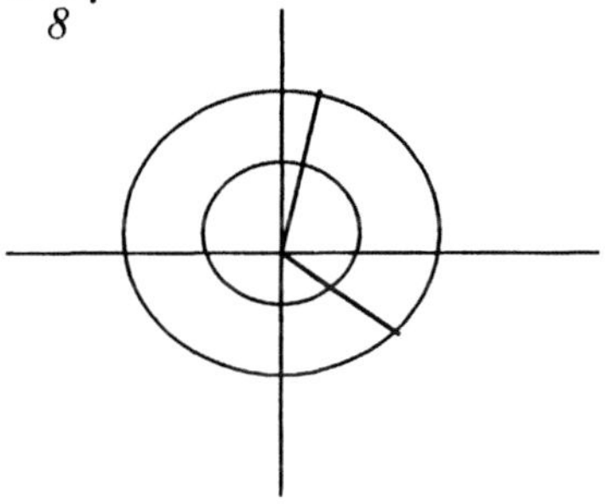

Hence $l = \frac{\lambda}{8} + n\frac{\lambda}{2} = \frac{\lambda}{8}(1 + 4n), n = 0,1,2........$

$(Z_{in})_{\max} = sZ_o = 2.618(60) = 157.08\Omega$

$(Z_{in})_{\min} = Z_o / s = 60 / 2.618 = \underline{\underline{22.92\ \Omega}}$

(does not exist if n = 0)

$l = \frac{62^o}{720^o}\lambda = 0.0851\lambda$

P.E. 11.6

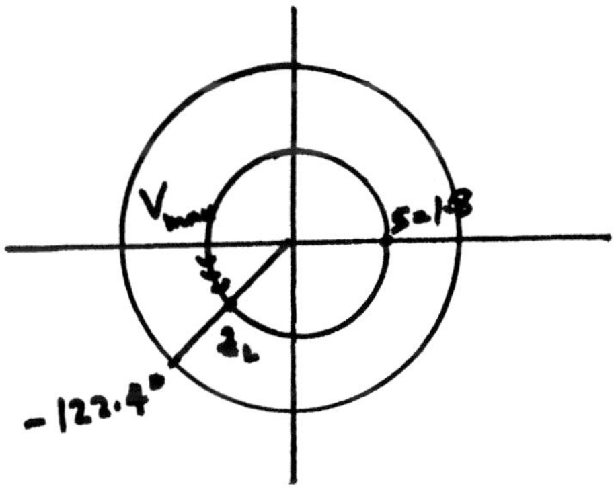

$\frac{\lambda}{2} = 37.5 - 25 = 12.5cm$ or $\lambda = 25cm$

$l = 37.5 - 35.5 = 2cm = \frac{2\lambda}{25}$

$l = 0.08\lambda \rightarrow 57.6^o$

$Z_L = 0.65 - j0.35$

$Z_L = Z_o z_L = 50(0.65 - j0.35)$

$\underline{\underline{= 33.5 - j17.5\Omega}}$

P.E. 11.7 See the Smith chart

$Z_L = \frac{100 - j80}{75} = 1.33 - j1.067$

$l_A = \frac{132^o - 65}{72}\lambda = \underline{\underline{0.093\lambda}}$

$l_B = \frac{132^o + 64^o}{720^o} = \underline{\underline{0.272\lambda}}$

$d_A = \frac{91}{720}\lambda = 0.126\lambda$

$d_B = 0.5\lambda - d_A \quad \underline{\underline{= 0.374\lambda}}$

$Y_s = \pm\frac{j0.95}{75} = \underline{\underline{\pm j12.67\ \text{mS}}}$

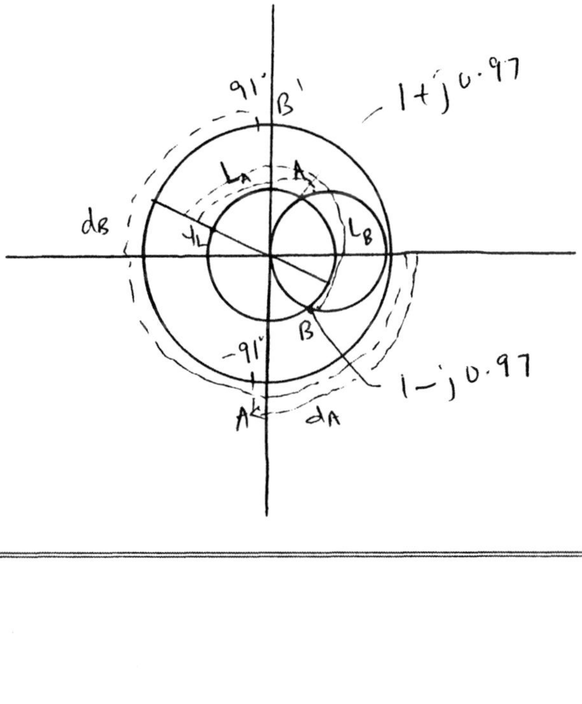

P.E. 11.8

(a) $\Gamma_G = \frac{1}{3}, \Gamma_L = \lim_{Z_L \to 0} \frac{Z_L - Z_o}{Z_L + Z_o} = -1$

$$V_\infty = \lim_{Z_L \to 0} \frac{Z_L}{Z_L + Z_g} V_g = 0, \qquad I_\infty = \lim_{Z_L \to 0} \frac{V_g}{Z_g + Z_g} = \frac{V_g}{Z_g} = \frac{12}{100} = 120mA$$

Thus the bounce diagrams for current and waves are as shown below.

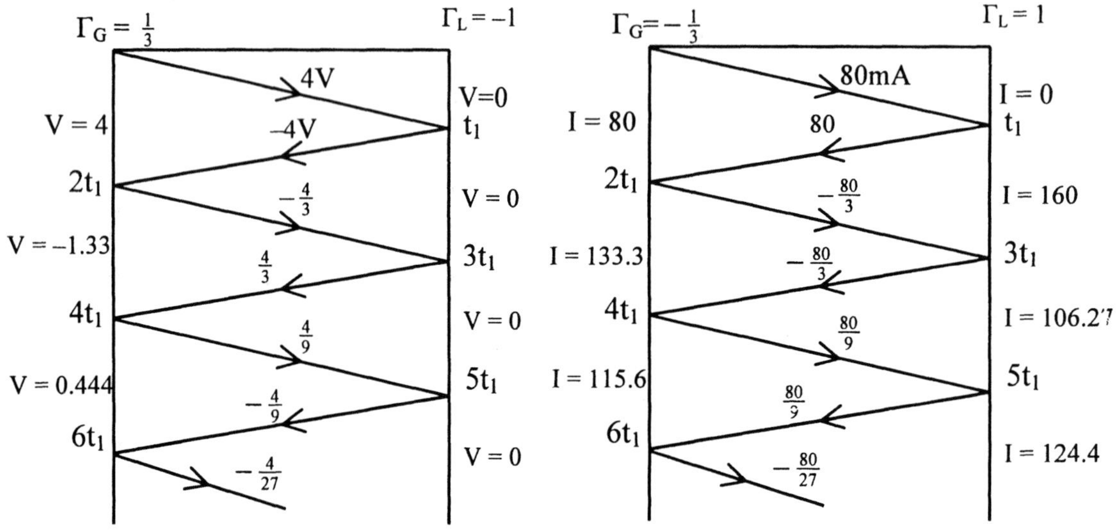

(Voltage) (Current)

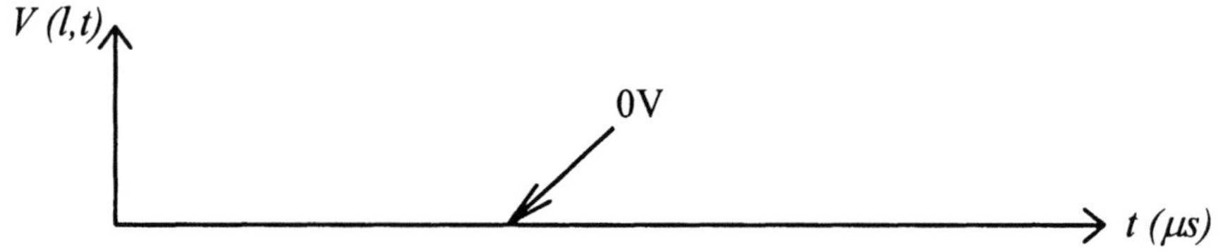

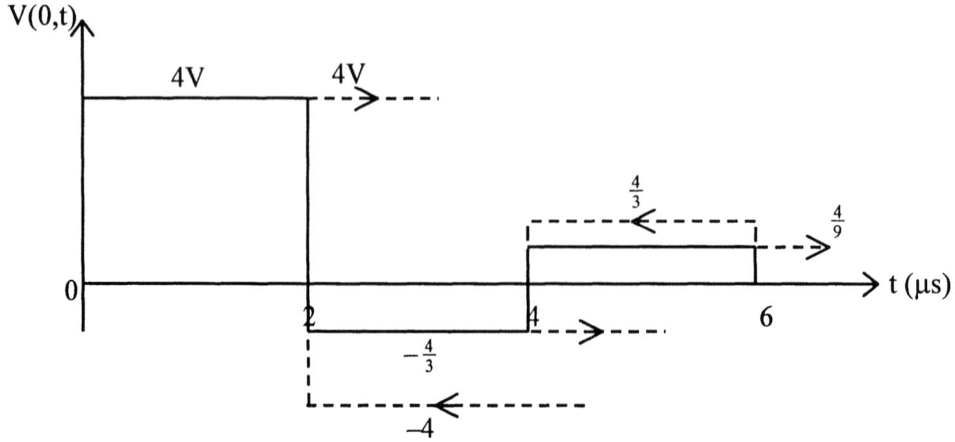

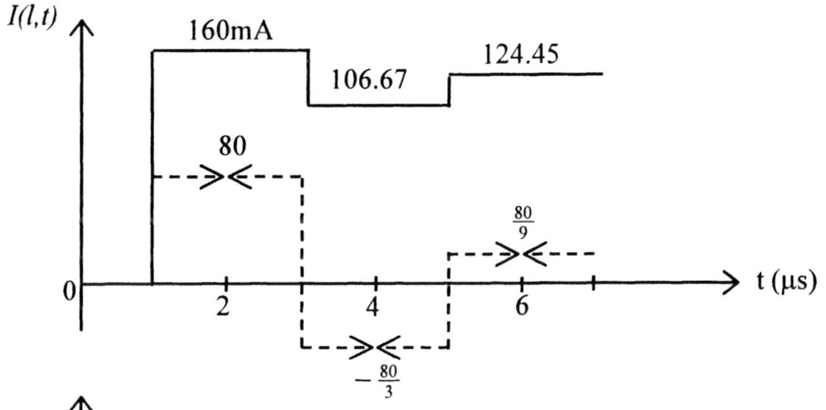

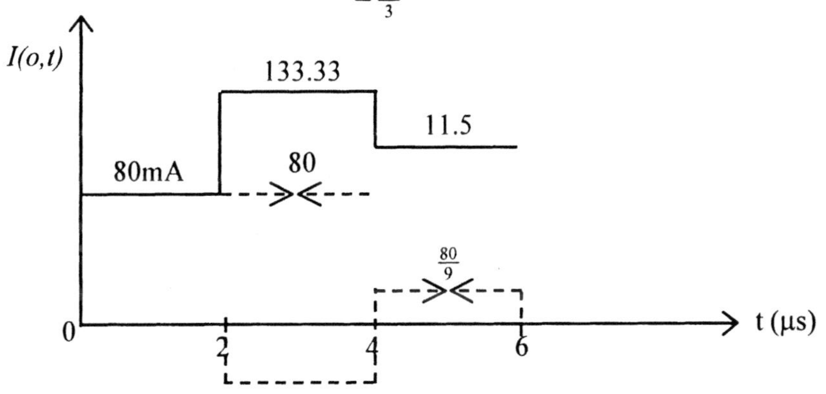

(b)) $\Gamma_G = \frac{1}{3}, \Gamma_L = \lim_{Z_L \to \infty} \frac{Z_L - Z_o}{Z_L + Z_o} = 1$

$$V_\infty = \lim_{Z_L \to \infty} \frac{Z_L}{Z_L + Z_g} V_g = V_g = 12V, \qquad I_\infty = \lim_{Z_L \to \infty} \frac{V_g}{Z_L + Z_g} = 0$$

The bounce diagrams for current and voltage waves are as shown below.

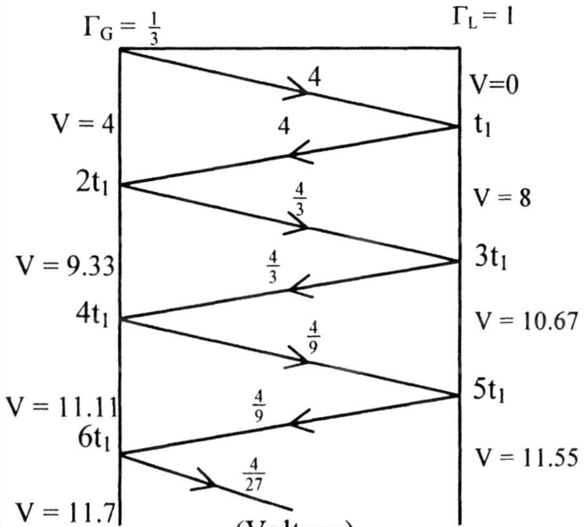

(Voltage)

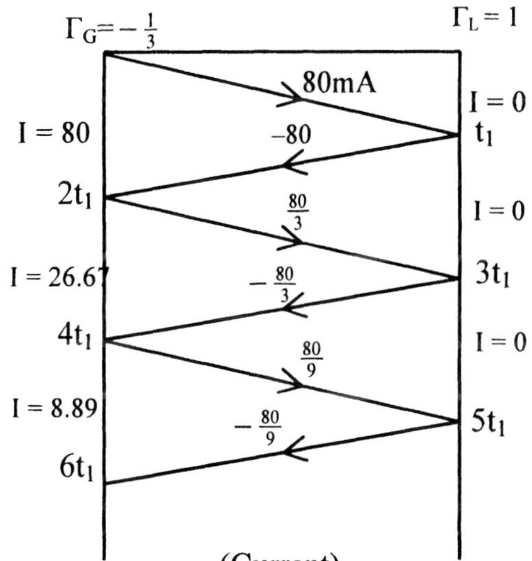

(Current)

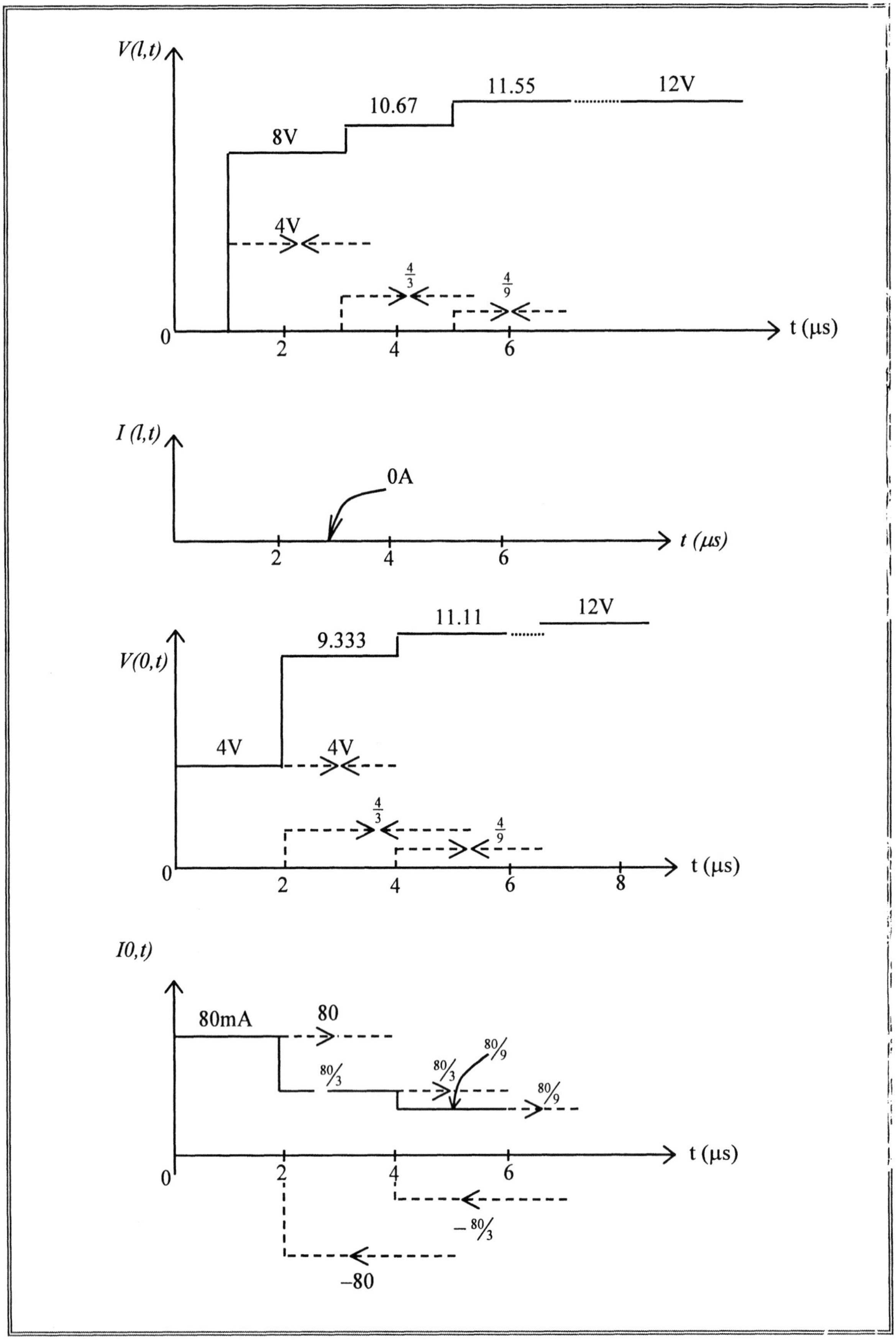
V(l,t)
8V
10.67
11.55
12V
4V
4/3
4/9
0
2
4
6
t (μs)
I (l,t)
0A
2
4
6
t (μs)
V(0,t)
4V
9.333
11.11
12V
4V
4/3
4/9
0
2
4
6
8
t (μs)
I0,t)
80mA
80
80/3
80/3
80/9
80/9
0
2
4
6
t (μs)
−80/3
−80

P.E. 11.9

$\Gamma_a = -\frac{1}{2}, \Gamma_L = \frac{1}{7}, t_l = 2\mu s$

$$(I_o)_{\max} = \frac{(V_g)_{\max}}{Z_g + Z_o} = \frac{10}{100} = 100mA$$

The bounce diagrams for maximum current are as shown below.

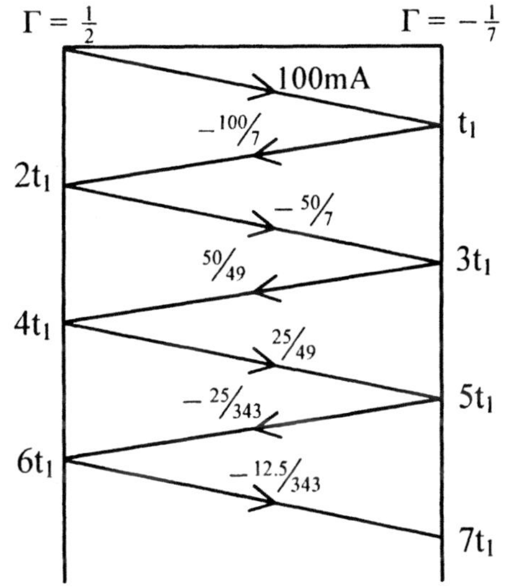

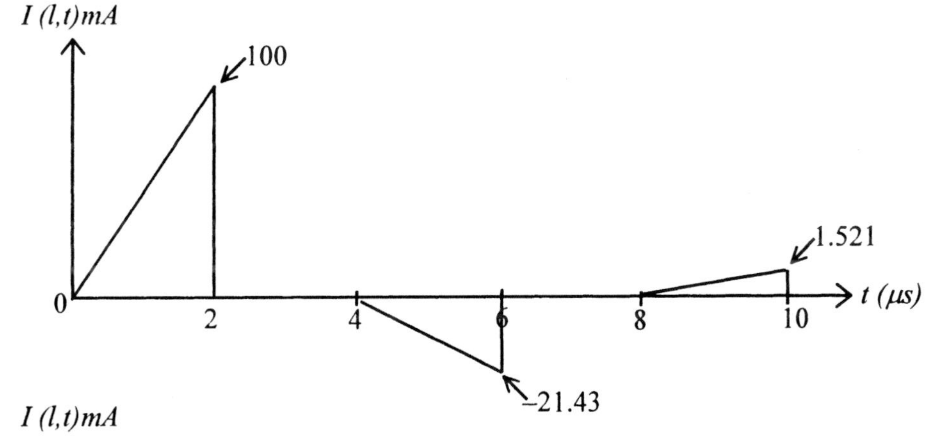

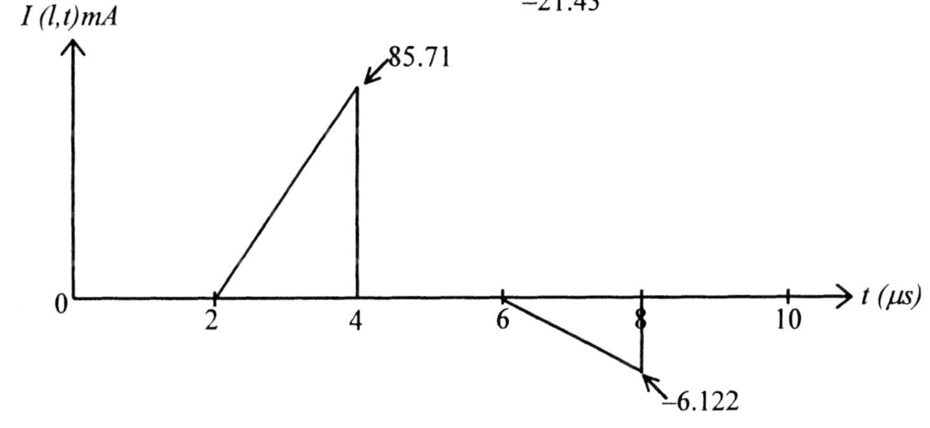

P.E. 11.10

(a) For $w/h = 0.8$, $\varepsilon_{eff} = \frac{4.8}{2} + \frac{2.8}{2}\left[1 + \frac{12}{0.8}\right]^{-1/2} = \underline{\underline{2.75}}$

(b) $Z_o = \frac{60}{\sqrt{2.75}} \ln\left(\frac{8}{0.8} + \frac{0.8}{4}\right) = 36.18 \ln 10.2 = \underline{\underline{84.03\Omega}}$

(c) $\lambda = \frac{3 \times 10^8}{10^{10}\sqrt{2.75}} = \underline{\underline{18.09 \text{ mm}}}$

P.E. 11.11

$$R_s = \sqrt{\frac{\pi f \mu_o}{\sigma_c}} = \sqrt{\frac{\pi \times 20 \times 10^9 \times 4\pi \times 10^{-7}}{5.8 \times 10^7}}$$

$$= 3.69 \times 10^{-2}$$

$$\alpha_c = 8.685 \frac{R_s}{wZ_o} = \frac{8.686 \times 3.69 \times 10^{-2}}{2.5 \times 10^{-3} \times 50}$$

$$= \underline{\underline{2.564 \text{ dB/m}}}$$

Prob. 11.1

$$\delta = \frac{1}{\sqrt{\pi F \mu \sigma}} = \frac{1}{\sqrt{\pi \times 5 \times 10^7 \times 4\pi \times 10^{-7} \times 6 \times 10^7}}$$

$$\delta = 9.19 \times 10^{-6}$$

$$R = \frac{2}{w\delta\sigma_c} = \frac{2}{0.3 \times 9.19 \times 10^{-6} \times 7 \times 10^7} = \underline{\underline{0.0104\Omega/\text{m}}}$$

$$L = \frac{\mu_o d}{w} = \frac{4\pi \times 10^{-7} \times 1.2 \times 10^{-2}}{0.3} = \underline{\underline{50.26 \text{ nH/m}}}$$

$$C = \frac{\varepsilon_o w}{d} = \frac{10^{-9}}{36\pi} \times \frac{0.3}{1.2 \times 10^{-2}} = \underline{\underline{221 \text{ pF/m}}}$$

Since $\sigma = 0$ for air,

$$G = \frac{\sigma w}{d} = 0$$

Prob. 11.2

$$C = \frac{\pi \varepsilon l}{\cosh^{-1}(d/2a)} \cong \frac{\pi \varepsilon l}{\ln(d/a)}$$

since $(d/2a)^2 = 11.11 >> 1$.

$$C = \frac{\pi x \frac{10^{-9}}{36\pi} x16x10^{-3}}{\ln(2/0.3)} = \underline{\underline{0.2342 \text{ pF}}}$$

$$\delta = \frac{1}{\sqrt{\pi f \mu \sigma}} = \frac{1}{\sqrt{\pi x10^7 x4\pi x10^{-7} x5.8x10^7}} = 2.09x10^{-5} \text{ m} << a$$

$$R_{ac} = \frac{l}{\pi a \delta \sigma} = \frac{16x10^{-3}}{\pi x0.3x10^{-3} x2.09x10^{-5} x5.8x10^7} = \underline{\underline{1.5x10^{-2}\ \Omega}}$$

Prob. 11.3

(a) Applying Kirchhoff's voltage law to the loop yields

$$V(z+\Delta z,t) + V(z,t) - R\Delta z I_1 - L\Delta z \frac{\partial I_1}{\partial t}$$

But $I_1 = I(z,t) - \frac{C}{2}\Delta z \frac{\partial V(z,t)}{\partial t} - \frac{G}{2}\Delta t V(z,t)$

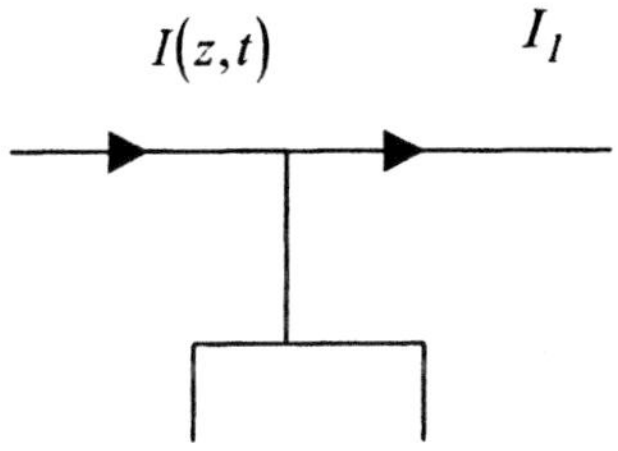

Hence,

$$V(z+\Delta z,t) = V(z,t) - R\Delta z\left[I(z,t) - \frac{C}{2}\Delta z\frac{\partial V}{\partial t} - \frac{G}{2}\Delta z V\right] - L\Delta z\left[\frac{\partial I}{\partial t} - \frac{C}{2}\Delta z\frac{\partial^2 V}{\partial t^2} - \frac{G}{2}\Delta z\frac{\partial V}{\partial t}\right]$$

Dividing by Δz and taking limits as $\Delta t \to 0$ give

$$\lim_{\Delta z \to 0} \frac{V(z+\Delta z,t) - V(z,t)}{\Delta z} = \lim_{\Delta z \to 0}\left[-RI - L\frac{\partial I}{\partial t} + \frac{RC}{2}\Delta z\frac{\partial V}{\partial t} + \frac{RG}{2}\Delta z V + \frac{LC}{2}\Delta z\frac{\partial^2 V}{\partial t^2} + \frac{LC}{2}\Delta z\frac{\partial V}{\partial t}\right]$$

or $-\frac{\partial V}{\partial z} = RL + L\frac{\partial I}{\partial t}$

Similarly, applying Kirchhoff's law to the node leads to

$$I(z+\Delta z,t)=I(z,t)-\frac{R}{2}\Delta zV(z,t)-\frac{C}{2}\Delta z\frac{\partial V}{\partial t}-\frac{G}{2}\Delta zV(z+\Delta z,t)-\frac{G}{2}\Delta z\frac{\partial V(z+\Delta z,t)}{\partial t}$$

$$\lim_{\Delta z\to 0}\frac{I(z+\Delta z,t)-I(z,t)}{\Delta z}=\lim_{\Delta z\to 0}\left[-\frac{R}{2}V(z,t)-\frac{C}{2}\frac{\partial V(z,t)}{\partial t}-\frac{G}{2}V(z+\Delta z,t)-\frac{C}{2}\frac{\partial V(z+\Delta z,t)}{\partial t}\right]$$

or $-\dfrac{\partial I}{\partial t}=GV+C\dfrac{\partial V}{\partial t}$

(b) Applying Kirchhoff's voltage law,

$$V(z,t)=R\frac{\Delta l}{2}I(z,t)+L\frac{\Delta l}{2}\frac{\partial I}{\partial t}(z,t)+V(z+\Delta l/2,t)$$

or

$$-\frac{V(z+\Delta l/2,t)-V(z,t)}{\Delta l/2}=RI+L\frac{\partial I}{\partial t}$$

As $\Delta l\to 0,\quad -\dfrac{\partial V}{\partial t}=RI+L\dfrac{\partial I}{\partial t}$

Applying Kirchhoff's current law,

$$I(z,t)=I(z+\Delta l,t)+C\Delta lV(z+\Delta l,t)+C\Delta l\frac{\partial V\left(z+\Delta l/2,t\right)}{\partial t}$$

or

$$-\frac{I(z+\Delta l,t)-I(z,t)}{\Delta l}=GV(z+\Delta l,t)+C\frac{\partial V(z+\Delta l,t)}{\partial t}$$

As $\Delta l\to 0,\quad -\dfrac{\partial I(z,t)}{\partial t}=GV(z,t)+C\dfrac{\partial V(z,t)}{\partial t}$

Prob. 11.4

$$Z_o=\sqrt{\frac{L}{c}}=\sqrt{\frac{\mu d}{w}\cdot\frac{d}{\varepsilon w}}=\frac{d}{w}\sqrt{\frac{\mu}{\varepsilon}}$$

$$Z_o=\eta_o\frac{d}{w}=78$$

$$Z_o^{'}=\eta_o\frac{d}{w^{'}}=75$$

$$\frac{78}{75}=\frac{w^{'}}{w}\rightarrow w^{'}1.04w$$

i.e. the width must be increased by 4%.

Prob. 11.5

(a) $R + jwL = 40 + j2\pi \times 10^7 \times 0.2 \times 10^{-6} = 41.93\angle 17.44^o$

$$R + jwc = 400 \times 10^{-6} + j2\pi \times 10^7 \times 0.5 \times 10^{-9} = 31.42 \times 10^{-2}\angle 89.89^o$$

$$Z_o = \sqrt{\frac{R + jwL}{G + jwc}} = \sqrt{\frac{41.93\angle 17.44^o}{31.42 \times 10^{-2}\angle 89.89^o}} = 13.34\angle -36.24$$

$$\underline{\underline{Z_o = 10.76 - j7.886\Omega}}$$

$$\gamma = \sqrt{(R + j\omega L)(G + j\omega L)} = \sqrt{(41.93\angle 17.44^o)(31.42 \times 10^{-2}\angle 89.89^o)}$$

$$= 3.63\angle 53.68^o = 2.15 + j2.925 = \alpha + j\beta$$

$$u = \frac{\omega}{\beta} = \frac{2\pi \times 10^7}{2.925} = \underline{\underline{2.148 \times 10^7 \text{ m/s}}}$$

(b) $\alpha = 2.15$ Np/m $= 2.15 \times 8.686$ dB/m $= 18.675$ dB/m

$$\alpha l = 30 \rightarrow l = \frac{30}{18.675} = \underline{\underline{1.606 \text{ m}}}$$

Prob. 11.6

(a) $\frac{R}{L} = \frac{G}{C} \rightarrow G = \frac{R}{L}C = \frac{20 \times 63 \times 10^{-12}}{0.3 \times 10^{-6}}$

$$G = 4.2 \times 10^{-3} \text{ S/m}$$

$$\alpha = \sqrt{RG} = \sqrt{20 \times 4.2 \times 10^{-3}} = 0.2898$$

$$\beta = \omega\sqrt{LC} = 2\pi \times 120 \times 10^6 \sqrt{0.3 \times 10^{-6} \times 63 \times 10^{-12}} = 3.278$$

$$\underline{\underline{\gamma = 0.2898 + j3.278 \text{ /m}}}$$

(b) Let V_o be its original magnitude

$$V_o e^{-\alpha z} = 0.2V_o \rightarrow e^{\alpha z} = 5$$

$$z = \frac{1}{\alpha}\ln 5 = \underline{\underline{5.554 \text{ m}}}$$

(c) $\beta l = 45^o = \pi/4 \rightarrow l = \frac{\pi}{4\beta} = \frac{4}{4 \times 3.278}$

$$\underline{\underline{l = 0.2396 \text{ m}}}$$

Prob. 11.7

(a) For a lossless line, R = 0 = G.

$$\gamma = j\omega\sqrt{LC} \longrightarrow \beta = \omega\sqrt{LC} = \omega\sqrt{\mu_o c_o} = \frac{\omega}{c}$$

$$u = \frac{\omega}{\beta} = c = \frac{1}{\sqrt{LC}}$$

(b) For lossless line, R = 0 =G

$$L = \frac{\mu}{\pi}\cosh^{-1}\frac{d}{2a}, C = \frac{\pi\varepsilon}{\cosh^{-1}\frac{d}{2a}}$$

$$Z_o = \sqrt{\frac{L}{C}} = \sqrt{\frac{\eta}{\pi}.\frac{1}{\pi\varepsilon}}\cosh^{-1}\frac{d}{2a} = \frac{120\pi}{\pi\sqrt{\varepsilon_r}}\cosh^{-1}\frac{d}{2a}$$

$$\underline{\underline{= \frac{120}{\sqrt{\varepsilon_r}}\cosh^{-1}\frac{d}{2a}}}$$

Prob. 11.8

$$L = \frac{\mu}{\pi}\cosh^{-1}\frac{d}{2a} = 4 \times 10^{-7}\cosh^{-1}\frac{0.32}{0.12}$$

$$\underline{\underline{L = 0.655 \ \mu\text{H/m}}}$$

$$C = \frac{\pi\varepsilon}{Cosh^{-1}\frac{d}{2a}} = \frac{\pi \times \frac{10^{-9}}{36\pi} \times 3.5}{Cosh^{-1} 2.667}$$

$$\underline{\underline{C = 59.4 \text{ pF/m}}}$$

$$Z_o = \sqrt{\frac{L}{C}} = \sqrt{\frac{0.655 \times 10^{-6}}{59.4 \times 10^{-12}}} = \underline{\underline{105.8\Omega}}$$

or

$$Z_o = \frac{120}{\sqrt{3.5}} \cosh^{-1} 2.667 = \underline{\underline{105\Omega}}$$

Prob. 11.9

Since R = 0 =G,

$$-\frac{\partial V}{\partial t} = L\frac{\partial L}{\partial t} \qquad (1)$$

$$-\frac{\partial I}{\partial t} = C\frac{\partial V}{\partial t} \qquad (2)$$

If $V = V_o \sin(wt - \beta t)$, from (1)

$$-\frac{\partial I}{\partial t} = V_o\beta \cos(wt - \beta z),$$

$$I = \frac{V_o}{L}\beta \cos(wt - \beta z)$$

Using (2)

$$\frac{V_o}{wL}\beta^2 \cos(wt - \beta z) = wcV_o \cos(wt - \beta z)$$

i.e. $$\frac{\beta^2}{wL} = wc \rightarrow \beta = w\sqrt{Lc}$$

But $Z_o = \sqrt{\frac{L}{c}}$, hence $Z_o = \frac{wL}{\beta}$ and $\underline{\underline{I_o = \frac{V_o}{Z_o}\sin(wt - \beta z)}}$

Prob. 11.10

(a) $\alpha = 0.0025$ Np/m, $\beta = 2$ rad/m,

$$u = \frac{\omega}{\beta} = \frac{10^8}{2} = \underline{\underline{5 \times 10^7 \text{ m/s}}}$$

(b) $$\Gamma = \frac{V_o^-}{V_o^+} = \frac{60}{120} = \frac{1}{2}$$

But $$\Gamma = \frac{Z_L - Z_o}{Z_L + Z_o} \rightarrow \frac{1}{2} = \frac{300 - Z_o}{300 + Z_o} \rightarrow \underline{\underline{Z_o = 100\Omega}}$$

$$I(l') = \frac{120}{Z_o} e^{0.0025l'} \cos(10^8 + 2l') - \frac{60}{Z_o} e^{-0.0025l'} \cos(10^8 t - zl')$$

$$= 0.12 e^{0.0025l'} \cos(10^8 + 2l') - 0.6 e^{-0.0025l'} \cos(10^8 t - zl') A$$

Prob. 11.11

(a) $$T_L = \frac{V_L}{V_o^+} = \frac{Z_L I_L}{\frac{1}{2}(V_L + Z_o I_L)} = \frac{2 Z_L I_L}{Z_L I_2 + Z_o I_2}$$

$$= \frac{Z_L I_L}{Z_L + Z_o}$$

$$1 + \Gamma_L = 1 + \frac{Z_L - Z_o}{Z_L + Z_o} = \frac{2 Z_L}{Z_L + Z_o}$$

(b) (i) $T_L = \dfrac{Z_n Z_o}{n Z_o + Z_o} = \dfrac{Z_n}{Z_n + 1}$

(ii) $T_L =_{Z_L} \xrightarrow{\lim} 0 = \dfrac{2}{1 + Z_o/Z_L} = 2$

(iii) $T_L =_{Z_L} \xrightarrow{\lim} 0 = \dfrac{2Z_L}{Z_L + Z_o} = 0$

(iv) $T_L = \dfrac{2Z_o}{2Z_o} = 1$

Prob. 11.12

$$R + j\omega L = 6.5 + j2\pi \times 2 \times 10^6 \times 3.4 \times 10^{-6} = 6.5 + j42.73$$

$$R + j\omega C = 8.4 \times 10^{-3} + j2\pi \times 2 \times 10^6 \times 21.5 \times 10^{-12} = (8.4 + j0.27) \times 10^{-3}$$

$$Z_o = \sqrt{\frac{R + j\omega L}{G + j\omega C}} = \sqrt{\frac{6.5 + j42.73}{(8.4 + j0.27) \times 10^{-3}}}$$

$$Z_o = 71.71\angle 39.75^o = 55.12 + j45.85\Omega$$

$$\gamma = \sqrt{(R + j\omega L)(G + j\omega C)} = \sqrt{(43.19\angle 81.34^o)(8.4 \times 10^{-3}\angle 1.84^o)}$$

$$= 0.45 + j0.39/\text{m}$$

↑ α ↑ β

$t = \frac{l}{u}$, but $u = \frac{w}{\beta}$,

$$t = \frac{\beta l}{\omega} = \frac{0.39 \times 5.6}{2\pi \times 2 \times 10^6} = \underline{\underline{0.1738 \mu s}}$$

Prob.11.13

$$Z_o = \sqrt{\frac{L}{c}}, \quad \gamma = j\beta = j\omega\sqrt{Lc}$$

$$Z_o\beta = \omega L \rightarrow \beta = \frac{\omega L}{Z_o} = \frac{2\pi \times 4.5 \times 10^9 \times 2.4 \times 10^6}{85}$$

$$= \underline{\underline{798.33 \text{ rad/m}}}$$

$$u = \frac{\omega}{\beta} = \frac{Z_o}{L} = \frac{85}{2.4 \times 10^{-6}} = \underline{\underline{3.542 \times 10^7 \text{ m/s}}}$$

Prob. 11.14

$$\Gamma_L = \frac{Z_L - Z_o}{Z_L + Z_o} = \frac{75 + j25 - 50}{75 + j25 + 50} = \underline{\underline{0.2773\angle 33.69^o}}$$

$$s = \frac{1 + |\Gamma|}{1 - |\Gamma|} = \frac{1.2773}{0.7227} = \underline{\underline{1.767}}$$

Prob. 11.15

From eq. (11.33)

$$Z_{sc} = Z_{in}\Big|_{Z_L = 0} = \tanh\gamma l$$

$$Z_{oc} = Z_{in}\Big|_{Z_L = \infty} = \frac{Z_o}{\tanh\gamma l} = Z_o \coth(\gamma l)$$

For lossless line, $\gamma = j\beta, \tan(\gamma l) = \tanh(j\beta l) = j\tan(\beta l)$

$$Z_{sc} = jZ_o \tan(\beta l), Z_{oc} = -jZ_o \cot(\beta l)$$

Prob. 11.16

$$Z_{in} = Z_{sc} = Z_o \tan\gamma l = Z_o \frac{\sinh(\gamma l)}{\cosh(\gamma l)}$$

But $\gamma l = (0.7 + j2.5)(0.8) = 0.56 + j2$

$$\sinh(x + jy) = \sinh(x)\cos(y) + j\cosh(x)\sin(y)$$

$$= \frac{\left(e^{o.56} - e^{-0.56}\right)}{2}\cos 2 + j\frac{\left(e^{o.56} + e^{-0.56}\right)}{2}\sin 2$$

$$= -0.245 + j0.0548$$

$$\cosh(x + jy) = \cosh(x)\cos(y) + j\sinh(x)\sin(y)$$

$$= -0.4831 + j0.5362$$

$$Z_{in} = \frac{(65 + j38)(-0.2454 + j1.0548)}{-0.4831 + j0.5362}$$

$$= \underline{\underline{113 + j2.726\Omega}}$$

Prob. 11.17

(a) $\Gamma = \frac{Z_L - Z_o}{Z_L + Z_o} = \frac{120 - 50}{170} = 0.4112$

$$\Gamma = \frac{1 + |\Gamma|}{1 - |\Gamma|} = 2.397$$

(b) $Z_{in} = Z_o \frac{Z_L + jZ_o \tan(\beta l)}{Z_o + jZ_L \tan(\beta l)}$

$$\beta l = \frac{2\pi}{\lambda}.\frac{\lambda}{6} = 60^o$$

$$Z_{in} = 50\left[\frac{120 + j50\tan(60^o)}{50 + j120\tan(60^o)}\right] = \underline{\underline{34.63\angle -40.65^o\,\Omega}}$$

Prob. 11.18

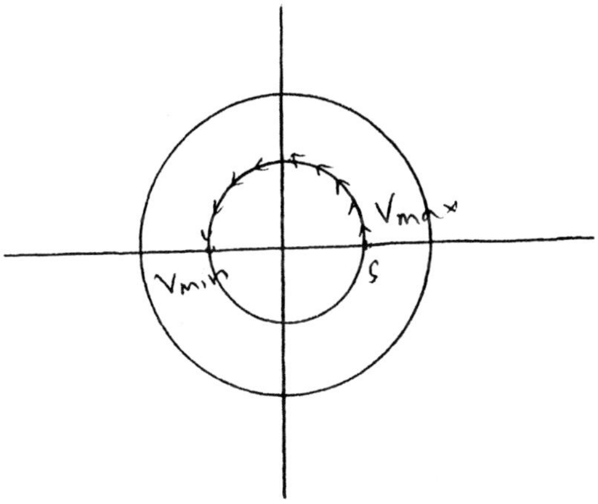

$$Z_L = \frac{Z_L}{Z_O} = \frac{210}{100} = 2.1 = s$$

$$\text{Or } \Gamma = \frac{Z_L - Z_O}{Z_L + Z_O} = \frac{110}{310},$$

$$s = \frac{1+|\Gamma|}{1-|\Gamma|} = 2-1$$

But $s = \frac{V_{\max}}{V_{\min}} \rightarrow V_{\max =} sV_{\min}$

Since the line is $\frac{\lambda}{4}$ long, $\frac{\lambda}{4} \rightarrow \frac{720^o}{4} = 120^o$

Hence the sending end will be $V_{\min}$,

while the receiving end at $V_{\min}$

$$V_{\max} = sV_{\min} = 1.2 \times 80 = \underline{\underline{96V}}$$

Prob. 11.19

$$I_l = \frac{V_L}{Z_L}, \Gamma = \frac{Z_L - Z_o}{Z_L + Z_o} = \frac{50e^{j30^o} - 50}{50e^{j30^o} + 50}$$

$$\approx j0.2679$$

From eq.(11.30),

$$V_o^+ = \frac{1}{2}(V_L + Z_o.\frac{V_L}{Z_L})e^{\gamma l} = \frac{V_L}{2Z_L}(Z_L + Z_o)e^{\gamma l}$$

$$V_o^- = \frac{V_L}{2Z_L}(Z_L - Z_o)e^{-\gamma l}$$

Substituting these in eq.(11.25),

$$I_s = \frac{V_L}{2Z_L Z_o}\left[(Z_L + Z_o)e^{\gamma l}e^{-\gamma z} - (Z_L - Z_o)e^{-\gamma l}e^{\gamma z}\right]$$

$$= \frac{V_L / Z_o}{1+\Gamma}\left[e^{-\gamma(z-l)} - \Gamma e^{\gamma(z-l)}\right]$$

But $l - z = \frac{\lambda}{8}$ or $z - l = -\frac{\lambda}{8}$

$$I_s = \frac{10\angle 25^o}{1.035\angle 15^o}\left(\frac{1}{50}\right)\left(e^{j\pi/4} - j0.2679e^{-j\pi/4}\right)$$

$$= \underline{\underline{0.2\angle 40^o\,A}}$$

or

$$\beta z = \frac{2\pi}{\lambda}.\frac{\lambda}{8} = \frac{\pi}{4}, \quad I_L = \frac{V_L}{Z_L} = \frac{10e^{j25^o}}{50e^{j30^o}} = 0.2e^{-j5^o}$$

$$I\left(z = \frac{\pi}{8}\right) = I_L e^{j\beta l} = 0.2e^{-j5^o}e^{j45^o}$$

$$= \underline{\underline{0.2e^{j40^o}\,A}}$$

Prob. 11.20

(*a*) $\beta l = \frac{1}{4}\times 100 = 25rad = 1432.4^o = 352.4^o$

$$Z_{in} = 60\left[\frac{j40 + j60\tan 352.4^o}{60 - 40\tan 352.4^o}\right] = \underline{\underline{j29.375\Omega}}$$

$$V(Z=0) = V_o = \frac{Z_{in}}{Z_{in} + Z_g}V_g = \frac{j29.375(10\angle 0^o}{j29.375 + 50 - j40}$$

$$= \frac{29.375\angle 90^o}{51.116\angle -12^o} = \underline{\underline{0.575\angle 102^o}}$$

(*b*) $Z_{in} = Z_L = \underline{\underline{j40\Omega.}}$

$V_L = V_s(Z = l), \quad V_o = V_L e^{j\beta l}$

$V_L = V_o e^{-j\beta l} = \left(0.575e^{j102^o}\right)\left(e^{-j352.4^o}\right)$

$= \underline{\underline{0.575\angle -250.4^o}}$

(c) $\beta l' = \frac{1}{4} \times 4 = 1 rad = 57.3^o$

$$Z_{in} = 60\left[\frac{j40 + j60\tan 57.3^o}{60 - 40\tan 57.3^o}\right] = \underline{\underline{-j3487.11\Omega.}}$$

$$V = V_L e^{j\beta l} = \left(0.575\angle -250.4^o\right)e^{j57.3^o}$$

$$= \underline{\underline{0.575\angle -193.1^o.}}$$

(d) 3m from the source is the same as 97m from the load., i.e.

$$l' = 100 - 3 = 97m, \quad \beta l' = \frac{1}{4} \times 97 = 24.25 rad = 309.42^o$$

$$Z_{in} = 60\left[\frac{j40 + j60\tan 309.42^o}{60 - 40\tan 309.42^o}\right] = \underline{\underline{-j18.2\Omega}}$$

$$V = V_L e^{j\beta l} = \left(0.575\angle -250.4^o\right)e^{j309.42^o}$$

$$= \underline{\underline{0.575\angle 59.02^o.}}$$

Prob. 11.21

$$\beta l = \frac{2\pi}{\lambda}(1.25\lambda) = \frac{\pi}{2} + 360^o,$$

$$\tan \beta l \to \infty$$

$$Z_{in} = \frac{Z_o^2}{Z_L} = \underline{\underline{46.875\Omega.}}$$

$$V_o = V(Z=0) = \frac{Z_{in}}{Z_{in} + Z_g} V_g = 48.39V.$$

for a loss less line,

$$|V_L| = |V(Z=0)| = \underline{\underline{48.39.}}$$

Prob. 11.22

Using the Smith chart, $Z_L = \dfrac{60 - j35}{100} = 0.6 - j0.35$

At C, $Z_{in} = Z_L = 60 - j35$

$$Z_L = \frac{60 - j35}{75} = 0.8 - j0.4667$$

$$l = \frac{3\lambda}{4} \rightarrow \frac{3}{4} \times 720^\circ = 540^\circ$$

At B, $Z_{in} = 75(0.95 + j0.54) = 71.25 + j40.5$

$$Z_L = \frac{71.25 - j40.5}{50} = 1.425 + j0.81$$

$$l = \frac{5\lambda}{8} \rightarrow 450^\circ = 360^\circ + 90^\circ$$

At A, $Z_{in} = 50(1.4 + j0.81) = \underline{\underline{70 + j40.5\Omega}}$

Prob. 11.23

$$V_1 = V_s(Z = 0) = V_o^+ + V_o^- \qquad (1)$$

$$V_2 = V_s(Z = l) = V_o^+ e^{-\gamma l} + V_o^- e^{\gamma l} \qquad (2)$$

$$I_1 = I_s(Z = 0) = \frac{V_o^+}{Z_o} - \frac{V_o^-}{Z_o} \qquad (3)$$

$$I_2 = -I_s(Z = l) = -\frac{V_o^+}{Z_o} e^{-\gamma l} + \frac{V_o^-}{Z_o} e^{\gamma l} \qquad (4)$$

$$(1) + (3) \rightarrow V_o^+ = \frac{1}{2}(V_1 + Z_o I_1)$$

$$(1) - (3) \rightarrow V_o^- = \frac{1}{2}(V_1 - Z_o I_1)$$

Substituting V_o^+ and V_o^- in (2) gives

$$V_2 = \frac{1}{2}(V_1 + Z_o I_1)e^{-\gamma l} + \frac{1}{2}(V_1 - Z_o I_1)e^{\gamma l}$$

$$= \frac{1}{2}(e^{\gamma l} + e^{-\gamma l})V_1 + \frac{1}{2}Z_o(e^{-\gamma l} - e^{\gamma l})I_1$$

$$V_2 = \cosh \gamma l V_1 + Z_o \sinh \gamma l I_1 \qquad (5)$$

Substituting V_o^+ and V_o^- in (4),

$$I_2 = -\frac{1}{2Z_o}(V_1 + Z_o I_1)e^{-\gamma l} + \frac{1}{2Z_o}(V_1 - Z_o I_1)e^{\gamma l}$$

$$= \frac{1}{2Z_o}(e^{\gamma l} - e^{-\gamma l})V_1 + \frac{1}{2}(e^{\gamma l} + e^{-\gamma l})I_1$$

$$I_2 = -\frac{1}{Z_o}\sinh\gamma l V_1 - \cosh\gamma l I_1 \qquad (6)$$

From (5) and (6)

$$\begin{bmatrix} V_2 \\ I_2 \end{bmatrix} = \begin{bmatrix} \cosh\gamma l & Z_o\sinh\gamma l \\ -\frac{1}{Z_o}\sinh\gamma l & -\cosh\gamma l \end{bmatrix}\begin{bmatrix} V_1 \\ I_1 \end{bmatrix}$$

But

$$\begin{bmatrix} \cosh\gamma l & Z_o\sinh\gamma l \\ -\frac{1}{Z_o}\sinh\gamma l & -\cosh\gamma l \end{bmatrix}^{-1} = \begin{bmatrix} \cosh\gamma l & Z_o\sinh\gamma l \\ -\frac{1}{Z_o}\sinh\gamma l & -\cosh\gamma l \end{bmatrix}$$

Thus

$$\begin{bmatrix} V_1 \\ I_1 \end{bmatrix} = \begin{bmatrix} \cosh\gamma l & Z_o\sinh\gamma l \\ \frac{1}{Z_o}\sinh\gamma l & \cosh\gamma l \end{bmatrix}\begin{bmatrix} V_2 \\ -I_2 \end{bmatrix}$$

Prob. 11.24

Method 1: $\quad Z_{in} = \dfrac{80 - j60}{50} = 1.6 - j1.2$

$$\lambda = \frac{u}{f} = \frac{0.8\times 3\times 10^8}{3\times 10^8} = 0.8m$$

$$l_1 = \frac{4.2}{2}m = 2.1m \rightarrow 720^\circ \times \frac{2.1}{0.8} = 5 \text{ revolutions} + 90^\circ$$

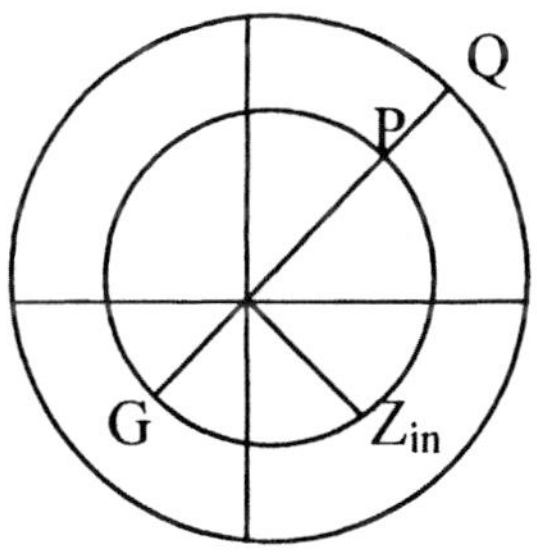

At G, $\quad Z_{in} = 0.44 - j0.4$

$$Z_{in} = Z_{in}Z_o = 50(0.44 - j0.4)$$

$$= \underline{\underline{22 - j20\Omega}}$$

$$|\Gamma| = \frac{OP}{OQ} = \frac{4.3cm}{9.3cm} = 0.4624, \theta_\Gamma = 50.5$$

$$\Gamma = \underline{\underline{0.4624\angle 50.5^\circ}}$$

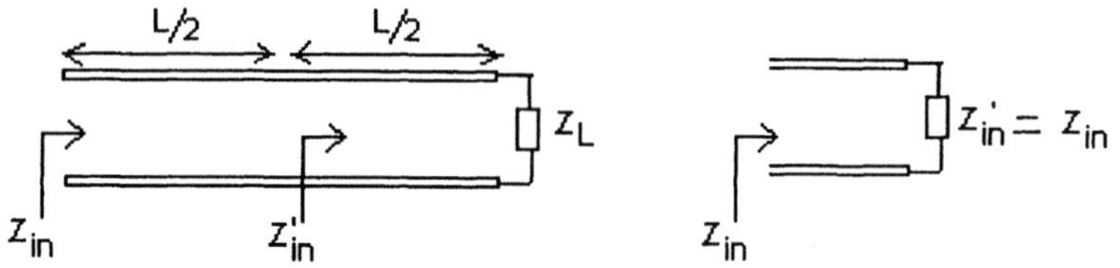

$$\tan\beta l = \tan\frac{\omega l}{u} = \tan\frac{2\pi\times 3\times 10^8}{0.8\times 3\times 10^8}(2.1)$$

$$= \tan\left(21\times\frac{\pi}{4}\right) = 1$$

$$Z_{in} = Z_o\left[\frac{Z_L' + jZ_o\tan\beta l}{Z_o + jZ_L'\tan\beta l}\right] = 50\left[\frac{80 - j60 + j50\times 1}{50 + j80 - j60\times 1}\right]$$

$$= 29.6\angle -43.152^\circ = \underline{\underline{21.6 - 20.2\Omega}}$$

$$\Gamma' = \frac{Z_L' - Z_o}{Z_L' + Z_o} = \frac{80 - j60 - 50}{80 - j60 + 50} = \frac{3 - j6}{13 - j6} = 0.4685\angle -38.66^\circ$$

$|\Gamma| = |\Gamma'| = 0.4685,$ but

$$\theta_\Gamma = \theta_{\Gamma'} + 2\times\frac{\pi}{4} = -38.66^\circ + 90^\circ = 51.34^\circ$$

$$\Gamma = \underline{\underline{0.4685\angle 51.34^\circ}}$$

Prob. 11.25

$$Z_{in} = \frac{Z_{in}}{Z_o} = \frac{90 + j150}{60} = 1.5 + j2.5$$

$$\lambda = \frac{u}{f} = \frac{3\times 10^8}{20\times 10^6} = 15m, \; l = 10m = \frac{2}{3}\lambda$$

If $\lambda \to 720^\circ$, then $\frac{2}{3}\lambda \to 480^\circ = 1$ revolution $+ 120^\circ$

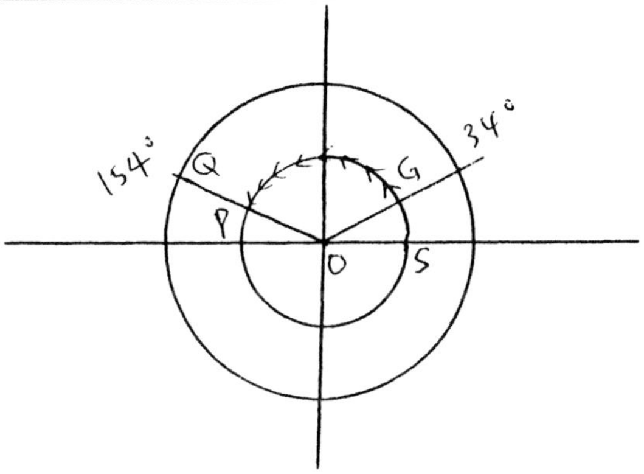

At the Load P, $z_L = 0.17 + j0.23$

$Z_L = Z_o z_L = 60(0.17 + j0.23) = \underline{\underline{10.2 + j13.8\Omega}}$

$$|\Gamma| = \frac{OP}{OQ} = \frac{6.5\,\text{cm}}{9\,\text{cm}} = 0.7222, \theta = 154°$$

$\Gamma = \underline{\underline{0.7222\angle 154°, s = 6.2}}$

Prob. 11.26

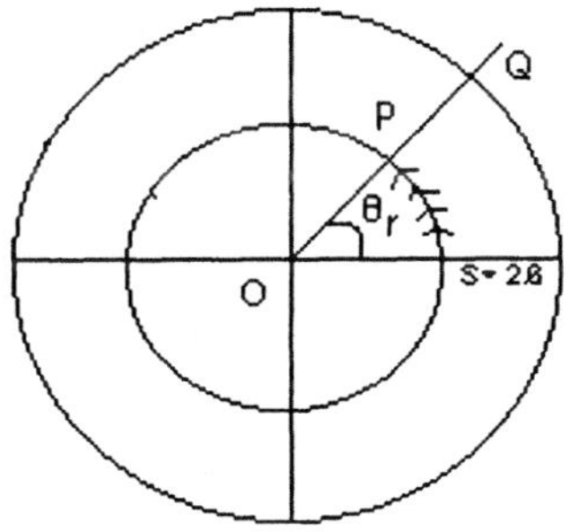

$(a) \quad Z_{in} = \frac{Z_{in}}{Z_o} = \frac{120 + j80}{75} = 1.6 + j1.067$

$|\Gamma| = \frac{OP}{OQ} = \frac{3.8\text{ cm}}{8.7\text{ cm}} = 0.4367, \quad \theta_\Gamma = 38^\circ$

$\Gamma = \underline{\underline{0.4367\angle 38^\circ}}, \quad s = \underline{\underline{2.6}}$

(b) The Load is purely resistive at s.

$\theta_\Gamma = 38^\circ$

But $\quad 720^\circ \rightarrow \lambda,$ hence $38^\circ \rightarrow \frac{38\lambda}{720} = \underline{\underline{0.053\lambda}}$ from the load

Prob. 11.27

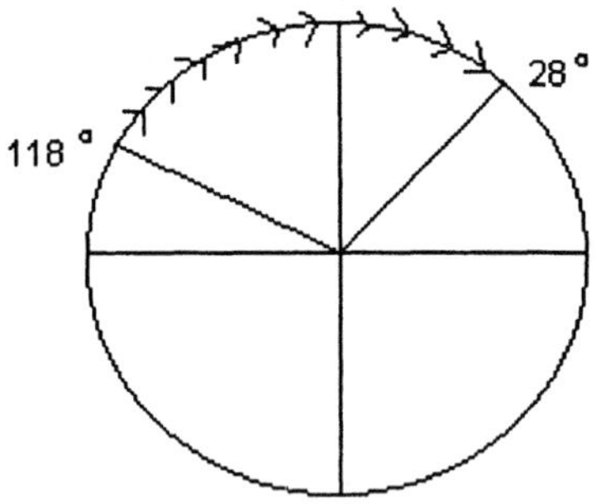

(a) If $\lambda \rightarrow 720°$, then $\frac{5\lambda}{8} \rightarrow \frac{5}{8} \times 720° = 450° \longrightarrow 90°$

$z_L = \frac{Z_L}{Z_o} = \frac{j45}{75} = j0.6$

$z_{in} = 0 + j4, \quad Z_{in} = Z_o Z_{in} = 75(j4) = \underline{\underline{j300\Omega}}$

(b) $\quad z_L = \frac{25 - j65}{75} = 0.333 - j0.867$

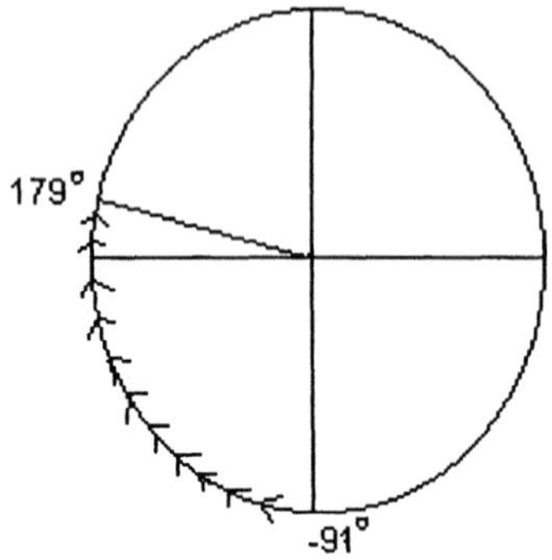

$z_{in} = 0.2 + j0.01$

$Z_{in} = 75(0.2 + j0.01) = \underline{\underline{15 + j0.75\Omega}}$

Prob. 11.28

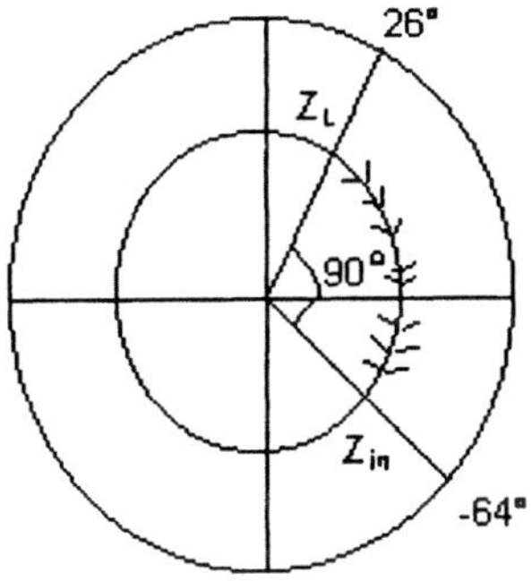

(a) $\lambda \rightarrow 720°$ so then $\frac{\lambda}{8} \rightarrow 90°$

$z_{in} = \underline{\underline{1 - j}}$

(b)

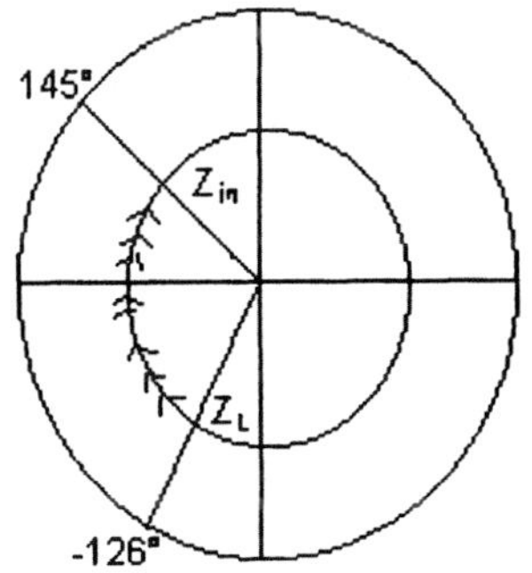

$z_{in} = 0.18 + j0.31$

(c)

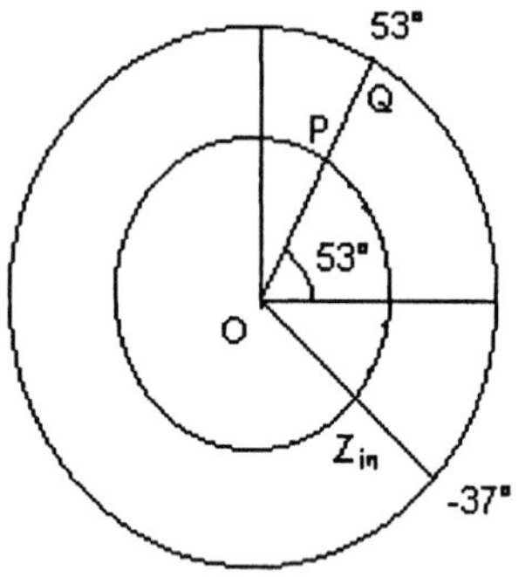

$$\Gamma = 0.3 + j0.4$$
$$= 0.5\angle 53.13°$$
$$\frac{OP}{OQ} = 0.5$$
$$z_{in} = \underline{\underline{1.7 + j1.35}}$$

Prob. 11.29

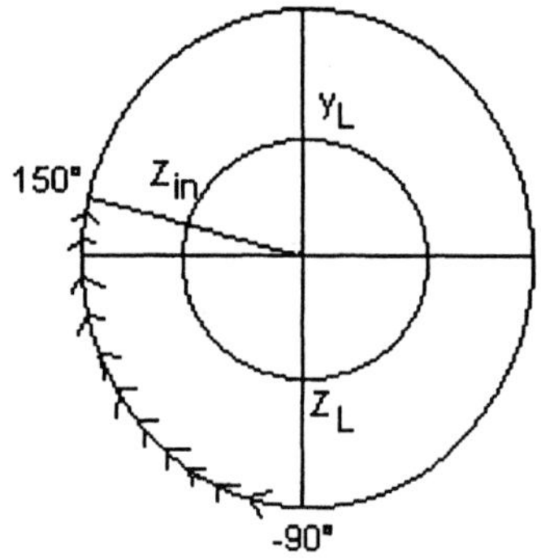

If $\lambda \to 270°$, then $\frac{\lambda}{6} \to 120°$

$$z_{in} = \underline{\underline{0.35 + j0.24}}$$

Prob. 11.30

(a) $Z_{in} = \frac{Z_{in}}{Z_o} = \frac{100 - j120}{80} = 1.25 - j1.5$

$\lambda = \frac{u}{f} = \frac{0.8 \times 3 \times 10^8}{12 \times 10^6} = 20\ m$

$l_1 = 22\ m = \frac{22\lambda}{20} = 1.1\lambda \to 720° + 72°$

$l_2 = 28\ m = \frac{28\lambda}{20} = 1.4\lambda \to 720° + 72° + 216°$

To locate P(the load), we move 2 revolution s plus 72° toward the load. At P,

$|\Gamma_L| = \frac{OP}{OQ} = \frac{5.1cm}{9.2cm} = 0.5543$

$\theta_\Gamma = 72° - 47° = 25°$

$\Gamma_L = \underline{\underline{0.5543 \angle 25°}}$

$Z_{in}, \max = sZ_o = 3.7(80) = \underline{\underline{296\ \Omega}}$

$Z_{in}, \min = \frac{Z_o}{s} = \frac{80}{3.7} = \underline{\underline{21.622\ \Omega}}$

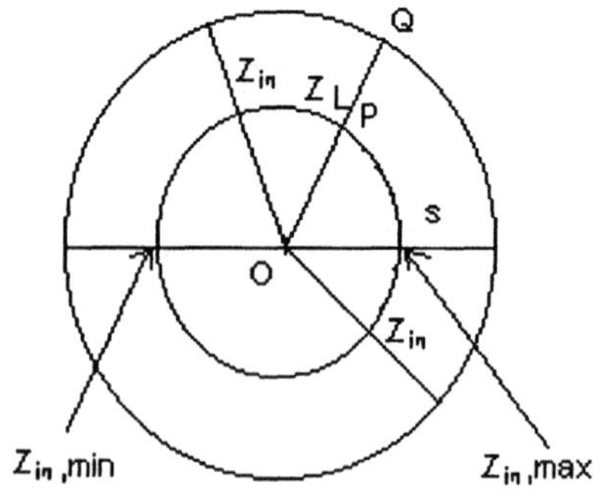

(b) Also, at P, $\quad Z_L = 2.3 + j1.55$

$$Z_L = 80(2.3 + j1.55) = \underline{\underline{184 + j124\Omega}}$$

At S, $\quad s = \underline{\underline{3.7}}$

To Locate $Z_{in}^{'}$, we move 216° from Z_{in} toward the geneator.

At $Z_{in}^{'}$,

$Z_{in}^{'} = 0.48 + j0.76$

$Z_{in}^{'} = 80(0.48 + j0.76) = \underline{\underline{38.4 + j60.8\Omega}}$

(c) Between Z_L and Z_{in}, we move 2 revolutions and 72°. During the movement, we pass through $Z_{in,\max}$ 3 times and $Z_{in,\min}$ twice.

Thus there are :

$$\underline{\underline{3\,Z_{in,\max} \text{ and } 2\,Z_{in,\min}}}$$

Prob. 11.31

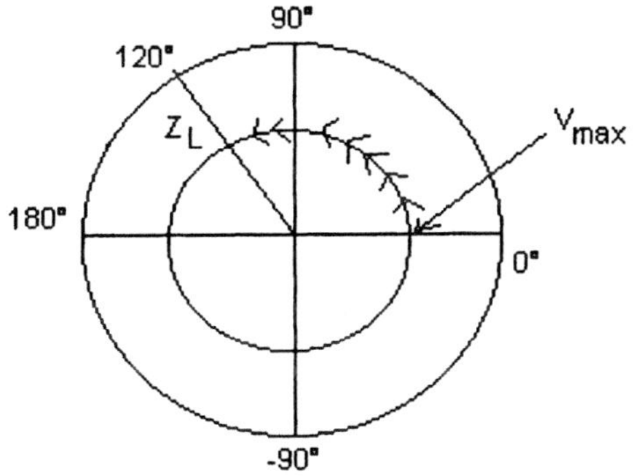

(a) $\quad \frac{\lambda}{2} = 120cm \rightarrow \lambda = 2.4m$

$u = f\lambda \rightarrow f = \frac{u}{\lambda} = \frac{3\times10^8}{2.4} = \underline{\underline{125MHz}}$

(b) $\quad 40cm = \frac{40\lambda}{240} = \frac{\lambda}{6} \rightarrow \frac{720°}{6} = 120°$

$Z_L = Z_o Z_L = 150(0.48 + j0.48$

$\quad = \underline{\underline{72 + j72}}$

(c) $\quad |\Gamma| = \frac{s-1}{s+1} = \frac{1.6}{3.9} = 0.444,$

$\quad \Gamma = \underline{\underline{0.444\angle 120°}}$

Prob. 11.32

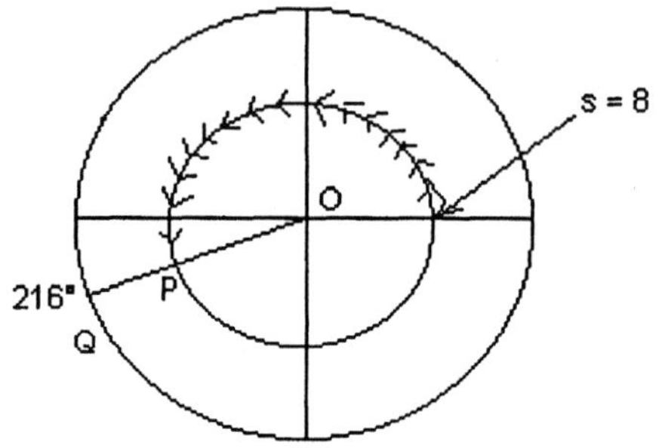

$0.3\lambda \rightarrow 720° \times 0.3 = 216°$

At P, $Z_L = 0.15 - j0.32$

$Z_L = Z_o Z_L = \underline{\underline{15 - j32\Omega.}}$

$$|\Gamma| = \frac{OP}{OQ} = \frac{7.2\text{ cm}}{9.3\text{ cm}} = 0.7742$$

$\Gamma = \underline{\underline{0.7742\angle 216°}}$

Prob. 11.33

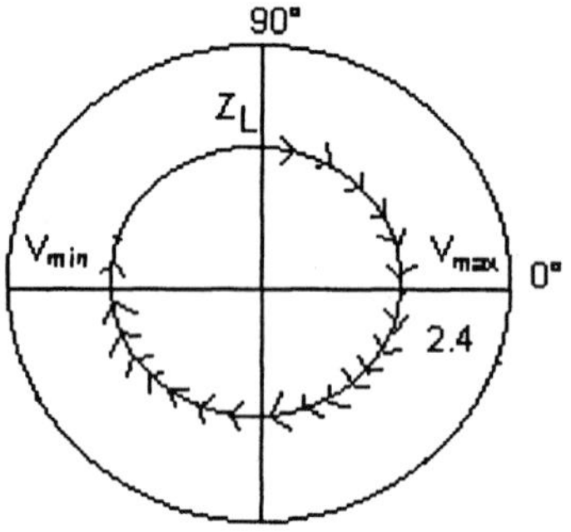

(a) If $\lambda \rightarrow 720°$, then $\dfrac{\lambda}{8} \rightarrow 90°$

$Z_L = 0.7 + j0.68$

$Z_L = 50(0.7 + j0.68) = \underline{\underline{35 + j34\Omega}}$

(b) $l = \dfrac{\lambda}{4} + \dfrac{\lambda}{8} = \underline{\underline{0.375\lambda}}$

Prob. 11.34

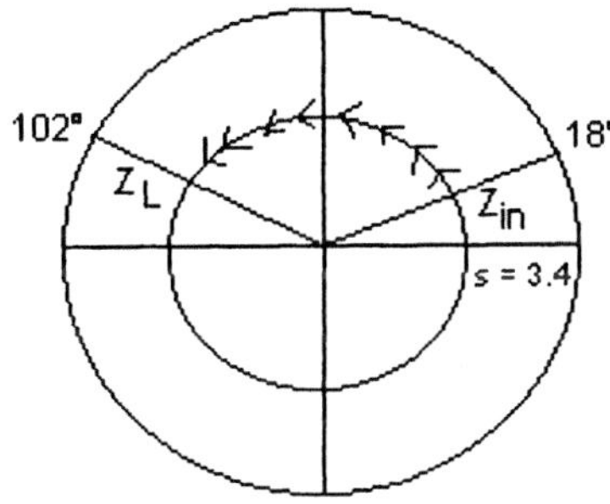

$l = 0.2\lambda \rightarrow 720° \times 0.2 = 144°$

$$Z_{in} = \frac{V_s}{I_s} = \frac{2+j}{10\times10^{-3}} = 200 + j100$$

$$Z_{in} = \frac{Z_{in}}{Z_L} = 2.667 + j1.33$$

$Z_L = 0.3 + j0.12$

$Z_L = 75(0.3 + j0.12) = \underline{\underline{22.5 + j9\Omega}}$, s = $\underline{\underline{3.4}}$

Prob. 11.35

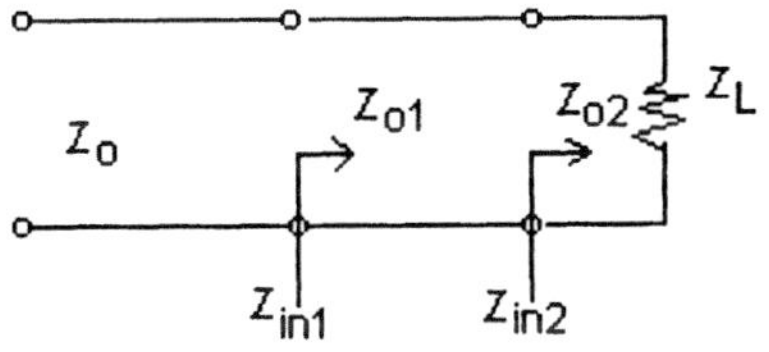

(a) From Eq. (11.43), $Z_{in2} = \dfrac{Z_{o2}^2}{Z_L}$

$$Z_{in1} = \frac{Z_{o1}^2}{Z_{in2}} = Z_o, \text{i.e. } Z_{in2} = \frac{Z_{o1}^2}{Z_o} = \frac{Z_{o2}^2}{Z_L}$$

$$Z_{o1} = Z_{o2}\sqrt{\frac{Z_o}{Z_L}} = 30\sqrt{\frac{50}{75}} = \underline{\underline{24.5\Omega.}}$$

(b) Also, $\dfrac{Z_o}{Z_{o1}} = \left(\dfrac{Z_{o2}}{Z_L}\right) \rightarrow Z_{o2} = \dfrac{Z_o Z_L}{Z_{o1}}$ (1)

Also, $\dfrac{Z_{o1}}{Z_{o2}} = \left(\dfrac{Z_{o2}}{Z_L}\right)^2 \rightarrow (Z_{o2})^3 = Z_{o1}Z_L^2$ (2)

From (1) and (2), $(Z_{o2})^3 = Z_{o1}Z_L^2 = \dfrac{Z_o^3 Z_L^3}{Z_{o1}^3}$ (3)

or $Z_{o1} = \sqrt[4]{Z_o^3 Z_L} = \sqrt[4]{(50)^3(75)} = \underline{\underline{53.33\Omega}}$

From (3), $Z_{o2} = \sqrt[3]{Z_{o1}Z_L^2} = \sqrt[3]{(55.33)(75)^2} = \underline{\underline{67.74\Omega.}}$

Prob. 11.36

$$\frac{\lambda}{4} \to 180°,\ Z_L = \frac{74}{50} = 1.48,\ \frac{1}{Z_L} = 0.6756$$

This acts as the Load to the left line. But there are two such loads in parallel due to the two lines on the right. Thus

$$Z_L^{'} = 50\frac{\left(\dfrac{1}{Z_L}\right)}{2} = 25(0.6756) = 16.892$$

$$Z_L^{'} = \frac{16.892}{50} = 0.3378,\ Z_{in} = \frac{1}{Z_L^{'}} = 2.96$$

$Z_{in} = 50(2.96) = \underline{\underline{148\Omega.}}$

Prob. 11.37

From the previous problem, $Z_{in} = 148\Omega$

$$I_{in} = \frac{V_g}{Z_g + Z_{in}} = \frac{120}{80+148} = 0.5263A$$

$$P_{ave} = \frac{1}{2}|I_{in}|^2 R_{in} = \frac{1}{2}(0.5263)^2(148) = 20.5W$$

Since the lines are lossless, the average power delivered to either antenna is $\underline{\underline{10.25W}}$

Prob. 11.38

(a) $\beta l = \dfrac{2\pi}{4}.\dfrac{\lambda}{4} = \dfrac{\pi}{2}$, $\tan\beta l = \infty$

$$Z_{in} = Z_o\left(\frac{Z_L + jZ_o\tan\beta l}{Z_o + jZ_L\tan\beta l}\right) = Z_o\frac{\left(\dfrac{Z_L}{\tan\beta l} + jZ_o\right)}{\left(\dfrac{Z_o}{\tan\beta l} + jZ_L\right)}$$

As $\tan\beta l \to \infty$,

$$Z_{in} = \frac{Z_o^2}{Z_L} = \frac{(50)^2}{100} = \underline{\underline{25\Omega}}$$

(b) If $Z_L = 0$,

$$Z_{in} = \frac{Z_o^2}{0} = \underline{\underline{\infty}} \qquad \text{(open)}$$

(c) $Z_L = 25 // \infty = \frac{25 \times \infty}{25 + \infty} = \frac{25}{1 + \frac{25}{\infty}} = 25\Omega$

$$Z_{in} = \frac{(50)^2}{25} = \underline{\underline{100\Omega}}$$

Prob. 11.39

$l_1 = \frac{\lambda}{4} \rightarrow Z_{in1} = \frac{Z_o^2}{Z_L} \quad or \quad y_{in1} = \frac{Z_L}{Z_o}$

$y_{in1} = \frac{200 + j150}{(100)^2} = 20 + j15 \text{ mS}$

$l_2 = \frac{\lambda}{8} \rightarrow Z_{in2} =_{Z_L} \underset{\longrightarrow 0}{\lim} Z_o \left(\frac{Z_L + jZ_o \tan\frac{\pi}{4}}{Z_o + jZ_L \tan\frac{\pi}{4}} \right) = jZ_o$

$y_{in2} = \frac{1}{jZ_o} = \frac{1}{j100} = -j10 \text{ mS}$

$l_3 = \frac{7\lambda}{8} \rightarrow Z_{in3} = Z_o \frac{\left(Z_i + jZ_o \tan\frac{7\pi}{4} \right)}{\left(Z_o + jZ_i \tan\frac{7\pi}{4} \right)} = \frac{Z_o(Z_i - jZ_o)}{(Z_o - jZ_i)}$

But

$$y_i = y_{in1} + y_{in2} = 20 + j5 \text{ mS}$$

$$z_i = \frac{1}{y_i} = \frac{1000}{20 + j5} = 47.06 - j11.76$$

$$y_{in3} = \frac{Z_o - jZ_o}{Z_o(Z_i - jZ_o)} = \frac{100 - j47.06 - 11.76}{100(47.06 - j111.76 - j100)}$$

$$= -6.408 + j5.1890 \text{ mS}$$

If the shorted section were often,

$y_{in1} = 20 + j15 \text{ mS}$

$y_{in2} = \frac{1}{Z_{in2}} = \frac{j \tan \pi/4}{Z_o} = \frac{1}{100} = \underline{\underline{j10 \text{ mS}}}$

$$l_3 = \frac{7\lambda}{8} \rightarrow Z_{in3} = Z_o \frac{\left(Z_i + jZ_o \tan\frac{7\pi}{4}\right)}{\left(Z_o + jZ_i \tan\frac{7\pi}{4}\right)} = \frac{Z_o(Z_i - jZ_o)}{(Z_o - jZ_i)}$$

$$y_i = y_{in1} + y_{in2} = 20 + j15 + j10 = 20 + j25 \text{ mS}$$

$$Z_i = \frac{1}{y_i} = \frac{1000}{20 + j25} = 19.51 - j24.39\Omega$$

$$y_{in3} = \frac{Z_o - jZ_i}{Z_o(Z_i - jZ_o)} = \frac{75.61 - \text{j}19.51}{100(19.51 - j124.39)}$$

$$= \underline{\underline{2.461 + j5.691 \text{ mS}}}$$

Prob. 11.40

$$z_L = \frac{Z_L}{Z_o} = \frac{60 - j50}{50} = 1.2 - j1$$

$$y_L = \frac{1}{2_L}$$

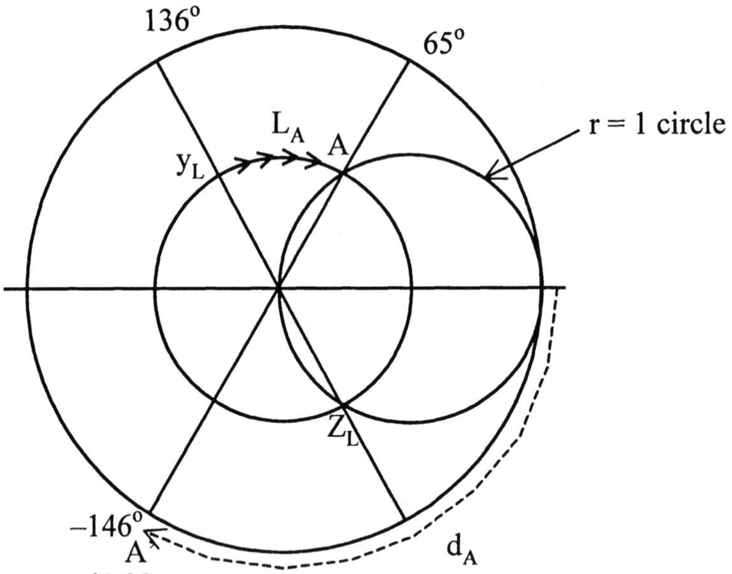

At A, y = 1 + j0.92, y_s = -j0.92

$$Y_s = Y_o y_s = \frac{-j0.92}{50} = \underline{\underline{-j18.4 \text{ mS}}}$$

$$L_A = (136^o - 65^o)\frac{\lambda}{720^o} = \underline{\underline{0.0986\lambda}}$$

$$d_A = \frac{146^o}{720^o} = \underline{\underline{0.2028\lambda}}$$

Prob. 11.41

$d_A = 0.12\lambda \rightarrow 0.12 \times 720^o = 86.4^o$

$l_A = 0.3\lambda \rightarrow 0.3 \times 720^o = 216^o$

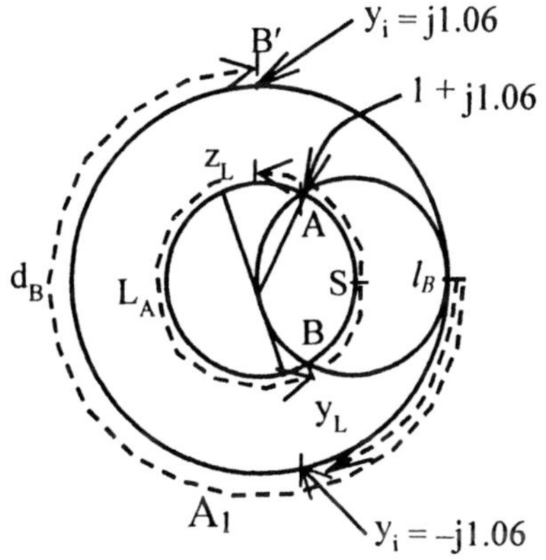

(a) From the Smith Chart,

$Z_L = 0.57 + j0.69$

$Z_L = 60(0.57 + j0.69)$

$= \underline{\underline{34.2 + j41.4\Omega}}$

(b) $d_B = \frac{360^o - 86.4^o}{720^o}\lambda = \underline{\underline{0.38\lambda}}$

$l_B = \frac{\lambda}{2} - \frac{(-62.4^o - -82^o)}{720^o}\lambda = \underline{\underline{0.473\lambda}}$

(c) $\underline{\underline{s = 2.65}}$

Prob. 11.42

$$\frac{\lambda}{4} \rightarrow \frac{720^o}{4} = 180^o$$

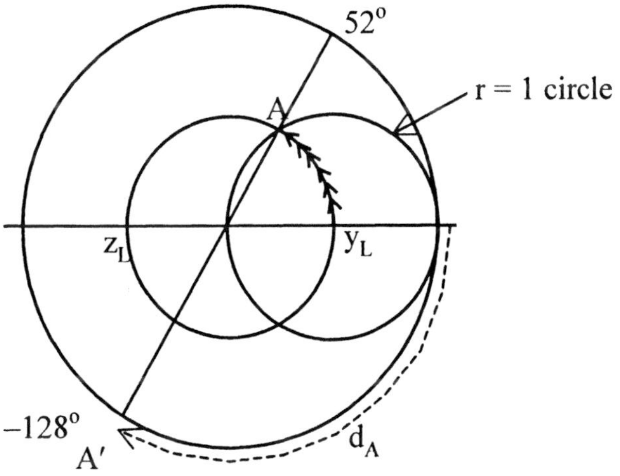

At A, $y = 1 + j1.5$, $y = -j1.5 \rightarrow Y_s = y_s Y_o = -j1.5 Y_o$

$$d_A = \frac{128^o \lambda}{720^o} = \underline{\underline{0.1778\lambda}}$$

$$L_A = \frac{52^o}{720^o}\lambda = \underline{\underline{0.0722\lambda}}$$

Prob. 11.43

$$s = \frac{V_{max}}{V_{min}} = \frac{4V}{1V} = \underline{\underline{4}}$$

$$|\Gamma| = \frac{s-1}{s+1} = \frac{3}{3} = 0.6$$

$$\frac{\lambda}{2} = 25\,\text{cm} - 5\,\text{cm} = 20\,\text{cm}$$

$$\rightarrow \quad \lambda = 40\,\text{cm}$$

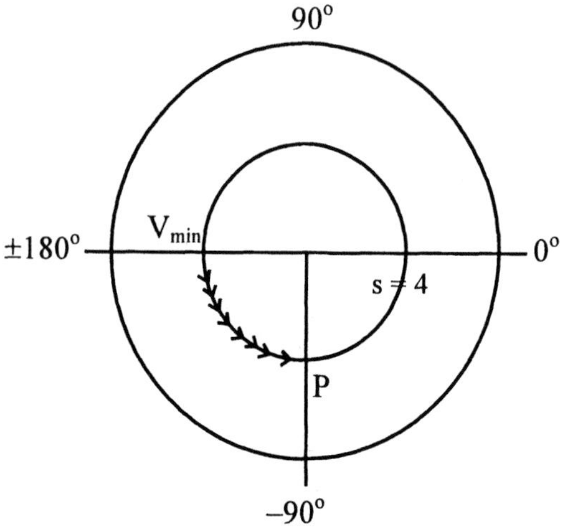

The load is l=5cm from V_{min}, i.e.

$$l = \frac{5\lambda}{40} = \frac{\lambda}{8} \quad \rightarrow \quad 90^o$$

On the s = 4 circle, move 90° from V_{min} towards the load and obtain $Z_L = 0.46 - j0.88$ at P.

$$Z_L = Z_o Z_L = 60(0.46 - j0.88) = \underline{\underline{27.6 - j52.8\ \Omega}}$$

$$\theta_\Gamma = 270^o \text{ or } 90^o$$

$$\Gamma = \underline{\underline{0.6\angle\text{-}90^o}}$$

Prob. 11.44

$$\frac{\lambda}{2} = 32 - 12 = 20cm \rightarrow \lambda = 40\,\text{cm}$$

$$f = \frac{u}{\lambda} = \frac{3\times10^8}{40\times10^{-2}} = \underline{\underline{0.75\,\text{GHz}}}$$

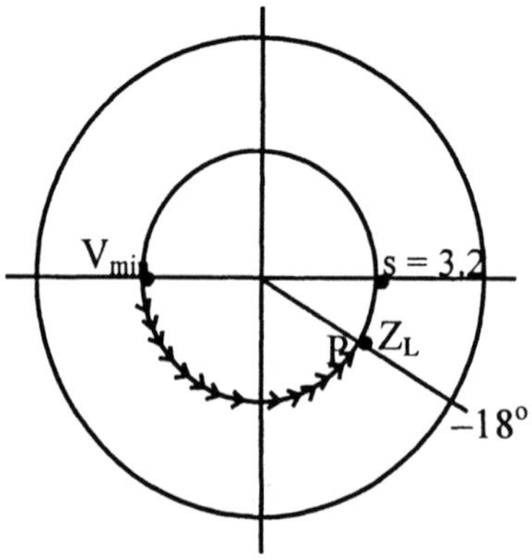

$$l = 21 - 12 = 9cm = \frac{9\lambda}{402} \rightarrow \frac{9}{40} \times 720^o = 162^o$$

At P, $z_L = 2.6 - j1.2$

$$Z_L = z_L Z_o = 50(2.6 - j1.2) = \underline{\underline{130 - j60\Omega}}$$

Prob. 11.45

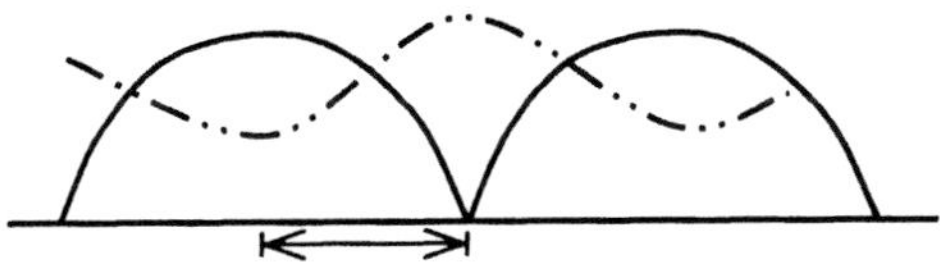

$$s = \frac{V_{max}}{V_{min}} = \frac{0.95}{0.45} = \underline{\underline{2.11}}$$

$$\frac{\lambda}{2} = 22.5 - 14 = 8.5 \quad \rightarrow \quad \lambda = 17\,\text{cm}$$

$$f = \frac{c}{\lambda} = \frac{3 \times 10^8}{0.17} = \underline{\underline{1.764\ \text{GHz}}}$$

$$l = 3.2\,\text{cm} = \frac{3.2}{17}\lambda \rightarrow 135.5^o$$

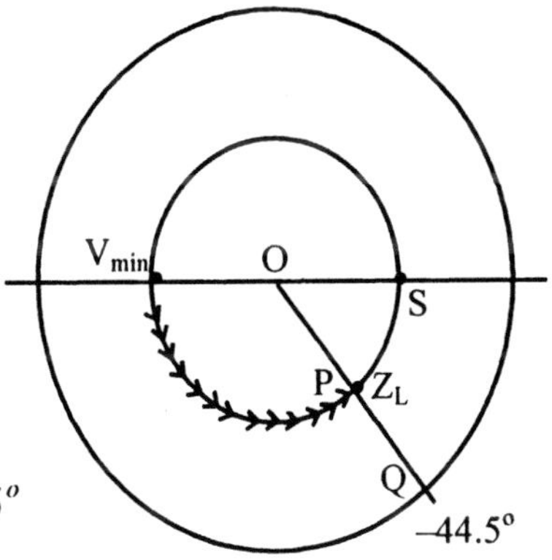

At P, $\quad Z_L = 1.4 - j0.8$

$$Z_L = 50(1.4 - j0.8) = \underline{\underline{70 - j40\Omega}}$$

$$|\Gamma| = \frac{s-1}{s+1} = \frac{1.11}{3.11} = 0.357, \quad \theta_\Gamma = -44.5^o$$

$$|\Gamma| = \underline{\underline{0.357\angle -44.5^o}}$$

Prob. 11.46

At $z = 0, t = 0^+, v_o = \dfrac{Z_o}{Z_o + Z_g} V_g$

$t_1 = \dfrac{l}{u} =$ transit time or time delay. Hence,

$V(l, t_1^+)$

$V(l, t_1^+) = V_o + \Gamma_L V_o$

$V(l, t_1^+) = V_o + \Gamma_L V_o$

$V(l, 3t_1^+) = V_o + \Gamma_L V_o + \Gamma_G \Gamma_L V_o$

$V(l, 5t_1^+) = V_o + \Gamma_L V_o + \Gamma_G \Gamma_L V_o + \Gamma_G \Gamma_L^2 V_o$

$V(l, 7t_1^+) = V_o(1 + \Gamma_L + \Gamma_G \Gamma_L + \Gamma_G \Gamma_L^2 + \Gamma_G^2 \Gamma_L^2)$

and so on. When $t >> \dfrac{l}{u}$

$$V(l,\infty)=V_o\left[1+\Gamma_G\Gamma_L+(\Gamma_G\Gamma_L)^2+(\Gamma_G\Gamma_L)^3+....\right]$$
$$+V_o\Gamma_L\left[1+\Gamma_G\Gamma_L+(\Gamma_G\Gamma_L)^2+(\Gamma_G\Gamma_L)^3+....\right]$$

But $1+x+x^2+x^3+.....=\dfrac{1}{1-x} \qquad |x|<1.$

Since $|\Gamma_G\Gamma_L|<1$,

$$V(l,\infty)=V_o\left[\frac{1}{1-\Gamma_G\Gamma_L}+\frac{\Gamma_L}{1-\Gamma_G\Gamma_L}\right]=V_o\frac{(1+\Gamma_L)}{1-\Gamma_G\Gamma_L}$$

$$=\frac{Z_oZ_g}{Z_g+Z_o}\left[\frac{1+\dfrac{Z_L-Z_o}{Z_L+Z_o}}{1-\dfrac{Z_L-Z_o}{Z_L+Z_o}\cdot\dfrac{Z_g-Z_o}{Z_g+Z_o}}\right]=\frac{V_gZ_L}{Z_L+Z_G}$$

Thus

$$V_\infty=\frac{V_gZ_L}{Z_L+Z_G}, \qquad I_\infty=\frac{V_\infty}{Z_L}=\frac{V_g}{Z_L+Z_G}$$

Prob. 11.47

$$t_1=\frac{l}{u}=\frac{6m}{3\times10^8}=2\mu s, \quad V_o=V_g\cdot\frac{Z_o}{Z_L+Z_g}=20\left(\frac{60}{100}\right)=12V,$$

$$\Gamma_g=\frac{Z_g-Z_o}{Z_g+Z_o}=\frac{40-60}{100}=-\frac{1}{5}, \quad \Gamma_L=\frac{Z_L-Z_o}{Z_L+Z_o}=\frac{100-60}{160}=\frac{1}{4}.$$

We only need the voltage bounce diagram because we can obtain *I(l, t)* from *V(l, t)/Z*$_L$.

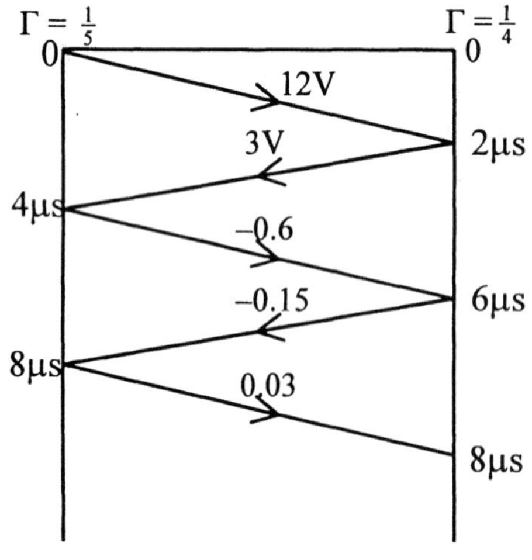

(Voltage bounce diagram)

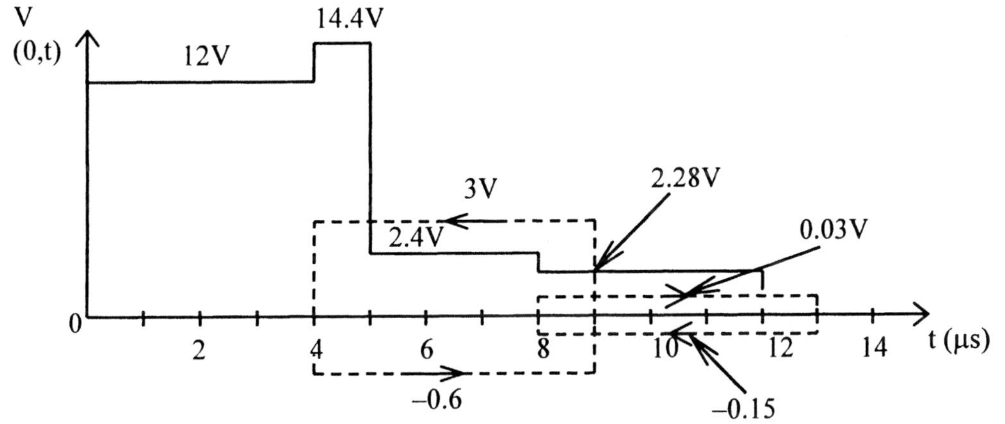

We obtain $V(1, t)$ from the bounce diagram and divide by $Z_L = 100\Omega$ to obtain $I(l, t)$.

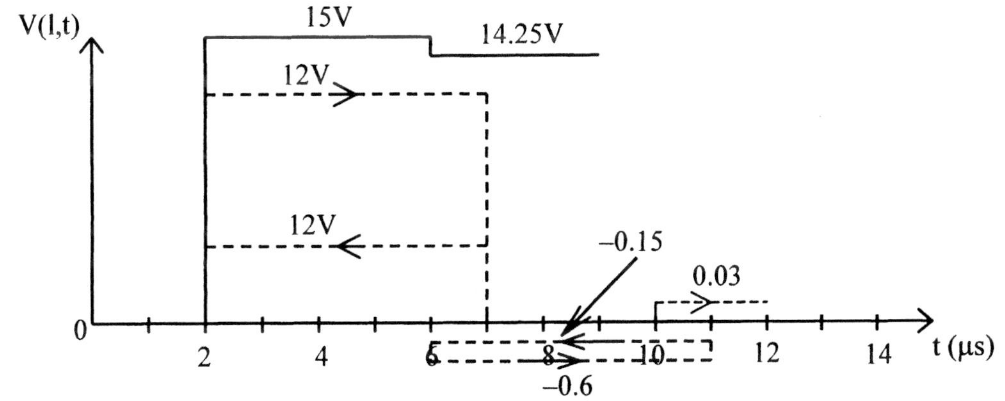

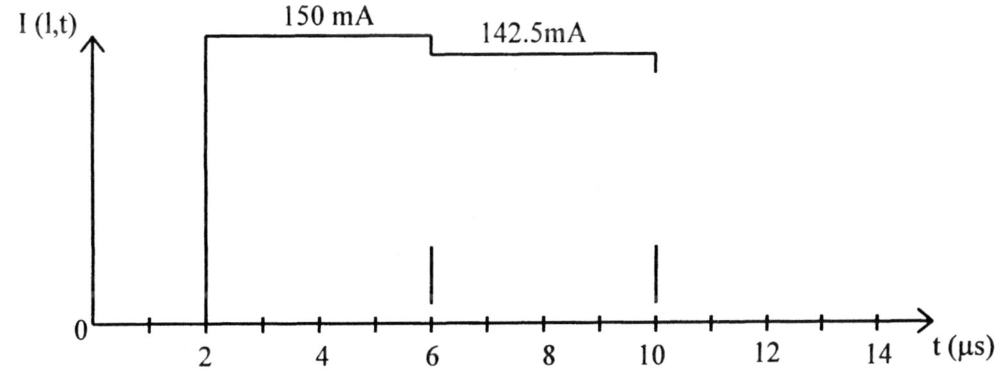

Prob. 11.48

$$\Gamma_L = \frac{Z_L - Z_o}{Z_L + Z_o} = \frac{0.5Z_o - Z_o}{1.5Z_o} = -\frac{1}{3}$$

$$\Gamma_g = \frac{Z_g - Z_o}{Z_g + Z_o} = \frac{Z_o}{3Z_o} = \frac{1}{3}$$

$$t_1 = \frac{l}{u} = 2\mu s, \quad V_o = \frac{Z_o}{3Z_o}(27) = 9\text{ V}, \quad I_o = \frac{V_o}{Z_o} = 180\text{ mA}$$

$$V_\infty = \frac{Z_L}{Z_g - Z_L}V_g = \frac{0.5}{2.5}(27) = 5.4\text{ V}, \quad I_\infty = \frac{V_\infty}{Z_L} = 216\text{ mA}$$

The voltage and current bounce diagram are shown below

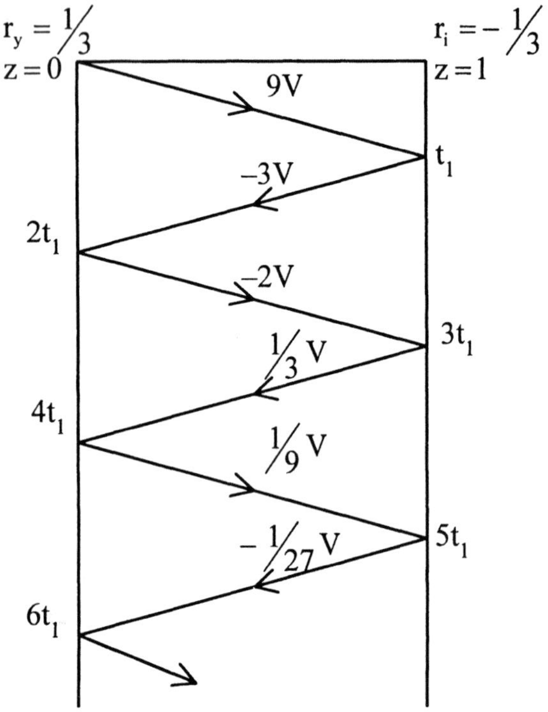

(Voltage bounce diagram)

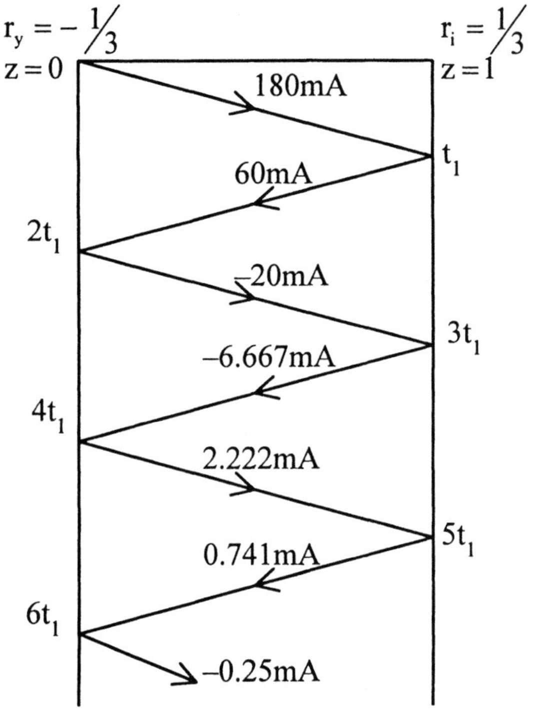

(Current bounce diagram)

From the bounce diagram, we obtain V(0,t) and I(0,t) as shown below:

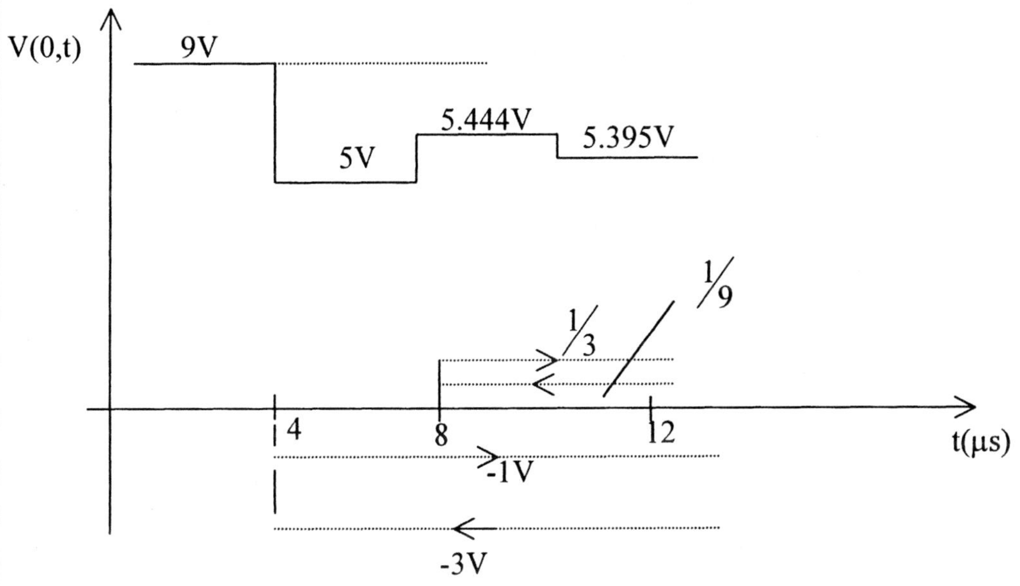

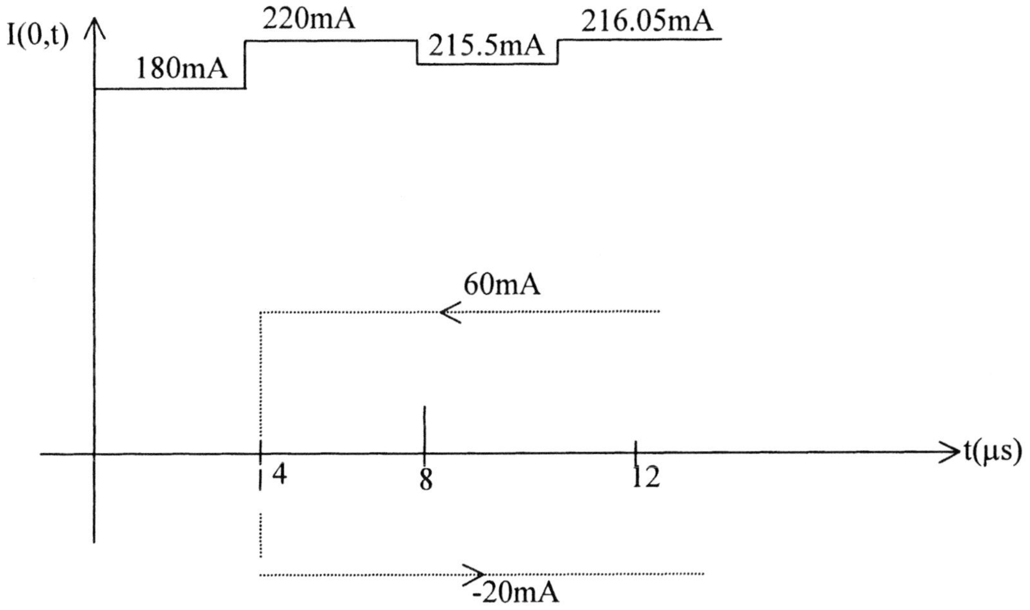

Prob.11.49 $V_0 = \frac{Z_0}{Z_0 + Z_9} V_g = \frac{75}{75 + 54}(100) = 60$

$$t_1 = \frac{l}{u} = \frac{200}{2 \times 10^8} = 1\mu s$$

$$\Gamma_9 = \frac{Z_9 - Z_0}{Z_9 + Z_0} = \frac{50 - 75}{50 + 75} = -\frac{1}{5}, \quad \Gamma_L = \frac{Z_l - Z_0}{Z_l + Z_0} = \frac{150 - 75}{150 + 75} = \frac{1}{3}$$

The voltage bounce diagram is shown below.

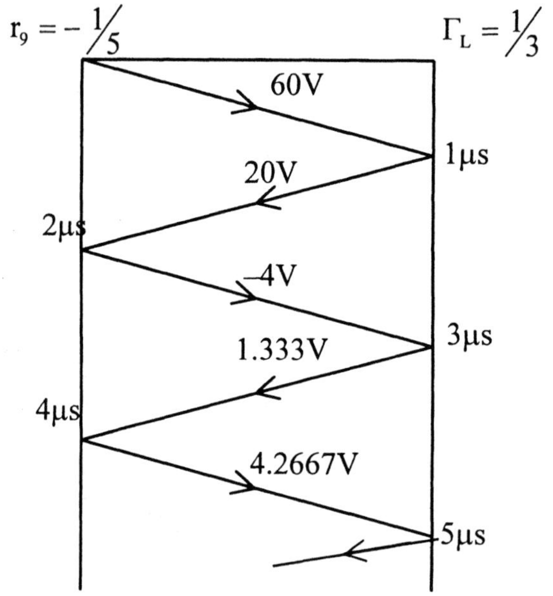

From the bounce diagram, we obtain V(l,t) as shown below.

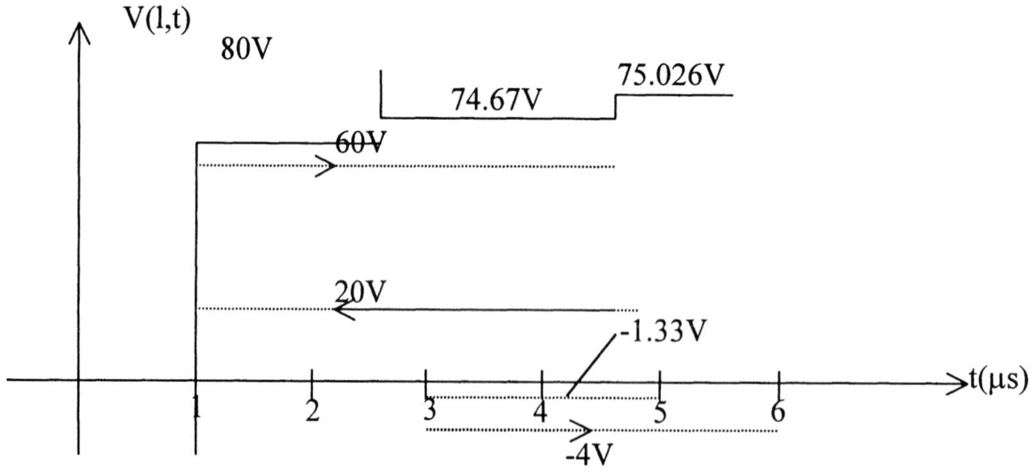

Since

$I(l,t) = \dfrac{V(l,t)}{150}$, we obtain I(l,t) by scaling V(l,t) down by 150.

The result is shown below.

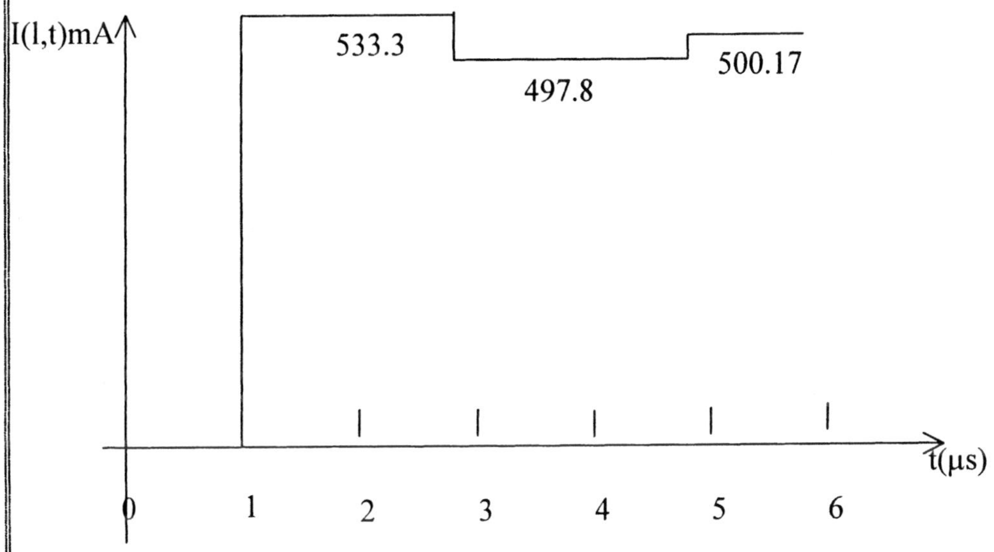

Prob. 11.50

(a) $t_1 = \dfrac{l}{u} = \dfrac{150}{3\times10^8} = 0.5\mu s,$

$$\Gamma_L = \frac{Z_L - Z_o}{Z_L + Z_o} = \frac{150-50}{150+150} = \frac{1}{2}, \quad \Gamma_g = \frac{Z_g - Z_o}{Z_g + Z_o} = \frac{25-50}{75} = -\frac{1}{3},$$

$$V_o = \frac{Z_o V_g}{Z_o + Z_g} = \frac{50(12)}{75} = 8V, \quad I_o = \frac{V_g}{Z_g + Z_o} = \frac{12}{75} = 160 \text{ mA}$$

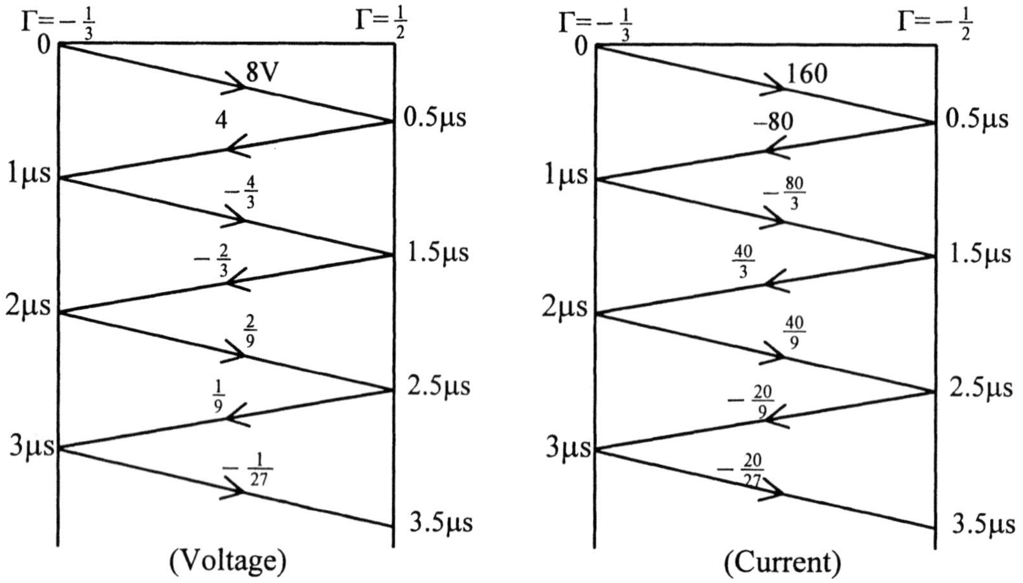
Γ=−1/3
Γ=1/2
0
8V
4
0.5μs
1μs
−4/3
−2/3
1.5μs
2μs
2/9
1/9
2.5μs
3μs
−1/27
3.5μs
(Voltage)
Γ=−1/3
Γ=−1/2
160
−80
−80/3
40/3
40/9
−20/9
−20/27
(Current)

(b)

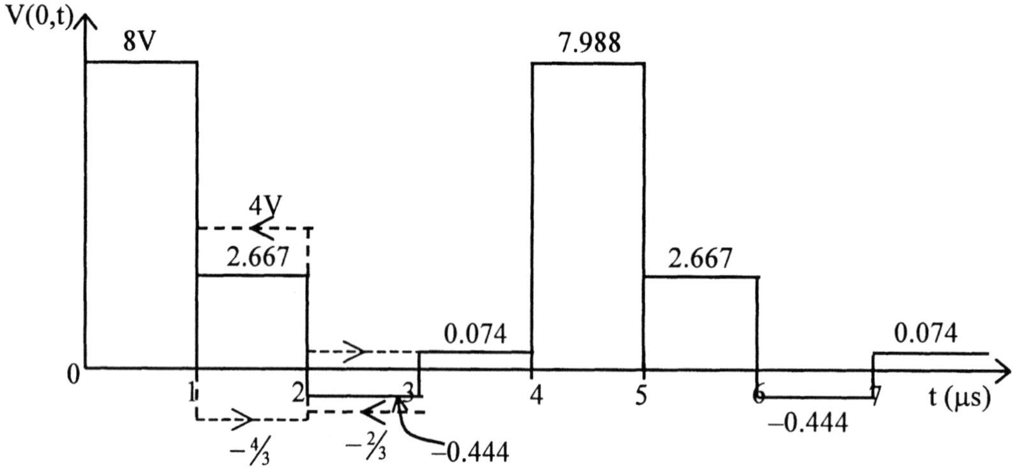
V(0,t)
8V
7.988
4V
2.667
2.667
0.074
0.074
−4/3
−2/3
−0.444
−0.444
t (μs)

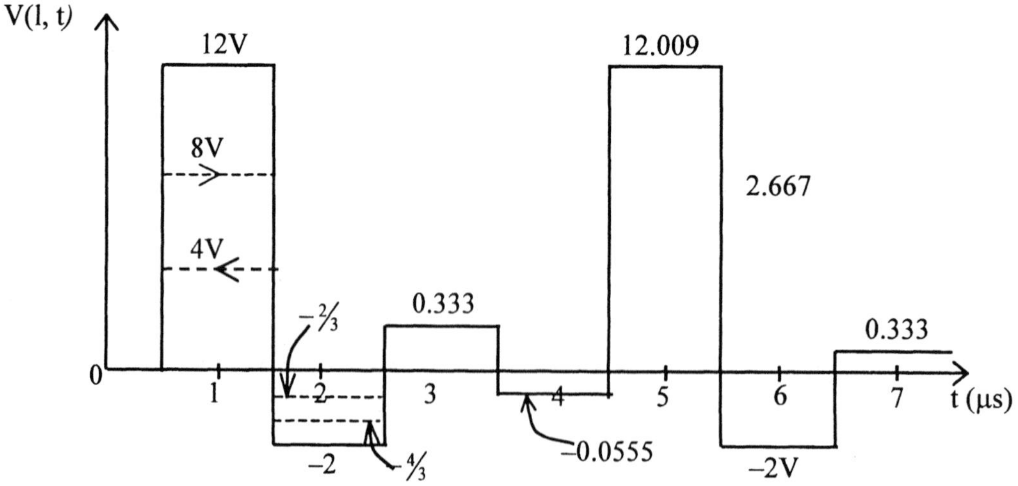
V(l, t)
12V
12.009
8V
4V
2.667
0.333
0.333
−2/3
−2
−4/3
−0.0555
−2V
t (μs)

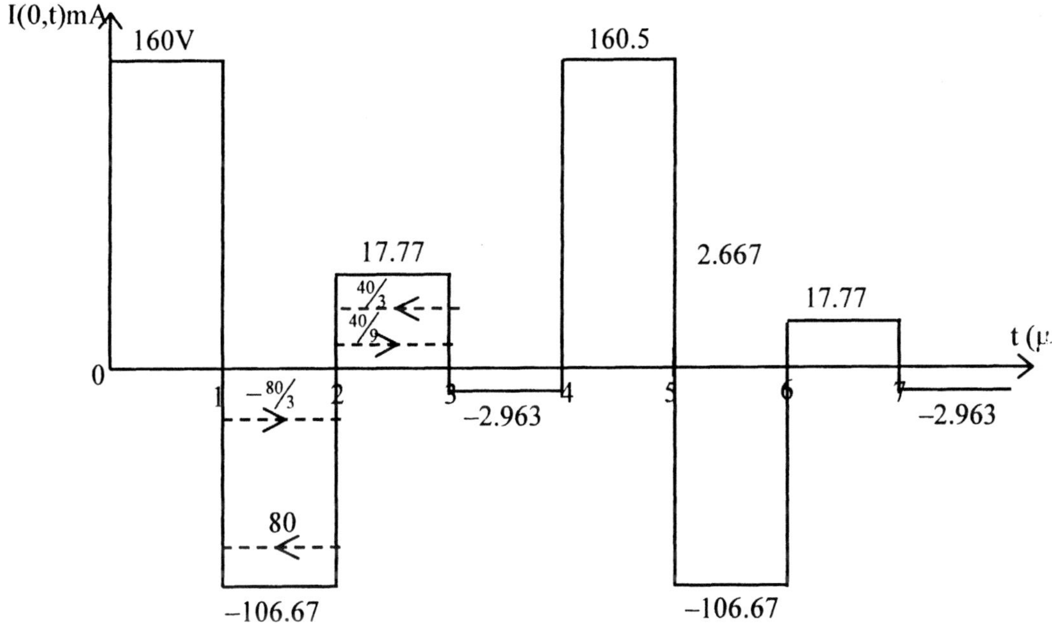
I(0,t)mA
160V
160.5
17.77
2.667
17.77
40/3
40/9
t (μs)
0
1
−80/3
2
3
4
5
6
7
−2.963
−2.963
80
−106.67
−106.67

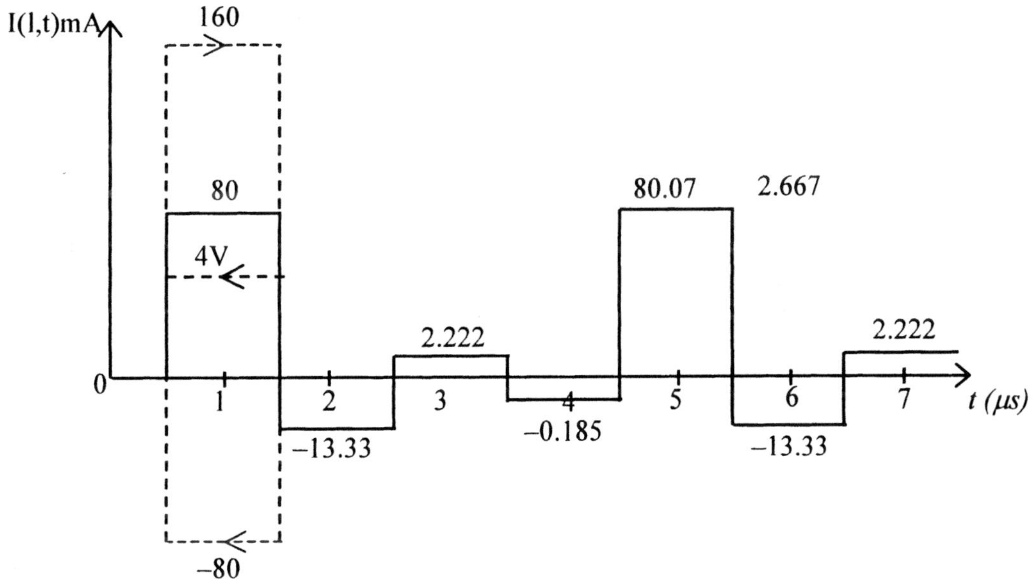
I(l,t)mA
160
80
80.07
2.667
4V
2.222
2.222
0
1
2
3
4
5
6
7
t (μs)
−13.33
−0.185
−13.33
−80

Prob.11.51

$$w = 1.5\text{cm},\ h = 1\text{cm},\ \frac{w}{h} = 1.5$$

(a) $$\varepsilon_{eff} = \left(\frac{6+1}{2}\right) + \frac{\varepsilon_r - 1}{2\sqrt{1+12h/w}} = 1.6 + \frac{0.6}{\sqrt{1+12/1.5}} = 1.8$$

$$Z_0 = \frac{377}{\sqrt{1.8}\ (1.5 + 1.393 + 0.667\ \text{In}\,(2.944))} = \frac{281}{3.613} = 77.77\mu$$

(b) $$\alpha_l = 8.686\frac{R_s}{wZ_0}$$

$$R_s = \frac{1}{\sigma_c\,\sigma} = \sqrt{\frac{\mu\pi f}{\sigma_c}} = \sqrt{\frac{19 \times 2.5 \times 10^9 \times 4\pi \times 10^{-3}}{1.1 \times 10^7}}$$

$$= 2.995 \times 10^{-2}$$

$$\alpha_l = \frac{8.686 \times 2.995 \times 10^{-2}}{1.5 \times 10^{-2} \times 77.77} = 0.223\text{dB/m}$$

$$u = \frac{c}{\sqrt{\varepsilon_{eff}}} \rightarrow \lambda = \frac{u}{f} = \frac{c}{f\sqrt{\varepsilon_{eff}}} = \frac{3 \times 10^8}{2.5 \times 10^9\sqrt{1.8}} = 8.944 \times 10^{-2}$$

$$\alpha_d = 27.3 \times \frac{0.8}{1.2}\frac{(2.2)}{1.8}\ \frac{2 \times 10^{-2}}{8.944 \times 10^{-2}} = \frac{96.096}{19.319} =$$

$$\alpha_d = 4.974\text{dB/m}$$

(c) $$\alpha = \alpha_l + \alpha_d = 5.197\text{dB/m}$$

$$\alpha l = 20\text{dB} \rightarrow l = \frac{20}{\alpha} = \frac{20}{5.197} = 3.848\text{m}$$

Prob 11.52

(a) Let $x = w/h$. If $x < 1$,

$$50 = \frac{60}{\sqrt{4.6}} In\left(\frac{8}{x} + x\right)$$

$$\sqrt[5]{4.6} \; - 6\text{In}\left(\frac{8}{\text{x}} + \text{x}\right) = 0$$

we solve for x (e.g using Maple) and get x = 2.027 or 3.945

which contradicts our assumptiom that $x < 1$. If $x > 1$,

$$50 = \frac{120\pi}{\sqrt{4.6\,(\text{x} + 1.393 + 0.667\,\text{In}(\text{x} \; + 1.444))}}$$

$$12\pi \; - 5\sqrt{4.6}\,(\text{x} \;\; 1.393\; 0.667\;\text{In}\,(\text{x} + 1.44))$$

solving for x, we obtain $x = 1.42 = \dfrac{\text{w}}{\text{h}}$

$w = 1.42 \times 8 = \underline{\underline{11.36\text{m}}}$

(b) $\quad \beta = \dfrac{\omega \varepsilon_{eff}}{c}$

$$\beta l = 45^0 = \frac{\pi}{4} = \frac{\text{wl}\varepsilon_{\text{eff}}}{c}$$

$$l = \frac{\pi\text{c}}{4\varepsilon_{\text{eff}}\, 2\pi f} = \frac{3 \times 10^8}{8 \times 4.6 \times 8 \times 10^9}$$

$\underline{\underline{l = 0.102\text{m}}}$

Prob. 11.53

For $\text{w} = 0.4$ mm, $\dfrac{w}{h} = \dfrac{0.4\text{ mm}}{2\text{ m}} = 0.2 \rightarrow$ narrow strip

$$A = \frac{12}{\sqrt{2(9.6+1)}} = 2.606, \quad B = \frac{1}{2}\left(\frac{8.6}{10.6}\right)\left(\ln\frac{\pi}{2} + \frac{1}{9.6}\ln\frac{4}{\pi}\right)$$

$$= 0.4057(0.4516+0.02516)$$

$$= 0.1934$$

$$C = \ln\frac{8}{0.2} + \frac{1}{32}(0.2)^2 = 3.69$$

$$Z_o = A(C-B) = 2.606(3.69-0.1934) = 9.112\Omega$$

For $w = 8\text{mm}, \quad \frac{w}{h} = \frac{8}{2} = 4 \rightarrow$ wide strip.

$$D = \frac{60\pi}{\sqrt{9.6}} = 60.84$$

$$E = 2.0 + 0.4413 + 0.08226 \times \frac{8.6}{(9.6)^2}$$

$$+ \frac{10.6}{2\pi(9.6)}(1.452 + \ln 2.94) = 2.449 + 0.4447$$

$$= 2.8936$$

$$Z_o = \frac{D}{E} = \frac{60.84}{2.8936} = 21.03.$$

Thus,

$$\underline{\underline{9.112\Omega < Z_o < 21.03\Omega}}$$

Prob 11.54

Suppose we guess that w/h < 2

$$A = \frac{75}{60}\sqrt{\frac{3.3}{2}} + \frac{1.3}{3.3}\left(0.23 + \frac{0.11}{2.3}\right) = 1.715$$

$$\frac{w}{h} = \frac{8e^{A}}{e^{2A} - 2} = \frac{44.453}{28.88} = 1.539 \rightarrow w = 1.539h = \underline{\underline{1.85\text{mm}}}$$

If we guess that w/h > 2,

$$\frac{60\pi^2}{Z_o\sqrt{\varepsilon_r}} = \frac{60\pi^2}{75\sqrt{2.3}} = 3.808$$

$$\frac{w}{h} = \frac{2}{\pi}\left[2.803 - In\,6.615 + \frac{1.3}{4.6}\left(In2.808 + 0.39 - \frac{0.61}{2.3}\right)\right]$$

$$= 0.793 \neq > 2$$

Thus $\frac{w}{h} = 1.539 < 2$

$$\varepsilon_{\text{eff}} = \frac{3.3}{22} + \frac{1.3}{2\sqrt{1 + \frac{12}{1.539}}} = 1.869$$

$$u = \frac{3 \times 10^8}{\sqrt{1.869}} = \underline{\underline{2.194 \times 10^8 \text{ m/s}}}$$

CHAPTER 12

P. E. 12.1 (a) For TE_{10}, $f_c = 3$ GHz,

$$\sqrt{1-(f_c/f)^2} = \sqrt{1-(3/15)^2} = \sqrt{0.96}, \quad \beta_o = \omega / u_o = 4\pi f / c$$

$$\beta = \frac{4\pi f}{c}\sqrt{0.96} = \frac{4\pi x15x10^9}{3x10^8}\sqrt{0.96} = \underline{\underline{615.6}} \text{ rad/m}$$

$$u = \frac{\omega}{\beta} = \frac{2\pi x15x10^9}{615.6} = \underline{\underline{1.531x10^8}} \text{ m/s}$$

$$\eta' = \sqrt{\frac{\mu}{\varepsilon}} = 60\pi, \quad \eta_{TE} = \frac{60\pi}{\sqrt{0.96}} = \underline{\underline{192.4\Omega}}$$

(b)For TM_{11}, $f_c = 3\sqrt{7.25}$ GHz, $\sqrt{1-(f_c/f)^2} = 0.8426$

$$\beta = \frac{4\pi f}{c}(0.8426) = \frac{4\pi x15x10^9(0.8426)}{3x10^8} = \underline{\underline{529.4}} \text{ rad/m}$$

$$u = \frac{\omega}{\beta} = \frac{2\pi x15x10^9}{529.4} = \underline{\underline{1.78x10^8}} \text{ m/s}$$

$$\eta_{TM} = 60\pi(0.8426) = \underline{\underline{158.8\Omega}}$$

P. E. 12.2 (a) Since $E_z \neq 0$, this is a TM mode

$$E_{zs} = E_o \sin(m\pi x / a)\sin(n\pi y / b)e^{-j\beta z}$$

$E_o = 20$, $\frac{m\pi}{a} = 40\pi$ ⟶ m=2, $\frac{n\pi}{b} = 50\pi$ ⟶ n=1

i.e. $\underline{\underline{TM_{21} \text{ mode}}}$.

(b) $f_c = \frac{u'}{2}\sqrt{(m/a)^2 + (n/b)^2} = \frac{3x10^8}{2}\sqrt{40^2 + 50^2} = 1.5\sqrt{41}$ GHz

$$\beta = \omega\sqrt{\mu\varepsilon}\sqrt{1-(f_c/f)^2} = \frac{2\pi f}{c}\sqrt{f^2 - f_c^2} = \frac{2\pi x10^9}{3x10^8}\sqrt{225 - 92.25} = \underline{\underline{241.3 \text{ rad/m.}}}$$

(c)

$$E_{xs} = \frac{-j\beta}{h^2}(40\pi)20\cos 40\pi x \sin 50\pi y e^{-j\beta z}$$

$$E_{ys} = \frac{-j\beta}{h^2}(50\pi)20\sin 40\pi x \cos 50\pi y e^{-j\beta z}$$

$$\frac{E_y}{E_x} = 1.25 \tan 40\pi x \cot 50\pi y$$

P. E. 12.3 If TE_{13} mode is assumed, f_c and β remain the same.

$f_c = 28.57$ GHz, $\beta = 1718.81$ rad/m, $\gamma = j\beta$

$$\eta_{TE13} = \frac{377/2}{\sqrt{1-(28.57/50)^2}} = 229.69\ \Omega$$

For m=1, n=3, the field components are:

$E_z = 0$

$$H_z = H_o \cos(\pi x/a)\cos(3\pi y/b)\cos(\omega t - \beta z)$$

$$E_x = -\frac{\omega\mu}{h^2}\left(\frac{3\pi}{b}\right) H_o \cos(\pi x/a)\sin(3\pi y/b)\sin(\omega t - \beta z)$$

$$E_y = \frac{\omega\mu}{h^2}\left(\frac{\pi}{a}\right) H_o \sin(\pi x/a)\sin(3\pi y/b)\sin(\omega t - \beta z)$$

$$H_x = -\frac{\beta}{h^2}\left(\frac{\pi}{a}\right) H_o \sin(\pi x/a)\cos(3\pi y/b)\sin(\omega t - \beta z)$$

$$H_y = -\frac{\beta}{h^2}\left(\frac{3\pi}{a}\right) H_o \cos(\pi x/a)\sin(3\pi y/b)\sin(\omega t - \beta z)$$

Given that $H_{ox} = 2 = -\frac{\beta}{h^2}(\pi/a)H_o$,

$$H_{oy} = -\frac{\beta}{h^2}(3\pi/b)H_o = 6a/b = 6(1.5)/8 = 11.25$$

$$H_{oz} = H_o = -\frac{2h^2 a}{\beta\pi} = \frac{-2x14.51\pi^2 x10^4 x1.5x10^{-2}}{1718.81\pi} = -7.96$$

$$E_{oy} = \frac{\omega\mu}{h^2}\left(\frac{\pi}{a}\right)H_o = -\frac{2\omega\mu}{\beta} = 2\eta_{TE} = -459.4$$

$$E_{ox} = -E_{oy}\frac{3a}{b} = 459.4(4.5/0.8) = 2584.1$$

$$E_x = 2584.1\cos(\pi x/a)\sin(3\pi y/b)\sin(\omega t - \beta z)\ \text{V/m},$$

$$E_y = -459.4\sin(\pi x/a)\sin(3\pi y/b)\sin(\omega t - \beta z)\ \text{V/m},$$

$E_z = 0$,

$$H_y = 11.25\cos(\pi x / a)\sin(3\pi y / b)\sin(\omega t - \beta z) \quad \text{A/m},$$

$$H_z = -7.96\cos(\pi x / a)\cos(3\pi y / b)\cos(\omega t - \beta z) \quad \text{A/m}$$

P. E. 12.4

$$f_{c11} = \frac{u'}{2}\sqrt{\frac{1}{a^2} + \frac{1}{b^2}} = \frac{3x10^8 x10^2}{2}\sqrt{1/8.636^2 + 1/4.318^2} = 3.883 \quad \text{GHz}$$

$$u_p = \frac{3x10^8}{\sqrt{1-(3.883/4)^2}} = \underline{\underline{12.5x10^8}} \quad \text{m/s},$$

$$u_g = \frac{9x10^{16}}{12.5x10^6} = \underline{\underline{7.203x10^7}} \quad \text{m/s}$$

P. E. 12.5 The dominant mode becomes TE_{01} mode

$$f_{c01} = \frac{c}{2b} = 3.75 \quad \text{GHz}, \quad \eta_{TE} = 406.7\Omega$$

From Example 12.2,

$$E_x = -E_o \sin(3\pi y / b)\sin(\omega t - \beta z), \quad \text{where } E_o = \frac{\omega \mu b}{\pi} H_o.$$

$$\mathcal{P}_{ave} = \int_{x=0}^{a}\int_{y=0}^{b} \frac{|E_{xs}|^2}{2\eta} dxdy = \frac{E_o^2 ab}{4\eta}$$

Hence $E_o = 63.77$ V/m as in Example 12.5.

$$H_o = \frac{\pi E_o}{\omega \mu b} = \frac{\pi x63.77}{2\pi x10^{10} x4\pi x10^{-7} x4x10^{-2}} = \underline{\underline{63.34}} \quad \text{mA/m}$$

P. E. 12.6 (a) For m=1, n=0, $f_c = u'/(2a)$

$$\frac{\sigma}{\omega \varepsilon} = \frac{10^{-15}}{2\pi x9x10^9 x2.6x10^{-9}/(36\pi)} = \frac{10^{-15}}{1.3} << 1$$

Hence,

$$u' \cong \frac{1}{\sqrt{\mu\varepsilon}} = c/\sqrt{2.6}, \qquad f_c = \frac{3x10^8}{2x2.4x10^{-2}\sqrt{2.6}} = 3.876 \quad \text{GHz}$$

$$\alpha_d = \frac{\sigma\eta'}{2\sqrt{1-(f_c/f)^2}} = \frac{10^{-15}x377/\sqrt{2.6}}{2\sqrt{1-(3.876/9)^2}} = 1.295x10^{-13} \text{ Np/m}$$

For n = 0, m=1,

$$\alpha_c = \frac{2R_s}{b\eta'\sqrt{1-(f_c/f)^2}}[\frac{1}{2}+\frac{b}{a}(f_c/f)^2]$$

$$= \frac{2\sqrt{2.6}\sqrt{\pi x9x10^9 x1.1x10^7 x4\pi x10^{-7}}}{377x1.5x10^{-2}x1.1x10^7\sqrt{1-(3.876/9)^2}}[0.5+(2.4/1.5)(3.876/9)^2] = \underline{\underline{3.148x10^{-2}}} \text{ Np/m}$$

(b)Since $\alpha_c >> \alpha_d, \alpha = \alpha_c + \alpha_d \cong \alpha_c = 3.148x10^{-2}$

loss = $\alpha l = 3.148x10^{-2}x0.4 = 1.259x10^{-2}$ Np = $\underline{\underline{0.1093 \text{ dB}}}$

P. E. 12.7 For TM_{11} , m = 1 = n,

$$E_{zs} = E_o \sin(\pi x/a)\sin(\pi y/b)e^{-\gamma z}$$

$$E_{xs} = -\frac{\gamma}{h^2}(\pi/a)E_o\cos(\pi x/a)\sin(\pi y/b)e^{-\gamma z}$$

$$E_{ys} = -\frac{\gamma}{h^2}(\pi/b)E_o\sin(\pi x/a)\cos(\pi y/b)e^{-\gamma z}$$

$$H_{xs} = \frac{j\omega\varepsilon}{h^2}(\pi/b)E_o\sin(\pi x/a)\cos(\pi y/b)e^{-\gamma z}$$

$$H_{ys} = -\frac{j\omega\varepsilon}{h^2}(\pi/a)E_o\cos(\pi x/a)\sin(\pi y/b)e^{-\gamma z}$$

$$H_{zs} = 0$$

For the electric field lines,

$$\frac{dy}{dx} = \frac{E_y}{E_x} = (a/b)\tan(\pi x/a)\cot(\pi y/b)$$

For the magnetic field lines

$$\frac{dy}{dx} = \frac{H_y}{H_x} = -(a/b)\cot(\pi x/a)\tan(\pi y/b)$$

Notice that $(\frac{E_y}{E_x})(\frac{H_y}{H_x}) = -1$

showing that the electric and magnetic field lines are mutually orthogonal. The field lines are as shown in Fig. 12.14.

P. E. 12.8

$$u' = \frac{1}{\sqrt{\mu\varepsilon}} = \frac{c}{\sqrt{\varepsilon_r}}$$

$$f_{TE101} = \frac{1.5x10^{10}}{\sqrt{3}}\sqrt{1/25 + 0 + 1/100} = \underline{\underline{1.936}} \text{ GHz}$$

$$Q_{TE101} = \frac{1}{61\delta}, \text{ where}$$

$$\delta = \frac{1}{\sqrt{\pi f_{101}\mu\sigma_c}} = \frac{1}{\sqrt{\pi x 1.936x10^9 x 4\pi x10^{-7} x 5.8x10^7}} = 1.5x10^{-6}$$

$$Q_{TE101} = \frac{10^6}{61x1.5} = \underline{\underline{10{,}929}}$$

Prob. 12.1 (a) For TM_{mn} modes, $H_z = 0$

$$E_{zs} = E_o \sin(\pi x / a)\sin(\pi y / b)e^{-\gamma z}$$

Using eq. (12.15), all field components vanish for TM_{01} and TM_{10}.

(b) See text.

Prob. 12.2 (a)

$$f_c = \frac{u'}{2}\sqrt{1/a^2 + 1/b^2} = \frac{3x10^8}{2\sqrt{4x10^{-2}}}\sqrt{1/2^2 + 1/3^2} = \underline{\underline{4.507}} \text{ GHz}$$

(b)

$$\beta = \beta'\sqrt{1 - (f_c / f)^2} = \frac{\omega}{u'}\sqrt{1 - (f_c / f)^2} = \frac{2\pi x 20x10^9\sqrt{4}}{3x10^8}\sqrt{1 - (4.508/20)^2}$$

$= \underline{\underline{816.2 \text{ rad/m}}}$

(c)

$$u = \omega / \beta = \frac{2\pi x 20x10^9}{816.21} = \underline{\underline{1.54x10^8}} \text{ m/s}$$

Prob. 12.3 (a)

$$f_c = \frac{u'}{2}\sqrt{(m/a)^2 + (n/b)^2} = \frac{3x10^8}{2x9x10^{-2}}\sqrt{(m/1)^2 + (n/2)^2} = \frac{15}{18}\sqrt{4m^2 + n^2} \text{ GHz}$$

Mode	F_c (GHz)
TE_{01}	0.8333
TE_{10}, TE_{02}	1.667
TE_{11}, TM_{11}	1.863
TE_{12}, TM_{13}	2.357
TE_{03}	2.5
TE_{13}, TM_{13}	3
TE_{04}	3.333
TE_{14}, TM_{14}	3.727
TE_{05}, TE_{23}, TM_{23}	4.167
TE_{15}, TM_{15}	4.488

(b)The highest possible mode is TE_{15} or TM_{15}.

$$\eta' = \frac{120\pi}{9} = 41.89, \quad \sqrt{1-(f_c/f)^2} = \sqrt{1-(4.488/4.5)^2} = 0.073$$

$$\eta_{TE15} = \frac{\eta'}{\sqrt{1-(f_c/f)^2}} = \frac{41.89}{0.073} = \underline{\underline{573.8\Omega}}$$

$$\eta_{TM15} = \eta'\sqrt{1-(f_c/f)} = \underline{\underline{3.058\Omega}}$$

(c) The lowest mode is TE_{01}

$$u' = c/9, \quad u_g = u'\sqrt{1-(f_c/f)^2} = \frac{3x10^9}{9}\sqrt{1-(0.8333/4.5)^2} = 3.276x10^8 \text{ m/s}$$

Prob. 12.4 a/b = 3 ⟶ a = 3b

$$f_{c10} = \frac{u'}{2a} \longrightarrow a = \frac{u'}{2f_{c10}} = \frac{3x10^8}{2x18x10^9} \text{ m} = 0.833\text{cm}$$

A design could be <u>a = 9mm, b = 3mm.</u>

Prob. 12.5 For the dominant mode,

$$f_c = \frac{c}{2a} = \frac{3x10^8}{2x8} = 18.75\,MHz$$

(a) It $\underline{\underline{\text{will not pass}}}$ the AM signal, (b) it $\underline{\underline{\text{will pass}}}$ the FM signal.

Prob. 12.6 (a) For TE_{10} mode, $f_c = \frac{u'}{2a}$

Or $a = \frac{u'}{2f_{c10}} = \frac{3x10^8}{2x5x10^9} = \underline{\underline{3\text{ cm}}}$

For TE_{01} mode, $f_c = \frac{u'}{2b}$

Or $b = \frac{u'}{2f_c} = \frac{3x10^8}{2x12x10^9} = \underline{\underline{1.25\text{ cm}}}$

(b) Since a > b, 1/a < 1/b, the next higher modes are calculated as shown below.

Mode	f_c (GHz)
TE_{10}	5
$*TE_{20}$	10
TE_{30}	15
TE_{40}	20
$*TE_{01}$	12
TE_{02}	24
$*TE_{11}$	13
TE_{21}	15.62

The next three higher modes are starred ones, i.e. $\underline{\underline{TE_{20}, TE_{01}, TE_{11}}}$

(c) $u' = \frac{1}{\sqrt{\mu\varepsilon}} = \frac{c}{\sqrt{2.25}} = 2x10^8$ m/s

For TE_{11} modes,

$$f_c = \frac{3x10^8}{2x10^{-2}\sqrt{2.25}}\sqrt{\frac{1}{3^2} + \frac{1}{1.25^2}} = \underline{\underline{8.67}}\text{ GHz}$$

Prob. 12.7

$$u = \frac{\omega}{\beta} = \frac{u'}{\sqrt{1-(f_c/f)^2}} = \frac{3x10^8}{\sqrt{1-(6.5/7.2)^2}} = 6.975x10^8\text{ m/s}$$

$$t = \frac{2l}{u} = \frac{300}{6.975x10^8} = \underline{\underline{430}} \text{ ns}$$

Prob. 12.8

$$f_c = \frac{u'}{2}\sqrt{(m/a)^2 + (n/b)^2}$$

$$f_{c11} = f_{c03} \longrightarrow \frac{u'}{2}\sqrt{(1/a)^2 + (1/b)^2} = \frac{u'}{2}\sqrt{9/b^2}$$

$$\frac{9}{b^2} = \frac{1}{a^2} + \frac{1}{b^2} \longrightarrow a = \frac{b}{\sqrt{8}}$$

$$f_{c03} = \frac{3u'}{2b} \longrightarrow b = \frac{3c}{2f_{c03}} = \frac{9x10^8}{2x12x10^9} = 3.75 \text{ cm}$$

$\underline{\underline{a = 1.32 \text{ cm}, \ b = 3.75 \text{ cm}}}$

Since a < b, the dominant mode is TE_{01}

$$f_{c01} = \frac{c}{2b} = \frac{3x10^8}{2x3.75x10^{-2}} = 4 \text{ GHz} < f = 8 \text{ GHz}$$

Hence, the dominant mode $\underline{\underline{\text{will proprogate}}}$.

Prob. 12.9 $E_z \neq 0$. This must be TM_{23} mode (m=2, n=3). Since a= 2b,

$$f_c = \frac{c}{4b}\sqrt{m^2 + 4n^2} = \frac{3x10^8}{4x3x10^{-2}}\sqrt{4+36} = 15.81 \text{ GHz}, \quad f = \frac{\omega}{2\pi} = \frac{10^{12}}{2\pi} = 159.2 \text{ GHz}$$

$$\eta_{TM} = \frac{1}{377\sqrt{1-(15.81/159.2)^2}} = \underline{\underline{375.1\ \Omega}}$$

$$\mathcal{P}_{ave} = \frac{|E_{xs}|^2 + |E_{ys}|^2}{2\eta_{TM}} a_z$$

$$= \frac{\beta^2 E_o^2}{2h^4 \eta_{TM}}\left[(2\pi/a)^2\cos^2(2\pi x/a)\sin^2(3\pi y/b) + (3\pi/b)^2\sin^2(2\pi x/a)\cos^2(3\pi y/b)\right]a_z$$

$$P_{ave} = \int \mathcal{P}_{ave}.dS = \int_{x=0}^{a}\int_{y=0}^{b} \mathcal{P}_{ave}.dxdya_z$$

$$= \frac{\beta^2 E_o^2}{2h^4\eta_{TM}}\frac{1}{4}\left[\frac{4\pi^2}{a^2}+\frac{9\pi^2}{b^2}\right] = \frac{\beta^2 E_o^2}{8h^2\eta_{TM}}$$

But

$$\beta = \frac{\omega}{c}\sqrt{1-(f_c/f)^2} = \frac{10^{12}}{3x10^8}\sqrt{1-(15.81/159.2)^2} = 3.317x10^3$$

$$h^2 = \frac{4\pi^2}{a^2}+\frac{9\pi^2}{b^2} = \frac{10\pi^2}{b^2} = 1.098x10^5$$

$$P_{ave} = \frac{(3.317)^2 x10^6 x25}{8x(1.098x10^5)^2 x375.4} = \underline{\underline{0.8347}} \text{ W}$$

Prob. 12.10 (a) Since m=2 and n=1, we have $\underline{\underline{\text{TE}_{21} \text{ mode}}}$

(b) $\beta = \beta'\sqrt{1-(f_c/f)^2} = \omega\sqrt{\mu_o\varepsilon_o}\sqrt{1-(\omega_c/\omega)^2}$

$$\beta c = \sqrt{\omega^2 - \omega_c^2} \longrightarrow \omega_c^2 = \sqrt{\omega^2 - \beta^2 c^2}$$

$$f_c = \frac{\omega_c}{2\pi} = \sqrt{f^2 - \frac{\beta^2 c^2}{4\pi^2}} = \sqrt{36x10^{18} - \frac{144x9x10^{16}}{4\pi^2}} = \underline{\underline{5.973}} \text{ GHz}$$

(c) $\eta_{TE} = \dfrac{\eta}{\sqrt{1-(f_c/f)^2}} = \dfrac{377}{\sqrt{1-(5.973/6)^2}} = \underline{\underline{3978\Omega}}$

(d)For TE mode,

$$E_y = \frac{\omega\mu}{h^2}(m\pi/a)H_o\sin(m\pi x/a)\cos(n\pi y/b)\sin(\omega t - \beta z)$$

$$H_x = \frac{-\beta}{h^2}(m\pi/a)H_o\sin(m\pi x/a)\cos(n\pi y/b)\sin(\omega t - \beta z)$$

$\beta = 12$, m = 2, n =1

$$E_{oy} = \frac{\omega\mu}{h^2}(m\pi/a)H_o, \quad H_{ox} = \frac{\beta}{h^2}(m\pi/a)H_o$$

$$\eta_{TE} = \frac{E_{oy}}{H_{ox}} = \frac{\omega\mu}{\beta} = \frac{2\pi x6x10^9 x4\pi x10^{-7}}{12} = 4\pi^2 x100$$

$$H_{ox} = \frac{E_{oy}}{\eta_{TE}} = \frac{5}{4\pi^2 x 100} = 1.267 \text{ mA/m}$$

$$\underline{H_x = -1.267 \sin(m\pi x / a)\cos(n\pi y / b)\sin(\omega t - \beta z)} \text{ mA/m}$$

Prob. 12.11 (a) Since m=2, n=3, the mode is $\underline{TE_{23}}$.

(b) $\beta = \beta' \sqrt{1 - (f_c / f)^2} = \frac{2\pi f}{c} \sqrt{1 - (f_c / f)^2}$

But

$$f_c = \frac{u'}{2} \sqrt{(m/a)^2 + (n/b)^2} = \frac{3x10^8}{2x10^{-2}} \sqrt{(2/2.86)^2 + (3/1.016)^2} = 46.19 \text{ GHz}, \text{ f = 50 GHz}$$

$$\beta = \frac{2\pi x 50 x 10^9}{3x10^8} \sqrt{1 - (46.19/50)^2} = 400.68 \text{ rad/m}$$

$$\gamma = j\beta = \underline{j400.7} \text{ /m}$$

(c) $\eta = \frac{\eta'}{\sqrt{1 - (f_c / f)^2}} = \frac{377}{\sqrt{1 - (46.19/50)^2}} = \underline{985.3\Omega}$

Prob. 12.12

$$P_{ave} = \frac{1}{2\eta} \int_{y=0}^{b} \int_{0}^{a} (|E_{xs}|^2 + |E_{ys}|^2) dx dy$$

But

$$E_{xs} = \frac{-j\beta}{h^2} (\pi / a) H_o \cos(\pi x / a) \sin(\pi y / b) e^{-j\beta z}$$

$$E_{ys} = \frac{-j\beta}{h^2} (\pi / b) E_o \sin(\pi x / a) \cos(\pi y / b) e^{-j\beta z}$$

$$P_{ave} = \frac{1}{2\eta_{TM11}} \frac{\beta^2 \pi^2}{h^4} E_o^2 [\frac{1}{a^2} \int_0^a \cos^2(\pi x / a) dx \int_0^b \sin^2(\pi x / b) dy$$

$$+ \frac{1}{b^2} \int_0^a \sin^2(\pi x / a) dx \int_0^b \cos^2(\pi x / b) dy]$$

$$= \frac{1}{2\eta_{TM11}} \frac{\beta^2 \pi^2}{h^4} E_o^2 [\frac{1}{a^2} + \frac{1}{b^2}](a/2)(b/2)$$

Note that $h^2 = \frac{\pi^2}{a^2} + \frac{\pi^2}{b^2} = \frac{a^2+b^2}{a^2b^2}\pi^2$

$$P_{ave} = \frac{\beta^2 E_o^2}{8\pi^2 \eta_{TM11}} \frac{a^3b^3}{a^2+b^2}$$

Prob. 12.13 (a)

$$f_c = \frac{u'}{2}\sqrt{(m/a)^2 + (n/b)^2}, \beta = \beta'\sqrt{1-(f_c/f)^2}$$

$$u = \omega/\beta = \frac{u'}{\sqrt{1-(f_c/f)^2}}, \ \lambda = 2\pi/\beta = \frac{\lambda'}{\sqrt{1-(f_c/f)^2}}$$

(b) If a = 2b = 2.5cm, $f_c = \frac{u'}{2a}\sqrt{m^2+4n^2}$. For TE_{11},

$$f_c = \frac{3x10^8}{2x2.5x10^{-2}}\sqrt{1+4} = 13.42 \text{ GHz}, \quad u = \frac{3x10^8}{\sqrt{1-(13.42/20)^2}} = \underline{\underline{4.06x10^8}} \text{ m/s}$$

$$\lambda = u/f = \frac{4.046x10^8}{200x10^8} = \underline{\underline{2.023}} \text{ cm}$$

For TE_{21},

$$f_c = \frac{3x10^8}{2x2.5x10^{-2}}\sqrt{4+4} = 16.97 \text{ GHz}, \quad u = \frac{3x10^8}{\sqrt{1-(16.97/20)^2}} = \underline{\underline{5.669x10^8}} \text{ m/s}$$

$$\lambda = u/f = \frac{5.669x10^8}{200x10^8} = \underline{\underline{2.834}} \text{ cm}$$

Prob. 12.14 (a)

$$f_c = \frac{u'}{2}\sqrt{(m/a)^2 + (n/b)^2} = \frac{3x10^8}{2x10^{-2}}\sqrt{1/1+4/9} = 18.03 \text{ GHz}$$

f = 1.2 f_c = $\underline{\underline{21.63 \text{ GHz}}}$

(b) $\sqrt{1-(f_c/f)^2} = \sqrt{1-(1/1.2)^2} = 0.5528$

$$u_p = \frac{c}{\sqrt{1-(f_c/f)}} = \frac{3x10^8}{0.5528} = \underline{\underline{5.427x10^8}} \text{ m/s}$$

$$u_g = u\sqrt{1-(f_c/f)^2} = 3x10^8 x0.5528 = \underline{\underline{1.658x10^8}} \text{ m/s}$$

Prob. 12.15

$$f_c = \frac{3x10^8}{2}\sqrt{(m/0.025)^2 + (n/0.01)^2} = 15\sqrt{n^2 + (m/2.5)^2} \text{ GHz}$$

f_{c10} = 6 GHz, f_{c20} = 12 GHz, f_{c01} = 15 GHz.
Since f_{c20} , f_{c10} > 11 GHz, only the dominant TE_{10} mode is propagated.

(a) $$\frac{u_p}{u} = \frac{1}{\sqrt{1-(f_c/f)^2}} = \frac{1}{\sqrt{1-(6/11)^2}} = \underline{\underline{1.193}}$$

(b) $$\frac{u_g}{u} = \sqrt{1-(6/11)^2} = \underline{\underline{0.8381}}$$

Prob. 12.16 Let $F = \sqrt{1-(f_c/f)^2} = \sqrt{1-(16/24)^2} = 0.7453$

$$u' = \frac{1}{\sqrt{\mu\varepsilon}} = \frac{3x10^8}{\sqrt{2.25}} = 2x10^8, \qquad u_p = \frac{u'}{F}, \qquad u_g = u'F = 2x10^8 x0.7453 = \underline{\underline{1.491x10^8}} \text{ m/s}$$

$$\eta_{TE} = \eta'/F = \frac{377}{1.5x0.7453} = \underline{\underline{337.2\Omega}}$$

Prob. 12.17 In free space,

$$\eta_1 = \frac{\eta_o}{\sqrt{1-(f_c/f)^2}}, \qquad f_c = \frac{c}{2a} = \frac{3x10^8}{2x5x10^{-2}} = 3 \text{ GHz}$$

$$\eta_1 = \frac{377}{\sqrt{1-(3/8)^2}} = 406.7$$

$$\eta_2 = \frac{\eta'_1}{\sqrt{1-(f_c/f)^2}}, \eta' = \frac{120\pi}{\sqrt{2.25}} = 80\pi, f_c = \frac{u'}{2a}, u' = \frac{c}{\sqrt{\varepsilon_r}}$$

$$f_c = \frac{3x10^8}{2x5x10^{-2}\sqrt{2.25}} = 2 \text{ GHz}, \quad \eta_2 = \frac{80\pi}{\sqrt{1-(2/8)^2}} = 82.62$$

$$\Gamma = \frac{\eta_2 - \eta_1}{\eta_2 + \eta_1} = \frac{82.62 - 406.7}{82.62 + 406.7} = -0.662$$

$$s = \frac{1+|\Gamma|}{1-|\Gamma|} = \frac{1.662}{0.338} = \underline{\underline{4.917}}$$

Prob. 12.18 Substituting $E_z = R\Phi Z$ into the wave equation,

$$\frac{\Phi Z}{\rho}\frac{d}{d\rho}(\rho R') + \frac{RZ}{\rho^2}\Phi'' + R\Phi Z'' + k^2 R\Phi Z = 0$$

Dividing by $R\Phi Z$,

$$\frac{1}{R\rho}\frac{d}{d\rho}(\rho R') + \frac{\Phi''}{\Phi\rho^2} + k^2 = -\frac{Z''}{Z} = -k_z^2$$

i.e. $Z'' - k_z^2 Z = 0$

$$\frac{1}{R\rho}\frac{d}{d\rho}(\rho R') + \frac{\Phi''}{\Phi\rho^2} + (k^2 + k_z^2) = 0$$

$$\frac{\rho}{R}\frac{d}{d\rho}(\rho R') + (k^2 + k_z^2)\rho^2 = -\frac{\Phi''}{\Phi} = k_\phi^2$$

or

$$\underline{\underline{\Phi'' + k_\phi^2\Phi = 0}}$$

$\rho\frac{d}{d\rho}(\rho R') + (k_\rho^2\rho^2 - k_\phi^2)R = 0$, where $k_\rho^2 = k^2 + k_z^2$. Hence

$$\underline{\underline{\rho^2 R'' + \rho R' + (k_\rho^2\rho^2 - k_\phi^2)R = 0}}$$

Prob. 12.19

$$\mathcal{P}_{ave} = \frac{|E_{xs}|^2 + |E_{ys}|^2}{2\eta}a_z = \underline{\underline{\frac{\omega^2\mu^2\pi^2}{2\eta b^2 h^4}H_o^2\sin^2\pi y/b a_z}}$$

$$P_{ave} = \int \mathcal{P}_{ave}.dS = \frac{\omega^2\mu^2\pi^2}{2\eta b^2 h^4}H_o^2\int_{x=0}^{a}\int_{y=0}^{b}\sin^2\pi y/b\,dxdy$$

$$P_{ave} = \frac{\omega^2\mu^2\pi^2}{2\eta b^2 h^4}H_o^2 ab/2$$

But $h^2 = (m\pi/a)^2 + (n\pi/b)^2 = \frac{\pi^2}{b^2}$,

$$P_{ave} = \frac{\omega^2 \mu^2 a b^3 H_o^2}{4\pi^2 \eta}$$

Prob. 12.20

$$R_s = \sqrt{\frac{\pi \mu f}{\sigma_c}} = \sqrt{\frac{\pi x 12x10^9 x 4\pi x 10^{-7}}{5.8x10^7}} = 2.858x10^{-2}$$

$$f_{c10} = \frac{u'}{2a} = \frac{3x10^8}{2\sqrt{2.6}x2x10^{-2}} = 4.651 \text{ GHz}$$

$$f_{c11} = \frac{u'}{2}\left[\frac{1}{a^2} + \frac{1}{b^2}\right]^{1/2} = 10.4 \text{ GHz}$$

$$\eta' = \sqrt{\frac{\mu}{\varepsilon}} = \frac{377}{\sqrt{2.6}} = 233.81\Omega$$

(a) For TE_{10} mode, eq.(12.57) gives

$$\alpha_d + j\beta_d = \sqrt{-\omega^2 \mu\varepsilon + k_x^2 + k_x^2 + j\omega\mu\sigma_d}$$

$$= \sqrt{-\omega^2 / u^2 + \frac{\pi^2}{a^2} + j\omega\mu\sigma_d}$$

$$= \sqrt{-\left(\frac{2\pi x 12x10^9}{3x10^8}\right)^2 (2.6) + \frac{\pi^2}{(2x10^{-2})^2} + j2\pi x 12x10^9 x 4\pi x 10^{-7} x 10^{-4}}$$

$$= 0.012682 + j373.57$$

$\alpha_d = 0.012682$ Np/m

$$\alpha_c = \frac{2R_s}{b\eta'\sqrt{1-(f_c/f)^2}}\left[\frac{1}{2} + \frac{b}{a}\left(\frac{f_c}{f}\right)^2\right]$$

$$= \frac{2x2.858x10^{-2}}{10^{-2}(233.81)\sqrt{1-(4.651/12)^2}}\left[\frac{1}{2} + \frac{1}{2}\left(\frac{4.651}{12}\right)^2\right] = \underline{0.1525} \text{ Np/m}$$

(b) For TE_{11} mode,

$$\alpha_d + j\beta_d = \sqrt{-\omega^2 / u^2 + 1/a^2 + 1/b^2 + j\omega\mu\sigma_d}$$

$$= \sqrt{-139556.21 + \frac{\pi^2}{(10^{-2})^2} + j9.4748} = 0.02344 + j202.14$$

$\alpha_d = 0.02344$ Np/m

$$\alpha_c = \frac{2R_s}{b\eta'\sqrt{1-(f_c/f)^2}}\left[\frac{(b/a)^3+1}{(b/a)^2+1}\right] = \frac{2x2.858x10^{-2}}{10^{-2}(233.81)\sqrt{1-(10.4/12)^2}}\left[\frac{(1/8)+1}{(1/4)\ +1}\right]$$

$\alpha_c = 0.0441$ Np/m

Prob. 12.21 $\varepsilon_c = \varepsilon' - j\varepsilon'' = \varepsilon - j\dfrac{\sigma}{\omega}$

Comparing this with

$$\varepsilon_c = 16\varepsilon_o(1 - j10^{-4}) = 16\varepsilon_o - j16\varepsilon_o x10^{-4}$$

$$\varepsilon\ = 16\varepsilon_o, \qquad \frac{\sigma}{\omega} = 16\varepsilon_o x10^{-4}$$

For TM_{21} mode,

$$f_c = \frac{u'}{2}\left[\frac{m^2}{a^2} + \frac{n^2}{b^2}\right]^{1/2} = 4.193 \text{ GHz}, \quad f = 1.1f_c = 4.6123 \text{ GHz}$$

$$\sigma = 16\varepsilon_o\omega x10^{-4} = 16x2\pi x4.6123x10^9 x\frac{10^{-9}}{36\pi}x10^{-4} = 4.1x10^{-4}$$

$$\eta' = \sqrt{\frac{\mu}{\varepsilon}} = 30\pi$$

$$\alpha_d = \frac{\sigma\eta'}{2\sqrt{1-(f_c/f)^2}} = \frac{4.1x10^{-4}x30\pi}{2\sqrt{1-1/1.12}} = \underline{\underline{0.04637}} \text{ Np/m}$$

$$E_o e^{-\alpha_d z} = 0.8E_o \longrightarrow z = \frac{1}{\alpha_d}\ln(1/0.8) = \underline{\underline{4.811}} \text{ cm}$$

Prob. 12.22 For TM_{21} mode,

$$\alpha_c = \frac{2R_s}{b\eta'\sqrt{1-(f_c/f)^2}}$$

$$R_s = \frac{1}{\sigma_c \delta} = \sqrt{\frac{\pi f \mu}{\sigma_c}} = \sqrt{\frac{\pi x 4.6123 x 10^9 x 4\pi x 10^{-7}}{1.5 x 10^7}} = 3.484 x 10^{-2}$$

$$\alpha_c = \frac{2x3.48x10^{-2}}{4\pi x 10^{-2} x 30\pi x 0.4166} = 0.04406 \text{ Np/m}$$

$$E_o e^{-\alpha_c z} = 0.7 E_o \longrightarrow z = \frac{1}{\alpha_c} \ln(1/0.7) = \underline{\underline{8.097}} \text{ m}$$

Prob. 12.23 For TE_{10} mode,

$$f_c = \frac{u'}{2a} = \frac{3x10^8}{2\sqrt{2.11}x4.8x10^{-2}} = 2.151$$

(a) loss tangent $= \dfrac{\sigma}{\omega\varepsilon} = d$

$$\sigma = d\omega\varepsilon = 3x10^{-4} x 2\pi x 4x10^9 x 2.11x \frac{10^{-9}}{36\pi} = 1.407x10^{-4}$$

$$\eta' = \frac{120\pi}{\sqrt{2.11}} = 259.53$$

$$\alpha_d = \frac{\sigma\eta'}{2\sqrt{1-(f_c/f)^2}} = \frac{1.4067x10^{-4}x259.53}{2\sqrt{1-(2.151/4)^2}} = \underline{\underline{2.165x10^{-2} \text{ Np/m}}}$$

(b) $$R_s = \sqrt{\frac{\mu f \pi}{\sigma_c}} = \sqrt{\frac{\pi x 4x10^9 x 4\pi x 10^{-7}}{4.1x10^7}} = 1.9625x10^{-2}$$

$$\alpha_c = \frac{2R_s}{b\eta'\sqrt{1-(f_c/f)^2}}\left[\frac{1}{2} + \frac{b}{a}(f_c/f)^2\right] = \frac{3.925x10^{-2}(0.5 + 0.5x0.2892)}{2.4x10^{-2}x259.53x0.8431}$$

$$= \underline{\underline{4.818x10^{-3} \text{ Np/m}}}$$

Prob. 12.24 (a) For TE_{10} mode,

$$f_c = \frac{u'}{2a}, \quad u' = \frac{c}{\sqrt{2.11}}$$

$$f_c = \frac{3x10^8}{\sqrt{2.11}(2x2.25x10^{-2})} = \underline{\underline{4.589 \text{ GHz}}}$$

(b) $$\alpha_{cTE10} = \frac{2R_s}{b\eta'\sqrt{1-(f_c/f)^2}}\left[\frac{1}{2}+\frac{b}{a}(f_c/f)^2\right]$$

$$R_s = \sqrt{\frac{\pi f \mu}{\sigma_c}} = \sqrt{\frac{\pi x 5x10^9 x 4\pi x 10^{-7}}{1.37x10^7}} = 3.796x10^{-3}$$

$$\eta' = \frac{377}{\sqrt{2.11}} = 259.54$$

$$\alpha_c = \frac{2x3.796x10^{-2}[0.5+\frac{1.5}{2.25}(4.589/5)^2]}{1.5x10^{-4}(259.54)\sqrt{1-(4.589/5)^2}} = \underline{\underline{0.05217}} \text{ Np/m}$$

Prob. 12.25 For TE_{10} mode,

$$\alpha_c = \frac{2R_s}{b\eta'\sqrt{1-(f_c/f)^2}}\left[\frac{1}{2}+\frac{b}{a}(\frac{f_c}{f})^2\right]$$

But a = b, $R_s = \frac{1}{\sigma_c \delta} = \sqrt{\frac{\pi f \mu}{\sigma_c}}$

$$\alpha_c = \frac{2\sqrt{\frac{\pi f \mu}{\sigma_c}}}{a\eta'\sqrt{1-(f_c/f)^2}}\left[\frac{1}{2}+(\frac{f_c}{f})^2\right] = \frac{k\sqrt{f}\left[\frac{1}{2}+(\frac{f_c}{f})^2\right]}{\sqrt{1-(f_c/f)^2}}$$

where k is a constant.

$$\frac{d\alpha_c}{df} = \frac{k[1-(\frac{f_c}{f})^2]^{1/2}[\frac{1}{4}f^{-1/2}-\frac{3}{2}f_c^2 f^{-5/2}] - \frac{k}{2}[\frac{1}{2}f^{1/2}+f_c^2 f^{-3/2}](2f_c^2 f^{-3})[1-(\frac{f_c}{f})^2]^{-1/2}}{1-(f_c/f)^2}$$

For minimum value, $\frac{d\alpha_c}{df} = 0$. This leads to $\underline{\underline{f = 2.962\, f_c}}$.

Prob. 12.26

$$\alpha = k\sqrt{\frac{f}{1-(f_c/f)^2}}$$, where k is a constant

$$\alpha = k\frac{f^{3/2}}{\sqrt{f^2 - f_c^2}}$$

$$\frac{d\alpha}{df} = k\frac{\sqrt{f^2 - f_c^2}\,\frac{3}{2}f^{1/2} - f^{3/2}\frac{1}{2}2f\frac{1}{\sqrt{f^2 - f_c^2}}}{f^2 - f_c^2}$$

For maximum α, $\frac{d\alpha}{df} = 0$ which implies that

$$(f^2 - f_c^2).\frac{3}{2}f^{1/2} - f^{5/2} = 0$$

or

$$\underline{\underline{f = \sqrt{3}f_c}}$$

Prob. 12.27 For the TE mode to z,

$$E_{zs} = 0, H_{zs} = H_o\cos(m\pi x/a)\cos(n\pi y/b)\sin(p\pi z/c)$$

$$E_{ys} = -\frac{\gamma}{h^2}\frac{\partial E_{zs}}{\partial y} + \frac{j\omega\mu}{h^2}\frac{\partial H_{zs}}{\partial x} = \underline{\underline{-\frac{j\omega\mu}{h^2}(m\pi/a)H_o\sin(m\pi x/a)\cos(n\pi y/b)\sin(p\pi z/c)}}$$

as required.

$$E_{xs} = -\frac{\gamma}{h^2}\frac{\partial E_{zs}}{\partial x} - \frac{j\omega\mu}{h^2}\frac{\partial H_{zs}}{\partial y} = \frac{j\omega\mu}{h^2}(n\pi/b)H_o\cos(m\pi x/a)\sin(n\pi y/b)\sin(p\pi z/c)$$

From Maxwell's equation,

$$-j\omega\mu\mathrm{H_s} = \nabla x\mathrm{E_s} = \begin{vmatrix} \frac{\partial}{\partial x} & \frac{\partial}{\partial y} & \frac{\partial}{\partial z} \\ E_{xs} & E_{ys} & 0 \end{vmatrix}$$

$$H_{xs} = \frac{1}{j\omega\mu}\frac{\partial E_{ys}}{\partial z} = \underline{\underline{-\frac{1}{h^2}(m\pi/a)(p\pi/c)H_o\sin(m\pi x/a)\cos(n\pi y/b)\cos(p\pi z/c)}}$$

Prob. 12.28 Maxwell's equation can be written as

$$H_{xs} = \frac{j\omega\varepsilon}{h^2}\frac{\partial E_{zs}}{\partial y} - \frac{\gamma}{h^2}\frac{\partial H_{zs}}{\partial x}$$

For a rectangular cavity,

$$h^2 = k_x^2 + k_y^2 = (m\pi / a)^2 + (n\pi / b)^2$$

For TM mode, $H_{zs} = 0$ and

$$E_{zs} = E_o \sin(m\pi x / a)\sin(n\pi y / b)\cos(p\pi z / c)$$

Thus

$$H_{xs} = \frac{j\omega\varepsilon}{h^2}\frac{\partial E_{zs}}{\partial y} = \underline{\underline{\frac{j\omega\varepsilon}{h^2}(n\pi / b)E_o \sin(m\pi x / a)\cos(n\pi y / b)\sin(p\pi z / c)}}$$

as required.

$$H_{xs} = -\frac{j\omega\varepsilon}{h^2}\frac{\partial E_{zs}}{\partial x} - \frac{\gamma}{h^2}\frac{\partial H_{zs}}{\partial y}$$

$$= -\frac{j\omega\varepsilon}{h^2}(m\pi / a)E_o \cos(m\pi x / a)\sin(n\pi y / b)\cos(p\pi z / c)$$

From Maxwell's equation,

$$j\omega\varepsilon \mathrm{E_s} = \nabla x \mathrm{H_s} = \begin{vmatrix} \frac{\partial}{\partial x} & \frac{\partial}{\partial y} & \frac{\partial}{\partial z} \\ H_{xs} & H_{ys} & 0 \end{vmatrix}$$

$$E_{ys} = \frac{1}{j\omega\varepsilon}\frac{\partial H_{xs}}{\partial z} = \underline{\underline{-\frac{1}{h^2}(n\pi / b)(p\pi / c)E_o \sin(m\pi x / a)\cos(n\pi y / b)\sin(p\pi z / c)}}$$

Prob. 12.29

$$f_r = \frac{u'}{2}\sqrt{(m/a)^2 + (n/b)^2 + (p/c)^2}$$

where for TM mode to z, m = 1, 2, 3,…, n=1, 2, 3, …., p = 0, 1, 2, ….

and for TE mode to z, m = 1, 2, 3,…, n=1, 2, 3, …., p = 1, 2, 3, ….

(a) If a < b < c, 1/a > 1/b > 1/c,

The lowest TM mode is TM_{110} with $f_r = \frac{u'}{2}\sqrt{\frac{1}{a^2} + \frac{1}{b^2}}$

The lowest TE mode is TE_{011} with $f_r = \frac{u'}{2}\sqrt{\frac{1}{b^2} + \frac{1}{c^2}} < \frac{u'}{2}\sqrt{\frac{1}{a^2} + \frac{1}{b^2}}$

Hence the dominant mode is TE_{011}.

(b) If a > b > c, 1/a < 1/b < 1/c,

The lowest TM mode is TM_{110} with $f_r = \frac{u'}{2}\sqrt{\frac{1}{a^2}+\frac{1}{b^2}}$

The lowest TE mode is TE_{101} with $f_r = \frac{u'}{2}\sqrt{\frac{1}{a^2}+\frac{1}{c^2}} < \frac{u'}{2}\sqrt{\frac{1}{a^2}+\frac{1}{b^2}}$

Hence the dominant mode is TM_{110}.

(c) If a = c > 1/b, 1/a = 1/c < 1/b,

The lowest TM mode is TM_{110} with $f_r = \frac{u'}{2}\sqrt{\frac{1}{a^2}+\frac{1}{b^2}}$

The lowest TE mode is TE_{101} with $f_r = \frac{u'}{2}\sqrt{\frac{1}{a^2}+\frac{1}{c^2}} < \frac{u'}{2}\sqrt{\frac{1}{a^2}+\frac{1}{b^2}}$

Hence the dominant mode is TE_{101}.

Prob. 12.30

$f_r = 1.5x10^{10}\sqrt{(m/3)^2+(n/2)^2+(p/4)^2}$ Hz

$f_{rTE011} = 15\sqrt{0+1/4+1/16} = 8.385$ GHz, etc.

The resonant frequencies are listed below.

Modes	Resonant frequencies (GHz)
TE_{101}	6.25
TE_{011}	8.38
TM_{110}	9.01
TM_{111}	9.76

Prob. 12.31 b = 2a, c = 3a

$$f_r = \frac{u'}{2}\sqrt{(m/a)^2+(n/2a)^2+(p/3a)^2}, u' = \frac{c}{\sqrt{2.5}}$$

$$\frac{u'}{2a} = \frac{3x10^8}{2\sqrt{2.5}x3x10^{-2}} = 3.162x10^9$$

$$f_r = 3.162\sqrt{m^2 + n^2/4 + p^2/9} \text{ GHz}$$

Mode	f_r (GHz)
011	1.9
110	3.535
101	3.333
102	3.8
120, 103	4.472
022	3.8

Thus the lowest five modes have resonant frequencies at

1.9, 3.333, 3.535, 3.8, and 4.472 GHz

Prob. 12.32

$$f_r = \frac{u'}{2}\sqrt{1/a^2 + 1/c^2}$$

For cubical cavity, a = b = c

$$f_r = \frac{u'}{2a}\sqrt{2} \longrightarrow a = \frac{u'}{\sqrt{2}f_r} = \frac{3x10^8}{\sqrt{2}x2x10^9} = 10.61 \text{ mm}$$

a = b = c = 1.061 cm

Prob. 12.33 (a)

$$f_r = \frac{u'}{2}\sqrt{(m/a)^2 + (n/b)^2 + (p/c)^2}$$

a = b = c = 3.2 cm, m=1, n=0, p=1, u' = c

$$f_r = \frac{3x10^8}{2x3.2x10^{-2}}\sqrt{1^2 + 0^2 + 1^2} = 6.629 \text{ GHz}$$

(b)

$$Q = \frac{a}{3}\sqrt{\pi f_{r101}\mu_o\sigma_c} = \frac{3.2x10^{-2}}{3}\sqrt{\pi x6.629x10^9 x4\pi x10^{-7} x1.37x10^7}$$

$$= 6.387$$

Prob. 12.34

$$f_r = \frac{c}{2a}\sqrt{m^2 + n^2 + p^2}$$

The lowest possible modes are TE_{101}, TE_{011}, and TM_{110}. Hence

$$f_r = \frac{c}{2a}\sqrt{2} \longrightarrow \quad a = \frac{c}{f_r\sqrt{2}} = \frac{3x10^8}{\sqrt{2}x3x10^9} = 7.071 \text{ cm}$$

a = b = c = 7.071 cm

Prob. 12.35 This is a TM mode to z. From Maxwell's equations,

$$\nabla x E_s = -j\omega\mu H_s$$

$$H_s = -\frac{1}{j\omega\mu}\nabla x E_s = \frac{j}{\omega\mu}\begin{vmatrix} \frac{\partial}{\partial x} & \frac{\partial}{\partial y} & \frac{\partial}{\partial z} \\ 0 & 0 & E_{zs}(x,y) \end{vmatrix} = \frac{j}{\omega\mu}\left(\frac{\partial E_{zs}}{\partial y}a_x - \frac{\partial E_{zs}}{\partial x}a_y\right)$$

But

$$E_{zs} = 200\sin 30\pi x \sin 30\pi y, \frac{1}{\omega\mu} = \frac{1}{6x10^9 x4\pi x10^{-7}} = \frac{10^{-2}}{24\pi}$$

$$H_s = \frac{j10^{-2}}{24\pi}x200x30\pi\left\{\sin 30\pi x\cos 30\pi y a_x - \cos 30\pi x \sin 30\pi y a_y\right\}$$

$$\mathbf{H} = \text{Re}\,(\mathbf{H}_s\, e^{j\omega t})$$

$$H = 2.5\left\{-\sin 30\pi x\cos 30\pi y a_x + \cos 30\pi x\sin 30\pi y a_y\right\}\sin 6x10^9\pi t \quad \text{A/m}$$

CHAPTER 13

P. E. 13.1

$$\mathbf{r}_{max} = \frac{2\mathbf{d}^2}{\lambda} = \frac{2\left(\lambda/100\right)^2}{\lambda} = \frac{\lambda}{5{,}000} \Longrightarrow \mathbf{r} = \frac{\lambda}{5} \text{ is in far field}$$

(a) $H_{\phi s} = \dfrac{jI_o\beta\partial l\sin\theta\, e^{j\beta r}}{4\pi r}$, $\beta\mathbf{r} = \dfrac{2\pi}{\lambda}.\dfrac{\lambda}{5} = 72^o$

$$\lambda = \frac{2\pi c}{\omega} = \frac{2\pi \times 3\times 10^8}{10^8} = 6\pi$$

$$H_{\phi s} = \frac{j(0.25)(\frac{2\pi}{\lambda})\frac{\lambda}{100}\sin 30^o e^{-j72^o}}{4\pi(6\pi/5)} = 0.1652e^{j18^o} \text{ mA/m}$$

$H = \text{Im}\left(H_{\phi s}e^{j\omega t}{}_{a_\phi}\right)$ Im is used since $I = I_o \sin wt$

$$= \underline{\underline{0.1628\sin(10^8 + 18^o)a_\phi}} \text{ mA/m}$$

(b) $\beta = \dfrac{2\pi}{\lambda}.200\lambda = 0^o$

$$H_{\phi s} = \frac{j(0.25)(\frac{2\pi}{\lambda})\frac{\lambda}{100}Sin60^o e^{-j0^o}}{4\pi(6\pi\times 200)} = 0.2871e^{j90^o} \quad \mu Am$$

$$H = \text{Im}\,(H_{\phi s}a_\phi e^{j\omega t}) = \underline{\underline{0.2671\sin(10^8 + 90^o)a_\phi}} \quad \mu Am.$$

P. E. 13.2

(a) $l = \dfrac{\lambda}{4} = \underline{\underline{1.5m}}$,

(b) $I_o = \underline{\underline{83.3\text{mA}}}$

(c) $P_{rad} = 36.56\,\lambda$, $P_{rad} = \dfrac{1}{2}(0.0833)^2\, 36.56$

$= \underline{\underline{126.8 \text{ mW}}}.$

(d) $Z_L = 36.5 + j21.25$,

$$\Gamma = \frac{36.5 + j21.25 - 75}{36.5 + j21.25 + 75} = 0.3874\angle 140.3^o$$

$$S = \frac{1 + 0.3874}{1 - 0.3874} = \underline{\underline{2.265}}$$

P. E. 13.3

$$D = \frac{4\pi U_{max}}{P_{rad}}$$

(a) For the Hertzian monopole

$$U(\theta,\phi) = \sin^2\theta \text{ , } 0 < \theta < \frac{\pi}{2}, \quad 0 < \phi < 2\pi \text{ , } U_{max} = 1$$

$$P_{rad} = \int_{\theta=0}^{\pi/2}\int_{\phi=0}^{2\pi} \sin^2 \sin\theta \, d\theta \, d\phi = \frac{4\pi}{3}$$

$$D = \frac{4\pi .1}{4\pi/3} = \underline{\underline{3}}$$

(b) For the $\frac{\lambda}{4}$ monopole,

$$U(\theta,\phi) = \frac{\cos^2(\frac{\pi}{2}\cos\theta)}{\sin^2\theta} \text{ , } U_{max} = 1$$

$$P_{rad} = \int_{\theta=0}^{\pi/2}\int_{\phi=0}^{2\pi} \frac{\cos^2(\frac{\pi}{2}\cos\theta)}{\sin^2\theta} \sin\theta \, d\theta \, d\phi = 2\pi(0.609)$$

$$D = \frac{4\pi(1)}{2\pi(0.609)} = \underline{\underline{3.28}}$$

P. E. 13.4

(a) $P_{rad} = \eta_r P_{in} = 0.95(0.4)$

$$D = \frac{4\pi U_{max}}{P_{rad}} = \frac{4\pi(0.5)}{0.4 \times 0.95} = \underline{\underline{16.53}}$$

(b) $D = \frac{4\pi(0.5)}{0.3} = \underline{\underline{20.94}}$

P. E. 13. 5

$$P_{rad} = \int_{\theta=0}^{\pi/2} \int_{\phi=0}^{2\pi} \sin\theta \sin\theta \, d\theta \, d\phi = \frac{\pi^2}{2}, \quad U_{max} = 1$$

$$D = \frac{4\pi(1)}{\pi^2/2} = \underline{\underline{2.546}}$$

P. E. 13. 6

(a) $$f(\theta) = |\cos\theta| \cos\left[\frac{1}{2}(\beta d \cos\theta + \alpha)\right]$$

where $\alpha = \pi$, $\beta d = \frac{2\pi}{\lambda} \cdot \frac{\lambda}{2} = \pi$

$$f(\theta) = |\cos\theta| \cos\left[\frac{1}{2}(\pi \cos\theta + \pi)\right]$$

$$\Downarrow \qquad\qquad \Downarrow$$

unit pattern group pattern

For the group pattern, we have nulls at

$$\pi/2(\cos\theta + 1) = \pi/2 \longrightarrow \theta = \pi/2$$

and maxima at

$$\pi/2(\cos\theta + 1) = 0 \longrightarrow \cos\theta = -1$$

Thus the group pattern and the resultant patterns are as shown in Fig.13.15(a)

(b) $$f(\theta) = |\cos\theta| \cos\left[\frac{1}{2}(\beta d \cos\theta + \alpha)\right]$$

where $\alpha = \pi/2$, $\beta d = \pi$

$$f(\theta) = |\cos\theta| \cos\left[\frac{1}{2}\left(\pi \cos\theta - \pi/2\right)\right]$$

$$\Downarrow \qquad\qquad \Downarrow$$

unit pattern group pattern

For the group pattern, the nulls are at

$$\pi/4(\cos\theta - 1) = \pm \pi/2, \pm 3\pi/2 \longrightarrow \theta = 180^o$$

and maxima at

$\cos\theta - 1 = 0 \longrightarrow \theta = 0$

Thus the group pattern and the resultant patterns are as shown in Fig.13.15(b)

P. E. 13.7

(a)

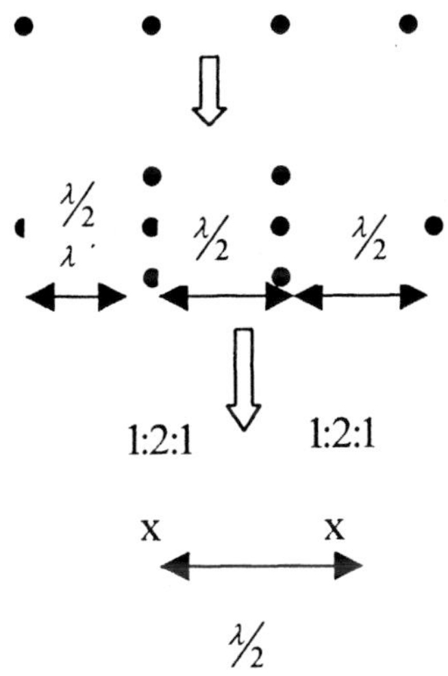

Thus, we take a pair at a time and multiply the patterns as shown below.

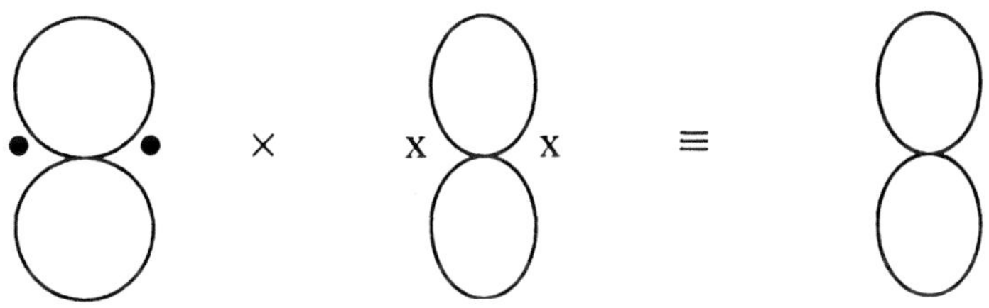

The group pattern is the normalized array factor, i.e.

$$(AF)_n = \frac{1}{\sum}\left|1 + Ne^{i\psi} + \frac{N(N-1)}{2!}e^{i2\psi} + \frac{N(N-1)(N-2)}{3!}e^{i3\psi} + \ldots\ldots\ldots\ldots + e^{i(N-1)\psi}\right|$$

where $\sum = \sum_{i-1}^{N-1}\binom{N}{i} = 1 + N + \frac{N-1}{2!} + \frac{N(N-1)(N-2)}{3!} + \ldots\ldots\ldots$

$$= (1+1)^{N-1} = 2^{N-1}$$

$$(AF)_n = \frac{1}{2^{N-1}}\left|1 + e^{j\psi}\right|^{N-1} = \frac{1}{2^{N-1}}\left|e^{j\psi/2}\right|\left|e^{-j\psi/2} + e^{j\psi/2}\right|^{N-1}$$

$$= \frac{1}{2^{N-1}}\left|2\cos\frac{\psi}{2}\right|^{N-1} = \underline{\underline{\left|\cos\frac{\psi}{2}\right|^{N-1}}}$$

P. E. 13.8

$$A_e = \frac{\lambda^2}{4\pi}G_d, \quad \lambda = \frac{c}{f} = \frac{3\times 10^8}{10^8} = 3m$$

For the Hertzian dipole,

$$G_d = 1.5\sin^2\theta$$

$$A_e = \frac{\lambda^2}{4\pi}(1.5\sin^2\theta)$$

$$A_{e,\max} = \frac{1.5\lambda^2}{4\pi} = \frac{1.5\times 9}{4\pi} = \underline{\underline{1.074\ \text{m}^2}}$$

By definition,

$$P_r = AeP_{ave} \longrightarrow P_{ave} = \frac{P_r}{A_e} = \frac{3\times 10^{-6}}{1.074}$$

$$= \underline{\underline{2.793\ \mu\text{W/m}^2}}$$

P. E. 13.9

(a) $$G_d = \frac{4\pi r^2 P_{ave}}{P_{rad}} = \frac{4\pi r^2 \frac{1}{2}\frac{E^2}{\eta}}{P_{rad}} = \frac{2\pi r^2 E^2}{\eta P_{rad}}$$

$$= \frac{2\pi\times 400\times 10^6\times 144\times 10^{-6}}{120\pi\times 100\times 10^3} = 2.16$$

$$G = 10\log_{10} G_d = \underline{\underline{3.34\ \text{dB}}}$$

(b) $$G = \eta_r G_d = 0.98\times 2.16 = \underline{\underline{2.117}}$$

P. E. 13.10

$$r = \left[\frac{\lambda^2 G_d{}^2\sigma}{(4\pi)^3}\frac{P_{rad}}{P_r}\right]^{1/4}$$

where $\lambda = \frac{c}{f} = \frac{3\times 10^8}{6\times 10^9} = 0.05m$

$$A_e = 0.7\pi a^2 = 0.7\pi(1.8)^2 = 7.125m^2$$

$$G_d = \frac{4\pi A_e}{\lambda^2} = \frac{4\pi(7.125)}{25\times 10^{-4}} = 3.581\times 10^4$$

$$r = \left[\frac{5 \times 10^{-4} \times (3.581)^2 \times 10^4 \times 5 \times 60 \times 10^3}{(4\pi)^3 \times 0.26 \times 10^{-3}} \right]^{1/4}$$

$$= 1270 \text{ m} = 0.857 \text{ nm}$$

At $r = {r_{\max}}/{2} = 635m,$

$$P = \frac{G_d P_{rad}}{4\pi r^2} = \frac{3.581 \times 10^4 \times 60 \times 10^3}{4\pi (635)^2} = \underline{\underline{42.4 \text{ W/m}^2}}$$

Prob. 13.1

Using vector transformation,

$$A_{rs} = A_{xs} \sin\theta \cos\phi \ , \ A_{\theta s} = A_{xs} \cos\theta \cos\phi \ , \ A_{\phi s} = A_{xs} \sin\phi$$

$$A_s = \frac{50e^{-j\beta r}}{r} (\sin\theta \cos\phi \, a_r + \cos\theta \cos\phi \, a_\theta - \sin\phi \, a_\phi)$$

$$\frac{\nabla \times A_s}{\mu} = H_s = \frac{-100 \cos\theta \sin\phi}{\mu r^2 \sin\theta} e^{-j\beta r} a_r - \frac{50}{\mu r^2} (\sin\theta + j\beta r) \sin\phi \, e^{-j\beta r} a_\theta$$

$$- \frac{50}{\mu r^2} \cos\theta \cos\phi (1 + j\beta r) e^{-j\beta r} a_\phi$$

At far field, only $\frac{1}{r}$ term remains. Hence

$$H_s = \frac{-j50}{\mu r} \beta e^{-j\beta r} (\sin\phi \, a_\theta + \cos\theta \cos\phi \, a_\phi)$$

$$E_s = -\eta \, a_r \times H_s = \frac{-j50\beta \eta e^{-j\beta r}}{\mu r} (\sin\phi \, a_\phi - \cos\theta \cos\phi \, a_\theta)$$

$$H = \text{Re}\left[H_s e^{jwt} \right] = \frac{-50}{\mu r} \beta \sin(\omega t - \beta r)(\sin\phi \, a_\theta + \cos\theta \cos\phi \, a_\phi)$$

$$E = \text{Re}\left[E_s e^{jwt} \right] = \frac{-50\eta \beta}{\mu r} \sin(\omega t - \beta r)(- \sin\phi \, a_\phi + \cos\theta \cos\phi \, a_\theta)$$

Prob. 13.2

$$\lambda = \frac{c}{f} = \frac{3\times 10^8}{10^7} = 30 \text{ m}$$

$$|E_{\theta s}| = \frac{\eta I_o \beta dl}{4\pi r}\sin\theta$$

At (100,0,0), r = 100m, $\theta = \frac{\pi}{2}$

$$|E_{\theta s}| = \frac{120(10)}{4\pi(100)}\frac{2\pi}{30}(0.2)(1) = \underline{\underline{0.04}} \text{ V/m}$$

Prob. 13.3

$$\lambda = \frac{c}{f} = \frac{3\times 10^8}{3\times 10^8} = 1m \qquad \beta = \frac{2\pi}{\lambda} = 2\pi$$

r = 10, $\theta = 30^o$, $\phi = 90^o$

$$H_{\phi s} = \frac{j(2)(2\pi)5\times 10^{-3}}{4\pi}\sin 30 e^{-j2\pi} = \underline{\underline{j0.25}} \text{ mA/m}$$

$$\eta = 120\pi = 377$$

$$E_{\theta s} = \eta H_{\phi s} = \underline{\underline{94.25}}\,\text{mV/m}$$

Prob. 13.4

(a) $$A_{zs} = \frac{e^{-j\beta r}}{4\pi r}\int_{-l/2}^{l/2} I_o(1-\frac{2|z|}{l})e^{j\beta z\cos\theta}\partial z$$

$$= \frac{e^{-j\beta r}}{4\pi r}I_o\left[\int_{-l/2}^{l/2}(1-\frac{2|z|}{l})e^{j\beta z\cos\theta}dz + j\int_{-l/2}^{l/2}(1-\frac{2|z|}{l})\sin(\beta z\cos\theta)dz\right]$$

$$= \frac{e^{-j\beta r}}{4\pi r}2I_o\int_0^{l/2}(1-\frac{2z}{l})\cos(\beta z\cos\theta)dz$$

$$= \frac{I_o e^{-j\beta r}}{2\pi r\beta^2\cos^2\theta}\cdot\frac{2}{l}\left[1-\cos(\beta l/2\,\cos\theta)\right]$$

$$E_s = -jw\mu A_s \longrightarrow E_{\theta s} = jw\mu\sin\theta A_{zs} = j\beta\eta\sin\theta A_{zs}$$

$$E_{\theta s} = \frac{j\eta I_o e^{-j\beta r}}{\pi r l} \frac{\sin\theta\left[1-\cos(\beta l/2 \cos\theta)\right]}{\beta \cos^2\theta}$$

If $\beta l/2 \ll 1$, $\cos(\beta l/2 \cos\theta) = 1 - \frac{(\frac{\beta l}{2}\cos\theta)^2}{2!}$. Hence

$$E_{\theta s} = \frac{j\eta I_o}{8\pi r} \beta l e^{-j\beta r} \sin\theta \ , \quad H_{\phi s} = \eta E_{\theta s}$$

$$P_{ave} = \frac{|E_{\theta s}|^2}{2\eta}, \qquad P_{rad} = \int P_{ave} dS$$

$$P_{rad} = \int_0^{2\pi}\int_0^{\pi} \frac{n}{2}\left(\frac{I_o \beta l}{8\pi}\right)^2 \frac{1}{r^2} \sin^2\theta r^2 \sin\theta d\theta d\phi$$

$$= 10\pi^2 I_o^2\left(\frac{l}{\lambda}\right)^2 = \frac{1}{2} I_o^2 R_{rad}$$

or $R_{rad} = 20\pi^2\left(\frac{l}{\lambda}\right)^2$

(b) $0.5 = 20\pi^2\left(\frac{l}{\lambda}\right)^2 \longrightarrow l = 0.05\lambda$

Prob. 13.5

$$\partial l = 5m, \quad \lambda = \frac{c}{f} = \frac{3\times10^8}{3\times10^6} = 100$$

$$\frac{\partial l}{\lambda} = \frac{5}{100} = \frac{1}{20} < \frac{1}{10}$$

$$R_{rad} = 80\pi^2\left(\frac{\partial l}{\lambda}\right)^2 = \frac{80\pi^2}{400} = \underline{\underline{1.974\Omega}}$$

Prob. 13.6

$$Z_{in} = 73 + j42.5$$

$$\Gamma = \frac{Z_{in} - Z_o}{Z_{in} + Z_o} = \frac{23 + j42.5}{123 + j42.5} = \underline{\underline{0.3713\angle 42.52^o}}$$

$$s = \frac{1+|r|}{1-|r|} = \frac{1.3713}{1-0.3713} = \underline{\underline{2.181}}$$

Prob. 13.7

This is a monopole antenna

$$\lambda = \frac{c}{f} = \frac{3\times 10^8}{1.5\times 10^6} = 200$$

$l \langle\langle \lambda$,hence it is a Hertzian monopole.

$$R_{rad} = \frac{1}{2} 80\pi^2 \left(\frac{dl}{\lambda}\right)^2 = 40\pi^2 \left(\frac{1}{200}\right)^2 = 9.87 \text{ m}\Omega$$

$$P_{rad} = P_t = \frac{1}{2} I_o R_{rad}$$

$$I_o^2 = \frac{2P_t}{R_{rad}} = \frac{8}{9.87\times 10^{-3}} = 810.54$$

$$\underline{\underline{I_o = 28.47\, A}}$$

Prob. 13.8

Change the limits in Eq. (13.16) to $\pm \frac{l}{2}$ i.e.

$$A_s = \frac{\mu I_o e^{j\beta z Cos\theta}}{4\pi r} \frac{(j\beta \cos\theta \cos\beta t + \beta \sin\beta t)}{-\beta^2 \cos^2\theta + \beta^2} \Bigg|_{-l/2}^{l/2}$$

$$= \frac{\mu I_o e^{j\beta r}}{2\pi r} \frac{1}{\beta \sin^2\theta} \left[\sin\frac{\beta l}{2} \cos\left(\frac{\beta l}{2}\cos\theta\right) - \cos\theta \cos\frac{\beta l}{2} \sin\left(\frac{\beta l}{2}\cos\theta\right) \right]$$

But $B = \mu H = \nabla \times A$

$$H_{\phi s} = \frac{1}{\mu r}\left[\frac{\partial}{\partial r}(rA_\theta) - \frac{\partial A_r}{\partial\theta}\right],$$

where $A_o = -A_z \sin\theta, A_r = A_z \cos\theta$

$$H_{\phi s} = \frac{I_o}{2\pi r}\frac{e^{-j\beta r}}{\beta}\left(\frac{j\beta}{\sin\theta}\right)\left[\sin\frac{\beta l}{2} \cos\left(\frac{\beta l}{2}\cos\theta\right) - \cos\theta \cos\frac{\beta l}{2} \sin\left(\frac{\beta l}{2}\cos\theta\right) \right] + \frac{I_o}{2\pi r^2} e^{-j\beta r}(......)$$

For far field, only the $\frac{1}{r}$-term remains. Hence

$$H_{\phi s} = \frac{jI_o}{2\pi r} e^{-j\beta r} \frac{\left[\sin\frac{\beta l}{2} \cos\left(\frac{\beta l}{2}\cos\theta\right) - \cos\theta \cos\frac{\beta l}{2} \sin\left(\frac{\beta l}{2}\cos\theta\right) \right]}{\sin\theta}$$

(b) $f(\theta) = \dfrac{\cos\left(\frac{\beta l}{2}\cos\theta\right) - \cos\frac{\beta l}{2}}{\sin\theta}$

For $l = \lambda, f(\theta) = \dfrac{\cos(\pi\cos\theta) + 1}{\sin\theta}$

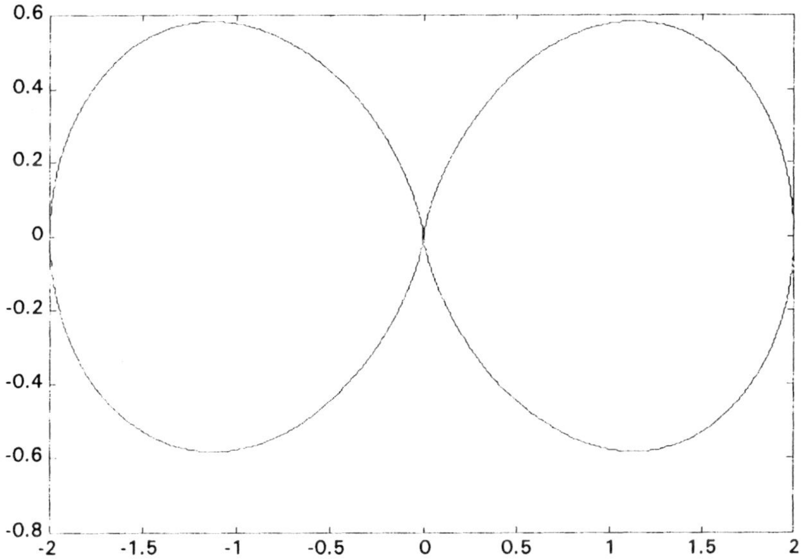

For $l = \frac{3\lambda}{2}$, $f(\theta) = \frac{\cos\left(\frac{3\pi}{2}\cos\theta\right)}{\sin\theta}$

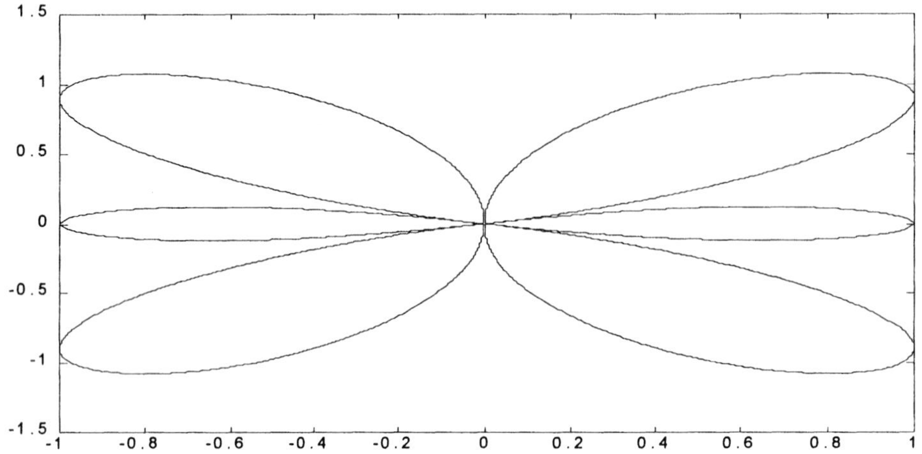

For $l = 2\lambda$, $f(\theta) = \frac{\cos(2\pi\cos\theta) - 1}{\sin\theta}$

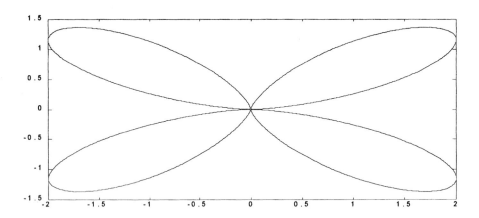

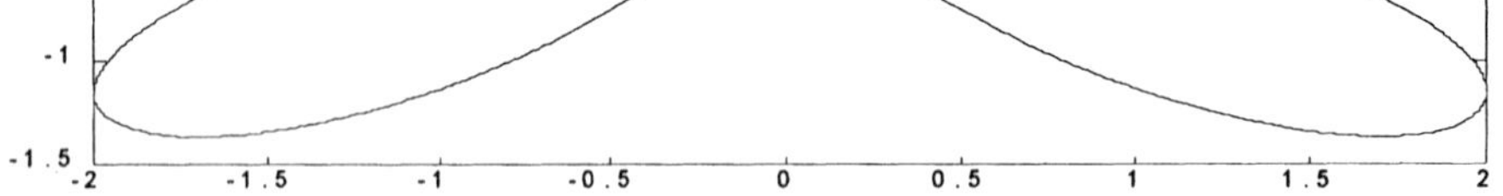

Prob. 13.9

(a) From Prob. 13.4,

$$E_{\theta s} = \frac{j\eta I_o}{8\pi r}\beta l e^{-j\beta r}\sin\theta\,, \quad H_{\phi s} = \eta E_{\theta s}$$

(b) $D = \frac{U_{\max}}{U_{ave}}$

$$U(\theta,\phi) = \sin^2\theta, \quad U_{\max} = 1$$

$$U_{ave} = \frac{P_{rad}}{4\pi} = \frac{1}{4\pi}\int_{\theta=0}^{\pi}\int_{\phi=0}^{2\pi}\sin^3\theta\, d\theta\, d\phi$$

$$= \frac{2\pi}{4\pi}\left(\frac{4}{3}\right) = \frac{2}{3}$$

$$D = \frac{1}{2/3} = \underline{\underline{1.5}}$$

Prob. 13.10

(a) $P_{rad} = \int P_{rad}.\partial s = P_{ave}.2\pi r^2$ (hemisphere)

$$P_{ave} = \frac{P_{rad}}{2\pi r^2} = \frac{200 \times 10^3}{2\pi(50 \times 10^6} = 12.73\mu W / m^2$$

$$P_{ave} = \underline{\underline{12.73 a_r \ \mu\text{W}/\text{m}^2}}.$$

(b) $P_{ave} = \frac{(E_{\max})^2}{2\eta}$

$$E_{\max} = \sqrt{2\eta P_{ave}} = \sqrt{240\pi \times 12.73 \times 10^{-6}}$$

$$= \underline{\underline{0.098 \text{ V/m}}}$$

Prob. 13.11

(a) $\lambda = \frac{c}{f} = \frac{3 \times 10^8}{100 \times 10^6} = 3m$

$$E_{\max} = \frac{\eta\pi I_o S}{r\lambda^2} \longrightarrow I_o = \frac{E_{\max} r\lambda^2}{\eta\pi S}$$

$$I_o = \frac{50 \times 10^{-3} \times 3 \times 3^2}{120\eta^2\pi(0.2)^2 100} = 90.71 \ \mu\text{A}$$

(b) $R_{rad} = \frac{320\pi^4 S}{\lambda^4} = 320\pi^4\pi^2(0.2)\times 10^4 = 60.77 k\Omega$

$$P_{rad} = \frac{1}{2}I_o^2 R_{rad} = \frac{1}{2}(90.71)^2 \times 10^{-12} \times 60.77 \times 10^3$$

$$= \underline{\underline{0.25 \text{ mW}}}$$

Prob. 13.12

(a) $f(\theta) = \left| \frac{\cos\left(\frac{\pi}{2}\cos\theta\right)}{\sin\theta} \right|$

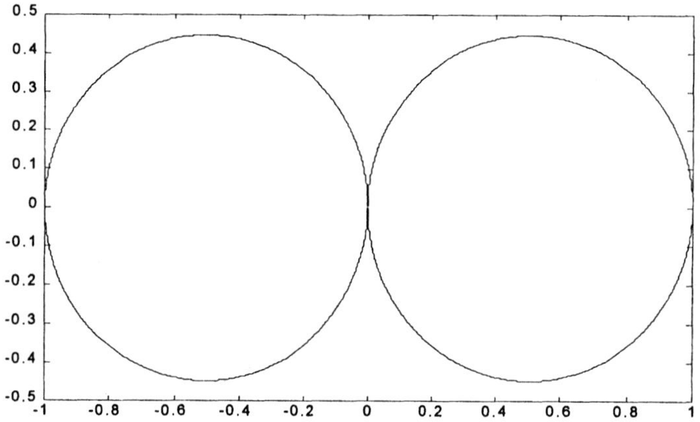

(b) The same as for $\frac{\lambda}{2}$ dipole except that the fields are zero for $\theta \rangle \frac{\pi}{2}$ as shown.

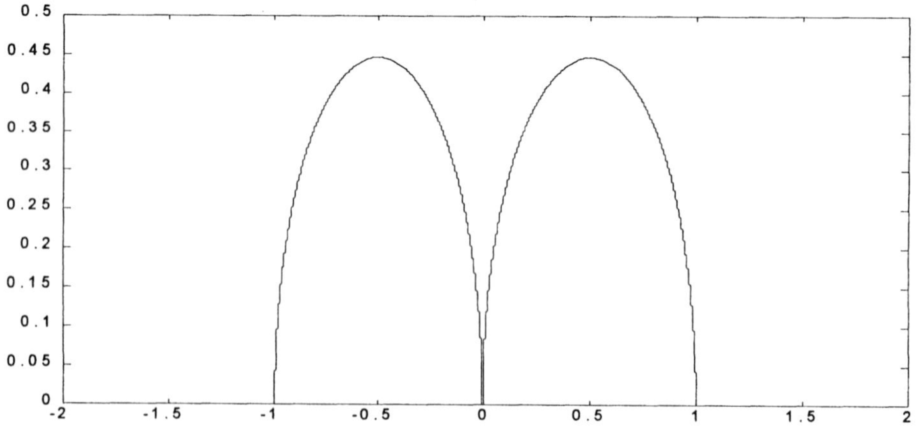

Prob. 13.13

For $l = 3\lambda/2$ and $l = \lambda$, the plots are the upper portions of those in Prob. 13.8(b). For $l = 5\lambda/8$, the plot is as shown below.

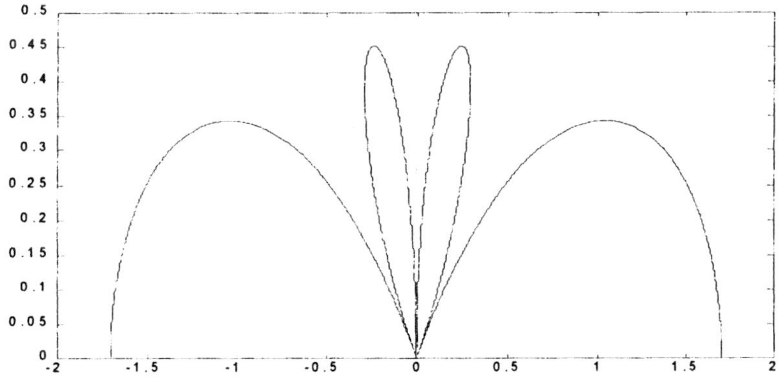

Prob. 13.14

$$P_{ave} = \frac{|E_s|^2}{2\eta} a_r = \frac{25\sin^2 2\theta}{2\eta r^2} a_r$$

$$P_{rad} = \frac{25}{2\eta} \iint (2\sin\theta\cos\theta)^2 \sin\theta\, d\theta\, d\phi$$

$$P_{rad} = \frac{25}{240\pi}(2\pi)\int_0^\pi 4\sin^2\theta\cos^2\theta\, d(-\cos\theta)$$

$$= \frac{25}{120}\int_0^\pi \left(\cos^4\theta - \cos^2\theta\right) d(-\cos\theta)$$

$$= \frac{25}{120}\left(\frac{\cos^5\theta}{5} - \frac{\cos^3\theta}{3}\right)\Bigg|_0^\pi = \frac{25}{120}\left(-\frac{2}{5} + \frac{2}{3}\right)$$

$P_{rad} = \underline{\underline{55.55 \text{ mW}}}$

Prob. 13.15

$f(\theta) = |\cos\theta\cos\phi|$

For the vertical pattern, $\phi = 0$ ⟶ $f(\theta) = |\cos\theta|$ which is sketched below.

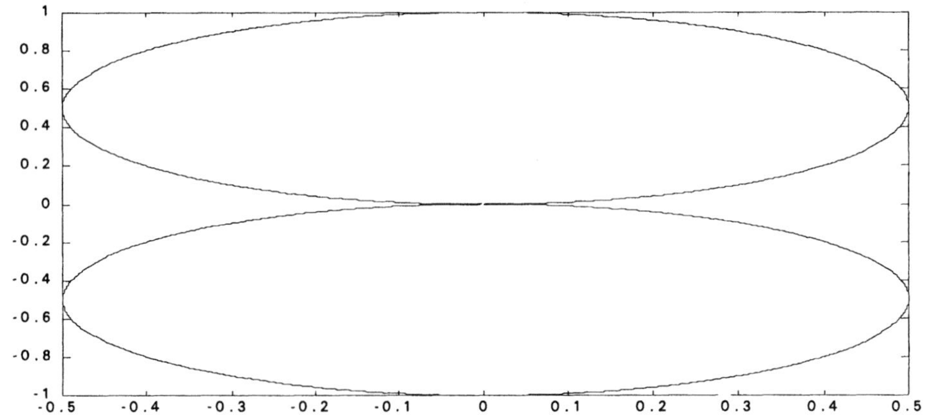

Prob. 13.16

$$P_{rad} = \frac{I_o^2 \eta \beta^2}{32\pi^2}(2\pi)\frac{4}{3} = \frac{I_o^2 \eta \beta^2 (dl)^2}{12\pi}$$

$$P_{ave} = \frac{I_o^2 \eta \beta^2 (dl)^2 \sin^2\theta}{32\pi^2 r^2}$$

$$\frac{P_{ave}}{P_{rad}} = \frac{\sin^2\theta}{32\pi^2 r^2} 12\pi = \frac{1.5\sin^2\theta}{4\pi r^2}$$

$$P_{ave} = \frac{1.5\sin^2\theta}{4\pi r^2} P_{rad}$$

Prob. 13.17

$$G_d = \frac{U}{U_{ave}} = \frac{4\pi r^2 P_{ave}}{\int P_{ave}.\partial s} = \frac{8\pi \sin\theta \cos\phi}{\int P_{ave}.dS}$$

But $\int P_{ave}.dS = \int_{\theta=0}^{\pi} \int_{\phi=0}^{\pi/2} 2\sin\theta \cos\phi \sin\theta \, d\theta \, d\phi$

$$= 2\int_0^{\pi/2} \cos\phi \, d\phi \int_0^{\pi} \sin^2\theta \, d\theta = 2\sin\phi \Big|_0^{\pi/2} \left(\frac{\pi}{2}\right) = \pi$$

$$G_d = \underline{\underline{8\sin\theta \cos\phi}}$$

Prob. 13.18

From Prob. 13.8, $E_{\theta s} = \dfrac{j\eta I_o e^{-j\beta r}\left[\cos\left(\dfrac{\beta l}{2}\cos\theta\right) - \cos\dfrac{\beta l}{2}\right]}{2\pi r\sin\theta}$

For $l = \lambda, \dfrac{\beta l}{2} = \dfrac{2\pi}{\lambda}.\dfrac{\lambda}{2} = \pi$

$$|E_{\theta s}| = \frac{\eta I_o[\cos(\pi\cos\theta) + 1]}{2\pi r\cos\theta}$$

$$f(\theta) = \frac{|E_{\theta s}|}{|E_{\theta s}|_{max}} = \underline{\underline{\frac{\cos(\pi\cos\theta) + 1}{\sin\theta}}}$$

It is sketched below.

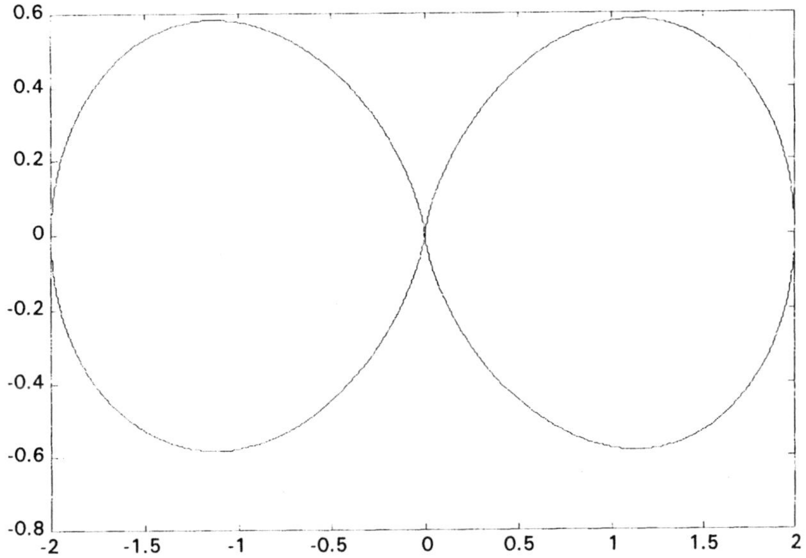

Prob. 13.19

(a) $E_{\theta s} = \dfrac{j\eta I_o \beta\, dl}{4\pi r}\sin\theta\, e^{-j\beta r}$

$$R_{rad} = 80\pi^2\left(\frac{dl}{\lambda}\right)^2$$

$$G_\phi = \frac{4\pi r^2 P_{ave}}{P_{rad}} = \frac{4\pi r^2 . \frac{1}{2\eta}|E_{\theta s}|^2}{\frac{1}{2} I_o^2 R_{rad}}$$

$$= \frac{4\pi r^2}{I_o^2} . \frac{1}{80\pi^2}\left(\frac{\lambda}{dl}\right)^2 . \frac{1}{\eta} \frac{\eta^2 I_o^2 \beta^2 (dl)^2 \sin^2\theta}{16\pi^2 r^2}$$

$$G_\phi = \underline{\underline{1.5 \sin^2\theta}}$$

(b) $D = G_{\phi,\max} = \underline{\underline{1.5}}$

(c) $A_e = \frac{\lambda^2}{4\pi} G_\phi = \underline{\underline{\frac{1.5\lambda^2 \sin^2\theta}{4\pi}}}$

(d) $R_{rad} = 80\pi^2 \left(\frac{1}{16}\right)^2 = \underline{\underline{3.084}}$

Prob. 13.20

(a) $$E_{\phi s} = \frac{120\pi^2 I_o}{r} \frac{S}{\lambda^2} \sin\theta\, e^{-j\beta r}$$

$$R_{rad} = \frac{320\pi^4 S^2}{\lambda^4}$$

$$G_d = \frac{4\pi U(\theta,\phi)}{P_{rad}} = \frac{4\pi r^2 P_{ave}}{\frac{1}{2} I_o R_{rad}} = \frac{8\pi r^2}{I_o^2} . \frac{1}{2\eta} \frac{|E_{\phi s}|^2}{R_{rad}}$$

$$= \frac{8\pi r^2}{I_o^2} . \frac{1}{2\eta} .14400\pi^4 \frac{I_o^2}{r^2} \frac{S^2}{\lambda^4} \sin^2\theta \frac{\lambda^2}{320\pi^4 S^2}$$

$$G_d = \underline{\underline{1.5 \sin^2\theta}}$$

(b) $\underline{\underline{D = 1.5}}$

(c) $A_e = \frac{\lambda^2 G_d}{4\pi} = \underline{\underline{\frac{\lambda^2}{4\pi} 1.5 \sin^2\theta}}$

(d) $S = \pi a^2 = \frac{\pi d^2}{4} = \frac{320\pi^6}{(576)^2}$

$$R_{rad} = \underline{\underline{0.927\Omega}}$$

Prob. 13.21

$$R_{ac} = \frac{l}{\sigma S}, \quad S = \pi a^2$$

$$R_{ac} = \frac{l}{\sigma \pi a^2}$$

$$R_l = R_{ac} = \frac{a}{2\delta} R_{dc} = \frac{a}{2\delta} \frac{l}{\sigma \pi a^2}$$

$$\text{Now } \delta = \sqrt{\frac{\pi f \mu}{\sigma}} = \sqrt{\frac{\pi \times 15 \times 10^6 \times 4\pi \times 10^7}{5.8 \times 10^7}} = 1.01 \times 10^{-3} m$$

Alternatively, since $\delta \langle\langle a$, current is confined to a cylindrical shell of thickness δ. Hence

$$R_l = R_{ac} = \frac{l}{\sigma (2\pi a)\delta}$$

$$l = \frac{\lambda}{2} = \frac{c}{2f} = \frac{3 \times 10^8}{2 \times 15 \times 10^6} = 10m$$

$$R_l = \frac{10}{2 \times 1.01 \times 5.8 \times 10^7 \times \pi \times 1.3 \times 10^{-2}} = 0.0209\Omega$$

$$R_{rad} = 73\Omega$$

$$\eta_r = \frac{R_{rad}}{R_{rad} + R_l} = \frac{73}{73.0209} = \underline{\underline{99.97\%}}$$

Prob. 13.22

(a) $U_{max} = 1$

$$U_{ave} = \frac{P_{rad}}{4\pi} = \frac{\int u d\Omega}{4\pi}$$

$$= \frac{1}{4\pi} \int\int \sin^2 2\theta \sin\theta \, d\theta \, d\phi$$

$$= \frac{1}{4\pi}(2\pi)\int_0^{\pi} (2\sin\theta\cos\theta)^2 d(-\cos\theta)$$

$$= 2\int_0^{\pi} (\cos^4\theta - \cos^2\theta) d(\cos\theta)$$

$$= 2\left[\frac{\cos^5\theta}{5} - \frac{\cos^3\theta}{3}\right]\Bigg|_0^\pi$$

$$= 2\left[-\frac{2}{5} + \frac{2}{3}\right] = \frac{8}{15}$$

$$U_{ave} = \underline{\underline{0.5333}}$$

$$D = \frac{U_{\max}}{U_{ave}} = \underline{\underline{1.875}}$$

(b) $U_{max} = 4$

$$U_{ave} = \frac{1}{4\pi}\int u\,d\Omega = \frac{4}{4\pi}\int\int \frac{1}{\sin^2\theta}\,d\theta\,d\phi$$

$$= \frac{1}{\pi}\int_0^\pi d\phi \int_{\pi/3}^{\pi/2} \frac{d(-\cos\theta)}{1-\cos^2\theta} = \frac{\pi}{\pi}\int \frac{dv}{u^2-1} = \ln\frac{1-u}{1+u}\Bigg|_{\pi/3}^{\pi/2}$$

$$= \ln 1 - \ln\frac{0.5}{1.5} = \ln 3$$

$$U_{ave} = \underline{\underline{1.099}}$$

$$D = \frac{U_{\max}}{U_{ave}} = \frac{4}{1.099} = \underline{\underline{3.641}}$$

(c) $U_{max} = 2$

$$U_{ave} = \frac{1}{4\pi}\int u\,d\Omega = \frac{1}{4\pi}\int\int 2\sin^2\theta\,\sin^2\phi\,\sin\theta\,d\theta\,d\phi$$

$$= \frac{1}{2\pi}\int_0^\pi \sin^2\phi\,d\phi \int_0^\pi \left(1-\cos^2\theta\right)d(-\cos\theta)$$

$$= \frac{1}{2\pi}\cdot\frac{\pi}{2}\left(\frac{\cos^3\theta}{3} - \cos\theta\right)\Bigg|_0^\pi = \frac{1}{4}\left[-\frac{2}{3} + 2\right] = \frac{1}{3}$$

$$U_{ave} = \underline{\underline{0.333}}$$

$$D = \frac{U_{\max}}{U_{ave}} = \underline{\underline{6}}$$

Prob. 13.23

(a) $$U_{ave} = \frac{1}{4\pi}\int u d\Omega$$

$$= \frac{1}{4\pi}\int\int \sin^2\theta \sin^2\phi \sin\theta d\theta d\phi$$

$$= \frac{1}{2}\left(\frac{\cos^3\theta}{3} - \cos\theta\right)\Bigg|_0^\pi = \frac{1}{2}\left(-\frac{2}{3}+2\right) = \frac{1}{2}\cdot\frac{4}{3} = \frac{2}{3}$$

$$U_{ave} = 0.6667$$

$$G_\phi = \frac{U}{U_{ave}} = \underline{\underline{1.5 Sin^2\theta}}$$

$$D = G_{\phi,\max} = \underline{\underline{1.5}}$$

(b) $$U_{ave} = \frac{1}{4\pi}\int\int 4\sin^2\theta\cos^2\phi\sin\theta d\theta d\phi$$

$$= \frac{1}{\pi}\int_0^\pi \cos^2\phi d\phi \int_0^\pi (1-\cos^2\theta) d(-\cos\theta)$$

$$= \frac{1}{\pi}\int_0^\pi \frac{1}{2}(1 - Cos2\phi)\partial\phi\left(\frac{Cos^3\theta}{3} - Cos\theta\right)\Bigg|_0^\pi$$

$$= \frac{1}{2\pi}\left(\phi + \frac{Sin2\phi}{2}\right)\Bigg|_0^\pi\left(\frac{4}{3}\right)$$

$$= \frac{1}{2\pi}(\pi)\left(\frac{4}{3}\right) = \frac{2}{3}$$

$$U_{ave} = 0.6667$$

$$G_{d,\max} = \frac{U}{U_{ave}} = \underline{\underline{6\sin^2\theta\cos^2\phi}}$$

$$D = G_{d,\max} = \underline{\underline{6}}$$

(c) $$U_{ave} = \frac{10}{4\pi}\int\int \cos^2\theta \sin^2\frac{\phi}{2}\sin\theta d\theta d\phi$$

$$= \frac{10}{4\pi}\int_0^{\pi/2}\sin^2\frac{\phi}{2}d\phi\int_0^{\pi}\left(\cos^2\theta\right)d(-\cos\theta)$$

$$= \frac{10}{4\pi}\int_0^{\pi/2}\frac{1}{2}(1-\cos\phi)d\phi\left(-\frac{\cos^3\theta}{3}\right)\Big|_0^{\pi}$$

$$= \frac{10}{4\pi}\left(\frac{2}{3}\right)\left(\frac{1}{2}\right)\left(\phi+\sin^2\phi\right)\Big|_0^{\pi} = \frac{10}{12\pi}\left(\frac{\pi}{2}-1\right)$$

$$U_{ave} = 0.1514$$

$$G_{d,\max} = \frac{U}{U_{ave}} = \underline{\underline{66.05\cos^2\theta\cos^2\frac{\phi}{2}}}$$

$$D = G_{d,\max} = \underline{\underline{66.05}}$$

Prob. 13.24

(a) $$P_{rad} = \int P_{ave}.dS = \frac{1}{2\eta}\int\left|E_{\phi s}\right|^2\partial S$$

$$= \frac{0.04}{16\pi^2}\left(\frac{1}{2\pi}\right)\int\int\frac{\cos^4\theta}{r^2}r^2\sin\theta d\theta d\phi$$

$$= \frac{0.04}{16\pi^2}\left(\frac{1}{240\pi}\right)(2\pi)\int_0^{\pi}\cos\theta d(-\cos\theta).10^6$$

$$= \frac{0.04}{16\pi^2}\frac{10^6}{120}\left(-\frac{\cos^5\theta}{5}\right)\Big|_0^{\pi} = \frac{10^4}{480\pi^2}.\frac{2}{5}$$

$$P_{rad} = \underline{\underline{0.8443\text{ W}}}$$

(b) $$G_d = \frac{4\pi U(\theta,\phi)}{P_{rad}} = \frac{4\pi r^2 P_{ave}}{P_{rad}}$$

$$= 4\pi r^2.\frac{0.04\cos^4\theta}{16\pi^2 r^2}.\frac{10^6}{240\pi}.\frac{12\pi^2}{100}$$

$$G_d = 5\cos^4\theta$$

Since $\cos 60^o = \frac{1}{2}$,

$$G_d = 5\left(\frac{1}{2}\right)^4 = \underline{\underline{0.625}}$$

Prob. 13.25

This is similar to Fig. 13.10 except that the elements are z-directed.

$$E_s = E_{s1} + E_{s2} = \frac{j\eta\beta I_o dl}{4\pi}\left[\sin\theta_1 \frac{e^{-j\beta r_1}}{r_1} a_{\theta 1} + \sin\theta_2 \frac{e^{-j\beta r_2}}{r_2} a_{\theta 2}\right]$$

where $r_1 \cong r - \frac{d}{2}\cos\theta, \quad r_2 \cong r + \frac{d}{2}\cos\theta, \quad \theta_1 \cong \theta_2 \cong \theta, \quad a_{\theta 1} \cong a_{\theta 2} = a_\theta$

$$E_s = \frac{j\eta\beta I_o dl}{4\pi}\sin\theta \; a_\theta\left[e^{j\beta d\cos\theta/2} + e^{-j\beta d\cos\theta/2}\right]$$

$$\underline{\underline{E_s = \frac{j\eta\beta I_o dl}{4\pi}\sin\theta \;\; \cos(\frac{1}{2}\beta d\cos\theta) a_\theta}}$$

Prob. 13.26

(a) $\text{AF} = 2\cos\left[\frac{1}{2}(\beta d\cos\theta + \alpha)\right], \quad \alpha = 0, \quad \beta d = \frac{2\pi}{\lambda}\lambda = 2\pi$

$$\text{AF} = \underline{\underline{2\cos(\pi\cos\theta)}}$$

(b)Nulls occur when

$\cos(\pi\cos\theta) = 0 \quad \longrightarrow \quad \pi\cos\theta = \pm\pi/2, \pm 3\pi/2, \ldots$

or

$$\theta = \underline{\underline{60^o, 120^o}}$$

(c) Maxima and minima occur when

$\frac{df}{d\theta} = 0 \quad \longrightarrow \quad \sin(\pi\cos\theta)\pi\sin\theta = 0$

i.e. $\sin\theta = 0 \quad \longrightarrow \quad \theta = 0^o, 180^o$

$\cos\theta = 0 \quad \longrightarrow \quad \theta = 90^o$

or

$$\theta = \underline{\underline{0^o, 90^o, 180^o}}$$

(d) The group pattern is sketched below.

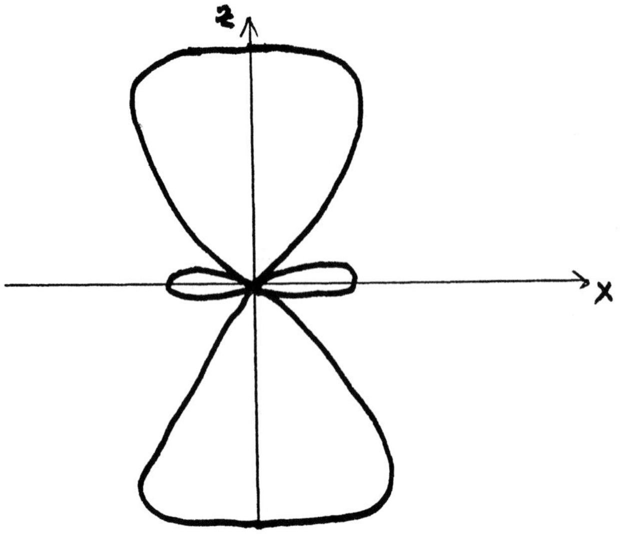

Prob. 13.27

(a) The group pattern is

$$f(\theta) = \cos\left[\frac{1}{2}(\beta d \cos\theta + \alpha)\right]$$

$$f(\theta) = \cos\left[\frac{1}{2}\left(\frac{2\pi}{\lambda}.\frac{\lambda}{2}\cos\theta + \frac{\pi}{2}\right)\right]$$

$$= \cos\frac{\pi}{4}\left(\cos\frac{\pi}{4}(\cos\theta + 1)\right)$$

$\cos\frac{\pi}{4}(\cos\theta + 1) = 0 \longrightarrow \qquad \frac{\pi}{4}(\cos\theta + 1) = \pm\frac{\pi}{2}, \pm\frac{3\pi}{2}$

or $\cos\theta = 1 \longrightarrow \qquad \theta = 0$

Maximum and minimum occur when

$$\frac{d}{d\theta}\left[\cos\frac{\pi}{4}(\cos\theta + 1)\right] = 0$$

$$\sin\theta \sin\frac{\pi}{4}(1 + \cos\theta) = 0$$

$$\sin\theta = 0 \qquad \theta = -1 \text{ or } \theta = 180^o$$

Alternatively $f(\theta)$ can be plotted using Matlab or Maple.
The group pattern is shown below.

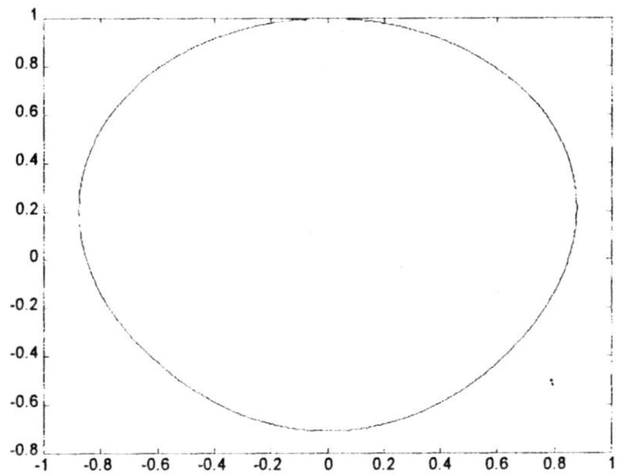

(b) For $d = \frac{\lambda}{2}, f(\theta) = \cos\left[\frac{1}{2}\left(\frac{2\pi}{\lambda}.\frac{\lambda}{2}\cos\theta + \frac{\pi}{2}\right)\right]$

$$= \cos\left(\frac{\pi}{2}\cos\theta + \frac{\pi}{4}\right)$$

$\cos\left(\frac{\pi}{2}\cos\theta + \frac{\pi}{4}\right) = 0 \longrightarrow \quad \frac{\pi}{2}\cos\theta + \frac{\pi}{2} = \pm\frac{\pi}{2}, \pm\frac{3\pi}{2}$

$\cos\theta = \frac{1}{2} \longrightarrow \quad \theta = 60^o$

For maximum or minimum,

$$\frac{d}{d\theta}\left[\cos\frac{\pi}{2}\left(\cos\theta + \frac{\pi}{4}\right)\right] = 0$$

$$\sin\theta \sin\left(\frac{\pi}{2}\cos\theta + \frac{\pi}{4}\right) = 0$$

$$\sin\theta = 0 \longrightarrow \quad \theta = 0^o, 180^o$$

$$\sin\left(\frac{\pi}{2}\cos\theta + \frac{\pi}{4}\right) = 0$$

$\frac{\pi}{2}\cos\theta + \frac{\pi}{4} = 0 \longrightarrow \quad \cos\theta = -\frac{1}{2} \longrightarrow \quad \theta = 120$

$\left|\cos\left(\frac{\pi}{2}\cos\theta + \frac{\pi}{4}\right)\right| = 1 \longrightarrow \quad \frac{\pi}{2}\cos\theta + \frac{\pi}{4} = 0, \pi, 2\pi \longrightarrow \quad \theta = 120^o$

The group pattern is sketched below.

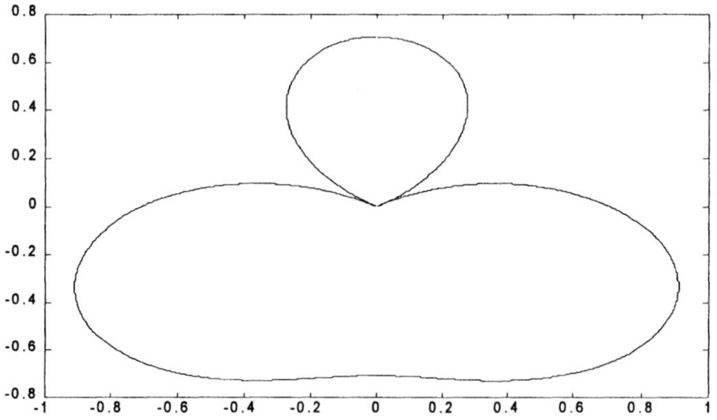

Prob. 13.28

$$f(\theta) = \cos\left[\frac{1}{2}\left(\beta d \cos\theta + \alpha\right)\right]$$

(a) $\alpha = \pi/2, \beta d = \frac{2\pi}{\lambda}.\lambda = 2\pi$

$$f(\theta) = \cos\left(\pi \cos\theta + \pi/4\right)$$

Nulls occur at $\pi \cos\theta + \pi/4 = \pm\frac{\pi}{2}, \pm\frac{3\pi}{2}, \ldots$ or $\theta = 75.5^o, 138.6^o$

Maxima occur at $\frac{\partial f}{\partial \theta} = 0 \longrightarrow \sin\theta = 0 \longrightarrow \theta = 0^o, 180^o$

Or $\sin\left(\pi \cos + \frac{\pi}{4}\right) = 0 \longrightarrow \theta = 41.4^o, 104.5^o$

With $f_{max} = 0.71, 1$.

Hence the group pattern is sketched below.

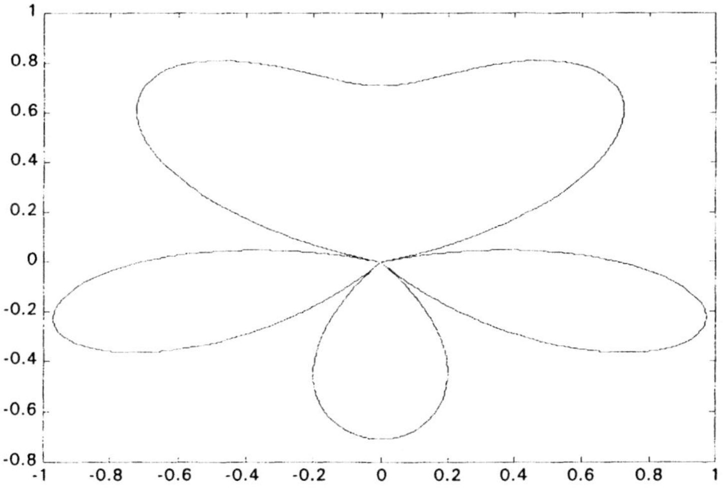

(b) $\alpha = \frac{3\pi}{4}, \beta d = \frac{2\pi}{\lambda}.\frac{\lambda}{4} = \frac{\pi}{2}$

$$f(\theta) = \left|\cos\left(\frac{\pi}{4}\cos\theta + \frac{3\pi}{8}\right)\right|$$

Nulls occur at $\frac{\pi}{4}\cos\theta + \frac{3\pi}{8} = \pm\frac{\pi}{2}, \pm\frac{3\pi}{2}, \ldots \longrightarrow \theta = 60^o$

Maxima and minima occur at $\sin\theta \sin\left(\frac{\pi}{4}\cos\theta + \frac{3\pi}{8}\right) = 0$

i.e. $\theta = 0^o, 180^o \rightarrow f(\theta) = 0.383, 0.924$

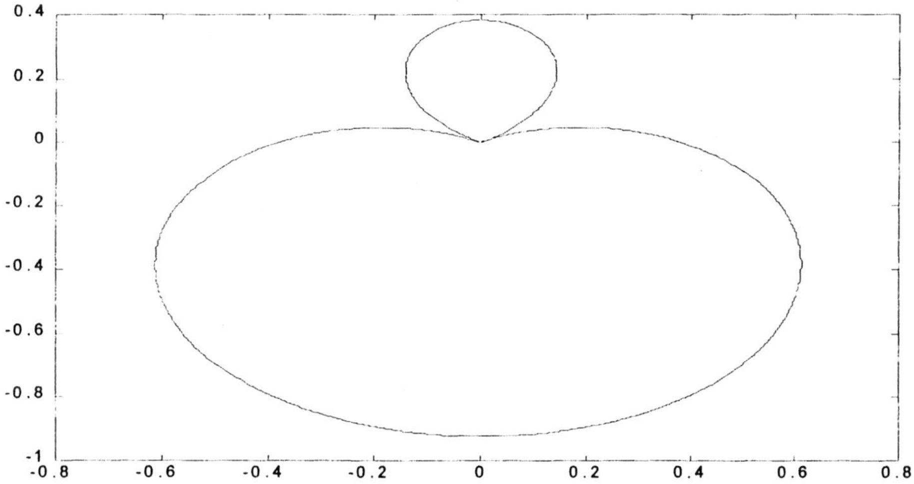

(c) $\alpha = 0, \beta d = \frac{2\pi}{\lambda} \cdot \frac{3\lambda}{4} = \frac{3\pi}{2}$

$$f(\theta) = \left| \cos\left(\frac{3\pi}{4} \cos\theta \right) \right|$$

It has nulls at $\frac{3\pi}{4}\cos\theta = \pm\frac{\pi}{2}, \pm\frac{3\pi}{2}, \ldots \rightarrow \theta = 48.2^o, 131.8^o$

It has maxima and minima at $\frac{df}{d\theta} = 0 \rightarrow \sin\theta \sin\left(\frac{3\pi}{4}\cos\theta \right) = 0$

i.e. $\theta = 0^o, 180^o \rightarrow f(\theta) = 0.71, 1$

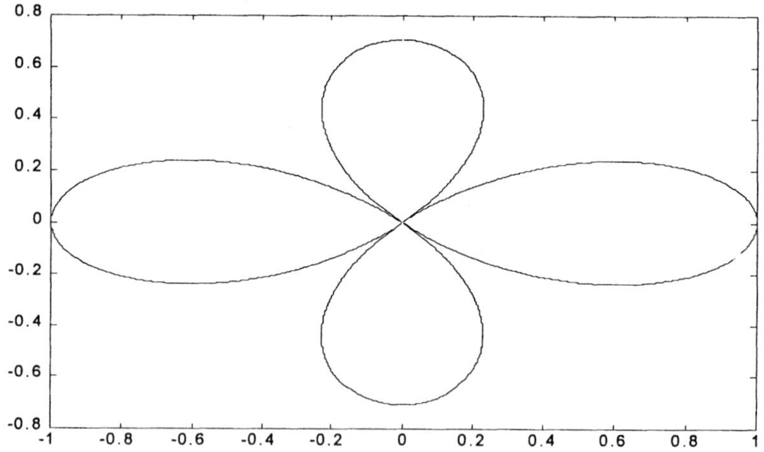

Prob. 13.29

(a) For N = 2, $f(\theta) = \cos\left[\frac{1}{2}(\beta d \cos\theta + \alpha) \right]$

$\alpha = 0, d = \frac{\lambda}{4}$

$$f(\theta) = \cos\left[\frac{1}{2}\left(\frac{2\pi}{\lambda} \cdot \frac{\lambda}{4}\cos\theta + 0 \right) \right] = \cos\left(\frac{\pi}{4}\cos\theta \right)$$

Maxima and minima occur at

$$\frac{d}{d\theta}\left[\cos\left(\frac{\pi}{4}\cos\theta \right) \right] = 0$$

$$\sin\theta \sin\left(\frac{\pi}{4}\cos\theta \right) = 0$$

$$\sin\theta = 0 \rightarrow \theta = \pi, 0 \text{ and } f(\theta) = 0.707$$

$$\sin\left(\frac{\pi}{4}\cos\theta\right) \rightarrow \cos = 0 \rightarrow \theta = 90^o, f(\theta) = 1$$

Nulls occur as $\frac{\pi}{4}Cos\theta = \pm\frac{\pi}{2}, \pm\frac{3\pi}{2}, \ldots$ (No Solution)

The group pattern is sketched below.

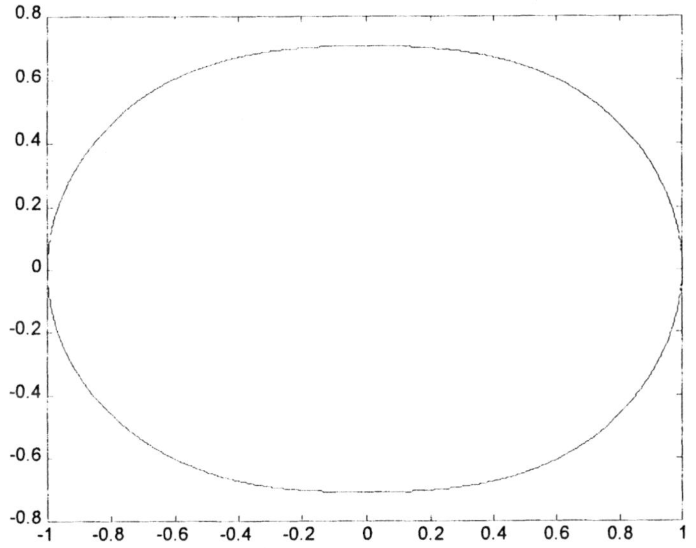

(b) For N = 4,

$$AF = \frac{\sin 2(\beta d\cos\theta + 0)}{\sin\frac{1}{2}(\beta d\cos\theta + 0)}$$

Now, $\frac{\sin 4\theta}{\sin\theta} = \frac{2\sin 2\theta\cos 2\theta}{\sin\theta} = 4\cos 2\theta\cos\theta$

$$AF = 4\cos(\beta d\cos\theta)\cos\left(\frac{1}{2}\beta d\cos\theta\right)$$

$$f(\theta) = \cos\left(\frac{2\pi}{\lambda}.\frac{\lambda}{4}\cos\theta\right)\cos\left(\frac{1}{2}\frac{2\pi}{\lambda}\frac{\lambda}{4}\right)\cos\theta$$

$$= \cos\left(\frac{\pi}{2}\cos\theta\right)\cos\left(\frac{\pi}{4}\cos\theta\right)$$

The plot is shown below.

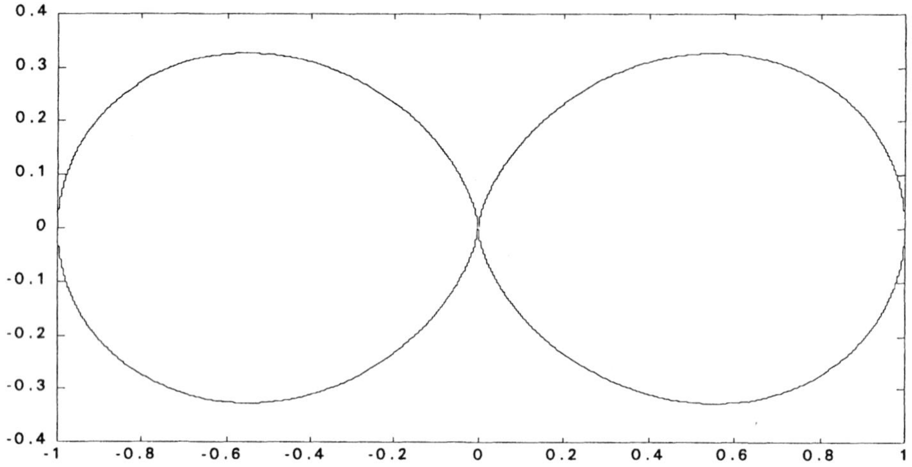

Prob. 13.30

(a) The given array is replaced by
where + represents

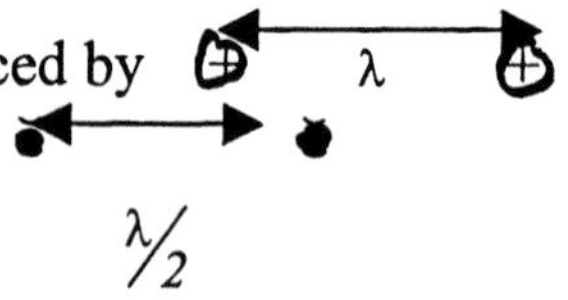

Thus the resultant pattern is obtained as follows.

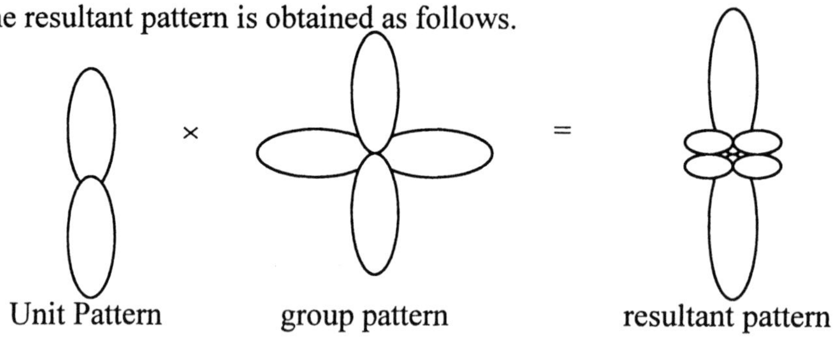

(b) The array is replaced by by

where + stands for

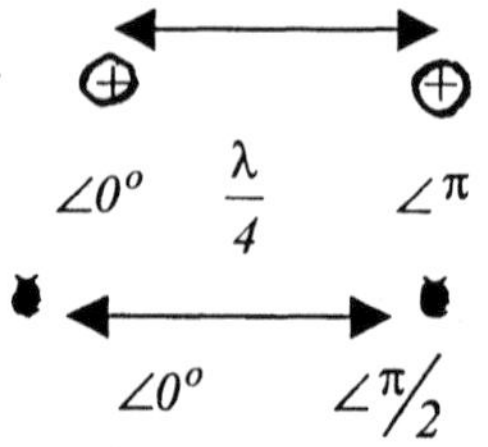

Thus the resultant pattern is obtained as shown.

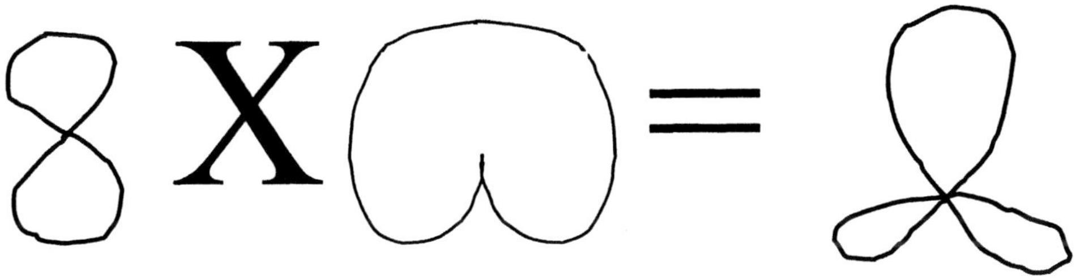

Prob. 13.31

$$A_e = \frac{\lambda^2}{4\pi} G_d$$

where $G_d = \dfrac{4\pi u}{U_{ave}} = \dfrac{4\pi U}{P_{rad}}$

But $E_{\phi s} = \dfrac{\eta \pi I_o S}{r\lambda^2} \sin\theta \, e^{-j\beta r}$

$$U = r^2 P_{ave} = \frac{r^2 \left| E_{\phi s} \right|^2}{2\eta} = \frac{\eta \pi^2 I_o^2 S^2 \sin^2\theta}{\lambda^4}$$

$$P_{rad} = \int P_{ave} dS = \frac{\eta \pi^2 I_o^2 S^2}{\lambda^4} \int\int \sin^3\theta \, \partial\theta \, \partial\phi$$

$$= \frac{\eta\pi^2 I_o^2 S^2}{\lambda^4} . (2\pi)\left(\frac{4}{3}\right)$$

$$G_d = 4\pi \frac{\dfrac{\eta \pi^2 I_o^2 S^2 \sin^2\theta}{\lambda^4}}{\dfrac{\eta \pi^2 I_o^2 S^2}{\lambda^4} . \dfrac{8\pi}{3}} = \frac{3}{2} \sin^2\theta$$

$$\lambda = \frac{c}{f} = \frac{3 \times 10^8}{10^8} = 3m,$$

$$A_e = \frac{3\lambda^2}{8\pi} \sin^2\theta = \frac{3 \times 9}{8\pi} \left(\frac{1}{2}\right)^2 = \underline{\underline{0.2686}}$$

Prob. 13.32

$$A_e = \frac{P_r}{P_{ave}} = \frac{P_r}{E_r/2\eta} = \frac{2\eta P_r}{E_r}$$

$$= \frac{2 \times 120\pi \times 2 \times 10^{-6}}{25 \times 10^2 \times 10^{-6}} = \frac{48\pi}{250} = \underline{\underline{0.6031}}$$

Prob. 13.33

(a) $A_{cr} = \frac{\lambda^2}{4\pi} G_{dr}, \qquad A_{ct} = \frac{\lambda^2}{4\pi} G_{dt}$

$$P_r = G_{dr} G_{dt} \left(\frac{\lambda^2}{4\pi r} \right) P_t = \left(\frac{4\pi}{\lambda^2} A_{er} \right) \left(\frac{4\pi}{\lambda^2} A_{et} \right) \left(\frac{\lambda^2}{4\pi r} \right) P_t$$

or $\frac{P_r}{P_t} = \frac{A_{er} A_{et}}{\lambda^2 r^2}$

(b) $P_{r,\max} = \frac{A_{cr} A_{ct}}{\lambda^2 r^2} P_t \quad , \quad A_{cr} = A_{ct} = \frac{\lambda^2}{4\pi}(1.68)$

$$\lambda = \frac{c}{f} = \frac{3 \times 10^8}{100 \times 10^6} = 3m,$$

$$P_{r,\max} = \frac{\left(0.13\lambda^2\right)^2 (80)}{\lambda^2 \left(10^3\right)^2} = \underline{\underline{12.8 \mu W}}$$

Prob. 13.34

$$P_r = P_i A_e = P_i \frac{\lambda^2}{4\pi} G_d$$

$$P_{r,\max} = P_i \frac{\lambda^2}{4\pi} G_{d,\max}$$

But $G_{d,\max} = D = 1.64$ and

$$P_i = \frac{E^2}{2\eta}, \ \lambda = \frac{c}{f} = \frac{3 \times 10^8}{60 \times 10^6} = 5m$$

$$P_{r,\max} = \frac{E^2 \lambda^2 D}{8\pi\eta} = \frac{9 \times 10^{-6} \times 25 \times 1.64}{8\pi(120\pi)}$$

$= \underline{\underline{38.9 \text{ nW}}}$

Prob. 13.35

$$G_{dt} = 10^4, G_{dr} = 10^{3.2} = 1585$$

$$\lambda = \frac{c}{f} = \frac{3\times 10^8}{15\times 10^9} = 0.02m = \frac{1}{50}$$

$$P_r = G_{dr}G_{dt}\left(\frac{\lambda}{4\pi r}\right)^2 P_t = 10^4(1585)\left(\frac{0.02}{4\pi\times 2.456741\times 10^7}\right)^2 320$$

$$= 2.128\times 10^{-11}\ \text{W} = \underline{\underline{21.28\ \text{pW}}}$$

Prob. 13.36

$$G_d = \frac{U}{U_{ave}} = \frac{4\pi r^2 P_{ave}}{P_{rad}} \quad \text{or} \quad P_{ave} = \frac{G_d P_{rad}}{4\pi r^2}$$

$$G_d = 10^{3.4} = 2511.9$$

$$P_{ave} = \frac{2511.9\times 7.5\times 10^3}{4\pi\left(40\times 10^3\right)^2}$$

$$= \underline{\underline{0.937\ \text{mW/m}^2}}$$

Prob. 13.37

$$30dB = \log\frac{P_t}{P_r} \rightarrow \frac{P_t}{P_r} = 10^3 = 1000$$

But $P_r = (G_d)^2\left(\frac{3}{50\times 4\pi\times 12}\right)^2 P_t = P_t\left(\frac{G_d}{800\pi}\right)^2$

$$\left(\frac{G_d}{800\pi}\right)^2 = \frac{P_r}{P_t} = \frac{1}{1000} = \left(\frac{1}{10\sqrt{10}}\right)^2$$

or $G_d = \frac{800\pi}{10\sqrt{10}} = 79.476$

$$G_d = 10\log 79.476 = \underline{\underline{19\ \text{dB}}}$$

Prob. 13.38

$$G_{dt} = 25 = 10\log_{10} G_{dt} \rightarrow G_{dt} = 10^{2.5} = 316.23$$

$$G_{dr} = 10^3 = 1000$$

$$P_r = 316.23 \times 10^3 \left(\frac{1}{4\pi \times 1.5 \times 10^3} \cdot \frac{3 \times 10^8}{1.5 \times 10^9} \right) = \underline{\underline{7.12 \text{ mW}}}$$

Prob. 13.39

(a) $P_i = \dfrac{|E|^2}{2\eta_o} = \dfrac{P_{rad} G_d}{4\pi r^2} \rightarrow |E_i| = \sqrt{\dfrac{240\pi P_{rad} G_d}{4\pi r^2}}$

$$|E_i| = \frac{1}{r}\sqrt{60 P_{rad} G_d} = \frac{1}{120 \times 10^6}\sqrt{60 \times 200 \times 10^3 \times 3500}$$

$$= \underline{\underline{1.708 \text{ V/m}}}$$

(b) $|E_s| = \sqrt{\dfrac{|E_i|^2 \sigma}{4\pi r^2}} = \sqrt{\dfrac{1.708^2 \times 8}{4\pi \times 14400 \times 10^6}} = \underline{\underline{11.36 \ \mu\text{V/m}}}$

(c) $P_c = P_i \sigma = \dfrac{1.708^2}{240\pi}(8) = \underline{\underline{30.95 \text{ mW}}}$

(d) $P_i = \dfrac{|E|^2}{2\eta_o} = \dfrac{(11.36)^2 \times 10^{-12}}{240\pi} = 1.712 \times 10^{-13} \ \text{W/m}^2$

$$\lambda = \frac{3 \times 10^8}{15 \times 10^8} = 0.2m, A_{2r} = \frac{\lambda^2 G}{4\pi} = \frac{0.04 \times 3500}{4\pi}$$

$$P_r = P_a A_{er} = 1.712 \times 10^{-13} \times 11.14 = 1.907 \times 10^{-12}$$

or $P_r = \dfrac{(\lambda G_d)^2 \sigma P_{rad}}{(4\pi)^3 r^4} = \dfrac{(0.2 \times 3500)^2 \times 8 \times 2 \times 10^5}{(4\pi)^3 \times 12^4 \times 10^{16}}$

$$= \underline{\underline{1.91 \times 10^{-12} \text{ W}}}$$

Prob. 13.40

$$\lambda = \frac{c}{f} = \frac{3 \times 10^8}{6 \times 10^6} = 0.5m$$

$$P_r = G_{dr}G_{dt}\left(\frac{\lambda}{4\pi r}\right)^2 P_t = (1)(1)\left(\frac{0.5}{4\pi \times 10^3}\right)^2 (80)$$

$$= \underline{\underline{0.1267\ \mu\text{W}}}$$

Prob. 13.41

$$P_r = \frac{(\lambda G_d)^2 \sigma P_{rad}}{(4\pi)^3 r^4} \rightarrow P_{rad} = \frac{(4\pi)^3 r^4 P_r}{(\lambda G_d)^2 \sigma}$$

$$\lambda = \frac{c}{f} = \frac{3 \times 10^8}{6 \times 10^9} = \frac{1}{20} \langle\langle r = 250m$$

$$40 = \log_{10} G_d \rightarrow G_d = 10^4$$

$$P_{rad} = \frac{(4\pi)^3 (0.25 \times 10^3)^4 \times 2 \times 10^{-6}}{\left(\frac{1}{20} \times 10^4\right)^2 \times 0.8} = \underline{\underline{77.52\ \text{W}}}$$

Prob. 13.42

$$P_{rad} = \frac{4\pi}{G_{dt}G_{dr}}\left(\frac{4\pi r_1 r_2}{\lambda}\right)^2 \frac{P_r}{\sigma}$$

But $G_{dt} = 36dB = 10^{3.6} = 3981.1$

$$G_{dt} = 20dB = 10^2 = 100$$

$$\lambda = \frac{c}{f} = \frac{3 \times 10^8}{5 \times 10^9} = 0.06$$

$$r_1 = 3km,\ r_2 = 5km$$

$$P_{rad} = \frac{4\pi}{3981.1 \times 100}\left(\frac{4\pi \times 15 \times 10^6}{6 \times 10^{-2}}\right)^2 \frac{8 \times 10^{-12}}{2.4}$$

$$= \underline{\underline{1.038\ \text{kW}}}$$

CHAPTER 14

P. E. 14.1

$$S_i = \frac{1+0.4}{1-0.4} = \frac{1.4}{0.6} = \underline{\underline{2.333}}$$

$$S_o = \frac{1+0.2}{1-0.2} = \frac{1.2}{0.8} = \underline{\underline{1.5}}$$

P. E. 14.2

(a) By Snell's law, $n_1 \sin\theta_1 = n_2 \sin\theta_2$. Thus

$\theta_2 = 90^o \longrightarrow \sin\theta_2 = 1$

$\sin\theta_1 = n_2/n_1$, $\theta_1 = \sin^{-1} n_2/n_1 = \sin^{-1} 1.465/1.48 = \underline{\underline{81.83^o}}$

(b) $NA = \sqrt{n_1^2 - n_2^2} = \sqrt{1.48^2 - 1.465^2} = \underline{\underline{0.21}}$

P. E. 14.3

$\alpha l = 10 \log P(0)/P(l) = 0.2 \text{ X } 10 = 2$

$P(0)/P(l) = 10^{0.2}$, i.e. $P(l) = P(0)\, 10^{-0.2} = 0.631\, P(0)$

i.e. $\underline{\underline{63.1\ \%}}$

Prob. 14.1 Microwave is used:

(1) For surveying land with a piece of equipment called the *tellurometer*. This radar system can precisely measure the distance between two points.
(2) For guidance. The guidance of missiles, the launching and homing guidance of space vehicles, and the control of ships are performed with the aid of microwaves.
(3) In semiconductor devices. A large number of new microwave semiconductor devices have been developed for the purpose of microwave oscillator, amplification, mixing/detection, frequency multiplication, and switching. Without such achievement, the majority of today's microwave systems could not exist.

Prob. 14.2 (a) In terms of the S-parameters, the T-parameters are given by

$T_{11} = 1/S_{21}$, $T_{12} = -S_{22}/S_{21}$, $T_{21} = S_{11}/S_{21}$, $T_{22} = S_{12} - S_{11} S_{22}/S_{21}$

(b) $T_{11} = 1/0.4 = 2.5$, $T_{12} = -0.2/0.4$,

$T_{21} = 0.2/0.4$, $T_{22} = 0.4 - 0.2 \text{ x } 0.2/0.4 = 0.3$

Hence,

$$T = \begin{bmatrix} 2.5 & 0.5 \\ -0.5 & 0.3 \end{bmatrix}$$

Prob. 14.3 Since $Z_L = Z_o$, $\Gamma_L = 0$.

$$\Gamma_i = S_{11} = \underline{0.33 - j0.15}$$

$$\Gamma_g = (Z_g - Z_o)/(Z_g + Z_o) = (2-1)/(2+1) = 1/3$$

$$\Gamma_o = S_{22} + S_{12}S_{21}\Gamma_g /(1 - S_{11}\Gamma_g)$$

$$= 0.44 - j0.62 + 0.56x0.56\,x(1/3)/[1 - (0.11 - j0.05)]$$

$$= \underline{0.5571 - j0.6266}$$

Prob. 14.4 The microwave wavelengths are of the same magnitude as the circuit components. The wavelength in air at a microwave frequency of 300 GHz, for example, is 1 mm. The physical dimension of the lumped element must be in this range to avoid interference. Also, the leads connecting the lumped element probably have much more inductance and capacitance than is needed.

Prob. 14.5

$$\lambda = c/f = \frac{3x10^8}{8.4x10^9} = \underline{3.571 \text{ mm}}$$

Prob. 14.6

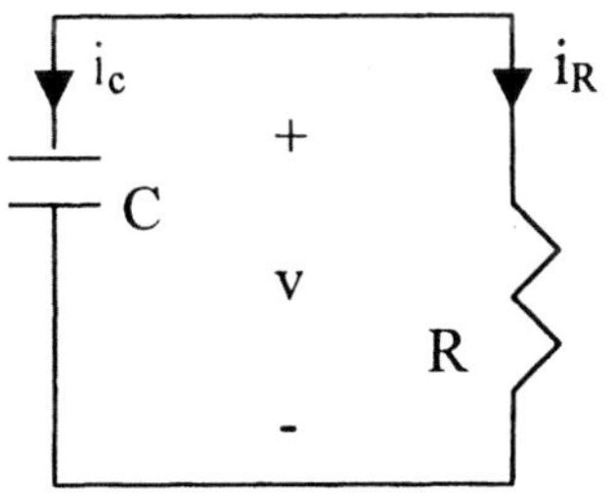

$i_c + i_R = 0$; hence $Cdv/dt + v/R = 0$

or $dv/v = -dt/RC$

so that $\ln v = -t/\tau + \ln v_o$, $\tau = RC = 125 \times 10^{-12} \times 2 \times 10^3 = 0.5\ \mu s$

$v = v_o e^{-t/\tau}$, $v(0) = v_o = 1500$

$i_c = C\,dv/dt = C(-1/\tau)\,v_o e^{-t/\tau} = \dfrac{-125x10^{-12}}{0.5x10^{-6}} \text{ x } 1500\, e^{-t/\tau}$

$= \underline{-0.375e^{-t/\tau}\text{ A},\ \tau = 0.5\ \mu s}$

Prob. 14.7

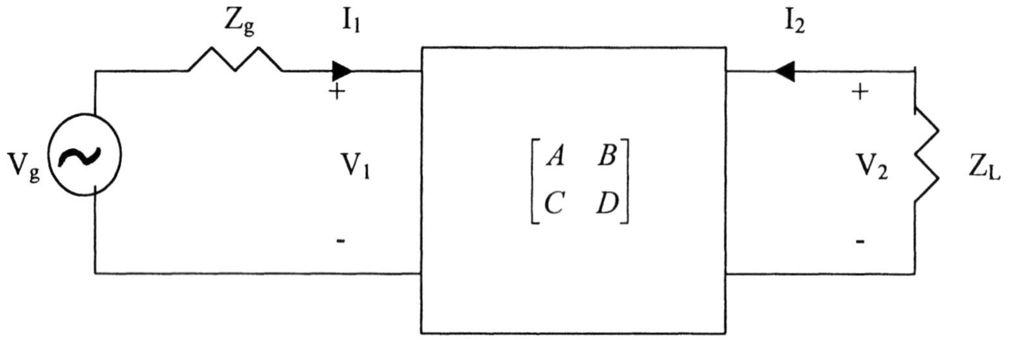

By definition,

$$V_1 = AV_2 - BI_2 \qquad (1)$$
$$I_1 = CV_2 - DI_2 \qquad (2)$$

We eliminate I_1 and I_2.

$$V_g = V_1 + Z_g I_1 \quad \text{or} \quad I_1 = (V_g - V_1)/Z_g \qquad (3)$$

$$V_2 = -Z_L I_2 \quad \text{or} \quad I_2 = -V_2/Z_L \qquad (4)$$

Substituting (3) and (4) into (1) and (2) and expressing V_1 and V_2 in terms of V_g, we obtain

$$IL = 20 \log V_1/V_2 = 20 \log_{10} \left| \frac{AZ_L + B + CZ_g Z_L + DZ_g}{Z_g + Z_L} \right|$$

Prob. 14.8

(a) $R_{dc} = \dfrac{l}{\sigma S} = \dfrac{10^3}{0.96x10^{-4}x6.1x10^7} = \underline{16.73\ m\Omega\ /km}$

(b) $R_{ac} = \dfrac{l}{\delta w \sigma}$, $\quad \pi a^2 = 0.8 \text{ x } 1.2 = 0.96$ or $a = 0.5528$

$$\delta = \frac{1}{\sqrt{\pi f \mu \sigma}} = \frac{1}{\sqrt{\pi X 6x10^6 x 4\pi x10^{-7} x6.1x10^7}} = \frac{1}{12\pi x10^3}$$

$$R_{ac} = \frac{1000x12\pi x10^3}{1.2x10^{-2}x6.1x10^7} = \underline{51.5\ \Omega}$$

Prob. 14.9

$$n = c/u_m = \frac{3x10^8}{2.1x10^8} = \underline{\underline{1.428}}$$

Prob. 14.10 When an optical fiber is used as the transmission medium, cable radiation is eliminated. Thus, optical fibers offer total EMI isolation because they neither emit nor pick up EM waves.

Prob. 14.11

(a) $NA = \sqrt{n_1^2 - n_2^2} = \sqrt{1.62^2 - 1.604^2} = \underline{\underline{0.2271}}$

(b) $NA = \sin\theta_a = 0.2271$ or $\theta_a = \sin^{-1} 0.2271 = \underline{\underline{13.13^o}}$

(c) $V = \dfrac{\pi d}{\lambda} NA = \dfrac{\pi x 50x10^{-6} x 0.2271}{1300x10^{-9}} = 27.441$

$N = V^2/2 = \underline{\underline{376 \text{ modes}}}$

Prob. 14.12

(a) $V = \dfrac{\pi d}{\lambda}\sqrt{n_1^2 - n_2^2} = \dfrac{\pi x 2.5x10^{-6} x 2}{1.3x10^{-6}}\sqrt{1.45^2 - 1^2} = \underline{\underline{12.69}}$

(b) $NA = \sqrt{n_1^2 - n_2^2} = \sqrt{1.45^2 - 1^2} = \underline{\underline{1.05}}$

(c) $N = V^2/2 = \underline{\underline{80 \text{ modes}}}$

Prob. 14.13

(a) $NA = \sin\theta_a = \sqrt{n_1^2 - n_2^2} = \sqrt{1.53^2 - 1.45^2} = 0.4883$
$\theta_a = \sin^{-1} 0.4883 = \underline{\underline{29.23^o}}$

(b) $P(l)/P(0) = 10^{-\alpha l/10} = 10^{-0.4X5/10} = 0.631$

i.e. $\underline{\underline{63.1\ \%}}$

Prob. 14.14

$P(l) = P(0)\, e^{-\alpha\, l/10} = 10e^{-0.5 \times 0.85/10}$ mW $= \underline{\underline{9.584 \text{ mW}}}$

Prob. 14.15 As shown in Eq. (10.35), $\log_{10} P_1/P_2 = 0.434 \ln P_1/P_2$,

$1\ Np = 20 \log_{10} e = 8.686$ dB or 1 Np/km $= 8.686$ dB/km, or 1Np/m = 8686 dB/km. Thus,

$$\alpha_{10} = \underline{\underline{8686\ \alpha_{14}}}$$

Prob. 14.16

$P(0) = P(l)\, e^{\alpha l/10} = 0.2\, e^{0.4 \times 30/10}$ mW $=$ $\underline{\underline{0.664 \text{ mW}}}$

Prob. 14.17 See text.

CHAPTER 15

P. E. 15.1 The program in Fig. 15.3 was used to obtain the plot in Fig. 15.5.

P. E. 15.2 For the exact solution,

$(D^2 + 1)\, y = 0 \quad \rightarrow \quad y = A \cos x + B \sin x$

$y(0) = 0 \rightarrow \quad A = 0$

$y(1) = 1 \rightarrow 1 = B \sin 1$ or $B = 1/\sin 1$

Thus, $y = \sin x / \sin 1$

For the finite difference solution,

$$y'' + y = 0 \quad \rightarrow \quad \frac{y(x+\Delta) - 2y(x) + y(x-\Delta)}{\Delta^2} + y = 0$$

or

$$y(x) = \frac{y(x+\Delta) + y(x-\Delta)}{2 - \Delta^2}, y(0) = 0, y(1) = 1, \Delta = 1/4$$

With the Fortran program shown below, we obtain the exact result y_e and FD result y.

```
      DIMENSION
      Y(0) = 0.0
      Y(4) = 1.0
      DEL = 0.25
      DO 10 N = 1,20  !  N = NO. OF ITERATIONS
      DO 10  I = 1,3
      Y(I) = ( Y(I+1) + Y(I-1) )/(2.0 – DEL*DEL)
      X = FLOAT (I)*DEL
      YE = SIN(X)/SIN(1.0)
      PRINT *, N, I, Y(I), YE
   10 CONTINUE
      STOP
      END
```

The results are listed below.

y(x)	N=5	N=10	N=15	N=20	Exact $y_e(x)$
y(0.25)	0.2498	0.2924	0.2942	0.2943	0.2941
y(0.5)	0.5242	0.5682	0.5701	0.5701	0.5697
y(0.75)	0.7867	0.8094	0.8104	0.8104	0.8101

P. E. 14.3 By applying eq. (15.16) to each node as shown below, we obtain the following results after 5 iterations.

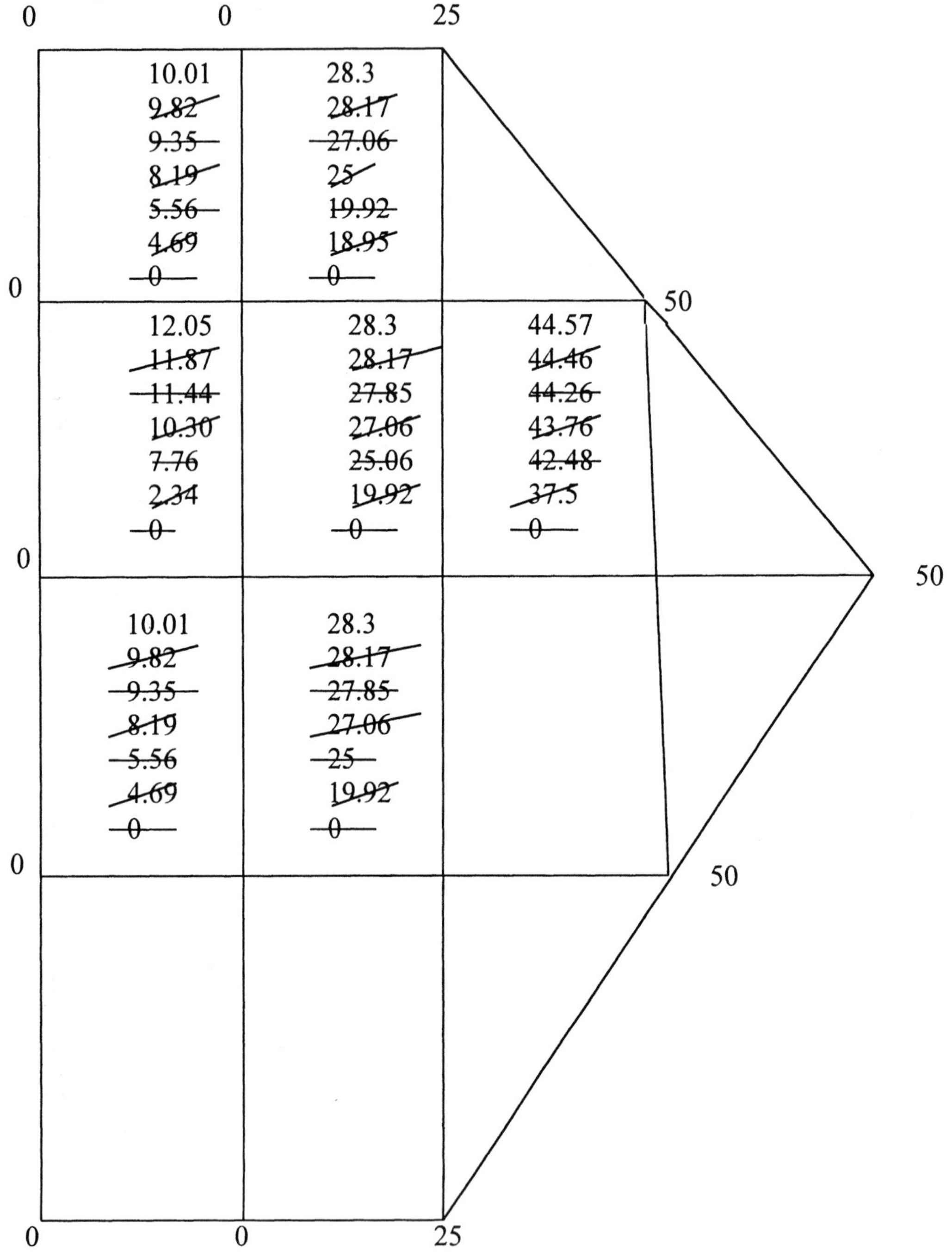

P. E. 15.4 (a) Using the program in Fig. 15.16 with NX = 4 and NY = 8, we obtain the

potential at center as

$V(2,4) = \underline{\underline{23.80\ \ V}}$

(b) Using the same program with NX = 12 and NY = 24, the potential at the center is

$V(6,12) = \underline{\underline{23.89\ V}}$

P. E. 15.5 By combining the ideas in Figs. 15.21 and 15.25, and dividing each wire into N segments, the results listed in Table 14.2 is obtained.

P. E. 15.6
(a)

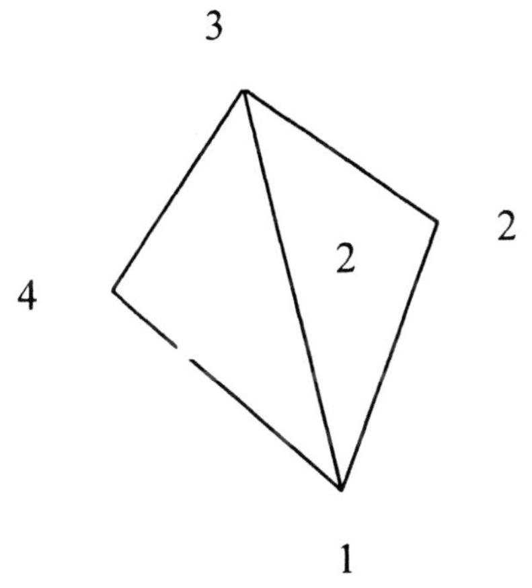

For element 1, local 1-2-3 corresponds with global 1-3-4 so that $A_1 = 0.35$,

$P_1 = 0.8,\ P_2 = 0.6,\ P_3 = -1.4,\ Q_1 = -0.5,\ Q_2 = 0.5, Q_3 = 0$

$$C^{(1)} = \begin{bmatrix} 0.6357 & 0.1643 & -0.8 \\ 0.1643 & 0.4357 & -0.6 \\ -0.8 & -0.6 & 1.4 \end{bmatrix}$$

For element 2, local 1-2-3 corresponds with global 1-2-3 so that $A_2 = 0.7$,

$P_1 = 0.1,\ P_2 = 1.4,\ P_3 = -1.5,\ Q_1 = -1,\ Q_2 = 0, Q_3 = 1$

$$C^{(2)} = \begin{bmatrix} 0.3607 & 0.05 & -0.4107 \\ 0.05 & 0.7 & -0.75 \\ -0.4107 & -0.75 & 1.1607 \end{bmatrix}$$

The global coefficient matrix is given by

$$C = \begin{bmatrix} C^{(1)}_{11} + C_{11}^{(2)} & C_{12}^{(2)} & C_{12}^{(1)} + C_{13}^{(2)} & C_{13}^{(1)} \\ C_{21}^{(2)} & C_{22}^{(2)} & C_{23}^{(2)} & 0 \\ C_{21}^{(1)} + C_{31}^{(2)} & C_{32}^{(2)} & C_{22}^{(1)} + C_{33}^{(2)} & C_{23}^{(1)} \\ C_{31}^{(1)} & 0 & C_{32}^{(2)} & C_{33}^{(1)} \end{bmatrix}$$

$$= \begin{bmatrix} 0.9964 & 0.05 & -0.2464 & -0.8 \\ 0.05 & 0.7 & 0.75 & 0 \\ -0.2464 & 0.75 & 1.596 & -0.6 \\ -0.8 & 0 & -0.6 & 1.4 \end{bmatrix}$$

(b)

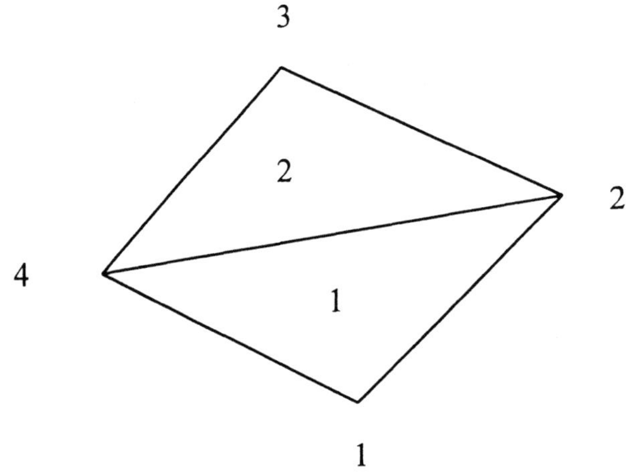

For element 1, local 1-2-3 corresponds with global 1-2-4 and $A_1 = 0.675$,

$P_1 = 0.8$, $P_2 = -0.9$, $P_3 = 0.1$, $Q_1 = -0.5$, $Q_2 = 1.5$, $Q_3 = -1.0$

$$C^{(2)} = \begin{bmatrix} 0.5933 & -0.9800 & 0.3867 \\ -0.9800 & 2.040 & -1.060 \\ 0.3867 & -1.060 & 0.6733 \end{bmatrix}$$

For element 2, local 1-2-3 corresponds with global 2-3-4 and $A_2 = 0.375$,

$P_1 = 0.1$, $P_2 = 1.4$, $P_3 = -1.5$, $Q_1 = -1$, $Q_2 = 0$, $Q_3 = 1$

$$C^{(1)} = \begin{bmatrix} 0.3607 & 0.05 & -0.4107 \\ 0.05 & 0.7 & -0.75 \\ -0.4107 & -0.75 & 1.1607 \end{bmatrix}$$

The global coefficient matrix is

$$C = \begin{bmatrix} C^{(1)}{}_{11} & C_{12}{}^{(1)} & 0 & C_{13}{}^{(1)} \\ C_{21}{}^{(1)} & C_{22}{}^{(1)} + C_{11}{}^{(2)} & C_{12}{}^{(2)} & C_{23}{}^{(1)} + C_{13}{}^{(2)} \\ 0 & C_{12}{}^{(2)} & C_{22}{}^{(2)} & C_{23}{}^{(2)} \\ C_{31}{}^{(1)} & C_{32}{}^{(1)} + C_{31}{}^{(2)} & C_{32}{}^{(2)} & C_{33}{}^{(1)} + C_{33}{}^{(2)} \end{bmatrix}$$

$$= \begin{bmatrix} 1.333 & -0.0777 & 0 & -1.056 \\ -0.0777 & 0.8192 & -0.98 & 0.2386 \\ 0 & -0.98 & 2.04 & -01.06 \\ -1.056 & 0.2386 & -1.06 & 1.877 \end{bmatrix}$$

P. E. 15.7 We use the FORTRAN program in Fig. 15.34. The input data for the region in Fig. 14.35 is a follows:

```
NE = 32;  ND =  26;  NP = 18;
NL = [ 1  2  5
    2  4  5
    2  3  5
    3  6  5
    4  5  9
    5  10  9
    5  6  10
    6  11  10
    7  8  12
    8  13  12
    8  9  13
    9  14  13
    9  10  14
    10  15  15
    10  11  14
    11  16  15
    12  13  17
    13  18  17
    13  14  18
    14  19  18
    14  15  19
    15  20  19
    15  16  20
    16  21  20
    17  18  22
```

```
18 23 22
18 19 22
19 24 23
19 20 24
20 25 24
20 21 25
21 26 25];
X = [ 1.0 2.5 2.0 1.0 1.5 2.0 0.0 0.5 1.0 1.5 2.0 0.0 0.5 1.0 1.5 2.0 0.0 0.5 1.0 1.5
1.5 0.0 0.5 1.0 1.5 2.0];
Y = [ 0.0 0.0 0.0 0.5 0.5 0.5 1.0 1.0 1.0 1.0 1.0 1.5 1.5 1.5 1.5 1.5 2.0 2.0 2.0
2.0 2.0 2.5 2.5 2.5 2.5 2.5 2.5];
NDP = [ 1 2 3 6 11 16 21 26 25 24 23 22 17 12 7 8 9 4];
VAL = [0.0 0.0 15.0 30.0 30.0 30.0 30.0 30.0 20.0 20.0 20.0 10.0 0.0 0.0 0.0
0.0 0.0 0.0];
```

With this data, the finite element (FEM) solution is compared with the finite difference (FD) solution as shown below.

Node #	x	y	FEM	FD
5	1.5	0.5	11.265	11.25
10	1.5	1.0	15.06	15.02
13	0.5	1.5	4.958	4.705
14	1.0	1.5	9.788	9.545
15	1.0	1.5	18.97	18.84
18	0.5	2.0	10.04	9.659
19	1.0	2.0	15.22	14.85
20	1.5	2.0	21.05	20.87

Prob. 15.1 (a) Using the Matlab code in Fig. 15.3, we input the data as:

```
>> plotit( [-1 2 1], [-1 0; 0 2; 1 0], 1, 1, 0.01, 0.01, 8, 2, 5 )
```

and the plot is shown below.

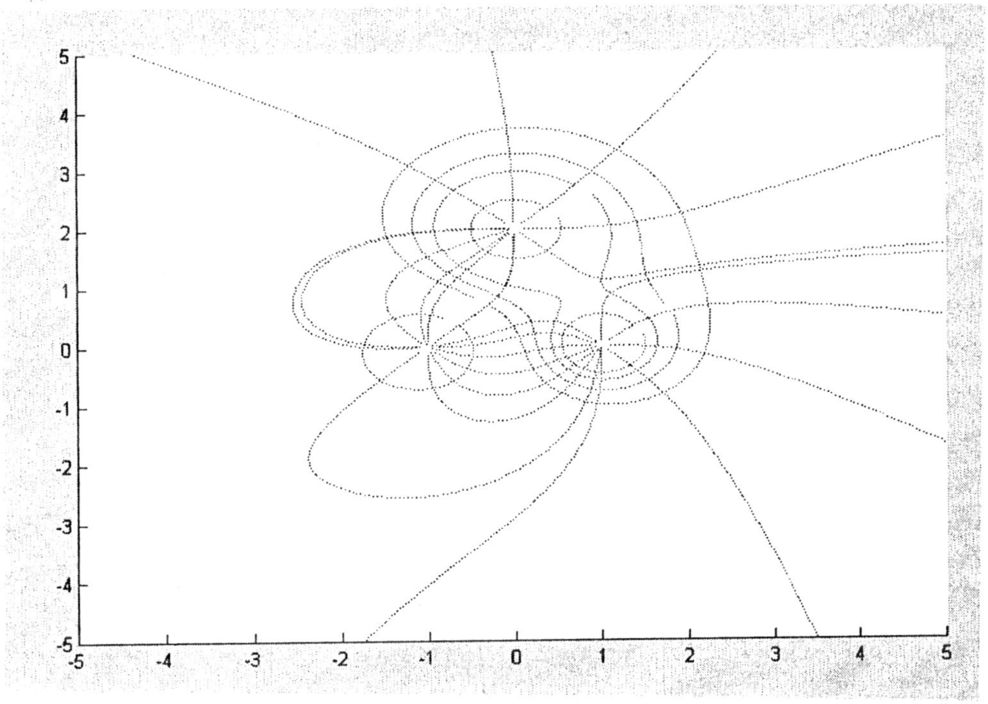

(b) Using the Matlab code in Fig. 15.3, we input the required data as:

>> plotit([1 1 1 1 1], [-1 -1; -1 1; 1 –1; 1 1; 0 0], 1, 1, 0.02, 0.01, 6, 2, 5)

and obtain the plot shown below.

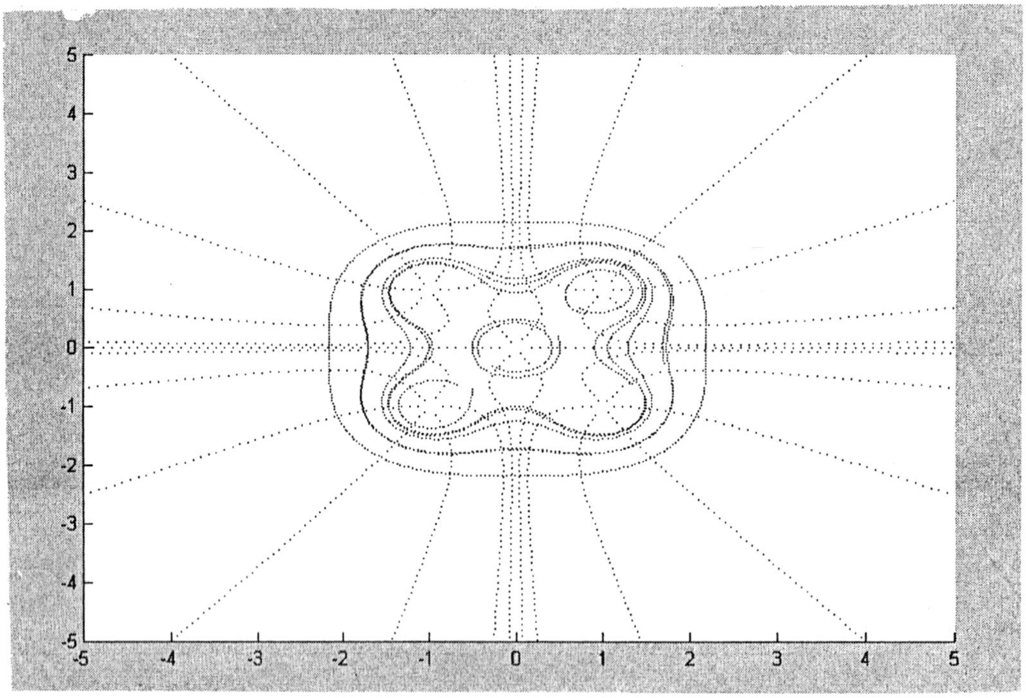

Prob. 15.2

Exact solution: y = Ax + B

x = 0, y = 0 ⟶ B = 0; x= 1, y = 10 ⟶ A = 10

y = 10x; y(0.25) = 2.5

Finite difference solution:

$$\frac{d^2y}{dx^2} \cong \frac{y(x+\Delta) - 2y(x) + y(x-\Delta)}{\Delta^2} = 0$$

or

$$y(x) = \frac{1}{2}[y(x+\Delta) + y(x-\Delta)], \Delta = 0.25$$

Using this scheme, we obtain the result shown below.

Iteration	0	0.25	0.5	0.75	1.0
0	0	0	0	0	10
1	0	0	0	5	10
2	0	0	2.5	7.5	10
3	0	1.25	5.0	8.75	10
4	0	2.5	5.625	7.5	10
5	0	2.8125	5.0	7.8125	10
6	0	2.5	5.3125	7.5	10
…	…	…	…	…	…

From this, we obtain y(0.25) = 2.5.

Prob. 15.3 (a)

$$\frac{dV}{dx} = \frac{V(x_o + \Delta x) - V(x \quad - \Delta x)}{2\Delta x}$$

For $\Delta x = 0.05$ and at x = 0.15,

$$\frac{dV}{dx} = \frac{2.0134 - 1.00}{0.05X2} = \underline{\underline{10.117}}$$

$$\frac{d^2V}{dx^2} = \frac{V(x + \Delta x) - 2V(x_o) + V(x_o - \Delta x)}{(\Delta x)^2} = \frac{2.0134 + 1.0017 - 2x1.5056}{(0.05)^2} = \underline{\underline{1.56}}$$

(b) V = 10 sinh x, dV/dx = 10 cosh x. At x = 0.15, dV/dx = $\underline{\underline{10.113}}$

which is close to the numerical estimate.

d^2V/dx^2 = 10 sinh x. At x = 0.15, d^2V/dx^2 = $\underline{\underline{1.5056}}$

which is lower than the numerical value.

Prob. 15.4

$$\nabla^2 V = \frac{\partial^2 V}{\partial \rho^2} + \frac{1}{\rho}\frac{\partial V}{\partial \rho} + \frac{\partial^2 V}{\partial z^2} = 0$$

The equivalent finite difference expression is

$$\frac{V(\rho_o + \Delta\rho, z_o) - 2V(\rho_o, z_o) + V(\rho_o - \Delta\rho, z_o)}{(\Delta\rho)^2} + \frac{1}{\rho_o}\frac{V(\rho_o + \Delta\rho, z_o) - V(\rho_o + \Delta\rho, z_o)}{2\Delta\rho}$$

$$+ \frac{V(\rho_o, z_o + \Delta z) - 2V(\rho_o, z_o) + V(\rho_o, z_o - \Delta z)}{(\Delta z)^2} = 0$$

If $\Delta z = \Delta\rho = h$, rearranging terms gives

$$V(\rho_o, z_o) = \frac{1}{4}V(\rho_o, z_o + h) + \frac{1}{4}V(\rho_o, z_o - h) + (1 + \frac{h}{2\rho_o})V(\rho + h, z_o)$$

$$+(1 - \frac{h}{2\rho_o})V(\rho - h, z_o)$$

as expected.

Prob. 15.5

$$\nabla^2 V = \frac{\partial^2 V}{\partial \rho^2} + \frac{1}{\rho}\frac{\partial V}{\partial \rho} + \frac{1}{\rho^2}\frac{\partial^2 V}{\partial \phi^2} = 0, \qquad (1)$$

$$\frac{\partial^2 V}{\partial \rho^2} = \frac{V_{m+1}{}^n - 2V_m{}^n + V_{m+1}{}^n}{(\Delta \rho)^2}, \quad (2)$$

$$\frac{\partial^2 V}{\partial \phi^2} = \frac{V_m{}^{n+1} - 2V_m{}^n + V_m{}^{n-1}}{(\Delta \phi)^2}, \quad (3)$$

$$\left.\frac{\partial V}{\partial \rho}\right|_{m,n} = \frac{V^n{}_{m+1} - V^n{}_{m-1}}{2\Delta \rho}. \quad (4)$$

Substituting (2) to (4) into (1) gives

$$\nabla^2 V = \frac{V^n{}_{m+1} - V^n{}_{m-1}}{m\Delta\rho(2\Delta\rho)} + \frac{V_{m+1}{}^n - 2V_m{}^n + V_{m+1}{}^n}{(\Delta\rho)^2} + \frac{V_m{}^{n+1} - 2V_m{}^n + V_m{}^{n-1}}{(m\Delta\rho\Delta\phi)^2}$$

$$= \frac{1}{(\Delta\rho)^2}\left[(1 - \frac{1}{2m})V_{m-1}{}^n - 2V_m{}^n + (1 + \frac{1}{2m})V_{m-1}{}^n + \frac{1}{(m\Delta\phi)^2}(V_m{}^{n+1} - 2V_m{}^n + V_m{}^{n-1})\right]$$

as required.

Prob. 15.6

$$V_o = \frac{V_1 + V_2 + V_3 + V_4}{4} = \frac{-10 + 0 + 30 + 60}{4} = \underline{\underline{20}} \text{ V}$$

Prob. 15.7

$V_1 = 0.25 (V_2 + 30 + 0 - 20) = V_2/4 + 2.5$ (1)

$V_2 = 0.25 (V_1 + 20 + 0 + 30) = V_1/4 + 12.5$ (2)

Substituting (2) into (1),

$V_1 = 2.5 + V_1/16 + 3.125 \longrightarrow V_1 = \underline{\underline{6\ V}}$

$V_2 = V_1/4 + 12.5 = \underline{\underline{14\ V}}$

Prob. 15.8

$$k = \frac{h^2 \rho_o}{\varepsilon_o} = \frac{10^{-2} x \dfrac{100}{\pi} x 10^{-9}}{\dfrac{10^{-9}}{36\pi}} = 36$$

$$V_1 = \frac{1}{4}(V_2 + 30 + 0 - 20 + k) = V_2/4 + 11.5 \qquad (1)$$

$$V_2 = \frac{1}{4}(V_1 + 20 + 0 + 30 + k) = V_1/4 + 21.5 \qquad (2)$$

Substituting (2) into (1) gives

$$V_1 = 11.5 + V_1/16 + 5.375 \longrightarrow \qquad V_1 = \underline{\underline{18 \text{ V}}}$$

$$V_2 = V_1/4 + 12.5 = \underline{\underline{26 \text{ V}}}$$

Prob. 15.9 (a)

$$V_1 = \frac{1}{4}(0 + 100 + V_3 + V_2), \; V_2 = \frac{1}{4}(0 + 100 + V_1 + V_4),$$
$$V_3 = \frac{1}{4}(0 + 0 + V_1 + V_4), \; V_4 = \frac{1}{4}(0 + 0 + V_2 + V_3)$$

We apply these iteratively n=5 times and obtain the result below.

n	0	1	2	3	4	5
V_1	0	25	34.375	36.72	37.305	37.45
V_2	0	31.25	35.937	37.11	37.403	37.475
V_3	0	6.25	10.937	12.11	12.403	12.475
V_4	0	9.375	11.917	12.305	12.45	12.487

(b) By band matrix method,

$$4V_1 - V_2 - V_3 = 100$$

$$-V_1 + 4V_2 - V_4 = 100$$

$$-V_1 + 4V_3 - V_4 = 0$$

$$-V_2 - V_3 + 4V_3 = 0$$

In matrix form,

$$\begin{bmatrix} 4 & -1 & -1 & 0 \\ -1 & 4 & 0 & -1 \\ -1 & 0 & 4 & -1 \\ 0 & -1 & -1 & 4 \end{bmatrix} \begin{bmatrix} V_1 \\ V_2 \\ V_3 \\ V_4 \end{bmatrix} = \begin{bmatrix} 100 \\ 100 \\ 0 \\ 0 \end{bmatrix}$$

$$A\,V = B \longrightarrow V = A^{-1} B$$

which yields $\underline{V_1 = 37.5 = V_2, \quad V_3 = 12.5 = V_4}$.

These values are more accurate than those obtained in part(a). Why? The average of the values should give 25 V which is the potential at the center of the region. The values in part(a) give 24.96 V while the value in part (b) gives 25 V.

Prob. 15.10

$$\begin{bmatrix} -4 & 1 & 0 & 1 & 0 & 0 \\ 1 & -4 & 1 & 0 & 1 & 0 \\ 0 & 1 & -4 & 0 & 0 & 1 \\ 1 & 0 & 0 & -4 & 1 & 0 \\ 0 & 1 & 0 & 1 & -4 & 1 \\ 0 & 0 & 1 & 0 & 1 & -4 \end{bmatrix} \begin{bmatrix} V_a \\ V_b \\ V_c \\ V_d \\ V_e \\ V_f \end{bmatrix} = \begin{bmatrix} -200 \\ -100 \\ -100 \\ -100 \\ 0 \\ 0 \end{bmatrix}$$

[A] [B]

(b)

$$\begin{bmatrix} -4 & 1 & 0 & 1 & 0 & 0 & 0 & 0 \\ 1 & -4 & 1 & 0 & 1 & 0 & 0 & 0 \\ 0 & 1 & -4 & 0 & 0 & 1 & 0 & 0 \\ 1 & 0 & 0 & -4 & 1 & 0 & 1 & 0 \\ 0 & 1 & 0 & 1 & -4 & 1 & 0 & 1 \\ 0 & 0 & 1 & 0 & 1 & -4 & 0 & 0 \\ 0 & 0 & 0 & 1 & 0 & 0 & -4 & 1 \\ 0 & 0 & 0 & 0 & 1 & 0 & 1 & -4 \end{bmatrix} \begin{bmatrix} V_1 \\ V_2 \\ V_3 \\ V_4 \\ V_5 \\ V_6 \\ V_7 \\ V_8 \end{bmatrix} = \begin{bmatrix} -30 \\ -15 \\ -30 \\ -15 \\ 0 \\ -15 \\ 0 \\ 0 \end{bmatrix}$$

[A] [B]

Prob. 15.11 (a) Matrix [A] remains the same. To each term of matrix [B], we add

$-h^2\rho_s / \varepsilon$.

(b) Let $\Delta x = \Delta y = h = 0.05$ so that $NX = 20 = NY$.

$$\frac{\rho_s}{\varepsilon} = \frac{x(y-1)10^{-9}}{10^{-9}/36\pi} = 36\pi x(y-1)$$

Modify the program in Fig. 15.16 as follows.

```
DO 40  I=1, NX -1
DO 40  J=1, NY-1
SAVE = V(I,J)
X = H*FLOAT(I)
Y=H*FLOAT(J)
RO = 36.0*PIE*X*(Y-1)
V(I,J) = 0.25*( V(I+1,J) + V(I-1,J) + V(I,J+1) + V(I,J-1) + H*H*RO )
40  CONTINUE
```

This is the major change. However, in addition to this, we must set

```
V1 = 0.0
V2 = 10.0
V3 = 20.0
V4 = -10.0
NX = 20
NY = 20
```

The results are:

$V_a = 4.276$, $V_b = 9.577$, $V_c = 11.126$, $V_d = -2.013$, $V_e = 2.919$,

$V_f = 6.069$, $V_g = -3.424$, $V_h = -0.109$, $V_i = 2.909$

Prob. 15.12

$$\frac{1}{c^2}\frac{\Phi^{j+1}_{m,n} + \Phi^{j-1}_{m,n} - 2\Phi^{j}_{m,n}}{(\Delta t)^2} = \frac{\Phi^{j}_{m+1,n} + \Phi^{j}_{m-1,n} - 2\Phi^{j}_{m,n}}{(\Delta x)^2}$$

$$+\frac{\Phi^{j}_{m,n+1} + \Phi^{j}_{m,n-1} - 2\Phi^{j}_{m,n}}{(\Delta z)^2}$$

If $h = \Delta x = \Delta z$, then after rearranging we obtain

$$\Phi^{j+1}{}_{m,n} = 2\Phi^{j+1}{}_{m,n} - \Phi^{j-1}{}_{m,n} + \alpha(\Phi^{j}{}_{m,n} + \Phi^{j}{}_{m-1,n} - 2\Phi^{j}{}_{m,n})$$
$$\alpha(\Phi^{j}{}_{m,n+1} + \Phi^{j}{}_{m,n-1} - 2\Phi^{j+1}{}_{m,n})$$

where $\alpha = (c\Delta t / h)^2$.

Prob. 15.13 Applying the finite difference formula derived above, the following programs was developed.

```
      DIMENSION V(0:50,0:50)

      U = 1.0
      DT = 0.1
      DX = 0.1
      NT = 4/DT
      NX =1/DX
      ALPHA = (U*DT/DX)**2
      DO 10 I=0,NX-1
      DO 10 J=0,NT-1
10    V(I,J) = 0.0
      DO 20 J=0,NT-1
      V(0,J) = 0
      V(10,J) = 0
20    CONTINUE
      DO 30 I=0,NT-1
      V(I,0) = SIN(FLOAT(I-1)*3.142/10.0)
      V(I,1) = V(I,0)
30    CONTINUE
      DO 40 J=1,NT-2
      DO 40 I=1,NX-2
      V(I,J+1) = ALPHA*( V(I-1,J) + V(I+1,J) ) + 2*(1.0 – ALPHA)*V(I,J)
     1  - V(I,J-1)
40    CONTINUE
      ...
      WRITE(6,*) V(I,J)
      ...
      STOP
      END
```

The results of the finite difference algorithm agree perfectly with the exact solution as shown below.

T	x	V(FD)	V(exact)
0.0	0.0	0.0	0.0
0.0	0.1	0.30903	0.30902
0.0	0.2	0.58779	0.58779
0.0	0.3	0.80902	0.80902
0.0	0.4	0.95106	0.95106
0.0	0.5	1.0	1.0
0.0	0.6	0.95106	0.95106
0.0	0.7	0.80902	0.80902
...	...	...	...

Prob. 15.14
(a)Points 1, 3, 5, and 7 are equidistant from O. Hence

$$V_o = \frac{1}{4}(V_1 + V_3 + V_5 + V_7) \qquad (1)$$

Also points 2, 4, 6, and 8 are equidistant from O so that

$$V_o = \frac{1}{4}(V_2 + V_4 + V_6 + V_8) \qquad (2)$$

Adding (1) and (2) gives

$$V_o = \frac{1}{4}(V_1 + V_2 + V_3 + V_4 + V_5 + V_6 + V_7 + V_8)$$

as required.

Prob. 15.15 Combining the ideas in the programs in Figs. 15.20 and 15.24, we develop a Matlab code which gives

N = 20 ⟶ C = 19.4 pF/m

N = 40 ⟶ C = 13.55 pF/m

N = 100 ⟶ C = 12.77 pF/m

For the exact value, d/2a = 50/10 = 5

$$C = \frac{\pi\varepsilon}{\cosh^{-1}\frac{d}{2a}} = \frac{\pi x10^{-9}/36\pi}{\cosh^{-1} 5} = \underline{\underline{12.12}} \text{ pF/m}$$

Prob. 15.16 To determine V and **E** at (-1,4,5), we use the program in Fig. 15.21.

$$V = \int_0^L \frac{\rho_L dl}{4\pi\varepsilon_o R}, \quad \text{where} \quad R = \sqrt{26 + (4 - y')^2}$$

$$V = \frac{\Delta}{4\pi\varepsilon} \sum_{k=1}^{N} \frac{\rho_k}{\sqrt{26 + (y - y_k)^2}}$$

$$E = \int_0^L \frac{\rho_L dl R}{4\pi\varepsilon_o R^3}$$

where **R = r – r'** = (-1, 4-y', 5), R = |**R**|

$$E_x \cong \frac{\Delta}{4\pi\varepsilon} \sum_{k=1}^{N} \frac{(-1)\rho_k}{[26 + (4 - y_k)^2\ ]^{3/2}}$$

$$E_y \cong \frac{\Delta}{4\pi\varepsilon} \sum_{k=1}^{N} \frac{(4 - y_k)\rho_k}{[26 + (4 - y_k)^2\ ]^{3/2}}$$

$$E_z = -5E_x$$

For N = 20, V_o = 1V, L = 1m, a = 1mm, the following lines are added to the program in the Fortran version of Fig. 15.21 after 90 CONTINUE statement. (See second edition.)

```
         V = 0.0
         EX = 0.0
         EY = 0.0
         FACTOR = DELTA/(4.0*PIE*EO)
         DO 100 K=1,N
         R = SQRT(26.0 + (4.0 – YY(K) )**2)
         V=V + RO(K)/R
         EX = EX – R0(K)/R**3
         EY = EY + (4.0 – YY(K))*RO(K)/R**3
100      CONTINUE
         V = V*FACTOR
         EX = EX*FACTOR
         EY= EY*FACTOR
         EZ = -5.0*EX
         PRINT *, V, EX, EY, EZ
         ...
```

The result is:

$V = 12.47$ mV, $\mathbf{E} = -0.3266\, \mathbf{a}_x + 1.1353\mathbf{a}_y + 1.6331\mathbf{a}_z$ mV/m

Prob. 15.17

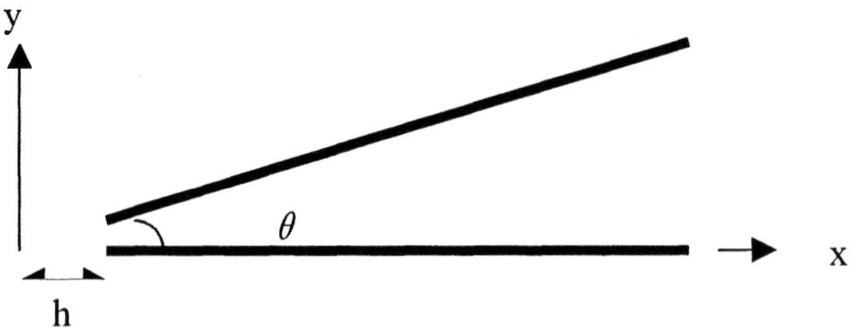

To find C, take the following steps:

(1)Divide each line into N equal segments. Number the segments in the lower conductor as 1, 2, …, N and segments in the upper conductor as N+1, N+2, …, 2N,

(2) Determine the coordinate (x_k, y_k) for the center of each segment.

For the lower conductor, $y_k = 0$, k=1, …, N, $x_k = h + \Delta\,(k-1/2)$, k = 1,2,… N

For the upper conductor, $x_k = [h + \Delta\,(k-1/2)] \sin\theta$, k=N+1, N+2, …,2 N,

$x_k = [h + \Delta\,(k-1/2)] \cos\theta$, k = N+1,N+2,… 2N

where h is determined from the gap g as

$$h = \frac{g}{2\sin\theta/2}$$

(3)Calculate the matrices [V] and [A] with the following elements

$$V_k = \begin{cases} V_o, k = 1,\ldots,N \\ -V_o, k = N+1,\ldots 2N \end{cases}$$

$$A_{ij} = \begin{cases} \dfrac{\Delta}{4\pi\varepsilon R_{ij}}, i \neq j \\ 2\ln\Delta / a, i = j \end{cases}$$

where $R_{ij} = \sqrt{(x_i - x_j)^2 + (y_i - y_j)^2}$

(4) Invert matrix [A] and find $[\rho] = [A]^{-1}\,[V]$.

(5) Find the charge Q on one conductor

$$Q = \sum \rho_k \Delta = \Delta \sum_{k=1}^{N} \rho_k$$

(6) Find $C = |Q|/V_o$

Taking N= 10, $V_o = 1.0$, a program was developed to obtain the following result.

θ	C (in pF)
10	8.5483
20	9.0677
30	8.893
40	8.606
50	13.004
60	8.5505
70	9.3711
80	8.7762
90	8.665
100	8.665
110	10.179
120	8.544
130	9.892
140	8.7449
150	9.5106
160	8.5488
170	11.32
180	8.6278

Prob. 15.18 We may modify the program in Fig. 15.25 and obtain $Z_o \cong 50\Omega$. For details, see M. N. O. Sadiku, "Numerical Techniques in Electromagnetics," (CRC Press, 1992), pp. 338-340.

Prob. 15.19 (a) Exact solution yields

$C = 2\pi\varepsilon / In(\Delta / a) = 8.02607x10^{-11}$ F/m and $\underline{\underline{Z_o = 41.559\Omega}}$

where a = 1cm and Δ = 2cm. The numerical solution is shown below.

N	C (pF/m)	$Z_o(\Omega)$
10	82.386	40.486
20	80.966	41.197
40	80.438	41.467
100	80.025	41.562

(b)For this case, the numerical solution is shown below.

N	C (pF/m)	$Z_o(\Omega)$
10	109.51	30.458
20	108.71	30.681
40	108.27	30.807
100	107.93	30.905

Prob. 15.20 We modify the Matlab code in Fig. 15.26 (for Example 15.5) by changing the input data and matrices [A] and [B]. We let

$x_i = h + \Delta\,(i-1/2),\ i = 1,2,\ldots N, \qquad \Delta = L/N$

$y_i = h/2,\ j = 1,2,\ldots N, \quad z_k = t/2,\ k = 1,2,\ldots N$

and calculate

$$R_{ij} = \sqrt{(x_i - x_j)^2 + (y_i - y_j)^2 + (z_i - z_j)^2}$$

We obtain matrices [A] and [B]. Inverting [A] gives

$[q] = [A]^{-1}\,[B],\quad [\rho_v] = [q]/(ht\Delta),\quad C = \dfrac{\sum_{i=1}^{N} q_i}{10}$

The computed values of $[\rho_v]$ and C are shown below.

i	$\rho_{vi}(x10^{-6})C/m^3$
1, 20	0.5104
2, 19	0.4524
3, 18	0.4324
4, 17	0.4215
5, 16	0.4144
6, 15	0.4096
7, 14	0.4063
8, 13	0.4041
9, 12	0.4027
10,11	0.4020

C = 17.02 pF

Prob. 15.21 From given figure, we obtain

$$\alpha_1 = \frac{A_1}{A} = \frac{1}{2A}\begin{vmatrix} 1 & x & y \\ 1 & x_2 & y_2 \\ 1 & x_3 & y_3 \end{vmatrix} = \frac{1}{2A}[(x_2y_3 - x_3y_2) + (y_2 - y_3)x + (x_3 - x_2)y]$$

as expected. The same applies for α_2 and α_3.

Prob. 15.22 (a) For the element in (a),

A = ½ (1 – 0.5x0.25) = 0.4375

$$\alpha_1 = \frac{1}{2A}[0.875 - 0.75x - 0.5y] = 1 - 0.8571x - 0.5714y$$

$$\alpha_2 = \frac{1}{2A}[0 + x - 0.5y] = 1.1428x - 0.5714y$$

$$\alpha_3 = \frac{1}{2A}[0 - 0.25x + y] = -0.2857x + 1.1429y$$

For the element in (b),

A = ½ [0.5x1.6 – (-1)x1.6] = 1.2

$$\alpha_1 = 1.25 - 0.625y$$

$$\alpha_2 = -1.5 + 0.667x + 0.4167y$$

$$\alpha_3 = 1.25 - 0.667x + 0.2083y$$

(b)For the element in (a),

$P_1 = -0.75$, $P_2 = 1.0$, $P_3 = -0.25$, $Q_1 = -0.5 = Q_2$, $Q_3 = 1.0$

$$C_{ij} = \frac{1}{4A}[P_jP_i + Q_jQ_j] = (\nabla\alpha_1.\nabla\alpha_2)A$$

Hence,

$$C^{(1)} = \begin{bmatrix} 0.4643 & -0.2857 & -0.1786 \\ -0.2857 & 0.7143 & -0.4286 \\ -0.1786 & -0.4286 & 0.6071 \end{bmatrix}$$

For the element in (b),

$P_1 = 0, P_2 = 1.6, P_3 = -1.6, Q_1 = -1.5, Q_2 = 1.0, Q_3 = 0.5$

Hence,

$$C^{(2)} = \begin{bmatrix} 0.4688 & -0.3125 & -0.1553 \\ -0.3125 & 0.7417 & -0.4292 \\ -0.1563 & -0.4292 & 0.5854 \end{bmatrix}$$

Prob. 15.23 (a)

$$2A = \begin{vmatrix} 1 & 1/2 & 1/2 \\ 1 & 3 & 1/2 \\ 1 & 2 & 2 \end{vmatrix} = 15/4$$

$$\alpha_1 = \frac{4}{15}[(6-1)+(-1\frac{1}{2})x+(-1)y] = \frac{4}{15}(5-1.5x-y)$$

$$\alpha_2 = \frac{4}{15}[(1-1)+\frac{3}{2}x-\frac{3}{2}y] = \frac{4}{15}(1.5x-1.5y)$$

$$\alpha_3 = \frac{4}{15}[(1/4-3/2)+0x+\frac{5}{2}y] = \frac{4}{15}(-1.25+2.5y)$$

$V = \alpha_1 V_1 + \alpha_2 V_2 + \alpha_3 V_3$

Substituting V=80, $V_1 = 100$, $V_2 = 50$, $V_3 = 30$, α_1, α_2, and α_3 leads to

20 = 7.5x + 10y + 3.75

Along side 12, y=1/2 so that

20 = 15x/2 + 5 + 15/4 $\longrightarrow$ x=3/2, i.e (1.5, 0.5)

Along side 13, x =y

20 = 15x/2 + 10x + 15/4 $\longrightarrow$ x=13/4, i.e. (13/14, 13/14)

Along side 23, y = -3x/2 + 5

20 = 15x/2 – 15 + 50 + 15/4 ⟶ x=-5/2 (not possible)

Hence intersection occurs at

(1.5, 0.5) along 12 and (0.9286, 0.9286) along 13

(b) At (2,1),

$$\alpha_1 = \frac{4}{15}, \quad \alpha_2 = \frac{6}{15}, \quad \alpha_3 = \frac{5}{15}$$

$V(1,2) = \alpha_1 V_1 + \alpha_2 V_2 + \alpha_3 V_3 =$ (400 + 300 + 150)/15 = 56.67 V

Prob. 15.24

$$2A = \begin{vmatrix} 1 & 0 & 0 \\ 1 & 2 & -1 \\ 1 & 1 & 4 \end{vmatrix} = 9$$

$$\alpha_1 = \frac{1}{9}[(0-0) + (4-0)x + (0-1)y] = \frac{1}{9}(4x - y)$$

$$\alpha_2 = \frac{1}{9}[(0-0) + (0+1)x + (2-0)y] = \frac{1}{9}(x + 2y)$$

$$\alpha_3 = \frac{1}{9}[(8+1) + (-1-4)x + (1-2)y] = \frac{1}{9}(9 - 5x - y)$$

$$V_e = \alpha_1 V_{e1} + \alpha_2 V_{e2} + \alpha_{31} V_{e3}$$

V(1,2) = 8(4-2)/9 + 12(1+4)/9 + 10(9-5-1)/9 = 96/9 = 10.667 V

At the center $\alpha_1 = \alpha_2 = \alpha_3 = 1/3$ so that

V(center) = (8 + 12 + 10)/3 = 10

Or at the center, (x, y) = (0 + 1 + 2, 0 + 4 –1)/3 = (1,1)

V(1,1) = 8(3)/9 + 12(3)/9 + 10(3)/9 = 10 V

Prob. 15.25

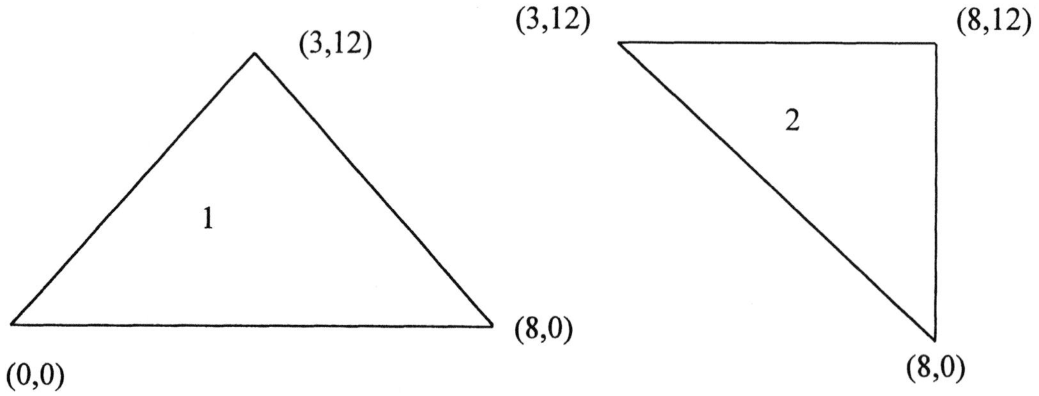

For element 1, local numbering 1-2-3 corresponds to global numbering 4-2-1.

$P_1 = 12, P_2 = 0, P_3 = -12, Q_1 = -3, Q_2 = 8, Q_3 = -5,$

A = (0 + 12 x 8)/2 = 48

$$C_{ij} = \frac{1}{4x48}[P_j P_i + Q_j Q_j]$$

$$C^{(1)} = \begin{bmatrix} 0.7956 & -0.1248 & -0.6708 \\ -0.1248 & 0.3328 & -0.208 \\ -0.6708 & -0.208 & 0.8788 \end{bmatrix}$$

For element 2, local numbering 1-2-3 corresponds to global numbering 2-4-3.

$P_1 = -12, P_2 = 0, P_3 = 12, Q_1 = 0, Q_2 = -5, Q_3 = 5,$

A = (0 + 60)/2 = 30

$$C_{ij} = \frac{1}{4x48}[P_j P_i + Q_j Q_j]$$

$$C^{(1)} = \begin{bmatrix} 1.2 & 0 & -1.2 \\ 0 & 0.208 & -0.208 \\ -1.2 & -0.208 & 1.408 \end{bmatrix}$$

$$C = \begin{bmatrix} C^{(1)}_{33} & C_{23}^{(1)} & 0 & C_{31}^{(1)} \\ C_{23}^{(1)} & C_{22}^{(1)} + C_{11}^{(2)} & C_{13}^{(2)} & C_{21}^{(1)} + C_{12}^{(2)} \\ 0 & C_{31}^{(2)} & C_{33}^{(2)} & C_{32}^{(2)} \\ C_{13}^{(1)} & C_{21}^{(1)} + C_{21}^{(2)} & C_{23}^{(2)} & C_{22}^{(2)} + C_{11}^{(1)} \end{bmatrix}$$

$$= \begin{bmatrix} 0.8788 & -0.208 & 0 & -0.6708 \\ -0.208 & 1.528 & -1.2 & -0.1248 \\ 0 & -1.2 & 1.408 & -0.206 \\ -0.6708 & -0.1248 & -0.208 & 1.0036 \end{bmatrix}$$

Prob. 15.26

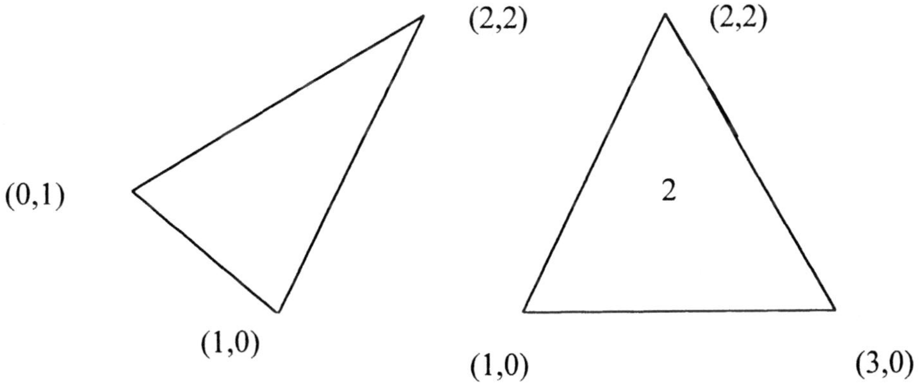

For element 1, local numbering 1-2-3 corresponds to global numbering 1-2-4.

P_1 = -2, P_2 = 1, P_3 = -1, Q_1 = 1, Q_2 = -2, Q_3 = 1,

A = ($P_2 Q_3$ - $P_3 Q_2$)/2 = 3/2, i.e. 4A = 6

$$C_{ij} = \frac{1}{4A}[P_j P_i + Q_j Q_j]$$

$$C^{(1)} = \frac{1}{6}\begin{bmatrix} 5 & -4 & -1 \\ -4 & 5 & -1 \\ -1 & -1 & 2 \end{bmatrix}$$

For element 2, local numbering 1-2-3 corresponds to global numbering 4-2-3.

P_1 = 0, P_2 = -2, P_3 = 2, Q_1 = 2, Q_2 = -1, Q_3 = -1,

A = 2, 4A = 8

$$C^{(2)} = \frac{1}{8}\begin{bmatrix} 4 & -2 & -2 \\ -2 & 5 & -3 \\ -2 & -3 & 5 \end{bmatrix}$$

The global coefficient matrix is

$$C = \begin{bmatrix} C^{(1)}{}_{11} & C_{12}{}^{(1)} & 0 & C_{13}{}^{(1)} \\ C_{12}{}^{(1)} & C_{22}{}^{(1)} + C_{22}{}^{(2)} & C_{23}{}^{(2)} & C_{23}{}^{(1)} + C_{21}{}^{(2)} \\ 0 & C_{23}{}^{(2)} & C_{33}{}^{(2)} & C_{31}{}^{(2)} \\ C_{13}{}^{(1)} & C_{23}{}^{(1)} + C_{21}{}^{(2)} & C_{31}{}^{(2)} & C_{33}{}^{(1)} + C_{11}{}^{(2)} \end{bmatrix}$$

$$= \begin{bmatrix} 0.8333 & -0.667 & 0 & -0.1667 \\ -0.6667 & 1.4583 & -0.375 & -0.4167 \\ 0 & -0.375 & 0.625 & -0.25 \\ -0.1667 & -0.4167 & -0.25 & 0.833 \end{bmatrix}$$

Prob. 15.27 We can do it by hand as in Example 15.6. However, it is easier to prepare an input files and use the program in Fig. 15.54. The Matlab input data is

```
NE = 2;
ND = 4;
NP = 2;
NL = [1  2  4
      2   3  4];
X = [ 0.0    1.0   3.0   2.0];
Y = [ 1.0  0.0  0.0  2.0];
NDP= [ 1  3 ];
VAL = [ 10.0   30.0]
```

The result is $V = \begin{bmatrix} 10 \\ 18 \\ 30 \\ 20 \end{bmatrix}$

From this,

$\underline{\underline{V_2 = 18 \text{ V}, \quad V_4 = 20 \text{ V}}}$

Prob. 15.28

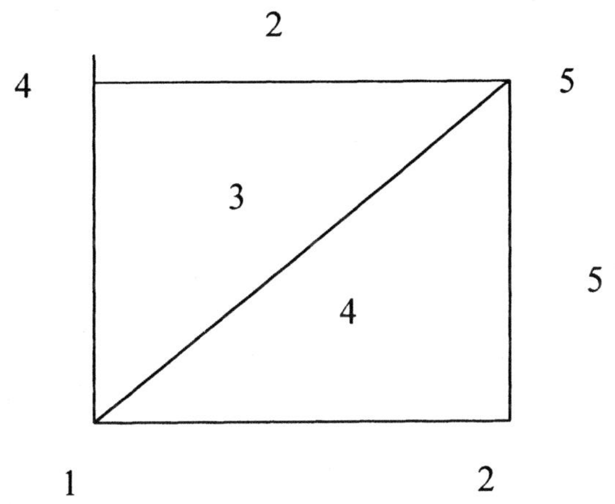

The local numbering 1-2-3 in element 3 corresponds with the global numbering 5-4-1, while the local number 1-2-3 in element 4 corresponds with the global numbering 5-1-2.

$C_{5,5} = C_{11}^{(2)} + C_{11}^{(3)} + C_{11}^{(4)} + C_{11}^{(5)}$, $A = 2$,

$C_{11}^{(2)} = (2 \times 2 + 2 \times 2)/8 = 1 = C_{11}^{(5)}$

$C_{11}^{(3)} = (2 \times 2 + 0)/8 = ½ = C_{11}^{(4)}$

$C_{5,5} = 1 + 1 + ½ + ½ = \underline{\underline{3}}$

$C_{5,1} = C_{31}^{(3)} + C_{21}^{(4)}$

But $C_{31}^{(3)} = \frac{1}{8}(P_3P_1 + Q_3 Q_1) = 0$ since $P_3 = 0 = Q_3$

$C_{21}^{(4)} = \frac{1}{8}(P_2P_1 + Q_2 Q_1) = 0$ since $P_3 = 0 = Q_3$

$\underline{\underline{C_{5,1} = 0}}$

Prob. 15.29 As in P. E. 14.7, we use the program in Fig. 15.34. The input data based on Fig. 15.56 is as follows.

NE =50; ND= 36; NP= 20;

NL = [1	8	7
1	2	8
2	9	8
2	3	9
3	10	9
3	4	10

```
        4    11    10
        4    5     11
        5    12    11
        5    6     12
        7    14    13
        7    8     14
        8    15    14
        8    9     15
        9    16    15
        9    16    16
        10   17    16
        10   11    17
        11   18    17
        11   12    18
        13   20    19
        13   14    20
        14   21    20
        14   15    21
        15   22    21
        15   16    22
        16   23    22
        16   17    23
        17   24    23
        17   18    24
        19   26    25
        19   20    26
        20   27    26
        20   21    27
        21   28    27
        21   22    28
        22   29    28
        22   23    29
        23   30    29
        23   24    30
        25   32    31
        25   26    32
        26   33    32
        26   27    33
        27   34    33
        27   28    34
        28   35    34
        28   29    35
        29   36    35
        29   30    36];
X = [0.0 0.2 0.4 0.6 0.8 1.0 0.0 0.2 0.4 0.6 0.8 1.0 0.0 0.2 0.4 0.6 0.8 1.0
     0.0 0.2 0.4 0.6 0.8 1.0 0.0 0.2 0.4 0.6 0.8 1.0 0.2 0.4 0.6 0.8 1.0];
```

Y = [0.0 0.0 0.0 0.0 0.0 0.0 0.2 0.2 0.2 0.2 0.2 0.2 0.4 0.4 0.4 0.4 0.4 0.4 0.6 0.6 0.6 0.6 0.6 0.6 0.8 0.8 0.8 0.8 0.8 0.8 1.0 1.0 1.0 1.0 1.0 1.0];
NDP = [1 2 3 4 5 6 12 18 24 30 36 35 34 33 32 31 25 19 13 7];
VAL = [0.0 0.0 0.0 0.0 0.0 0.0 0.0 0.0 0.0 0.0 50.0 100.0 100.0 100.0
100.0 50.0 0.0 0.0 0.0];

With this data, the potentials at the free nodes are compared with the exact values as shown below.

Node no.	FEM Solution	Exact Solution
8	4.546	4.366
9	7.197	7.017
10	7.197	7.017
11	4.546	4.366
14	10.98	10.66
15	17.05	16.84
16	17.05	16.84
17	10.98	10.60
20	22.35	21.78
21	32.95	33.16
22	32.95	33.16
23	22.35	21.78
26	45.45	45.63
27	59.49	60.60
28	59.49	60.60
29	45.45	45.63

Prob. 15.30 We use exactly the same input data as in the previous problem except that the last few lines are replaced by the following lines.

VAL = [0.0 0.0 0.0 0.0 0.0 0.0 0.0 0.0 0.0 0.0 58.8 100.0 95.1 95.1
58.8 0.0 0.0 0.0];

The potential at the free nodes obtained with the input data are compared with the exact solution as shown below.

Node no.	FEM Solution	Exact Solution
8	3.635	3.412
9	5.882	5.521
10	5.882	5.521
11	3.635	3.412
14	8.659	8.217
15	14.01	13.30
16	14.01	13.30
17	8.659	8.217
20	16.99	16.37
21	27.49	26.49
22	27.49	26.49
23	16.69	16.37
26	31.81	31.21
27	51.47	50.5
28	51.49	50.5
29	31.81	31.21

Prob. 15.31

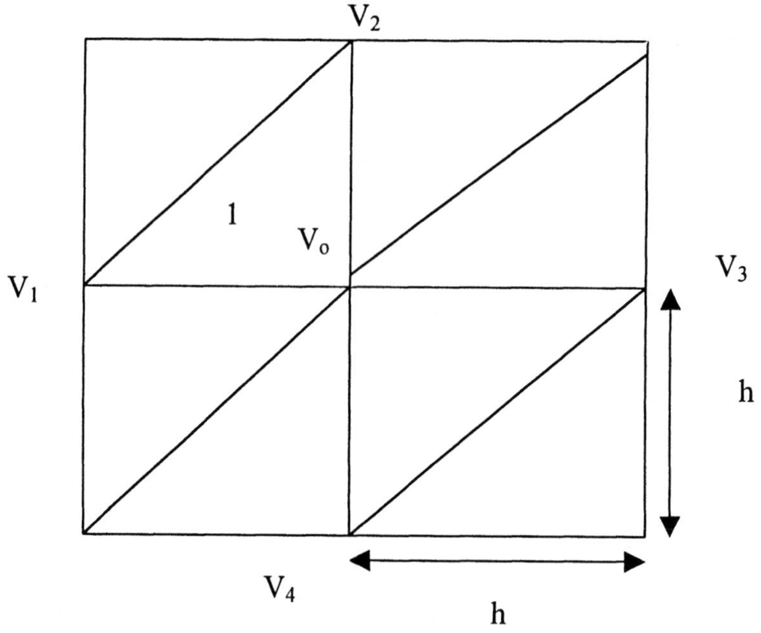

For element 1, the local numbering 1-2-3 corresponds witn nodes with V_1 , V_2 , and V_3.

$$V_o = -\frac{1}{C_{oo}}\sum_{i=1}^{4} V_i C_{io}$$

$$C_{oo} = \sum_{j=1}^{4} C_{oj}^{(e)} = \frac{1}{4h^2/2}(hh + hh)x2 + \frac{1}{4h^2/2}(hh + 0)x4 = 4$$

$$C_{o1} = \frac{2x1}{2h^2}[P_3P_1 + Q_3Q_1] = \frac{2}{2h^2}[-hh - 0] = -1$$

$$C_{o2} = \frac{2x1}{2h^2}[P_1P_2 + Q_1Q_2] = \frac{2}{2h^2}[-hx0 + hx(-h)] = -1$$

Similarly, $C_{03} = -1 = C_{04}$. Thus

$V_o = (V_1 + V_2 + V_3 + V_4)/4$

which is the same result obtained using FDM.